Developments in Geotectonics 13

RECENT CRUSTAL MOVEMENTS, 1977

Developments in Geotectonics 13

RECENT CRUSTAL MOVEMENTS, 1977

Proceedings of the Sixth International Symposium on
Recent Crustal Movements,
Stanford University, Palo Alto, California, July 25—30, 1977

Edited by

C.A. WHITTEN
9606 Sutherland Road, Silver Spring, Md. 20901 (U.S.A.)

R. GREEN
*Department of Geophysics, University of New England, Armidale,
N.S.W. 2351 (Australia)*

and

B.K. MEADE
5516 Bradley Boulevard, Alexandria, Va. 22311 (U.S.A.)

Reprinted from Tectonophysics Volume 52

ELSEVIER SCIENTIFIC PUBLISHING COMPANY
Amsterdam — Oxford — New York   1979

ELSEVIER SCIENTIFIC PUBLISHING COMPANY
335 Jan van Galenstraat
P.O. Box 211, 1000 AE Amsterdam, The Netherlands

Distributors for the United States and Canada:

ELSEVIER/NORTH-HOLLAND INC.
52, Vanderbilt Avenue
New York, N.Y. 10017

Library of Congress Cataloging in Publication Data

International Symposium on Recent Crustal Movements, 6th,
    Stanford University, 1977.
    Recent crustal movements, 1977.

    (Developments in geotectonics ; 13)
    "Reprinted from Tectonophysics, vol. 52."
    Bibliography: p.
    Includes index.
    1. Earth movements--Congresses. 2. Geodynamics--
Congresses. 3. Seismology--Congresses. I. Whitten,
Charles A. II. Green, Robert, 1930-   III. Meade,
B. K. IV. Series.
QE598.I57 1977   551.1'3   79-82
ISBN 0-444-41783-4

ISBN 0-444-41783-4 (Vol. 13)
ISBN 0-444-41714-1 (Series)

Printed in The Netherlands

PREFACE

This special issue of Tectonophysics contains the Proceedings of the Sixth International Symposium on Recent Crustal Movements held at Stanford University, Palo Alto, California, July 25—30, 1977. A brief report, including the resolutions adopted at the Symposium, is given on the following pages.

The volume contains 50 papers and 38 abstracts of the presentations at the nine technical sessions. In addition, a special report on the RCM Symposium held in Japan in 1976 and the report of the Fennoscandia Subcommission are included.

The papers are arranged by the scientific division of the sessions. An author index is provided.

Ronald Green of Armidale, Australia, Buford K. Meade of Alexandria, Virginia, U.S.A., and Charles A. Whitten of Silver Spring, Maryland, U.S.A., were the co-editors. They express their appreciation to the authors and the local organizing committee for their cooperation and also thank the editorial staff of Tectonophysics for their continued help and support of the International Commission for Recent Crustal Movements by publishing these proceedings.

C.A.W.

THE SIXTH INTERNATIONAL SYMPOSIUM ON RECENT CRUSTAL
MOVEMENTS, PALO ALTO, CALIFORNIA, JULY 25–30, 1977

The Sixth International Symposium on Recent Crustal Movements was convened under the auspices of the Commission on Recent Crustal Movements (CRCM) of the International Association of Geodesy (IAG). The meetings were held on the campus of Stanford University at Palo Alto, California during the last week of July, 1977.

The Symposium was sponsored by The National Aeronautics and Space Administration, the National Oceanic and Atmospheric Administration (National Geodetic Survey), and the U.S. Geological Survey. Members of the organizing committee were Charles A. Whitten, George A. Thompson, Wayne R. Thatcher, James C. Savage, Robert D. Nason, Sanford, R. Holdahl, and Robert O. Castle. The Symposium was convened by Buford K. Meade, Robert L. Kovach, Robert O. Burford, and Clarence R. Allen. Patricia H. Johnson and Robert O. Burford served as coordinator and general chairman, respectively, for a local arrangements committee composed of most of the California members of the organizing committee. They were assisted by Gregg A. Rice, Pamela M. Rowbotham-Castle, Gretchen O. Burford, and Thelma R. Snider.

Three field trips were offered to symposium participants before, during, and after the week of technical sessions. The trips were arranged by Robert D. Nason, with the support of Kenneth R. Lajoie, Thomas H. Rogers, Gerald E. Weber, Charles G. Bufe, Spencer H. Wood, Clarence R. Allen, Glenn R. Roquemore, and Kerry Sieh. Field trip A was conducted on Saturday and Sunday, July 23–24, along a route to include a site on the San Andreas fault trace showing a surface rupture that occurred during the 1906 earthquake, several sites to view various aspects of the deformation of uplifted marine terraces along the Pacific coast from Halfmoon Bay to Monterey Bay, and a number of sites to examine evidence of active fault creep on the San Andreas and Calaveras faults near San Juan Bautista and Hollister, east of Monterey Bay. Field trip B was conducted on Wednesday, July 27, affording the symposium participants an opportunity to see various natural and cultural features of the San Francisco Bay area, especially the San Andreas fault on the San Francisco peninsula, the city of San Francisco, several municipalities east of San Francisco Bay, the Hayward fault, and the Wiebel Winery. Field trip C was held from Sunday through Wednesday, July 31–August 3, following a route across California to include visits to Hollister, Pacheco Pass, Yosemite Valley, Tioga Pass, Owens Valley, the 1872 earthquake fault scarp on the Owens Valley fault near Lone Pine, the Garlock fault, the San Andreas fault at Palmdale, and the 1971 San Fernando earthquake area.

The symposium meetings were attended by 182 scientists and engineers from 19 nations, representing various disciplines within the fields of geology, geophysics, geodesy, civil engineering, structural engineering, and engineering seismology. The participants were welcomed to Stanford University by Dr. William F. Miller, University Vice President and Provost, who also briefly summarized the history, policies, and educational philosophy of that institution. Dr. Clarence R. Allen then bade the participants welcome to the week of technical sessions on behalf of the organizing committee. A review of the history and purpose of CRCM prepared by Prof. Yuri D. Boulanger, President of the Commission, was read by R.O. Burford *. The opening session concluded with a special presentation by Prof. Keichi Kasahara, Chairman of the North Pacific Subcommission, on the results of the 1976 Symposium on the Recent Crustal Movements of Japan. Proceedings volumes for the symposium in Japan were distributed to the participants. A special report from the Fennoscandian Subcommission prepared by Andrei A. Nikonov was also distributed during the opening session.

The program for the technical sessions was arranged by a program committee under the chairmanship of James C. Savage. The diverse subjects covered by the 82 oral presentations ** were grouped into 9 technical sessions distinguished by the following general themes: Crustal deformation measurements using extra-terrestrial geodesy; Measurement of strain, tilt, and gravity; Observed vertical crustal deformation (2 sessions); Geologic studies of Holocene deformation (2 sessions); Observed horizontal crustal deformation; Seismology; and Experimental and theoretical models of crustal movement and deformation. In addition to the technical sessions, meetings of the CRCM Executive Committee (Prof. Yuri D. Boulanger, Chairman, and Dr. Pavel Vyskočil, Secretary), The Working Group on Instruments and Methods (Dr. Peter L. Bender, Chairman), and Study Group 1 of Working Group 10 of the International Geodynamics Project (Dr. N. Pavoni, Chairman) were held during the symposium.

During the closing session of the symposium, in addition to other congratulatory and farewell remarks, Patricia H. Johnson and Gregg A. Rice were thanked by the organizing committee and by those assembled for their capable handling of all the details necessary for the smooth conduct of the meetings. As a final item of official symposium business, the following resolutions presented by the CRCM Executive Committee were discussed and approved by those in attendance:

(1) The Commission on Recent Crustal Movements, *recognizing* the need for detecting areas of rapid distortion in national geodetic networks due to large-scale crustal movements, both for the support of surveying and map-

---

* The text for Dr. Allen's presentation is included in this volume.
** Additional abstracts or papers prepared by several contributors who were unable to attend the meetings are also included in this volume.

making activities, and for improved understanding of tectonic processes within the countries, and *recognizing* the development of satellite techniques and of long baseline interferometry techniques capable of determining the three-dimensional geometric coordinates of several hundred points throughout the world with centimeter to decimeter accuracy, *recommends*:

(a) That a Worldwide Geometrical Position Reference Network be established with such techniques;

(b) That such points be well monumented and reliably connected to national networks;

(c) That gravity also be measured with high precision at such points; and

(d) That particular urgency be given to establishing such points in plate tectonic convergence zones and surrounding known seismic zones.

(2) The CRCM, *recognizing* the wealth of information available in existing and ongoing precise point position determinations for the future study of crustal movements and for earthquake prediction research, and *recognizing* that crustal movements amounting to several centimeters per year may occur across seismic zones and cause substantial distortions of control networks for mapping, *recommends*:

(a) That attention be paid to maintaining monumentation and documentation of such points for purposes of future reference;

(b) That an essential element of control survey programs be the resurvey of such networks to determine the location, rate, and nature of crustal deformations with their associated uncertainties; and

(c) That these movements be monitored to improve knowledge of the long-term effects.

(3) The CRCM, *recognizing* the need for crustal movement measurements of extremely high accuracy for earthquake prediction research and other purposes, and *recognizing* the need for earth rotation and polar motion measurements with subdecimeter accuracy in order to support such crustal movement measurements, and *recognizing* the capability of new techniques to achieve such accuracy with a moderate number of stations, *recommends* that the IAG and IUGG encourage and endorse international cooperative efforts for determining the earth's rotation and polar motion with subdecimeter accuracy.

(4) This symposium records its deep appreciation and extends its thanks to Stanford University for hosting the meetings, to the U.S. Geological Survey, the National Oceanic and Atmospheric Administration, and the National Aeronautics and Space Administration for financial support, and to the organizing committee who in their many ways have produced a most effective and interesting program on this relaxed and pleasant campus. Once again the Recent Crustal Movements of the State of California have proved most active and exciting.

R.O.B.

CONTENTS

LIST OF AUTHORS

Abdullah, Sh., 343, 344, 345
Allen, C.R., 1
Alt, J.N., 239, 533
Anderson, R.E., 417
Angelier, J., 267
Atkinson, L., 381
Barrows, A.G., 376
Beavan, R.J., 121
Beeby, D.J., 376
Bell, E.J., 561, 571
Belokopytov, V.A., 203
Bender, P.L., 69
Bennett, J.H., 303
Bilham, R.G., 121
Bill, M.G., 497
Bousquet, J-C., 277
Brander, J.L., 497
Brown, L.D., 181, 191, 223
Bucknam, R.C., 417
Bufe, C.G., 600, 603
Burford, R.O., 481, 603
Carter, W.E., 39
Castle, R.O., 287, 301
Chmyriov, V.M., 343, 344
Citron, G.P., 191, 223
Claflin, E.S., 49
Cluff, L.S., 431, 521, 533
Coleman, R., 15
Darrow, A.C., 389
Davidenko, Ya., 203
Davis, P., 83
Donabedov, A.T., 347
Donnelly, T.W., 431
Dracup, J.F., 49
Dragert, H., 87
Dronov, V.I., 343
Dunn, P.J., 59
Elliott, M.R., 249, 287
Evans, K., 83
Faller, J.E., 69, 107
Farrell, W.E., 97
Flemming, N.C., 177
Franzen, W., 97

Fuis, G.S., 601
Gerstenecker, C., 157
Ghisetti, F., 361
Gohn, G.S., 183
Golubev, A.N., 605
Goodkind, J.M., 99
Greiner, G., 349
Grigoriev, A.S., 347
Groten, E., 157
Guha, S.K., 549
Guseva, T.V., 203
Harada, T., 469
Hardy, R.L., 139
Harsh, P.W., 519, 603
Hein, G., 157
Higgins, B.B., 183
Hill, R.L., 303
Holdahl, S.R., 139
Holzer, T.L., 304
Hosfall, J., 83
Huggett, G.R., 84
Illies, J.H., 349
Itô, H., 629
Johnston, M.J.S., 85, 520
Kahle, J.E., 376
Kailasam, L.N., 211
Kasahara, K., 3, 329
Kato, T., 305
King, C-Y., 120
King, G., 83
Kolenkiewicz, R., 59
Kolotov, B.A., 344
Kumar, M., 75
Lajoie, K.R., 378, 380, 407
Lambert, A., 87
Lawrence, M.B., 181
Lensen, G.J., 317
Levine, J., 69, 77
Liard, J., 87
Lindh, A.G., 601
Lyttle, P.T., 183
MacDoran, P.F., 47, 49
Mälzer, H., 167
Marks, S.M., 600
Mason, R.G., 497

Masters, E.G., 15
Mather †, R.S., 15
McConnell, J.R., 505
McCrory, P.A., 407
McHugh, S., 520
Medovikov, A.S., 605
Michailova, A.V., 347
Moody, S.E., 69, 77
Morabito, D.D., 49
Mortensen, C.E., 85
Mueller, I.I., 75
Nason, R., 604
Niell, A.E., 49
Nikonov, A.A., 5, 644, 647
Odinev, N.N., 203
Ong, K.M., 49
O'Rourke, J.E., 505
Osokina, D.N., 647
Page, W.D., 533
Pavoni, N., 363
Pearlman, M.R., 69
Pelton, J., 179
Pevnev, A.K., 203
Plafker, G., 533
Prilepin, M.T., 605
Reilinger, R.E., 191
Resch, G.M., 49
Ridley, A.P., 319, 505
Rinker, R.L., 107
Rizos, C., 15
Rogers, T.H., 521
Roquemore, G.R., 409
Rundle, J.B., 627
Sanders, C.O., 585
Sarna-Wojcicki, A., 380
Schäfer, K., 118
Schmitt, G., 167
Schubert, C., 447
Schwartz, D.P., 431
Seeley, M.W., 319
Shachmuradova, Z.E., 347
Shimura, M., 469
Sidorov, V.A., 347
Silberman, M.N., 561

XX

Opening Remarks and Special Reports

WELCOMING REMARKS

CLARENCE R. ALLEN

Seismological Laboratory, California Institute of Technology, Pasadena, Calif. 91125 (U.S.A.)

On behalf of the Organizing Committee, I extend to each of you a cordial welcome to California, to Stanford University, and to the 1977 Recent Crustal Movements Symposium. We hope that you will find the Symposium both enjoyable and scientifically stimulating. We are encouraged by the variety of papers to be delivered and by the diversity of disciplines represented by the attendees. This multidisciplinary approach to the study of recent crustal movements has been a prime factor in making the field one of such great worldwide interest today.

For many of us from California, the most interesting aspect of the Symposium will be the opportunity to compare the recent crustal movements in this region with those that many of you will be describing from other parts of the world. A nagging question that repeatedly confronts us here in California is that of the degree to which crustal movements here are typical or atypical. To what degree are studies carried out here applicable elsewhere, and vice versa?

Not so many years ago, the alleged dominance of horizontal movements in California was thought to be unique, if not somehow peculiarly aberrant. Similar movements are now recognized in many parts of the world and are, of course, fully consistent with current concepts of plate tectonics. Indeed, the San Andreas fault now appears to be a remarkably simple transform fault as compared to those of many other regions. At the same time, vertical movements are being increasingly recognized in California as being more significant than we once thought. The rapidity of movement of the Southern California Uplift (or "Palmdale Bulge") has admittedly surprised all of us in California, but to what degree is it unique — even in the western United States? It is hard to believe that similar episodes are not typical of many other localities throughout the world.

Why is it that fault creep, which is so spectacular along parts of the San Andreas fault in California, is seen almost nowhere else? Is it because we simply haven't looked hard enough elsewhere, or is California truly unique in this respect? I — along with many of you — have looked hard for similar evidence of fault creep in tectonically active areas such as Japan, New Zealand, and the Middle East, and yet I am aware of only one isolated area in Turkey where anything comparable is taking place. Perhaps we have been

unduly influenced by one particular segment of the San Andreas fault in
California that is in fact somewhat unique in terms of the rock types and
very shallow tectonism along it. Assuredly this is a subject that will be
receiving attention at this Symposium and in future research efforts.

One unique aspect of California tectonism that seems to have stood the
test of time is the almost universal shallowness of earthquakes here; very few
occur at depths greater than 20 km. Such does not seem to be the case in
most other tectonically active regions of the world, including areas such as
New Zealand that are dominated by strike-slip faults otherwise remarkably similar to those of California. In this sense, perhaps crustal movements
in California really do represent a somewhat atypical tectonic environment.

Those of us who live in California are proud of our recent crustal movements, however typical or atypical they may be, and however ominous they
may appear to some on the outside. We do truly hope, however, that none of
these occurs as a rapid displacement on the San Andreas fault near Palo Alto
within the next five days. To our knowledge, no one has predicted such an
event so the Organizing Committee feels confident in welcoming you to a
safe as well as a productive meeting, and we wish you well for the days
ahead.

Tectonophysics, 52 (1979) 3

© Elsevier Scientific Publishing Company, Amsterdam — Printed in The Netherlands

3

THE SYMPOSIUM ON THE RECENT CRUSTAL MOVEMENTS, JAPAN, 1976

KEICHI KASAHARA

Earthquake Research Institute, Tokyo (Japan)

A symposium on recent crustal movements was held on October 14—15, 1976 at the International Latitude Observatory of Mizusawa. The Sub-Committee for Recent Crustal Movements, the National Committe for Geodesy and Geophysics of Japan and the Geodetic Society of Japan were the joint promoters.

Its principal subject was similar to that of the previous meeting in Tokyo, 1964 (J. Geod. Soc. Japan, 10 (1964): 111—214). However, the decade since then has seen remarkable progress in the scope and intensity of the present research field. Many factors may account for this; for example, the concepts on plate tectonics have provided us with a global view of crustal movement studies, and the increasing activity in earthquake prediction research is beginning to accumulate a great amount of data on crustal movements. Reorganization of international coordination, which was initiated at the last CRCM meeting (Grenoble, 1975), may also account for the expansion of this field. In order to respond to these circumstances, the symposium was convened with special interest in the methodological topics aside from field problems, i.e., formulation of empirical rules, instrumentation and data-filing problems.

There were five invited lectures and twenty individual contributions in the symposium. In the final session we also had a general discussion for the purpose of summarizing the symposium and of outlining the future scope of research. The Proceedings of the Symposium (Sub-Committee for the Recent Crustal Movements, 1977), present the talks at the meeting, with the invited lectures as papers and the individual contributions as letters. For editorial convenience, they are not necessarily arranged in the same order as they were given at the meeting. The general discussion in the closing session is also published (in Japanese text), in the same volume of the Journal.

K. KASAHARA (Tokyo)

REFERENCES

Sub-Committee for the Recent Crustal Movements, Japan, 1977. The Symposium on the Recent Crustal Movements. In: I. Nakagawa (Editor), J. Geol. Soc. Jap., 22: 225—318.

Tectonophysics 52 (1979) 5—14   5
© Elsevier Scientific Publishing Company, Amsterdam — Printed in The Netherlands

INTERIM REPORT 1975—1976 OF THE FENNOSCANDIAN SUBCOMMISSION OF THE COMMISSION ON RECENT CRUSTAL MOVEMENTS

A.A. NIKONOV

Academy of Sciences, Institute of Physics of the Earth, Moscow (U.S.S.R.)

INTRODUCTION

The activity of the Fennoscandian Sub-Commission dates back as far as the 3rd International Symposium on Recent Crustal Movements (Leningrad, U.S.S.R.) held in 1968. In 1971, shortly after the 15th General IUGG Assembly in Moscow it was expanded and given organizational status. Sub-Commission Progress Reports, compiled by its Chairman, Prof. T.J. Kukka-mäki, were published in 1969 (T.J. Kukkamäki, Report on the work of Fennoscandian Sub-Commission. In: Problems of Recent Crustal Movements. 3rd Int. Symp. Leningrad, U.S.S.R., 1968; Moscow, 1969), and 1975 (T.J. Kukkamäki, Report on the work of the Fennoscandian Sub-Commission. In: Problems of Recent Crustal Movements. 4th Int. Symp. Moscow, U.S.S.R., 1971; Tallin, 1975.

The last Sub-Commission Report including a list of references was presented to the 16th General IUGG Assembly in Grenoble, 1975 and distributed among the delegates. The Report covered the period 1971—1974.

The present Report is a continuation of the preceding one and covers the period 1975—1976. Being an interim report it does not claim to cover all the studies carried out in the region. It is compiled on the basis of information from representatives of each country at the request of the Sub-Commission secretary, and a bibliography for 1975—1976 by countries and regions is included.

U.S.S.R.

Kola peninsula and Karelia (according to Dr. G.D. Panasenko)

Geodetic measurements. In 1975 regular levelling was performed in the Khibine Mountains along the lines of Tikozero—Kirovsk, Titan—Rasvum-chorr, in the Rasvumchorr pit and in the Yukspor tunnel. At the tide-gauge stations in Kandalaksha and Umba (the Bay of Kandalaksha in the White Sea) geodetic investigations on measuring the zero position of the stations were accomplished. Similar measurements were carried out in 1976 at tide-gauge stations of Kandalaksha, Umba and Kovda. Levelling in the Kukisvum-chorr and Yukspor tunnels was repeated in 1976 and additional bench-marks were set up along the Kandalaksha—Kovda line.

Gravimetric investigations included measurements on secular gravimetric points in 1975 and installation of new points along the profile from Zasheeck to the Kovda settlements in 1976.

Tiltmeter measurements were carried out on the following temporary points: Murmansk (1974), Lovozero (1974—1975), Rikolatva (1975—1976), Neblogora (1976). Constant measurements are going on in the Kukisvum-chorr pit.

Geomorphological studies were conducted along the White Sea shores with the aim of making more precise the heights and age of Post-Glacial shorelines and to reveal their deformations.

Publications

1971—74 (Supplement)

Kolomyts, A.S. and Panasenko, G.D., 1973. First results of observation of measuring the tilt of the station 'Kukisvumchorr' Peninsula of Kol'sk. In: Comtemporary Movements of the Earth of the Geodynamical Polygon. Alma-Ata (in Russian).

Koshechkin, B.I. and Strelkov, S.A., 1974. Evidence of the Newest Tectonics in the North-eastern Baltic Shield. Petrozavodsk (in Russian).

Lukashov, A.D., 1974. Basic outlines of the Karelian neo-tectonics. In: Newest and Modern Movements of the Earth's Crust of the Eastern Part of the Baltic Shield. Petrozavodsk (in Russian).

Markov, G.A., Turchaninov, I.A., Ivanov, V.I. and Kozyrev, A.A., 1973. The field of tectonic high tension according to data in the Hibinsk massif. In: High Tension Conditions of the Earth Crust. Nauka, Moscow (in Russian).

Markov, G.A., Turchaninov, I.A. and Lovchinkov, A.V., 1973. High tension condition type in the massif of Lovosersk. In: Exploration of Ore Deposits in the Kola Peninsula Apatity (in Russian).

Panasenko, G.D., Bogdanov, V.I., Zhamaletdinov, A.A., Kolomyts, A.O. and Obolenskaia, L.M., 1972. Basic scientific findings of research in physics of the earth crust. A collected work. In: Geophysical Research in the Polar Lights. Apatity (in Russian).

Rubinraut, G.S., 1974. The Most Recent Movement of the Earth's Crust in the Upper-Ponoy Depression and its Relationship with the Technical Development of Subregion Keivsk. Petrozavodsk. (in Russian).

1975—1976

Koshechkin, B.I., Markov, G.A., Nikonov, A.A., Panasenko, G.D. and Strelkov, S.A., 1975. Postglacial and recent crustal movements in the northeast of the Baltic shield. Tectonophysics, 29 (1—4) 339—344.

Lilienberg, D., Sétounskaja, L., Blagovoline, N., Bylinskaja, L., Gorelov, S., Nikonov, A., Rozanov, L., Serebrjanny, L. and Filkine, V., 1971. L'analise morphostructurale des mouvements verticaux actuels de la partis européenne de l'URSS. Problems of Recent Crustal Movements. 4th Int. Symp. Moscow, USSR, Tallin, 1975.

Baltic region

Iakubovskii, O.V. The possible reasons for discrepancies of the vertical movements of the earth's crust of the Prebaltic. In: Contemporary Movements of the Prebaltic Territory, Tartu. (In Russian, English summary).
Pobedonostsev, S.V., 1975. Analysis of the vertical movements of the coast of the Prebaltic (English summary). In: Contemporary Movements of the Prebaltic Territory, Tartu. (In Russian, English summary).
Pobedonostsev, S.V., 1975. Determination of the recent vertical movements of the sea coast of the European territory of the U.S.S.R. by the oceanographical method. In: Problems of Recent Crustal Movements. 4th Int. Symp., Moscow, U.S.S.R., 1971. Tallin (Russian summary).
Razhinkas, A.K. Geopotential field of the Prebaltic and its correlation with the contemporary movements of the earth's crust. In: Contemporary Movements of the Prebaltic Territory, Tartu. (In Russian, English summary).
Serebriany, L.R. and Setunskaya, L.E., 1975. Of the geological—geomorphological interpretation of the maps of the contemporary vertical movements of the earth's crust of Eastern Europe in the Prebaltic region. In: Contemporary Movements of the Prebaltic Territory, Tartu. (In Russian, English summary).

ESTONIAN SOVIET SOCIALIST REPUBLIC (according to Dr. G.A. Zhelnin)

Major work was accomplished by the Institute of Astrophysics and Physics of the Atmosphere as well as by the Institute of Geology, Acad. of Sci., Est. S.S.R. and other organizations.

Geodetic measurements: (1) Studies of recent vertical movements of the earth's surface on geodynamic test-sites of Navesti (central part of the Republic) and Polukula (Hiiumaa Island) were continued. Repeated levelling on these test-sites has not shown any displacements for the past 3—5 years.

(2) Studies of horizontal movements at the geodynamic test-sites of Viru Nigula on the Azern fault. Two opposite sides of a geodetic quadrangle were measured by invar tape with first-order accuracy. All sides and diagonals of the quadrangle are measured by geodimeter SG-3, whereas the fault is measured directly on two cross-sections (150 and 250 m) by invar tapes and an optical deformometer. Processing of materials is proceeding.

(3) Measurements in large cities of the republic were carried out to reveal the earth's surface.

Theoretical studies: (1) Block structure of movements was studied using data from repeated levelling.

(2) Systematic errors of levelling caused by refraction in the near-earth layer of the atmosphere and the effect of temporary atmosphere loads were studied, as well as the errors depending on the length of the directional ray, the levelling method and the spike type.

(3) The method of differential hydrostatic levelling was elaborated and respective instruments were constructed.

Publication

1975
Pobul, E.A. and Sil'dvee, H.H., 1975. About the blocks of crystal formation of the Estonian foundation. Tartu (In Russian, English summary).

8

Sil'dvee, H., 1975. Repeated measurements of the intensity of gravity in the territory of Estonia. Proceed. Akad. Nauk. Estonia S.S.R., Ser. Chem. Geol., 24 (4). (In Russian).

Vallner, L.A., Torim, A. and Sil'dvee, H., 1975. Geodynamical minipolygons in the territory of Estonia. In: Contemporary Movements of Prebaltic Territory. Tartu (In Russian, English summary).

Vallner, L.A. and Zhelnin, G.A., 1975. The new map of isobases of the territory of the Republic of Estonia (In Russian, English summary).

Zhelnin, G.A., 1975. Sources of known charts (schematical) of the speed of contemporary movements of the earth's crust (for instance, the charts of the territory of the Baltics (In Russian, English summary).

Zhelnin, G.A., Vallner, L.A. and Sil'dvee, H., 1975. The results of the first stage of research of contemporary movements of the earth's surface in the territory of Estonia (In Russian, English summary).

1976

Riandarv, IU. The vertical movements of the earth's crust in the Prebaltic according to precise levelling data, 1930—70. In: Geodetic Measurements on Farms and Buildings. Elgava. (In Russian).

Sil'dvee, H. and Miidel, A. The results of the interpretation of the contemporary movements of the earth's crust in Estonia. Proc. Sci. Tech. Conf. Highways and Geodesy, Tallin. (In Russian).

Tamm, IU.A. Hydrostatical levelling under winter conditions. Proc. Sci. Tech. Conf. Highways and Geodesy, Tallin. (In Russian).

Torim, A.A. Invar rods used in measurements by the Academy of Science of Estonia. Proc. Sci. Tech. Conf. Highways and Geodesy, Tallin. (In Russian).

Vallner, L.A. Errors in levelling due to change of conditions in atmospheric pressure. Proc. Sci. Tech. Conf. Highways and Geodesy, Tallin. (In Russian).

Zhelnin, G.A. Development of research in the study of contemporary movements in the territory of Estonia. In: Geodetic Measurements on Farms and Buildings. Elgada. (In Russian).

FINLAND (according to Prof. T.J. Kukkamäki)

Relevelling. In Lapland a single levelling line was relevelled during the period 1971—75 (Takalo, 1977). This line was first levelled in 1949—61. The land uplift was computed from these levellings and shown in Fig. 2 of Mäkinen (1977).

High-precision traverse. In order to discover possible horizontal movements, two southern sides of the 890-km-high precision traverse (Parm, 1976), measured in 1974, were remeasured in 1976. No significant movements were discovered. Remeasurements will be continued.

Land uplift gravity lines. In order to observe secular variation in gravity, three high-precision gravity lines have been observed, in cooperation with Swedish, Norwegian and West German expeditions, across the Fennoscandian land uplift area in the period 1966—76 along the 60, 63 and 65° parallels. The 63° parallel will be measured in 1977 (Kiviniemi, 1974, 1977).

Long pipe levels. Recordings of earth tides with a 160-m pipe level have been continued.

Recording gravimeters. The tidal variation in gravity has been recorded with gravimeters in cooperation with the proper institutions at seven stations in Finland: Helsinki, Vaajakoski, Joensuu, Vaasa, Oulu, Sodankylä, Kevo; at four stations in Sweden: Vitvattnet, Kramfors, Gävle, Boxholm; and at one station in Norway: Trondheim. The observations have been analysed in the Centre International des Marées Terrestres in Brussels.

Satellite geodesy. A satellite laser has been constructed and is operational at present.

Doppler observations in the frame of EDOC II were performed in April— May 1977 at the terminal points of the 890-km-high precision traverse and several additional observations are proceeding. VLBI observations of the terminal points have been planned for 1978.

The Geodynamic Working Group of the Nordic Geodetic Commission has been active on land uplift gravity lines, on recording gravity and on vertical and horizontal movements. The Finnish Geodetic Institute has participated in the work of the group, which has arranged meetings in 1976 in Gävle and 1977 in Helsinki.

References

Kiviniemi, A., 1974. High-precision measurements for studying the secular variation of gravity in Finland. Publ. Finn. Geod. Inst., 78.
Kiviniemi, A., 1977. The Finnish measurements at the Fennoscandian land uplift gravity lines. Paper at the Non-tidal Gravity Variations and Methods for Their Study Symposium in Trieste.
Mäkinen, J., 1977. Pohjois-Suomen maannousu ja Toinen vaaitus. Geofysiikan päivät, Helsinki.
Parm, T., 1976. High precision traverse. Publ. Finn. Geod. Inst., 79.
Takalo, M., 1977. Bench mark list II of the second levelling of Finland. Publ. Finn. Geod. Inst., 83.

Supplement

Honkasalo, T., 1975. Report on the application of geodetic measurements to the determination of recent crustal movements in Finland. Tectonophysics, 29 (1—4): 345— 347.
Kääriäinen, J., 1975. A long pipe tiltmeter. Problems of Recent Crustal Movements. 4th Int. Symp. Moscow, U.S.S.R., Valgus, Tallin.
Kukkamäki, T., 1975. Report on the work of the Fennoscandian Subcommission. Problems of Recent Crustal Movements. 4th Int. Symp., Moscow, U.S.S.R., Valgus, Tallin.

SWEDEN (according to L. Pettersson)

Repeated high-precision levellings across fault lines in order to investigate any land uplift which might be in progress in blocks with old fault lines as boundaries between the blocks. The levellings were commenced in 1974 at four places and are to be repeated at intervals of one or two years. The mean square of one double levelling is about ± 0.20 mm/$\sqrt{}$ km.

Repeated gravity measurements in collaboration with Finland and Norway along three east—west lines across Fennoscandia in order to investigate the relation between the change of gravity and land uplift. This can give information concerning the mass movement associated with land uplift. First observations of the three lines took place in 1967, 1975 and 1976, respectively. The observations are to be repeated at five-year intervals. This means that hitherto only one line has been measured twice. The mean square of a gravity difference is 2—3 μgal when observed eight times with four LaCoste and Romberg gravimeters.

Studies are based on potential theories.

Publications

Ekman, M., 1976. Undersökning av lutningsändringar i jordskorpan vid förkastningslinjer i södra Sverige 1974—1976. Statens Lantmäteriverk, Gävle.

Pettersson, L., 1974. Studium av sekulär ändring i tyngdkraften utefter latitud 63° mellan Atlanten och Bottenhavet. Paper presented at N.G.C. Congress, København.

The second High Precision Levelling of Sweden, 1951—1967. Rikets allmänna Kartverk, Meddelande, No. A 40, Stockholm, 1974.

Ussisoo, I., 1975. Computation of land uplift and mean sea level in Sweden. Statens Lantmäteriverk, Vällingby.

DENMARK (according to Dr. O.B. Andersen)

Current observations at the tide-gauge stations continued. Preparations were made for measurements along high-precision gravimetric profiles within the Nordic block.

For list of publications see the 1971—74 report.

NORWAY (according to Dr. S. Bakkelid)

Studies carried out

Studies of recent crustal movements in Norway have hitherto been limited to vertical movements. Studies have been made in three different fields:

(1) Yearly rates of regular vertical movement of the land relative to sea level (RLU).

(2) Vertical displacements over short time periods caused by volcanic activity.

(3) Post-Glacial shore-level displacements.

Determination of rates of land uplift (RLU)

RLU-values are determined using data from the following sources:

(1) Long-time operation of tide gauges.

(2) Reobservation of the height relative to MSL of old watermarks cut into steep faces of rocks.

(3) Relevellings of precision levelling lines.

(4) Repeated observations of high-precision gravity profiles.

TABLE I

	Latitude	Longitude	RLU (mm/year)
Oslo	59°54'	10°45'	3.9 ± 0.3
Oscarsborg	59°40.5'	10°36.7'	2.7 ± 0.2
Nevlungshavn	58°58'	9°53'	1.9 ± 0.5
Tregde	58°00'	7°34'	0.0 ± 0.3
Stavanger	58°58'	5°44'	−0.3 ± 0.2
Bergen	60°24'	5°18'	0.1 ± 0.2
Kjølsdal	61°55'	5°38'	0.0 ± 0.6
Heimsjø	63°26'	9°07'	1.6 ± 0.5
Narvik	68°26'	17°25'	2.9 ± 0.4

Long-time operation of tide gauges

The Geographical Survey of Norway (N.G.O.) has continued the operation of twelve permanent gauges and has established 4 more that will be in operation in a few years.

The Norwegian Hydrographic Office (N.S.K.V.) and other governmental agencies operate 10 permanent or semipermanent gauges.

RLU-values, computed from tide-gauge data, have been reported in earlier reports. Updated RLU-values for some gauges are given in Table I. Table I includes data for gauges that have been in operation more than 40 years. In 10 years time reliable values of RLU can also be determined for the greater number of other gauges.

Observations of the height of old watermarks

Heights relative to MSL of five old watermarks which in 1839 were cut into steep faces of rock projecting from the sea at appropriate places along the coast were observed for the fourth time in 1974. Previous observations were made in 1865, 1890 and 1939.

In 1975 and 1976 fourteen more such marks were observed for the fourth time. The height relative to MSL of the upper rim of the growth of bladder wrack (*Fucus vesiculosus*) was observed first in 1890 and repeat were made in 1976. The new observations have not yet been computed.

Repeated precise levelling

The Alta—Kautokeino line (line A), levelled first in 1954, was relevelled in 1974—75. The Haman—Elverum—Alvdal line (line B), levelled first in 1935—36 and was relevelled in 1976. The scope of the relevelling of line A was primarily to locate a supposed error in the 1935 levelling; therefore, the difference in the rate of land uplift between Alta and Kautokeino is unreliable and is not reported. The levellings of line B give approximately 0.5 mm/yr as the rate of land uplift relative to Hamar. From Elverum northwards to Alvdal only insignificant variations of land uplift could be detected.

Repeated observations of precision gravity profiles

The northern profile, Korgenfjellet—Kuhmo, was observed first in 1975 and the southern, Bergen—Virolahti, was first observed in 1976 as a joint Finnish, Swedish and Norwegian enterprise.

Results of observations of the middle profile, Joensu—Vägstranda, was reported by L. Petterson at the meeting of the Nordic Geodetic Commission in Copenhagen in 1974. The gravity changes at the different stations along the profile seem to disagree with the rates of land uplift determined by repeated precise levellings (Petterson, 1974; Kiviniemi, 1974).

The data from tide gauges and from repeated height observations of old watermarks give absolute values of RLU for points on the coast. Repeated precise levelling and repeated precise gravity observations give relative values of RLU for points inland.

Crustal displacements by volcanic and earthquake activity

At Jan Mayen, the small volcanic island at 71°N 8°30'W, the three-station seismograph network equipped with Geotech Model S-13 seismometers and a network of six simple surface tiltmeters have been in operation since 1973, as has the tide gauge. Relatively large tilts are recorded at sites located on 'bedrock' and lesser tilts are located on sand foundations (Sylvester, 1975).

A gravity station network established in 1974 and comprising 19 stations was again observed in 1975.

The tide-gauge registrations and the gravity data have not yet been analysed to determine vertical crustal movements.

Plans have been made for additional crustal strain measurements on Jan Mayen, such as triangulation and laser-ranging. From a global net of seismological stations, as indicated from seismograms, a clear strike—slip solution has been obtained for an intra-plate earthquake on January 18, 1976 located just offshore West Storfjorden on Svalbard (Bungum, 1976).

Post-Glacial shore-level displacements

In Norway there is ample opportunity for, and also a long tradition in, studying ancient shore-level changes, i.e., primarily the relative shore-level changes, or resultant difference between the eustatic sea-level changes and isostatic crustal movements. Modern methods are based on pollen and diatom analysis and radiocarbon datings. The results obtained expose very clearly the great difference in Late Pleistocene shore-level changes existing from one region to another, in particular the profound differences between the peripheral coast regions and the central and eastern part of the country (Hafsten, 1975).

This difference concerns both the highest marine limit, which in the eastern part (Oslo area, at about long. 11°E) reached a height of about 220 m above present sea-level. The general trend of the shore-level displacement, particularly the Post-Weichselian sea-level (Tapes) transgressions, affected only the less depressed, peripheral, parts, i.e., the western and extreme southern coast regions.

Within the strongly depressed, eastern part of South Norway (viz. the Oslofjord and Trondheim regions) the early Post-Weichselian drop in shore level occurred so rapidly that no less than two-thirds of the total Post-Weichselian shore displacement had taken place before the transition to the Atlantic period (viz. prior to 8200 B.P.), i.e., at a rate of displacement of 7—8 m per century in average for Pre-Boreal (10,300—9400 B.P.) and Boreal time (9400—8200 B.P.). Due to the oblique, Late Pleistocene uplift of the land mass the marine levels (viz. the isobases) generally expose a rising gradient from the western and extreme southern coast regions towards the central part of South Norway. This is seen very clearly along the extensive fjord systems of western Norway, where, for example, the shore-level isobases during the deglaciation period (Younger Dryas/Pre-Boreal) drop from altitudes of 120—130 m at the head of the fjord systems to 20 m or less in the outer coast regions. Within the Oslofjord region the isobases show a rising gradient from south to north of about 0.8 m/km during early Pre-Boreal time.

Flooding or an actual rise in shore level, due to the Post-Weichselian sea-level (Tapes) transgressions, occurred only in the peripheral, western and extreme southern coast regions. The rise in shore level since the time of the Boreal transgression minimum on the extreme southwest coast (Jaeren-Fonnes) may of the order 6—7 m. At Fonnes the mid-Atlantic transgression maximum reached a height above present sea-level of 10.9 m, whereas the Boreal regression minimum lay well below present sea-level, probably at a depth of about 4 m. Beach ridges as well as stratigraphic evidence of the Post-Weichselian transgression up to a height of about 7 m above present sea-level suggest a maximum rise in shore level here of about 11 m.

The southwest coast, which became deglaciated during early Late Weichselian time (around 13,000 B.P.), also shows evidence of a Late Weichselian transgression, viz. during pre-Allerød time.

References

Bungum, H., 1976. Two focal mechanism solutions for earthquakes from Iceland and Svalbard. NORSAR Contrib., No. 201.

Hafsten, U., 1975. Late and Post-Weichselian Shore Level Changes in South Norway. Uppsala University, Uppsala.

Kiviniemi, A., 1974. High precision measurements for studying the secular variation in gravity in Finland. Publ. Finn. Geod. Inst., 78.

Petterson, L., 1974. Studium av sekulär ändring i tyngdkraften utefter latitud 63° mellan Atlanten och Bottenhavet. Presented in Copenhagen at the meeting of the Nordic Geodetic Commission.

14

Sylvester, A.G., 1975. History and surveillance of volcanic activity on Jan Mayen Island. Bull. Volcanol., 39 (2): 1—23.

Supplementary list of Swedish publications

Lundqvist, J. and Lagerbäck, R., 1976. The Pärve Fault: a late-glacial fault in the Precambrian of Swedish Lapland. Geol. Fören. Stockh. Förh., 98: 45—51.

Mörner, N.-A., 1973. Eustatic changes during the last 300 years. Palaeogr. Palaeoclimatol. Palaeoecol., 13: 1—14.

Mörner, N.-A., 1973. New method of separating glacio-isostatic and tectonic components in Scandinavian uplift. Abstr., INQUA 9th Congr., N.Z., 1973, p. 255; Proc. Symp. Jakarta, 1973, on Recent Crustal Movements and Associated Seismological and Volcanic Activity.

Mörner, N.-A., 1975. Postglacial earth movements. Nat. Rep. Swed. G.D.P. Comm., Grenoble, p. 1—10.

Mörner, N.-A., 1975. Double-nature of the Fennoscandian uplift. Abstr., 16th General Assembly, I.U.G.G., Grenoble, p. 45.

Tectonophysics, 52 (1979) 15—37
© Elsevier Scientific Publishing Company, Amsterdam — Printed in The Netherlands

Crustal Deformation Using Extra-Terrestrial Geodesy

GEODETIC REFERENCE SYSTEMS FOR CRUSTAL MOTION STUDIES

R.S. MATHER †, C. RIZOS, R. COLEMAN and E.G. MASTERS

Department of Geodesy, University of New South Wales, Sydney N.S.W. 2033 (Australia)

(Accepted for publication March 24, 1978)

ABSTRACT

Mather, R.S., Rizos, C., Coleman, R. and Masters, E.G., 1979. Geodetic reference systems for crustal motion studies. In: C.A. Whitten, R. Green and B.K. Meade (Editors), Recent Crustal Movements, 1977. Tectonophysics, 52: 15—37.

The behaviour of geodetic reference systems as a function of time cannot be ignored in determinations of position for secular geodynamic modelling. While non-specialist users are likely to appreciate the uncertainty surrounding the concept of a "fixed" point in regional surveys for crustal motion, its extension to global surveys based on techniques in dynamic geodesy is not widely understood. Such solutions are related to the natural coordinate system defined by the instantaneous geocentre (earth's centre of mass) and the instantaneous rotation axis whose location in earth space cannot be expected to be time invariant.

The effect of such movements of the natural system of reference on the coordinates of points at the earth's surface is computed for a plausible model of the former. The resulting changes in solutions for three-dimensional position from dynamic considerations are discussed. It is shown that for the model adopted, the earth-space variations of the natural reference system with time produce coordinate changes with magnitudes not dissimilar to those resulting from reasonable variations in the earth's figure, though with different wavelengths, when computed from three-dimensional considerations alone.

The resulting displacements, along with an accepted model for plate motions, are used to study changes in the shape of the level surfaces of the earth's gravity field with time. It is shown that plausible models for the deformation of the earth's figure produce significant changes in the geopotential and the shape of the geoid. However, the transfer of mass implied in extreme models of plate motion produce no significant changes in the datum level surface for crustal motion studies over periods as long as 10^2 yr.

1. THE CONCEPT OF GEODETIC POSITION IN FOUR DIMENSIONS

The basic objective of geodetic surveying is the determination of point positions at the earth's surface. All major geodetic determinations attempted to date have pursued this objective in three dimensions. The precision achieved at present appears to be 1 part per million (ppm) or ±6 m in each coordinate, both on a regional (e.g., Bomford, 1967) as well as a global scale (e.g., Anderle, 1974; Smith et al., 1976). Factors such as earth tides, plate motions, vertical crustal motions, etc., which cause variations in position are not capable of discrimination in such determinations. In the case of non-periodic phenomena, a measure of resolution may be obtained if continuity in geodetic concepts were maintained over periods as long as 10^2 yr. The situation is more favourable in the case of vertical crustal motion studies, where the relevelling of relatively large areas using first-order techniques appears to have defined vertical crustal motions to ± 0.2 mm yr^{-1} (e.g., RAK, 1974). All such deductions of vertical crustal motion rates assume that the radial component of the position of level surfaces of the earth's gravity field is invariant with time. This hypothesis is examined more closely in sections 5 and 7.

The realization of the potential for significant improvement in the precision of geodetic measurements (e.g., Bender et al., 1973; Ong et al., 1976; Smith et al., 1976) can provide the basis for the determination of position to ± 10 cm under certain conditions (Mather, 1974). The redetermination of position at the jth station P_j in a n-station network would give two different sets of coordinates $X_{ij}(t_1)$ at epoch ($\tau = t_1$) and $X_{ij}(t_2)$ at epoch ($\tau = t_2$). The term "epoch" is used in this paper to refer to a time span of up to one year, during which a system of observations is established for the purpose of determining position, ideally, on a global basis. The usefulness of such determinations for crustal motion studies is dependent on whether the system of reference used in each case is the same. Alternatively, it is necessary to establish the relationship between the earth-space locations of the two systems of reference used.

The present development deals with the consequences of interpreting geodetic results obtained for each of the epochs ($\tau = t_1$) and ($\tau = t_2$) using *three-dimensional considerations alone* without due regard for the behaviour of the implied systems of reference between epochs. Section 2 deals with the concept of a *simple* system of reference for geodetic use in four dimensions. Section 3 deals with the motion of the geocentre in earth space, which is the space having the same rotational and galactic motion as the earth (Mather, 1972). In geodetic practice calling for the highest precision, the concept of "earth rotation" is defined in the average sense as sampled from the network of geodetic observatories at the surface of the earth which generate the system of observations used in defining position. Section 4 considers the effect of changes in the shape of the reference model which best fits the geoid, defined as the level surface of the earth's gravity field in relation to which mean sea-level has zero mean on global sampling (Mather, 1977a). Section 5 deals with the role of the geoid in vertical crustal motion studies.

2. THE SYSTEM OF REFERENCE IN FOUR DIMENSIONS

Observations taken during the epoch of measurement are converted to notions of position by appropriate reference to a system of geodetic reference. The earth-space location of the most convenient system of reference may vary from epoch to epoch. The selection of an appropriate system of reference is influenced by the nature of the observations to be used in the determination of position. For example, the rates of change of observations made to extra-terrestrial sources which do not rotate with the earth, are sensitive to the rotation vector of the earth relative to the observing station. In addition, such data collected in relation to sources in earth orbit, contain information on the location of the instantaneous geocentre (earth's centre of mass) which is one focus of the osculating elliptical orbit.

Consequently, a *simple* "natural" coordinate system for observations to extra-terrestrial sources is the three-dimensional Cartesian coordinate system x_i in earth space with its origin at the instantaneous geocentre, the x_3 axis coincident with the instantaneous axis of rotation and the x_1x_3 plane (plane of zero longitude) defined by a *single* point. Present-day systems like the 1968 B.I.H. system (Guinot and Feissel, 1968) have their x_1x_3 plane defined by the point of intersection of the plane of zero longitude and the equator as maintained *in an average sense* by a network of astronomical observatories.

As a result of the motion of the rotation pole in relation to the earth's crust (e.g., Rochester, 1973), the instantaneous natural system of reference x_i at epoch ($\tau = t$) does not necessarily coincide with that (X_i) for the selected epoch of reference ($\tau = t_0$). The latter is similar to the 1968 B.I.H. system, hereafter called the *C.I.O.—Greenwich* system of reference, to which positions should be referred, irrespective of epoch. The coordinates of the same point P_j at the earth's surface on the two systems of reference are related by equations of the form:

$$X_{ij} = x_{ij} - \epsilon_{ikl}x_{kj}\omega_l + o\{\omega_l^2 X_{ij}\} \tag{1}$$

where:

$$\epsilon_{ikl} = \begin{cases} 0 & \text{if } i = k, \, k = l, \text{or } i = l \\ 1 & \text{if subscripts are ordered 12312 ...} \\ -1 & \text{if subscripts are ordered 13213 ...} \end{cases} \tag{2}$$

and ω_i are counterclockwise rotations about the X_i axes, being related to the conventional components x and y of polar motion (Guinot and Feissel, 1968, p. 9) by the relations:

$$\omega_1 = y, \qquad \omega_2 = x; \qquad \omega_3 = 0 \tag{3}$$

Equations similar to (1) and (2) can be used to relate coordinates on any regional geodetic system to a system of the C.I.O.—Greenwich type provided: (1) the displacement of the regional origin from the geocentre; and (2) the appropriate rotations ω_i are known.

Furthermore, any global geodetic position determination in three dimensions can be referred to the X_i axes system irrespective of epoch of observation, provided:

(1) Its relation to the instantaneous natural system of reference were known.

(2) Both polar motion and geocentre motion were defined between the epoch of reference ($\tau = t_0$) and the epoch of observation ($\tau = t$).

Obviously, the notion of geocentre and polar motion is relative to earth space *as seen from the n observatories P_j at the earth's surface*. Changes in the configuration of the set of observatories may cause an apparent change in the position of the geocentre and of the pole (e.g., see Mather et al., 1977). This is unlikely to cause any serious problems if the set of observatories is fixed and has a global distribution. The use of eqs. 1—3 implies that the coordinates of *one* point — either real or conceptual — are held fixed and do not vary with time. This is equivalent to all coordinate changes being referred to this single point.

3. THE MOTION OF THE GEOCENTRE

Sensitivity to the motion of the geocentre is exhibited only by systems of observations to a source/responder *in earth orbit*, i.e., near-earth satellites or the moon. It does not apply in the case of observations to quasars or stars. A simple method is available for the determination of geocentre motion from measurements of secular changes in absolute gravity (Mather, 1973, pp. 198 et seq.). A study based on simulated data indicates that geocentre motion with rates in excess of ± 1 cm yr^{-1} can be recovered from such changes in the presence of random station noise of ± 10 μGal as observed from at least 25 well-distributed stations from over 10 years' observations, even if oceanic and polar sites were not occupied (Mather et al., 1977). These 25 sites are marked with an asterisk in Table I.

Independent three-dimensional position determinations related to the natural system of reference in each instance will produce apparent position changes of observing stations due to geocentre motion. Let δr_g be the rate of geocentre motion in the direction (ϕ_g, λ_g) from the geocentre at epoch ($\tau = t_1$). This produces components δX_g in the direction X_i between epochs ($\tau = t_1$) and ($\tau = t_2$) defined by:

$$\delta X_{gi} = \delta r_g l_i (t_2 - t_1) \tag{4}$$

where:

$$l_1 = \cos \phi_g \cos \lambda_g; \qquad l_2 = \cos \phi_g \sin \lambda_g; \qquad l_3 = \sin \phi_g \tag{5}$$

The resulting changes $\delta\phi_j$, $\delta\lambda_j$, δh_j in the latitude ϕ_j, longitude λ_j and ellipsoidal elevation h_j, respectively, at P_j are given by:

$$\delta\phi_j = \delta r_g (t_2 - t_1)(-l_1 \sin \phi_j \cos \lambda_j - l_2 \sin \phi_j \sin \lambda_j + l_3 \cos \phi_j)/R$$

$$\delta\lambda_j = \delta r_g (t_2 - t_1)(l_2 \cos \lambda_j - l_1 \sin \lambda_j)/(R \cos \phi_j) \tag{6}$$

$$\delta h_j = \delta r_g (t_2 - t_1)(l_1 \cos \phi_j \cos \lambda_j + l_2 \cos \phi_j \sin \lambda_j + l_3 \sin \phi_j)$$

where R, for all practical purposes, is the mean radius of the earth.

The last three columns of Table I give magnitudes of the coordinate

TABLE I

Stations sites used for simulating changes in position due to geocentre motion. Model for geocentre motion: 1 cm yr^{-1} in direction 36°N 45°E.

TOTAL STATION COMPLEMENT = 54

NO	STATION	IDENTIFICATION	LATITUDE DEG N	LONGITUDE DEG E	HEIGHT M	PLATE IDENTIFICATION	Dφ	Dλ cm	Dh cm **
1	6052 MAWSONCNST	PCLAR-SOUTH	-67.6010	62.8730	43.60	ANTARCTIC	93.6	-24.8	-25.0
2	6052 MILKESST *	PCLAR-SOUTH	-66.2790	110.5350	-3.30	ANTARCTIC	54.3	-73.6	-40.3
3	6053 MCMURDO	PCLAR-SOUTH	-77.8450	166.6420	-41.20	ANTARCTIC	-29.1	-68.9	-66.4
4	6050 PALMERST	PCLAR-SOUTH	-64.7740	295.9480	-37.90	ANTARCTIC	89.1	-76.5	-64.4
5	1 CAPETOWN	AFRICA-SOUTH	-33.9170	18.4170	145.00	AFRICAN	98.8	10.2	27.2
6	2 CARLON TS	OCEANIC ISLE	-4.9170	37.7500	80.00	AFRICAN *		2	11.9
7	6043 HEARD IS	OCEANIC ISLE	-53.0190	73.4930	49.60	ANTARCTIC	92.2	-38.5	-8.6
8	6032 CAVERSHM *	AUSTRALIA NZ	-31.8400	115.9750	-3.20	AUSTRALIAN	63.8	-76.5	-50.1
9	1030 RRROA	AUSTRALIA NZ	-35.6260	148.9750	958.40	AUSTRALIAN	36.4	-78.5	-50.1
10	6013 INVERCAR *	AUSTRALIA NZ	-44.4160	168.3520	3.40	PACIFIC	8.3	-67.6	-73.2
11	3 CHATHAM IS	OCEANIC ISLE	-44.0000	183.5000	2.00	PACIFIC	0.2	-53.6	-84.4
12	5 PACIFIC	HYPOTHETICAL	-45.0000	225.0000	2.00	PACIFIC	-15.0	-0.0	-98.0
13	4 PACIFIC	HYPOTHETICAL	-45.0000	255.0000	2.00	ANTARCTIC	-8.0	40.5	-91.1
14	5 PACIFIC	STH AMERICA	-3.1500	289.3320	72.20	ANTARCTIC	30.0	72.9	-61.5
15	108 SANTIAGO	STH AMERICA	-54.2840	323.5060	16.80	STH AMERICAN	44.0	80.0	-40.7
16	6061 TRINDEOIS	OCEANIC ISLE	-37.0650	347.6850	4.20	AFRICAN	73.2	68.1	-0.6
17	6067 TRISTNIS *	OCEANIC ISLE	-25.8840	27.0700	1552.20	AFRICAN	86.6	24.0	-43.8
18	1036 UCHBUANG	OCEANIC ISLE	-1.0090	47.3000	137.20	AFRICAN	81.9	-3.2	57.7
19	103 MADAGASC	OCEANIC ISLE	-7.3520	72.4720	-62.90	AUSTRALIAN	65.5	-37.3	63.7
20	6072 DIEGO SC	AUSTRALIA NZ	-10.5840	113.7160	1.40	AUSTRALIAN	55.9	-75.4	-20.8
21	704 CARNARVN *	OCEANIC ISLE	-1.6520	168.3070	122.90	PACIFIC	42.5	-80.3	-60.2
22	6078 THRSDYIS	OCEANIC ISLE	-14.3320	189.2860	31.50	PACIFIC	40.7	-67.6	-78.2
23	6078 NEWHEBRD	OCEANIC ISLE	-25.0690	250.5730	21.40	PACIFIC	19.1	-47.2	-97.4
24	6032 AMSTAMCA	OCEANIC ISLE	-25.1170	288.5070	249.70	NAZCA	46.1	3.6	-91.8
25	6039 PITCAIRN	STH AMERICA	-16.4660	324.8350	28.20	STH AMERICAN	59.9	72.4	-51.3
26	6020 ADELAIDE **	STH AMERICA	-59.2700	345.5930	8.50	STH AMERICAN	63.0	79.7	-7.7
27	9007 AREQUIPA	OCEANIC ISLE	12.1320	15.0350	311.40	STH AMERICAN	42.7	69.6	32.6
28	6061 NATALBRZ **	AFRICA	-6.7480	38.9590	1915.10	AFRICAN	45.9	40.4	80.9
29	6054 CENASOL	INDIC PAK-SBC	25.3600	79.4580	1885.90	AUSTRALIAN	18.5	-45.4	87.0
30	604 AFTICAPY	ASIA/F EAST	16.7000	98.9670	26.90	EURASIAN	40.3	-65.4	251.1
31	908 ACCISARA *	OCEANIC ISLE	6.9220	122.0690	74.30	PACIFIC	81.6	-78.9	-20.6
32	900 NAINITAL *	OCEANIC ISLE	20.2910	166.6110	305.60	PACIFIC	69.5	-68.9	-20.7
33	6072 CHANGMA	HYPOTHETICAL	20.7070	203.1430	5.00	PACIFIC	79.4	-29.3	-49.7
34	6047 ABIDANG	OCEANIC ISLE	18.7300	249.0450	-8.00	CARIBBEAN	67.2	33.0	-51.1
35	6012 WAKE-IS	OCEANIC ISLE	-0.0000	285.0000	-25.00	STH AMERICAN	67.2	70.1	-23.9
36	9012 HAWAII	AFRICA	14.7450	304.7950	50.20	AFRICAN	59.9	79.6	-8.1
37	6033 PACIFIC	STH AMERICA	-18.0790	342.5170	96.60	EURASIAN	-0.3	71.7	51.1
38	603 SCCOROIS	AFRICA	36.2400	233.9330	410.50	EURASIAN	1.1	-20.4	94.3
39	11 CARIBBEAN	MIDDLE EAST	41.3830	59.6290	898.60	EURASIAN	-4.1	-32.9	57.2
40	6009 PARAMART *	SOVIET UNION	36.9060	169.0000	5.10	EURASIAN	2.8	-76.7	57.6
41	6003 CAARAR	HYPOTHETICAL	36.7140	139.1920		NTH AMERICAN	51.0	-80.7	29.8
42	6005 DINASLCS *	OCEANIC ISLE	54.0000	174.1240	235.00	NTH AMERICAN	98.4	-62.8	15.8
43	603 MASHHAD *	NTH AMERICA	44.5000	236.9000		NTH AMERICAN	89.4	44.1	5.7
44	7 MASHKENT	OCEANIC ISLE	32.3480	295.3460	-76.00	NTH AMERICAN	88.4	16.7	-32.0
45	9025 DCCAIRA	OCEANIC ISLE	37.4630	307.0940	-75.10	NTH AMERICAN	64.2	76.2	8.5
46	902 STEMYA	EUROPE RUSSIA	65.6620	355.1940	124.10	AFRICAN	47.7	80.2	36.2
47	906 EUROPIEN	PCLAR-NORTH	64.9770	218.4410	29.90	EURASIAN	-47.7	63.1	75.7
48	10 VANCOUVR *	PCLAR-NORTH		212.4750		NTH AMERICAN	96.4	-17.5	80.4
49	901 ATHODKR *								19.9

* Preferred 25 Station Set Effect of geocentre motion (per century)

** Representation of geodetic coordinate changes as changes at the local Laplacian triad.

changes which result on neglecting the effect of geocentre motion modelled as follows:

$$\delta r_g = 1 \text{ cm yr}^{-1}; \qquad \phi_g = 36°\text{N}; \qquad \lambda_g = 45°\text{E}; \qquad (t_2 - t_1) = 100 \text{ yr}$$

Such apparent changes in position could be inferred from *differences* in coordinates if determined using the natural system of reference for each epoch after correction for polar motion. Coordinate systems of this type are widely used in dynamic satellite geodesy.

4. EFFECT OF EARTH MODEL CHANGES ON GEODETIC COORDINATES

The geodetic coordinates $X_{ij}(t)$ of the point P_j in a n-station geodetic network at the earth's surface, can undergo changes δX_{ij} as a function of time for the following reasons:

(1) Changes in the earth-space location of the natural coordinate system with time (Mather, 1973).

(2) Relative motions between the n stations at the earth's surface.

Changes at (1) due to geocentre motion have already been considered in section 3. The motion of the instantaneous axis of rotation in earth space can be modelled and the model improved during the analysis of observations. There is no possibility that its neglect is a potential source of confusion to the uninitiated user. The changes at (2) can be due to the following causes: (a) plate motions; (b) local crustal movements; and (c) changes in the shape of the earth.

The motion of the jth plate in a system of N plates can be represented by the three components ω_{rij} of the rotation rate vector ω_{rj} defining the plate motion, in the directions X_i (e.g., see Solomon and Sleep, 1974). From purely geodetic considerations it is convenient if all such rotations are referred to the African plate, as the origin defining the $X_1 X_3$ plane of the C.I.O.—Greenwich system of reference is located on this plate, as discussed in section 2.

There is no evidence at present to support the existence of secular changes in the shape of the ellipsoid which best fits the geoid. Arguments can be advanced against placing bounds for polar wandering (e.g., Goldreich and Toomre, 1969). Estimates of secular polar motion have also been made on the basis of astronomical observations (e.g., Markowitz, 1974). The secular increase in the length of day, currently estimated to be 0.1 ms yr^{-1} (B.I.H., 1975) may presage a change in the earth's meridional flattening after an appropriate delay related to the rigidity of the earth. While corresponding changes in the gravitational attraction are only 10^{-3} μGal yr^{-1} (Mather et al., 1977), a rate of 1 cm yr^{-1} is modelled in the present study for the quantity $a\, df$, where a is the equatorial radius of the earth and df is the rate of change in the meridional flattening. This would be compatible with any tendency of mass transfers due to plate motions having characteristics of second-degree zonal harmonics on global consideration.

The resulting changes δX_{ijf} produced in the coordinates X_{ij} of the jth station P_j between epochs $(\tau = t_1)$ and $(\tau = t_2)$ are given by:

$$\delta X_{ijf} = -R \, df \, \sin^2\phi_j l_{ij}(t_2 - t_1) + o\{f\delta X_{ijf}\} \tag{7}$$

where:

$$l_{1j} = \cos\phi_j \cos\lambda_j; \qquad l_{2j} = \cos\phi_j \sin\lambda_j; \qquad l_{3j} = \sin\phi_j \tag{8}$$

Similarly, it is possible to define the changes $\delta X_{ijf'}$ in X_{ij} resulting from changes df' in the equatorial flattening f' of the reference ellipsoid using a relation of the form (e.g., Bomford, 1971, p. 477):

$$\delta X_{ijf'} = \tfrac{1}{2}R \, \cos^2\phi_j \, df' \cos 2(\lambda_j - \theta) \, l'_{ij}(t_2 - t_1) + o\{f \, \delta X_{ijf'}\} \tag{9}$$

where θ is the longitude of the principal axis of inertia corresponding to the semi-major axis of the equatorial ellipse, and

$$l'_{1j} = \cos\lambda_j; \qquad l'_{2j} = \sin\lambda_j; \qquad l'_{3j} = 0 \tag{10}$$

Variations of θ with time have not been considered in the present study.

Table II lists changes in earth-space coordinates due to earth expansion with rate δr of 0.1 cm yr^{-1}, the effects δX_{ije} on X_{ij} at P_j being given by

$$\delta X_{ije} = \delta r l_{ij}(t_2 - t_1) \tag{11}$$

along with changes δX_{ijf} and $\delta X_{ijf'}$ for meridional and equatorial flattening at rates of 1 cm yr^{-1} in $R \, df$ and $R \, df'$, plate motions using the model of Solomon and Sleep (1974, p. 2558) and secular polar motion computed using eq. 1, with the quantities ω_i being modelled using eq. 3 in the form (Gaposchkin, 1972, p. 22):

$$\omega_\alpha = \omega_{1\alpha} + \omega_{2\alpha}(t_2 - t_1) \tag{12}$$

Row 5 in Table II gives the coordinate changes which occur at five sample stations with a wide spread in position if the motion of the natural system of reference has been correctly allowed for, being the total of rows 1—4. Rows 6 and 7 in Table II give the coordinate variations which occur if the motion of the natural system of reference was not allowed for when computing geodetic coordinates. In the case of polar motion (row 6), the gross Chandler period and seasonal effects have been excluded as not influencing average considerations over an epoch of observation. The coordinate changes listed in row 8 of Table II would be the values obtained if the tectonic motions described in rows 1—4 occurred and were referred to the instantaneous natural system of reference without allowance for the earth-space motion of the latter.

The similarity in the order of magnitudes of the totals in rows 5—7 of Table II illustrates the importance of considering the motion of the natural system of reference when providing coordinates for geodynamic use.

TABLE II

Variations in earth-space coordinates due to changes in the earth's figure, plate motions and natural coordinate system movements (in centimetres per century) *

Factors causing variations	Row	Tromsø (ϕ = 69.66°; λ = 18.94°)			Addis Ababa (ϕ = 8.75°; λ = 38.96°)		
		ΔX_1	ΔX_2	ΔX_3	ΔX_1	ΔX_2	ΔX_3
Earth expansion	1	3	1	9	8	6	2
Meridional flattening change	2	−29	−10	−83	−2	−1	−0
Equatorial flattening change	3	2	1	0	−11	−9	0
Plate motion	4	158	225	−82	—	—	—
Total variation due to earth figure changes (1 + 2 + 3 + 4)	5	122	213	−155	−44	−36	1
Motion of principal axis of greatest moment of inertia	6	−192	1004	−54	−31	163	−508
Geocentre motion	7	−57	−57	−59	−57	−57	−59
Total variation if natural reference system motion is ignored	8	−127	1160	−268	−132	70	−566

* For rates used, see section 4.

5. THE REFERENCE SYSTEM FOR VERTICAL CRUSTAL MOTION STUDIES

Geodetic levelling has played a significant role in vertical crustal motion studies (e.g., Meade, 1973; RAK, 1974). The levelling operation is a differential process where the observed quantity between adjacent bench-marks P_i and P_{i+1} is the increment in orthometric height dz_i. Current practice calls for this elemental data to be treated in one of two ways when estimating the normal displacement above the datum level surface.

Method 1

Here, dz_i is converted to an equivalent difference in geopotential δW_i, using the relation:

$$\delta W_i = -g\, dz_i \tag{13}$$

where g is the local value of gravity. The differences δW_i form a holonomic system of increments and the resulting network is amenable to loop adjustment procedures. This results in a network of geopotential differences ΔW_i at the bench-mark P_i in relation to the adopted datum level surface.

Method 2

In this approximate method, the gravity field of the earth is modelled by that of a rotational equipotential ellipsoid. The dz_i are converted to a holo-

Invercargill ($\phi = -46.42°$; $\lambda = 168.35°$)			Vancouver ($\phi = 49.50°$ $\lambda = 236.90°$)			Natal ($\phi = -5.93°$; $\lambda = 324.84°$)		
ΔX_1	ΔX_2	ΔX_3	ΔX_1	ΔX_2	ΔX_3	ΔX_1	ΔX_2	ΔX_3
−7	1	−7	−3	−5	8	8	−6	−1
36	−7	38	21	32	−44	−1	1	0
−23	5	0	9	14	0	31	−21	0
−283	489	358	−132	227	101	−208	−299	16
−254	484	389	−84	303	64	−211	−297	15
148	−775	−287	−155	814	510	21	−111	779
−57	−57	−59	−57	−57	−59	−57	−57	−59
−163	−348	43	−296	1060	515	−247	−465	735

nomic dynamic height equivalent dh_i for purposes of adjustment, using a relation of the form:

$$dh_i = dz_i \left(1 + \frac{1 + \beta \sin^2\phi}{1 + \beta \sin^2\phi_s} \right) \tag{14}$$

where ϕ is the latitude at which dz_i was observed, ϕ_s is the standard latitude to which the dynamic heights refer and β is the dynamic flattening defining the formula for normal gravity (e.g., Heiskanen and Moritz, 1967, p. 79). The dynamic heights are reconverted to orthometric "heights" after adjustment on reversing eq. 14.

In both cases the concept of "height" as a normal displacement in relation to some datum level surface is ambiguous at the ±1—2-m level if based solely on data generated using eqs. 13 and 14.

Consider the following procedure for using levelling data collected at epochs ($\tau = t_1$) and ($\tau = t_2$) to establish vertical crustal movement. The "elevation" h_i of the bench-mark P_i is re-established and the difference δh_i given by:

$$\delta h_i = h_i(t_2) - h_i(t_1) \tag{15}$$

is considered to define vertical crustal movement at P_i. In levelling data processed by method 1, the difference is given by:

$$\delta h_i = \frac{1}{\gamma}[\Delta W_i(t_1) - \Delta W_i(t_2)] \tag{16}$$

where, for all practical purposes, γ is the global mean value of normal gravity. Eqs. 15 and 16 are valid only if (1) the earth space location of the elevation datum, and (2) the shape of the level surfaces have not changed between epochs. The validity of this assumption requires verification in the case of a non-rigid earth undergoing deformation with associated mass redistribution. For a more detailed treatment, see the Appendix (section 7).

The following factors were modelled in investigating gravity changes and their effect on the shape of the datum level surface as a function of time:

(1) Earth expansion.
(2) Geocentre motion.
(3) Changes in the earth's figure.
(4) Mass transfers due to plate motion.

Changes in gravity due to the causes mentioned at (1)—(3) can be represented by the following equations (Mather et al., 1977):

(1) The change δg_a in gravity due to earth expansion at a rate δr between epochs $(\tau = t_1)$ and $(\tau = t_2)$, given by:

$$\delta g_a = -2\frac{GM}{R^3}\, \delta r(t_2 - t_1) + o\{f\delta g_a\} \tag{17}$$

where G is the gravitational constant and M the mass of the earth.

(2) The change δg_g in gravity due to geocentre motion defined by eqs. 4 and 5, according to the relation:

$$\delta g_{gj} = -2\frac{GM}{R^3}\, \delta r_g(t_2 - t_1) \sum_{i=1}^{3} l_i l_{ij} \tag{18}$$

the quantities l_i and l_{ij} being defined by eqs. 5 and 8, the subscript j referring to P_j.

(3) The gravity change δg_{fi} due to the rate of change df in the meridional flattening at P_j, given by:

$$\delta g_{fj} = \gamma(1 - \sin^2\phi_j)\, df(t_2 - t_1) + o\{f\delta g_{fj}\} \tag{19}$$

where ϕ_j is the geodetic latitude of P_j.

(4) The change $\delta g_{f'j}$ in gravity at P_j due to the change df' in the equatorial flattening f', given by:

$$\delta g_{f'j} = \gamma\, df'\, \cos^2\phi_j \cos 2(\lambda_j - \theta)(t_2 - t_1) + o\{f'\, \delta g_{f'j}\} \tag{20}$$

θ being defined as in eq. 9.

(5) The gravity change $\delta g_{\omega j}$ at P_j due to secular movement of the principal axis of greatest moment of inertia, as represented by the rotations ω_i about the X_i axes in eq. 3, given by:

$$\delta g_{\omega j} = \gamma\beta \sin 2\phi_j(\omega_2 \cos \lambda_j - \omega_1 \sin \lambda_j) \tag{21}$$

If $\delta\Delta g$ is the change in the gravity anomaly between epochs $(\tau = t_1)$ and $(\tau = t_2)$, it follows that $\delta\Delta g$ is related to values of observed gravity at the two epochs by the relation:

$$\delta \Delta g = g(t_2) - g(t_1) \tag{22}$$

if the equipotential ellipsoidal gravity model is unchanged between epochs.
In this case, if $\delta \Delta g$ were also modelled as:

$$\delta \Delta g = \delta g_a + \delta g_g + \delta g_f + \delta g_{f'} + \delta g_\omega \tag{23}$$

where the subscript has been suppressed, the resulting change δN_g in the
radial displacement of the geoid from the reference ellipsoid can be obtained
from Stokes' integral (e.g., Heiskanen and Moritz, 1967, p. 94) as appropri-
ately corrected for the contribution of zero degree:

$$\delta N_g = \frac{R}{4\pi\gamma} \iint f(\psi) \, \delta \Delta g \, d\sigma - \frac{R}{\gamma} M\{\delta \Delta g\} \tag{24}$$

where ψ is the angular distance between the element of surface area $d\sigma$ at
which the gravity change is $\delta \Delta g$, and the point of computation, $f(\psi)$ is Stokes'
function (Heiskanen and Moritz, 1967) and $M\{\delta \Delta g\}$ is the global mean value
of $\delta \Delta g$. As Stokes' function contains no harmonics of degree 1 and 0, the
effect of δg_a is contained entirely in the second term of eq. 24. Eq. 24
excludes the effect of harmonics of degree 1. Consequently, the contribution
of δg_g has to be correctly interpreted. It follows that the solutions at each of
the epochs $(\tau = t_1)$ and $(\tau = t_2)$ are referred to ellipsoids with their centres
collocated with the instantaneous geocentre. Thus, eq. 24 is insensitive to
the gravitational effect of geocentre motion. This can be obtained directly
from the first-degree harmonic in the analysis of the global secular change in
gravity (Mather et al., 1977, p. 8).

The change δW_m in the geopotential at the earth's surface, due to mass
transfers δm at the element of surface area $d\sigma$ on account of plate motion,
can be computed using the relation:

$$\delta W_m = \frac{G}{2R} \iint \text{cosec} \tfrac{1}{2} \psi \, \delta m \tag{25}$$

There is no clear-cut model for the global distribution of δm on a geodetic
time scale. A tendency exists for subduction zones to be correlated with
geoidal highs. Gravity anomalies at mid-oceanic ridges are also correlated
with anomalies in the elevation of the sea-floor (Sclater et al., 1975). Neither
of these observations provides a basis for modelling mass transfers due to
plate motions over geodetic time scales (i.e., $<10^2$ yr). For simplicity in computations, it was decided to use two models for mass transfer based on
widely differing concepts, and estimate δW_m globally in each case.

Model 1

It is assumed that mass is conserved within a plate over the short period of
time involved. This type of model is questionable over longer time spans.

Present-day estimates of plate motion over the last 5—10 m.y. show that the rate of growth of plates is highly variable and mass transfer of some sort should occur between colliding plates. It has been pointed out by Garfunkel (1975, p. 4428) that in view of the absence of local balance between sub-duction zones and spreading zones, there can be a net flux of mass from the shrinking plates to the expanding ones through sublithospheric flow. In the current situation, this will occur from the northern hemisphere to the south-ern, with overall odd-degree zonal harmonic characteristics in the global dis-tribution of δW_m. Any tendency of this type will show up clearly in a planned global programme of secular gravity variation measurements. A feel for the magnitudes involved is best obtained by studying the gravitational effects of the meridional and equatorial flattening changes (see Table IV, rows 3 and 4). However, it was felt that a more sophisticated modelling of mass transfer on this basis over geodetic time scales was not possible at the present time owing to the difficulty of defining an alternate mechanism for mass transfer in the form of a global model for δm.

The change in position ΔX_{ij} at the point P_j on the boundary of the kth plate which is rotating with component ω_{rik} about the X_i axes, after a time interval dt is given by:

$$\Delta X_{ij} = \epsilon_{ilm} \omega_{rlk} X_{mj} \, dt + o\{\omega_{rlk}^2 X_{ij}\} \tag{26}$$

The components $(\delta N_j, \delta E_j)$ of this motion in the north and east directions in the local horizon at P_j are given by:

$$\delta N_j = -\Delta X_{1j} \sin \phi_j \cos \lambda_j - \Delta X_{2j} \sin \phi_j \sin \lambda_j + \Delta X_{3j} \cos \phi_j \tag{27}$$

and:

$$\delta E_j = -\Delta X_{1j} \sin \lambda_j + \Delta X_{2j} \cos \lambda_j \tag{28}$$

The component δl_j of motion orthogonal to the plate boundary is given by:

$$\delta l_j = \delta N_j \sin \alpha_j - \delta E_j \cos \alpha_j \tag{29}$$

where α_j is the azimuth of the plate boundary at P_j.

The mass δm accumulated or lost along a straight section of the plate boundary of length l can be modelled as:

$$\delta m = \rho l \, \delta l T \tag{30}$$

where ρ is the density of matter accumulated or lost and T is the thickness of the plate. In the current set of computations of δW_m, the density ρ_s of matter accumulating at subduction zones over short periods of time ($<10^2$ yr) was adopted as 2.67 g cm^{-3}. Values of $T = 5$ km and $T = 35$ km were adopted for the thicknesses of the oceanic and continental plates, respec-tively. On assuming conservation of mass within the plate for model 1, the average density of matter being "lost" at spreading zones (ρ_{sp}) can be estimated by the relation:

$$\rho_{sp} = -\frac{\rho_s \displaystyle\sum_{\text{subduction zones}} l\,\delta lT}{\displaystyle\sum_{\text{spreading zones}} l\,\delta lT} \tag{31}$$

Model 1 assumes that all plate boundaries other than subduction zones and spreading zones are not subject at any mass change.

Model 2

In this second model, mass conservation is enforced by introducing a negative mass Δm at the point whose coordinates are $\overline{X}_i$. These quantities are defined as follows:

$$\Delta m = -\rho_s \sum_{\text{subduction zones}} l\,\delta lT \tag{32}$$

and:

$$\overline{X}_i = -\frac{\rho_s \displaystyle\sum_{\text{subduction zones}} l\,\delta l\,TX_{il}}{\Delta m} \tag{33}$$

where X_{il} is the distance of the midpoint of l from the X_i axis. Model 2 acknowledges the temporary mass accumulation that occurs at subduction

TABLE III

Density of mass "lost" at spreading centres using model 1 for the transfer of mass due to plate motion (mass conservation per plate). Density of mass accumulating at subduction zones = 2.67 g cm^{-3}.

Plate identification	ρ_{sp} for model 1 (eq. 31) (g cm^{-3})
1. Arabian	13.9
2. African	—
3. Antarctic	0.1
4. Australian	6.6
5. Caribbean	0.0
6. Cocos	2.7
7. Eurasian	33.5
8. Nazca	2.5
9. North American	14.2
10. Pacific	3.4
11. Philippines	0.0
12. South American	13.9

TABLE IV

Gravity changes in microgal per century at five selected sites due to changes in the earth's figure and movements of the natural system of reference in earth space (for details of models used, see section 5)

Factors causing gravity variations	Row	Tromsø ($\phi = 69.66°$, $\lambda = 18.94°$)	Addis Ababa ($\phi = 8.75°$, $\lambda = 38.96°$)	Invercargill ($\phi = -46.42°$, $\lambda = 168.35°$)	Vancouver ($\phi = 49.50°$, $\lambda = 236.90°$)	Natal ($\phi = -5.93°$, $\lambda = 324.84°$)
Earth expansion	1	−31	−31	−31	−31	−31
Geocentre motion	2	250	270	−230	−21	24
Meridional flattening change	3	19	150	73	65	150
Equatorial flattening change	4	7	−45	73	−52	120
Motion of principal axis of greatest moment of inertia	5	−1	−1	3	6	−1

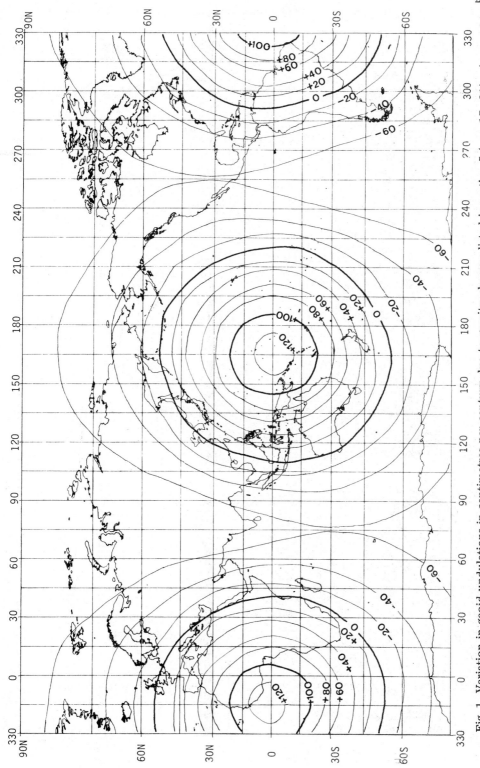

Fig. 1. Variation in geoid undulations in centimetres per century due to gravity changes listed in section 5 (eqs. 17—21) using eq. 24 and the rates specified therein (zero-degree contribution of −30 cm not included). Contour interval 20 cm per century.

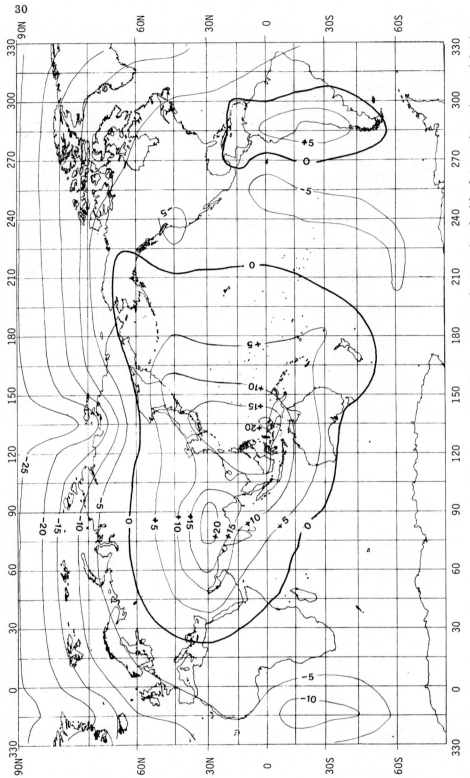

Fig. 2. Variation in geoid undulations in centimetres per century due to mass transfers associated with plate motion using model 1 for mass "loss" (section 5). Contour interval 5 mm per century.

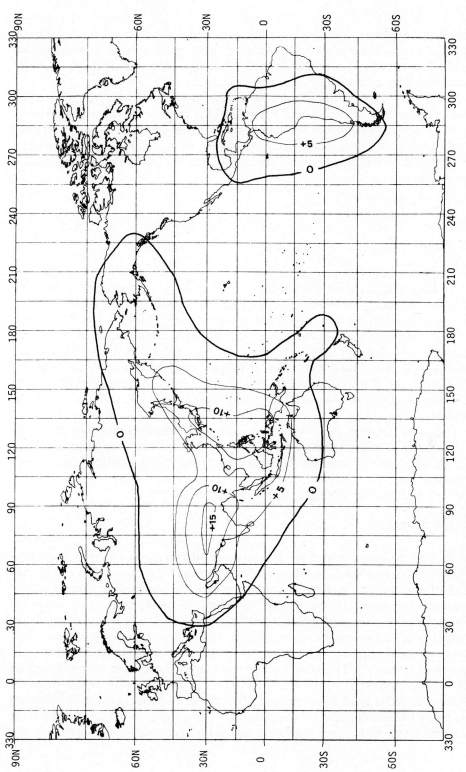

Fig. 3. Variation in geoid undulation in centimetres per century due to mass transfers associated with plate motion using model 2 for mass "loss" (section 5). Contour interval 5 mm per century.

zones and replaces the large number of sources from which mass is lost in model 1 by a single source for want of more information. It follows that model 2 is exactly equivalent to model 1 if the mass loss in the latter were represented by a first-degree harmonic model.

The model used for plate motions in the present calculations is based on that proposed by Solomon and Sleep (1974). This model was augmented by additional information on spreading zones from Chase (1972, p. 120) and Kaula (1975, p. 246). Table III gives values of ρ_{sp} computed for model 1 from eq. 31, while the values of Δm and $\overline{X}_i$ for the plate motion model used are as follow:

$$\Delta m = -0.139 \times 10^{20} \text{ g} \tag{34}$$

and:

$$\overline{X}_1 = -508.4 \text{ km}; \qquad \overline{X}_2 = 718.9 \text{ km}; \qquad \overline{X}_3 = 523.0 \text{ km} \tag{35}$$

Table IV lists the contributions of the various constituents listed in eqs. 17—21 to $\delta\Delta g$. Fig. 1 shows the variations in the shape of the geoid over a century as computed from eq. 24. It should be noted that geocentre motion makes no contribution to this figure for the reasons given above. Figs. 2 and 3 illustrate the change in the shape of the geoid computed from eq. 25 on adopting models 1 (eq. 31) and 2 (eqs. 34 and 35) respectively, for the distribution of mass "loss". The similar pattern of contours in Figs. 2 and 3 is a reflection of the mass accumulation at subduction zones and the dominant first-degree characteristics of the mass "loss" which are common to both models.

The mass transfers due to plate motion do not make significant contributions to the change in the shape of the geoid over short time intervals up to 10^2 yr, as illustrated in Figs. 2 and 3. This would not be the case if the mass transfers had the characteristics of low-degree harmonics, as illustrated in Fig. 1 for harmonics of degree 2, and discussed earlier.

6. CONCLUSIONS

Station positions based on a system of observations to extra-terrestrial sources in earth-orbit, should be verified for the accommodation of the motion of the natural system of reference in earth space before use for geodynamic purposes. If the entire spectrum of polar motion is correctly allowed for, as expected in orbit analysis, the effect of such movements on station coordinates can have magnitudes which are not negligible in comparison with those due to plate motions if the secular geocentre motion is in excess of 1 cm yr^{-1}. The latter can be adequately monitored by a network of 25 well-distributed absolute gravity stations in the presence of station noise at the ± 10 μGal level from at least 10 years of observations. Such measurements of secular changes in absolute gravity will also be of use in detecting mass transfers with zonal harmonic characteristics due to any change in the shape of the earth or as a result of the majority of expanding plates being in the southern hemisphere.

Changes in the shape of the geoid associated with the two models for mass transfer in plate motions, described in section 5, are too small to be of concern in vertical crustal motion studies over time intervals less than 10^2 yr, where phenomena are implied from changes in either geopotential or orthometric heights as defined in eqs. 15 and 16. This contention requires review if it were established that a steady rate of mass transfer substantially greater than that modelled in this study, were taking place from the shrinking plates to the expanding ones and if the latter were concentrated in a particular part of the globe.

The effect of geocentre motion on vertical crustal motion studies can be considered as follows. The true change in height dh_i at the general point P_i is related to the measured geopotential differences $\Delta W_i(t_1)$ and $\Delta W_i(t_2)$ at the two epochs of measurement by the relation:

$$dh_i = \frac{1}{\gamma}[\Delta W_i(t_1) - \Delta W_i(t_2)] + \delta N_i - \delta \zeta_{sr} + \delta \zeta_s \tag{36}$$

where $\delta \zeta_{sr}$ is the apparent rise in MSL at the regional elevation datum, and $\delta \zeta_s$ is the change in the height of the stationary sea surface topography between epochs as determined from independent global solutions (Mather, 1977a, section 7). The inclusion of the term δN_i from a solution of the geodetic boundary-value problem effectively filters out the geocentre motion from vertical crustal motion determinations. Here, δN_i, for all practical purposes, can be interpreted as the change in the quantity δN_g in eq. 24, supplemented by a contribution using eq. 25, if necessary. Eq. 36 provides the absolute definition of height change not available in eq. 16. For a more detailed treatment, see the Appendix, (section 7).

The last three correction terms have varying significance for regional studies. Both $\delta \zeta_{sr}$ and $\delta \zeta_s$ are constant for data related to the same height datum. It is unlikely that the higher frequency end of the spectrum of mass changes as a function of position will generate sufficient power to vitiate results obtained using eq. 16. However, it is not difficult or overly expensive to include the price of a precise absolute gravity survey in the cost of a levelling survey for vertical crustal movement and thereby minimize the uncertainty on this account. A global network of absolute gravity stations engaged in a programme of secular monitoring of gravity would assist considerably in defining δN_i without ambiguity.

7. APPENDIX

The influence of the stability of the levelling datum on vertical crustal motion studies

The relation between the datum for elevations D, the mean sea surface, the geoid, the reference ellipsoid and the general bench-mark P in the regional network of levelling

34

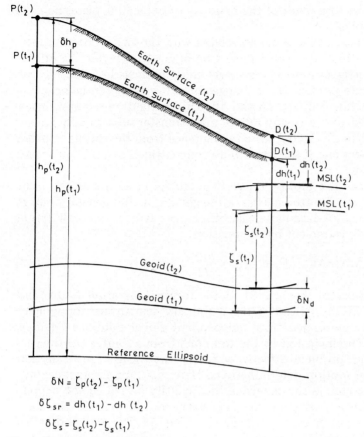

$$\delta N = \zeta_p(t_2) - \zeta_p(t_1)$$

$$\delta \zeta_{sr} = dh(t_1) - dh(t_2)$$

$$\delta \zeta_s = \zeta_s(t_2) - \zeta_s(t_1)$$

Fig. 4. Relationships between the levelling datum, mean sea-level, the geoid and the ellipsoid and the general bench-mark.

related to D, is shown in Fig. 4 for epochs ($\tau = t_1$) and ($\tau = t_2$). The observed data consist of the following:

(1) Levelled differences of geopotential $\Delta W(t_1)$ at epoch ($\tau = t_1$) between the earth-space locations of $P(t_1)$ and $D(t_1)$, and $\Delta W(t_2)$ between $P(t_2)$ and $D(t_2)$ at ($\tau = t_2$).

(2) The apparent rise in mean sea-level between epochs in relation to $D(\delta\zeta_{sr})$.

(3) The global change $\delta\zeta_{ss}$ of mean sea-level between epochs.

(4) The change $\delta\zeta_s$ in the sea surface topography determined from a global analysis of satellite altimeter and world-wide gravity data (Mather, 1977a, sec. 7) as evaluated at D between epochs.

(5) Determinations of the height anomaly ζ at both P and D with values $\zeta_p(t_1)$ and $\zeta_d(t_1)$ at epoch ($\tau = t_1$) and $\zeta_p(t_2)$ and $\zeta_d(t_2)$ at epoch ($\tau = t_2$).

If the reference ellipsoid were not changed between epochs, the height h_p of P above the ellipsoid is given by:

$$h_p = h_{np} + \zeta_p \tag{37}$$

h_{np} being the normal elevation given by the well-known relation (e.g., Heiskanen and

Moritz, 1967, p. 171)

$$h_n = -\frac{\Delta W'}{\gamma}(1 + c_\phi) \tag{38}$$

where:

$$c_\phi = o\{f\} = -\frac{\Delta W'}{a\gamma}\left[1 + m + f - 2f\sin^2\phi + \left(\frac{\Delta W'}{a\gamma}\right)^2\right] + o\{f^3\} \tag{39}$$

γ being the value of normal gravity computed for the ellipsoid defined by the parameters (a, f) and

$$m = a\omega^2/\gamma_e \tag{40}$$

the subscript e referring to the value of γ on the equator of the ellipsoid while ω is the angular velocity of rotation of the earth. In eqs. 38 and 39 $\Delta W'$ is the difference in geopotential between P and the geoid *at the epoch of measurement*. The latter is defined as the level surface of the earth's gravity field which corresponds to mean sea-level (MSL) as sampled globally (Mather 1977a, sec. 1). The geoid height at D for epoch $(\tau = t_1)$ can differ from that for epoch $(\tau = t_2)$ due to the following reasons:

(1) A secular change in global MSL $(\delta\zeta_{ss})$ causing a different level surface to be selected as the geoid. The new geoid will be higher than the old geoid by $\delta\zeta_{ss}$ at D on this account.

(2) The shape of the geoid can change due to mass redistribution. Such radial changes can be computed using eq. 24 from measurements of secular changes in gravity.

(3) Computations using eq. 24 are insensitive to geocentre motion. The position of the reference ellipsoid at each epoch is located with its centre at the instantaneous geocentre.

Geoid height variations cannot be determined independent of sea surface topography determinations (Mather, 1977b). Here, $\Delta W'$ can be related to the observed difference in geopotential ΔW by the relation:

$$\Delta W' = \Delta W - \gamma\,\delta\zeta_d \tag{41}$$

where $\delta\zeta_d$ is the height of D above the geoid. It follows from Fig. 4 that the change $d\delta\zeta_d$ in $\delta\zeta_d$ between epochs $(\tau = t_1)$ and $(\tau = t_2)$ is given by:

$$d\delta\zeta_d = -\delta\zeta_{sr} + \delta\zeta_s \tag{42}$$

As the effect of geocentre motion is effectively filtered out in vertical crustal motion studies on including a term based on a solution of the geodetic boundary-value problem, the consideration of eqs. 37, 38, 41 and 42 defines the true vertical crustal movement dh_p at P as

$$dh_p = \frac{1}{\gamma}[\Delta W(t_1) - \Delta W(t_2)] + \delta N_p - \delta\zeta_{sr} + \delta\zeta_s \tag{43}$$

as given in eq. 36, where:

$$\delta N_p = \zeta_p(t_2) - \zeta_p(t_1) \tag{44}$$

and is defined for all practical purposes by using the secular variations in gravity in eq. 24.

ACKNOWLEDGEMENTS

Valuable computer assistance for this project was provided by Bernd Hirsch. The authors are grateful to Peter Angus-Leppan for presenting this paper on their behalf at the 1977 R.C.M. Conference at Stanford University, Palo Alto, California held between July 25 and 31, 1977.

REFERENCES

Anderle, R.S., 1974. Transformation of terrestrial survey data to Doppler satellite datum. J. Geophys. Res., 79: 5319—5331.
Bender, P.L., Currie, D.G., Dicke, R.H., Eckhardt, D.H., Faller, J.E., Kaula, W.M., Mulholland, J.D., Plotkin, H.H., Poultney, S.K., Silverberg, E.C., Wilkinson, D.T., Williams, J.G. and Alley, C.O., 1973. The lunar laser ranging experiment. Science, 182: 229—238.
B.I.H., 1975. Rapport sur 1975. Bureau International de l'Heure, Paris.
Bomford, A.G., 1967. The geodetic adjustment of Australia 1963—1966. Surv. Rev., 144: 52—71.
Bomford, G., 1971. Geodesy. Oxford University Press, London, 561 pp.
Chase, C.G., 1972. The N plate problem of plate tectonics. Geophys. J. R. Astron. Soc., 29: 117—122.
Gaposchkin, E.M., 1972. Analysis of pole position from 1846 to 1970. In: P. Melchior and S. Yumi (Editors), Rotation of the Earth. Proc., Int. Astron. Union Symp. No. 48, Reidel, Dordrecht, pp. 19—32.
Garfunkel, Z., 1975. Growth, shrinking and long-term evolution of plates and their implications for the flow pattern in the mantle. J. Geophys. Res., 80 (32): 4425—4432.
Goldreich, P. and Toomre, A., 1969. Some remarks on polar wandering. J. Geophys. Res., 74: 2555—2569.
Guinot, B. and Feissel, M., 1968. Annual Report for 1968. Bureau International de l'Heure, Paris.
Heiskanen, W.A. and Moritz, H., 1967. Physical Geodesy. Freeman, San Francisco, Calif., 364 pp.
Kaula, W.M., 1975. Absolute plate motions by boundary velocity minimizations. J. Geophys. Res., 80 (2): 244—248.
Markowitz, W., 1974. Astronomical programs for the study of secular variations in position. In: Proc., A.A.S./I.A.G. Symp. Earth's Gravitational Field and Secular Variations in Position. Univ. N.S.W., Sydney, pp. 11—19.
Mather, R.S., 1972. Earth space. Unisurv G, 17: 1—41, Univ. N.S.W., Sydney.
Mather, R.S., 1973. Four-dimensional studies in earth space. Bull. Geod., 108: 187—209.
Mather, R.S., 1974. Geodetic coordinates in four dimensions. Can. Surv., 28 (5): 574—581.
Mather, R.S., 1977a. The role of the geoid in four-dimensional geodesy. Mar. Geod., in press.
Mather, R.S., 1977b. A geodetic basis for ocean dynamics. Marussi Septennial Issue, Boll. Geod. Sci. Affini, in press.
Mather, R.S., Masters, E.G. and Coleman, R., 1977. The role of non-tidal gravity variations in the maintenance of reference systems for secular geodynamics. Int. Symp. on Non-Tidal Gravity Variations and Methods for their Study, Trieste, June 20—24, 1977. Unisurv G (Aust. J. Geod., photogramm. and surv.), 26: 1—25.
Meade, B.K. (Ed.), 1973. Reports on Geodetic Measurements of Crustal Movement, 1906—1971. U.S. National Geodetic Survey, N.O.A.A., Stock No. 0317-00167, U.S. Government Printing Office, Washington D.C.
Ong, K., MacDoran, P.F., Thomas, J.B., Fuegel, H.F., Skjerve, L.J., Spitzmesser, D.S., Batelaan, P.D., Paine, S.R. and Newstead, M.G., 1976. A demonstration of a transportable radio interferometric surveying system with 3 cm accuracy on a 307 m baseline. J. Geophys. Res., 81 (20): 3587—3593.

RAK, 1974. Sveriges andra precisions — avvägning 1951—1967. Meddelande No. A40, Rikets Allmänna Kartwerk, Stockholm, 91 pp.

Rochester, M.G., 1973. The earth's rotation. EOS (Trans. Am. Geophys. Union), 54 (8): 769—780.

Sclater, J.G., Lawver, L.A. and Parsons, B., 1975. Comparison of long wavelength residual elevations and free air gravity anomalies in the North Atlantic and possible implications for the thickness of the lithospheric plate. J. Geophys. Res., 80 (8): 1031—1052.

Smith, D.E., Lerch, F.J., Wagner, C.A., Kolenkiewicz, R. and Khan, M.A., 1976. Contributions to the National Geodetic Satellite Program by Goddard Space Flight Center. J. Geophys. Res., 81 (5): 1006—1026.

Solomon, S.C. and Sleep, N.H., 1974. Some simple physical models for absolute plate motions. J. Geophys. Res., 79: 2557—2567.

Tectonophysics, 52 (1979) 39—46
© Elsevier Scientific Publishing Company, Amsterdam — Printed in The Netherlands

THE NATIONAL GEODETIC SURVEY PROJECT "POLARIS"

WILLIAM E. CARTER and WILLIAM E. STRANGE

National Geodetic Survey, National Ocean Survey, Rockville, Md. 20852 (U.S.A.)

ABSTRACT

Carter W.E. and Strange, W.E., 1979. The National Geodetic Survey project "Polaris". In: C.A. Whitten, R. Green and B.K. Meade (Editors), Recent Crustal Movements 1977. Tectonophysics, 52: 39—46.

The National Geodetic Survey (N.G.S.) of the National Ocean Survey has undertaken a new project called POLar-motion Analysis by Radio Interferometric Surveying (POLA-RIS). The premier goal of the project is to establish a fully operational polar motion monitoring network, with a spatial resolution and accuracy of 10 cm and a temporal resolution of 24 hours. The new network will utilize radio interferometric techniques. The fixed POLARIS observatories may also serve as base stations to be used in conjunction with small-aperture, 3—10-m, mobile units for positional surveys for geodetic and geodynamic applications. The N.G.S. plans to cooperate fully with other governmental organizations having related responsibilities and interest in polar motion, earth rotation and geodynamics [e.g., National Aeronautics and Space Administration (NASA), United States Geological Survey (U.S.G.S.), United States Naval Observatory (U.S.N.O.)] as well as academic and private researchers.

INTRODUCTION

At present there are two international organizations that determine and distribute polar motion data as a service to the user community — the International Polar Motion Service (I.P.M.S.) and the Bureau International de l'Heure (B.I.H.). Both services combine data collected at several tens of observatories scattered around the world and regularly publish positions of the pole having temporal resolutions of several days (B.I.H. 5 days, I.P.M.S. 0.05 year) and smoothed spatial resolutions of 30 cm. The observational data are quite noisy and the final accepted values have been greatly "smoothed" to constrain or eliminate any "apparent" short-term erratic motions that may occur. Systematic differences in the position of the pole as determined by the I.P.M.S. and B.I.H. often are as large as 1 m and persist for periods of months.

A large segment of the user community has traditionally consisted of geodetic surveying agencies that utilize the polar motion data to reduce astronomic observations, made over time spans of several decades, to a common epoch. The present services provide the required polar coordinates with an order of magnitude of better accuracy than the typical position determination made by classical geodetic astronomical methods, and therefore provide very adequate service to this user group.

Another, less well-defined, group of users may be described as scientific researchers. The polar motion data have been used by researchers to study a variety of questions dealing with the structure and composition of the physical earth, the response of the earth to disturbances such as earthquakes, and the long-range prediction of weather trends. The spatial and temporal resolutions required by these users vary according to the particular area of study, and, in general, the scientific users have generated very little impetus for improvement of the existing system. Nonetheless, the monitoring of polar motion has been recognized to be of such scientific value that extraordinary efforts have been made by participating nations to maintain the system, even during periods of world wars.

The United States has participated in the I.P.M.S. and its predecessor, the International Latitude Service (I.L.S.), since its inception. The N.G.S. has operated two observatories, at Gaithersburg, Maryland, and Ukiah, California, for more than seven decades.

The space age has brought requirements for polar position information with a spatial resolution of 10 cm and temporal resolution of 8—24 hours. The present techniques and instruments simply cannot meet these requirements. New methods, based on more modern technology, are required. The international scientific community has recognized this problem, and appropriate resolutions have been passed by both the International Astronomical Union (Commission 19, August 1973) and the International Union of Geophysics and Geodesy (Resolution 4, September 1975). In response to these international resolutions and in awareness of the maturing of several new techniques that could provide the desired improvements in temporal and spatial resolution, the N.G.S. began to develop a comprehensive plan to establish an improved pilot polar motion service.

After a review of several candidate methods, we have concluded that the next generation polar motion service should utilize Very Long Baseline Interferometry (VLBI). The following sections of this paper briefly outline our reasons for selecting VLBI, and our preliminary plans for the establishment of a minimally configured operational polar motion VLBI network.

SELECTION OF VLBI

A NASA-funded study, technically monitored jointly by NASA and N.G.S. scientists, to compare the relative benefits and costs of measuring polar motion and earth rotation by VLBI and satellite laser ranging methods was recently completed by the Systems Planning Corporation (Schultheis et al., 1977). The study concluded that a VLBI network would be the best way to perform routine operational measurements of polar motion and earth rotations.

Clearly, VLBI has certain inherent advantages over other known methods that make it the most desirable. The primary advantages are:

(1) VLBI has an almost total all-weather capability.

(2) VLBI measurements are made in the most nearly inertial frame of reference presently known.

(3) VLBI uses natural sources and does not rely on active transmitters or satellites having finite lifetimes.

The almost total all-weather capability of VLBI is a tremendous advantage for an operational system. It permits the development of a highly reliable system with a minimum of redundancy. Six to ten regularly participating radio observatories, devoting approximately one-third of their observing time to VLBI observations, would provide very adequate redundancy for an international polar-motion, earth-rotation monitoring system. Rather complex observing schedules, designed to optimize the scientific productivity of the participating observatories, could be formulated months in advance, with a very high probability that the desired data would be successfully collected. The record of optical systems, including visual, photographic and laser ranging techniques, attests to the severe disadvantage of systems that must have clear, cloud-free, stable atmospheric conditions.

Since VLBI uses very distant celestial objects, such as quasars, which are expected to exhibit very small proper motions (preliminary determinations have not been able to detect any proper motion of quasars) many of the problems associated with the classical reference frame are greatly reduced or eliminated.

Since VLBI sources are natural, rather than man-made, and have practically infinite life, many other complexities are also eliminated. No reflectors need be placed in space. No orbit ephemerides need be generated and constantly updated. No hazard, such as that presented by the laser pulses as they propagate through the atmosphere, is created. The system is equally as passive as any other classical astronomy program, simply collecting the natural energy that falls on the antennae from the quasars.

PROJECT POLARIS

The initial conceptual design and a preliminary project plan for Project POLARIS were developed by W.E. Carter, N.G.S., during January 1977. Under this initial plan, the N.G.S. would establish a minimal three-station VLBI network within the United States to study polar motion and earth rotation, and to serve as a base network for the activities of transportable VLBI systems for geodetic positioning and geodynamic investigations.

A very important ingredient in the POLARIS plan is the location of one of the network stations at the Haystack Observatory Complex (Westford Antenna). This will provide the N.G.S. with the use of an existing moderate-aperture radio telescope, time-sharing use of a Mark III correlator and, most importantly, will facilitate direct intercourse between the N.G.S. operational staff and the M.I.T./Haystack engineering and research staff. A study to determine the best location of the remaining two stations has begun.

COMPUTER SIMULATIONS OF VLBI POLAR MOTION NETWORK

Researchers at M.I.T./Haystack Observatory VLBI facilities recently performed computer simulations for the N.G.S. to investigate the predicted accuracy attainable with a three-station network located completely within the United States. Three different configurations have been investigated. Westford, Massachusetts, and Fort Davis, Texas, were included in all configurations, along with a third site in Alaska or California or Florida. Fig. 1 shows the three simulated networks. The reasons for investigating these particular networks are discussed later in this paper.

In formulating the simulations, it was assumed that the network stations would be equipped with the Mark III VLBI Data Acquisition System, presently under development by the M.I.T./Haystack/Goddard research team. Actual sources were used to develop realistic schedules of observations. Other constraints and uncertainties of parameters were set to appropriate

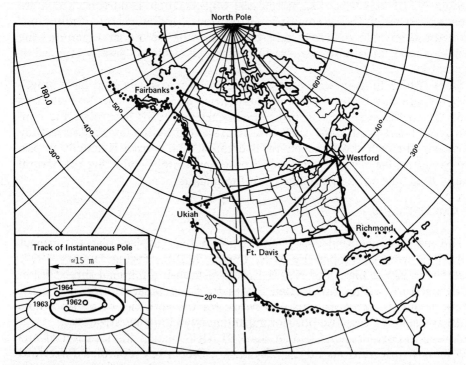

Fig. 1. Map (polar azimuthal orthographic projection) showing three possible configurations of the POLARIS network. The ●symbols indicate locations of earthquakes of magnitude 7.0 or greater on the Richter scale.

values for the anticipated network instrumentation. The simulations indicated that uncertainties in the X and Y components of the polar motion of less than ± 10 cm, and in the determination of UT1 of ± 0.1 msec were achievable in 8-hour schedules. The homogeneity of distribution of sources was such that the particular 8-hour observing window used has little impact on the predicted results.

The network of Westford—Fort Davis—Alaska yields significantly better (by roughly a factor of 2) results than the other two configurations. However, all three configurations yield results of the order of which we are seeking, and therefore these simulations do not rule out any of the configurations. Other factors, such as the cost of operation, possible use of existing facilities and site stability, will have significant weight in the selection of the final network.

SITE SELECTION

The Harvard Radio Astronomy Station (H.R.A.S.) located near Fort Davis, Texas, has some significantly desirable attributes. It is located far enough from the buffer zone between the North American and Pacific Plates that displacement velocities associated with regional crustal deformations are expected to be reasonably small. A detailed crustal deformation study of this region by the University of Texas and the N.G.S., under NASA funding, in support of the McDonald Observatory Lunar Ranging Experiment is presently in progress. Indeed, the Harvard Radio Astronomy Station is located less than 10 km line-of-sight from McDonald Observatory, not only making the crustal deformation studies applicable to both sites, but making possible direct comparison of data collected at the two locations. The principal antenna at H.R.A.S. is approximately 26 m in diameter, and with slight modifications and additions would be quite adequate for POLARIS observations. Exploratory discussions between the N.G.S. and Harvard College have begun.

The N.G.S. and the U.S. Naval Observatory (U.S.N.O.) have had a long history of cooperative programs in the areas of polar motion, earth rotation and time determinations. U.S.N.O. operates a Time Service substation near Richmond, Florida, which is under consideration as a possible location of one of the stations in the POLARIS network, and is also interested in VLBI as a technique to transfer time between widely distant locations, such as between the east and west coasts of the United States (Johnston et al., 1977). Desirable attributes of the Florida location include a very quiet seismic and tectonic setting, excellent vertical stability (Holdahl and Morrison, 1974) and the opportunity to develop a cooperative program that could increase the scientific productivity of the project without increasing the cost. The undesirable factors are the relatively weak geometric configuration of the net, as compared to the inclusion of a station in Alaska, a possible degradation of the accuracy with which the tropospheric correction can be made due to the relatively high humidity in Florida and the necessity to construct a new radio telescope facility.

The National Oceanic and Atmospheric Administration (N.O.A.A.) has a 26-m diameter antenna located near Fairbanks, Alaska, that has already been used by the M.I.T./Haystack/Goddard VLBI team (Whitney et al., 1976). The facility is in fairly heavy use for satellite communications and time-sharing may not be easy. The most desirable attribute of an Alaskan site is the very strong geometry it provides to a U.S. network. The disadvantages include the remoteness of the site (which affects operating costs, staffing and maintenance), the severity of the Alaskan weather conditions, and the close proximity of the site to the active North American and Pacific Plate boundary. Several earthquakes of magnitude 7 or greater have been recorded in the immediate vicinity of the N.O.A.A. Alaskan site, see Fig. 1. The close proximity of such large earthquakes is of concern not only because of the displacements they could cause in the location of the antenna, but also because of possible extensive damage to the facilities, resulting in costly repairs and highly undesirable gaps in the data records. Vertical velocities, probably related to the Pacific Plate subduction process, are known to exceed 1 cm per year over large portions of Alaska.

Consideration is also being given to locating a POLARIS observatory at the N.G.S. Ukiah, California IPMS site. Doppler satellite data are already being collected by the N.G.S. on a regular basis at Ukiah to support the Groupe de Richerches de Geodesie Spatiale (G.R.G.S.) MEDOC Experiment, i.e., the determination of polar motion by Doppler tracking of artificial satellites. The N.G.S. also has a cooperative program planned with U.S.N.O. to install a photographic zenith tube (PZT) at Ukiah in the very near future. Any potential benefits to be derived by collocating the new VLBI system with the optical and Doppler systems must be considered in conjunction with the other important site selection parameters. Ukiah is located very near the active North American—Pacific Plate boundary, causing concern about the site stability. It may well prove more desirable to include the Ukiah site in the points to be visited repeatedly by a transportable VLBI system.

More detailed studies of the advantages and disadvantages of potential locations for POLARIS stations are presently in progress.

NASA PARTICIPATION

VLBI has been developed by a number of research groups, primarily associated with universities and radio observatories, under funding from such agencies as the National Science Foundation (N.S.F.) and NASA. Thus far, VLBI has been widely considered as a "strictly research" program which has been the province of a very limited number of researchers. In order for the technology to be transferred from the realm of research to an operational system, the knowledge and experience accumulated by the researchers must be passed on to the operational organization. The complexities of the theory, hardware and computer software associated with VLBI are such that this transition from research to operational status is a formidable under-taking. One attempt directly to address this problem was the early recogni-tion and selection of the Haystack/Westford facilities as the premier station in the network. A second approach to the problem is presently being explored with the NASA Office of Applications. N.G.S. and NASA personnel have formulated a preliminary plan that, if approved and implemented, would result in a joint N.G.S.–NASA program to accomplish the transfer of technology and founding of the fully operational POLARIS network. Once the network reached operational status, N.G.S. would assume full responsi-bility for routine data collection, reduction, analysis and the dissemination to the user community.

At the time of preparation of this paper, no firm commitments had been made between N.G.S. and NASA concerning Project POLARIS.

TRANSPORTABLE VLBI GEODETIC SURVEYING SYSTEMS

The Jet Propulsion Laboratory (J.P.L.), with NASA funding, has had a continuing program over the past several years to develop transportable VLBI systems for use in geodetic and geodynamic applications (Ong et al., 1976). Depending upon other components of the system, an antenna 3—10 m in diameter is adequate. Such transportable systems, when used in conjunc-tion with a fixed network such as the anticipated POLARIS network, could prove to be a very powerful geodetic technique. At present the N.G.S. has only a supporting role in the development of an operational transportable VLBI system, but this role may be expanded as the capabilities and appli-cations of the system become better defined.

CONCLUSIONS

The National Geodetic Survey has begun a new project intended to esta-blish a greatly improved pilot polar motion monitoring program by use of VLBI techniques. Exploratory discussions have already begun with a number of private and governmental organizations having related responsibilities or interests. At the appropriate time, discussions will be initiated with foreign nations to explore the possibilities for international cooperation in the devel-opment of a new-generation polar motion service. The greatly improved spatial and temporal resolution of the polar motion and earth-rotation information by the new method will not only allow present theories and models to be tested but can be expected to produce new questions to chal-lenge earth scientists.

REFERENCES

Holdahl, S.R. and Morrison, N.L., 1974. Regional investigations of vertical crustal movements in the U.S., using precise relevelings and mareograph data. Tectonophysics, 23: 373—390.

Johnston, K.J., Kaplan, G.H., Klepczynski, W.J., Knowles, S.H., Mayer, C.H., Smith, C.A., Spencer, J.H. and Winkler, G.M.R., 1977. 1st Rep. Joint USNO/NRL Working Group on Radio Astronometry. U.S. Naval Observatory, Washington, D.C., 40 pp.

Ong, K.M., MacDoran, P.F., Thomas, J.B., Fliegel, H.F., Skjerve, L.J., Spitzmesser, D.J., Batelaan, P.D., Paines, S.R. and Newsted, M.G., 1976. A demonstration of a transportable radio interferometric surveying system with 3-cm accuracy on a 307-m baseline. J. Geophys. Res., 81 (20): 3587—3593.

Schultheis, A.C., Sullivan, R.J. and Harris, R.L., 1977. Estimates of Benefits and Costs of Measuring Polar Motion and Universal Time Using Very Long Baseline Interferometry on Satellite Laser Ranging. Rep. 289, Systems Planning Corporation, Arlington, Va.

Whitney, A.R., Rogers, A.E.E., Hinteregger, H.F., Knight, C.A., Levine, J.I., Lippincott, S., Clark, T.A., Shapiro, I.I. and Robertson, D.S., 1976. A very-long-baseline interferometer system for geodetic applications. Radio Sci., 11 (5): 421—432.

Tectonophysics, 52 (1979) 47
© Elsevier Scientific Publishing Company, Amsterdam — Printed in The Netherlands

HIGH MOBILITY RADIO INTERFEROMETRIC GEODETIC MONITORING OF CRUSTAL MOVEMENTS

P.F. MacDORAN

Jet Propulsion Laboratory, California Institute of Technology, Pasadena, Calif. 91103 (U.S.A.)

(Accepted for publication March 24, 1978)

ABSTRACT

Based upon actual transportable radio interferometric geodetic experience using the 9-m diameter ARIES (Astronomical Radio Interferometric Earth Surveying) system, it has been possible to redesign the elements to achieve an order of magnitude more mobility. A new 4-m-diameter ARIES station is now being implemented which will be capable of achieving 4—7 cm three-dimensional accuracy on baselines up to 500 km in virtually all weather conditions. Since the 4-m station remains essentially assembled during redeployments, the elapsed time between data acquisitions is estimated to be three days for sites a few hundred kilometers apart. There is also the capability for rapid deployments to epicentral regions where post-earthquake crustal deformation measurements are desirable.

Although the 4-m ARIES station is presently being designed to operate in combination with the 40-m diameter telescope of the Caltech Owens Valley Radio Observatory, further instrumentation improvements in radio sensitivity and higher digital recording rates will allow a 9-m-diameter antenna to function as a transportable base station. This future system will be fully transportable to any global land-based location and be capable of 3-cm three-dimensional relative position measurements on a baseline up to 1000 km with only 8 hours of on-site data acquisition.

The cost effectiveness of ARIES technology on geodetic networks of 100-km spacing density is substantially superior by comparison with monitoring such regional scale networks by conventional horizontal and vertical geodetic methods.

Tectonophysics, 52 (1979) 49—58
© Elsevier Scientific Publishing Company, Amsterdam — Printed in The Netherlands

COMPARISON OF A RADIO INTERFEROMETRIC DIFFERENTIAL BASELINE MEASUREMENT WITH CONVENTIONAL GEODESY

A.E. NIELL [1], K.M. ONG [1], P.F. MacDORAN [1], G.M. RESCH [1], D.D. MORABITO [1], E.S. CLAFLIN [1] and J.F. DRACUP [2]

[1] *Jet Propulsion Laboratory, California Institute of Technology, Pasadena, Calif. 91103 (U.S.A.)*
[2] *National Geodetic Survey, Rockville, Md. 20852 (U.S.A.)*

(Accepted for publication March 24, 1978)

ABSTRACT

Niell, A.E., Ong, K.M., MacDoran, P.F., Resch, G.M., Morabito, D.D., Claflin, E.S. and Dracup, J.F., 1979. Comparison of a radio interferometric differential baseline measurement with conventional geodesy. In: C.A. Whitten, R. Green and B.K. Meade (Editors), Recent Crustal Movements, 1977. Tectonophysics, 52: 49—58.

The ARIES (Astronomical Radio Interferometric Earth Surveying) 9-m transportable station has been deployed at two sites selected to be intervisible. By observing on approximately 380-km baselines from ARIES to the 40-m telescope of the Owens Valley Radio Observatory, we derived the three-dimensional position of each site relative to the OVRO telescope. By differencing the components of the positions of the two sites, we obtained the vector from Malibu to Palos Verdes — a 42-km path across Santa Monica Bay in southern California — with an estimated accuracy of better than 10 cm. The National Geodetic Survey has directly measured both the intersite distance and azimuth by conventional first-order horizontal geodetic control methods.The two techniques differ by 6 ± 10 cm in baseline length, and 0.5 ± 1.2 arc sec in azimuth (corresponding to 10 ± 20 cm).

INTRODUCTION

As one phase of demonstrating the feasibility of using mobile VLBI (Very Long Baseline Interferometry) stations for accurate baseline measurements, an experiment was structured, in cooperation with the National Geodetic Survey (N.G.S.), which has allowed an independent assessment of the accuracy of the VLBI system. Two intervisible sites were selected such that their separation was significant to possible future geophysical studies while still measureable in a single laser distance determination and accessible to existing geodetic control. The sites for this demonstration were at Palos Verdes and Malibu on Santa Monica Bay in southern California, with a separation of approximately 42 km.

As a base station for the VLBI measurements we used the 40-m diameter telescope at the Caltech Owens Valley Radio Observatory (OVRO); the transportable 9-m diameter ARIES (Astronomical Radio Interferometric Earth Surveying) station was deployed at the Palos Verdes (PV) site in March and October 1976 and at Malibu (MAL) in February 1977. The three-dimensional positions of the PV and MAL sites were then determined with OVRO as a common reference site.

By differencing the PV—OVRO and MAL—OVRO vectors, the PV—MAL vector was determined and then supplied to the N.G.S. for verification. The N.G.S. then assigned a geodetic survey party to measure directly the chord distance from PV to MAL, these sites then being connected to the nearby geodetic control. The result of this radio interferometry and conventional geodesy intercomparison is the subject of this paper.

VERY LONG BASELINE INTERFEROMETRY

The methods of VLBI are given elsewhere (see Thomas et al., 1976, for a summary and further references). We will give only a very brief description here.

The radio signal from an extragalactic radio source is recorded on magnetic tape at two radio telescopes (see Fig. 1). By cross-correlating the signals from the magnetic tapes, the travel time (or delay) of the radio wave from one telescope to the other can be measured. This delay is proportional to the

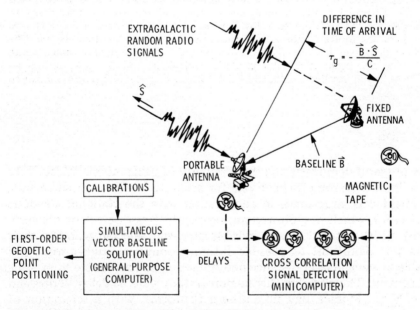

Fig. 1. A schematic outline of the ARIES VLBI system.

projection of the vector baseline between the two telescopes onto the direction of the radio source. By observing many different radio sources over the sky, both the directions to the radio sources and the vector baseline between the telescopes can be determined. If the radio source positions are known, then the baseline vector alone can be solved for with fewer observations. The baseline vector thus determined is a strictly geometric quantity, giving the position of one telescope relative to the other in a coordinate frame defined by the instantaneous spin axis and equatorial plane of the earth. Thus, it does not give an absolute position of both telescopes relative to the geocenter. It can, however, give the relative position independent of peripheral measurements, such as gravity, so that locating one site in an absolute sense serves to define the "absolute" locations of all other sites whose relative positions are determined by VLBI.

Since it is desirable to relate the positions of many different locations, the development of portable antennas is a necessity if VLBI is to be useful for geodetic programs. At J.P.L. within Project ARIES we have developed a 9-m-diameter transportable antenna which is being used in conjunction with a fixed radio telescope to monitor the relative locations of six sites in California (Fig. 2). (A complete description of the ARIES system is given by Ong et al., 1976). The fixed site which serves as the basic reference point

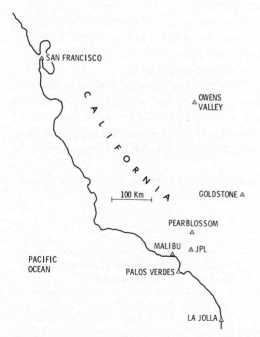

Fig. 2. A map of California indicating the sites to which the ARIES station has made baseline measurements. The other antennas involved are at the Owens Valley Radio Observatory (OVRO) and the NASA Deep Space Network at Goldstone.

for these experiments is the 40-m radio telescope of Caltech's Owens Valley Radio Observatory (OVRO) near Bishop, California. For most of the six sites visited by ARIES we have simultaneous measurements (not reported here) at one of the Goldstone antennas of the NASA Deep Space Network near Barstow, California. Inclusion of the third station gives three baselines around a triangle, which allows useful tests of closure.

THE EXPERIMENTS

To verify the accuracy of a new technique it must be shown that the results are both precise, by demonstrating repeatability, and accurate, by comparing results with an accepted standard. As another step in demonstrating the achievements of VLBI by these criteria, we have measured the relative locations of the two intervisible sites of Malibu (MAL) and Palos Verdes (PV) (Fig. 2).

Since we have only one transportable VLBI geodetic station, we deployed it first to Palos Verdes to determine the OVRO—PV baseline, then to Malibu for the OVRO—MAL baseline. The Palos Verdes—Malibu vector baseline was obtained by differencing the components of the baseline vectors to OVRO. The Palos Verdes—Malibu distance is the length of the difference vector, and is the distance between the locations of the intersection of axes of the ARIES antenna at the time of each experiment.

The data at Palos Verdes were taken on two separate occasions six months apart. Between these visits to Palos Verdes the ARIES antenna was moved to the Pearblossom and the J.P.L. (Pasadena) sites (Fig. 2). The dates of the experiments which we are reporting are shown in Table I.

Since the antenna cannot easily be relocated to exactly the same position when returning to a site, it is simpler to measure the offset from a reference position on the ground. This has been done by the N.G.S. with an error of less than 1 cm. Using these measurements, the Palos Verdes baselines from March were corrected to refer to the position of the ARIES antenna for the October experiment.

It is important to note that the test we are reporting, that of obtaining a

TABLE I

Summary of experiments

ARIES site	Experiment date	Duration (hours)	Water-vapor radiometer at ARIES
Palos Verdes	Mar. 4, 1976	12	No
	Mar. 5, 1976	12	Yes
	Oct. 10, 1976	12	Yes
	Feb. 23, 1977	12	Yes
Malibu	Feb. 24, 1977	26	Yes

vector baseline by differencing the results from two separate baselines, is much stronger than measuring a single baseline for direct comparison with conventional technique. The baselines to OVRO are almost 400 km long, so these results are a test of accuracies over several hundred kilometers, rather than just the 42 km of the differential baseline. Equally important, the experiments at Palos Verdes were performed on two separate occupations of the site several months apart, and thus demonstrate the ability to obtain repeatable results having dismantled and reassembled the antenna, under different atmospheric conditions and with different earth orientation parameters (UT1 and polar motion). Finally, note that the time involved in the experiments themselves is small, less than 40 hours per site.

DISCUSSION OF UNCERTAINTIES

The much simplified description of VLBI given above does not include the factors which contribute the major sources of uncertainty in our results. They are:

(1) Errors in the directions to the radio sources.

(2) Uncertainty in the propagation time of the radio wave as it traverses the ionosphere and troposphere to each site.

(3) Limited signal-to-noise ratio.

(4) Instabilities of the frequency systems and electronics at both stations.

(5) Errors in the Earth orientation parameters, UT1 and polar motion.

For most of the radio sources observed, the a priori positions used were averages of results obtained with other radio interferometers (Clark et al., 1976; Elsmore and Ryle, 1976; Wade and Johnston, 1977). The uncertainties in these positions contribute an uncertainty in the delay measurement of each source in the range 6—15 cm. For two of the radio sources the positions were determined from ARIES measurements on different baselines and have only slightly larger uncertainties.

The troposphere has two components, which are commonly referred to as the wet and dry components. The delay due to the dry component is larger, approximately 2 m at the zenith, and can be calculated with an uncertainty of less than 2 cm from measurements of the surface pressure and temperature. The wet component is smaller, up to a few tens of centimeters equivalent delay, but is more variable. For all but one of the experiments, a water-vapor radiometer (Winn et al., 1976) was used at the ARIES sites to measure directly the zenith water-vapor content (the wet component). At OVRO for all experiments and at ARIES for one experiment the atmospheric water-vapor content was estimated from surface measurements of temperature and relative humidity using a model developed by Berman (1976).

The largest error associated with the water-vapor radiometer is a possible bias of less than 2 cm, which is not elevation angle dependent. Although the uncertainty in modeling the water-vapor content from surface measurements may be large, the total content for these experiments was so small that the error is unlikely to be greater than 2 cm. A systematic error in the troposphere zenith delay estimate is amplified by a factor of about 2.5 and absorbed almost completely as a change in the local vertical component of the baseline at the site where the error is made. For the baselines we are measuring this error is nearly perpendicular to both the lengths and azimuths, so the PV—MAL results are insensitive to such possible systematic errors. All uncertainty in the calibration of the troposphere is included in the baseline error estimates for each experiment.

We have tried to estimate the effect of the ionosphere by using Faraday rotation measurements of the total electron content over the Goldstone Deep Space Network site near Barstow, California. Using a simple model to map the ionosphere in longitude to the OVRO and ARIES sites, we found the largest change on the PV—MAL baseline components to be 4 cm, and a length difference of 5 cm. Because we have no means of verifying the adequacy of our ionosphere correction, we have not included any corrections in the values of the baseline components or lengths. We have, however, increased the uncertainty in the PV—MAL length by adding 5 cm quadratically to the 7-cm formal uncertainty which is determined by quadratically combining the uncertainties of the individual experiments.

In the course of an experiment about 16 different radio sources are observed at the rate of 4 per hour for 12—24 hours. The delay precision for each observation, determined primarily by signal-to-noise ratio, is 0.1—0.5 nsec (1 nsec = 10^{-9} sec), corresponding to 3—15 cm.

The primary frequency reference at each station was a hydrogen maser from the NASA Goddard Space Flight Center. The frequency stability was expected to cause unmodeled time variations of less than 1 nsec. The actual unmodeled noise is estimated simultaneously with the baseline components, instrumental parameters and source positions, and is found to be less than about 0.3 nsec.

The values of UT1-UTC and polar motion provided by the Bureau International de l'Heure have an accuracy of better than 0.02 arc sec over a one-year interval (Chao and Fliegel, 1970; Fliegel and Lieske, 1970) which gives an uncertainty of less than 3 cm in the PV—MAL baseline length and less than 0.1 arc sec in its azimuth.

The cross-correlation of the videotapes from the two telescopes was done on the Caltech/J.P.L. Mk II VLBI Processor in Pasadena, California at various times between January and September 1977. To insure consistency of the cross-correlation results and all subsequent computer processing, two types of tests are made. As a continuing check on the correlator, one 4-min observation from an experiment on January 15, 1976 is correlated and its delay determined at least once per month. This delay has not varied by more than 0.01 nsec, corresponding to about 3 mm, over the 10 months the test has been run. To eliminate the possibility that some change would escape detection by checking only one scan, the entire January 1976 experiment was recorrelated in September 1977, nine months after the first correlation.

The fitted baselines agreed within 2 cm in all three geocentric components. Small differences were expected because both the correlator and the post-correlation computer programs were being optimized in the intervening months.

RESULTS

The results for the individual experiments are shown in Fig. 3. For these geocentric coordinates the Z-axis is parallel to the earth's spin axis; X and Y lie in the equatorial plane and form a right-handed system with X pointing toward Greenwich. All baselines have been referred back to the Conventional International Origin (mean pole of 1901–1905).

The repeatability of results for separate occupations of a site is illustrated in Fig. 3 by the March and October 1976 OVRO—PV results. The maximum spread for any component is 8 cm, and all measurements agree well with the weighted averages.

The weighted averages of the baseline components for each site and the differenced baseline components are given in Table II. Notice that the components of the baseline to OVRO are roughly a factor of 10 larger than the PV—MAL differential baseline components, so the fractional uncertainty in each direct baseline component is better than 2 parts in 10^7.

CONVENTIONAL GEODETIC MEASUREMENTS

Distance measurements were obtained on two different occasions in July 1977, under extremely poor observing conditions (smoke, fog and smog)

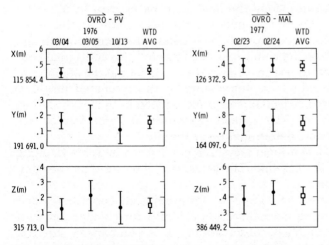

Fig. 3. The ARIES/OVRO baseline measurements by experiment for Palos Verdes and Malibu. The date of the experiment is given at the top of each column of bars. The uncertainty represented by the bars is one standard deviation.

TABLE II

Santa Monica Bay differential baseline

	OVRO→Palos Verdes (m)	OVRO →Malibu (m)	Palos Verdes→Malibu (m)
X	115,854.461 ± 0.027	126,372.388 ± 0.032	10,517.926 ± 0.042
Y	191,691.155 ± 0.042	164,097.746 ± 0.047	−27,593.400 ± 0.063
Z	315,713.146 ± 0.050	286,449.408 ± 0.060	−29,263.738 ± 0.078
		Length VLBI	41,573.902 ± 0.086
		N.G.S.	41,573.846 ± 0.06

compounded by a temperature inversion. Nine measurements were secured on the first occupation and five measurements on the second. The maximum difference between the 14 observations is 7.2 cm. Four distances were measured with a Model 8 Geodimeter. The remainder were observed with a modified Model 4 Geodimeter. Both instruments employ a red laser light source as the carrier beam. The modified instrument is equipped with a 10-mW laser which is considerably larger than that used in the model 8 geodimeter. Vertical angles were observed prior to and following the measurements at each end of the line. These observations were utilized to determine a refined refractive index for correcting the measurements (Meade, 1969).

The length accuracy estimate of 6 cm (one standard deviation) is based on several factors. Observed temperature differences ranged from 6 to 8°C. Investigations have shown that the use of end-point temperatures alone to estimate the mean temperature along the line could be in error by 1.5°C or about 6 cm over the line. Although unlikely, the magnitude of this error could be twice that stated. This estimate also takes into consideration the difficult and abnormal observing conditions encountered. An additional contributing factor is the need to use an estimated geoid height at the Malibu site in the reductions. Admittedly, the uncertainty in the estimated height is quite small. Should it approach 1 m, the error introduced in the lenght reduction would be about 0.5 cm. The geoidal separation at Palos Verdes was computed from an astronomic determination at the site.

The length from this conventional measurement is given in Table II, along with the VLBI length of the baseline. The agreement is better than the combined uncertainty of 10 cm.

The azimuth of the line was computed from adjusted coordinates and reduced to the corresponding normal section azimuth for comparison with the value determined from VLBI observations. This comparison shows a difference of only 0.5 arc sec. The accuracy of the azimuth of the line, as determined from adjusted coordinates derived from a computation of the High Precision Transcontinental Traverse network in California, should not be worse than 1.2 arc sec. In fact, it is reasonable to place the error estimate for this component at 0.8 arc sec or better.

CONCLUSIONS

The radio interferometric geodetic system incorporating a transportable antenna has been used to measure a three-dimensional baseline 42 km long with a demonstrated accuracy of better than 10 cm in length and 20 cm in azimuth, as determined by comparison with direct measurements by the National Geodetic Survey. The 42-km baseline was obtained by differencing two baseline vectors approximately 10 times as long, each of which is thought to be accurate in three dimensions to better than 10 cm.

To achieve a mobile VLBI system accuracy significantly better than 10 cm will require operation at a higher radio frequency to reduce the errors due to the ionosphere, water-vapor radiometers at both stations for tropospheric calibrations and better estimates of the earth rotation parameters. With these improvements we believe that the VLBI technique can provide an accuracy of 2—5 cm on baselines of 500—1000 km.

ACKNOWLEDGEMENTS

Because of the relative complexity of these experiments, the ARIES Project team is indebted to many persons for their assistance in making these demonstrations a success. Thanks are due to: H. Hardebeck, A. Moffet and the staff of the Caltech Owens Valley Radio Observatory (OVRO) for their cooperation in having the 40-m telescope function as the base station for these experiments; A. Rogers of Haystack Observatory and T. Clark of the NASA Goddard Space Flight Center (G.S.F.C.) for the use of the S/X-band receiver system deployed at OVRO; D. Kaufmann and R. Coates of G.S.F.C. for arranging the loan of the hydrogen maser frequency systems at OVRO and ARIES; K. Johnston of the Naval Research Laboratory for radio source positions prior to publication; T. Bandy of Rancho Palos Verdes Estates City Hall and Captain Zimmermann of the Los Angeles County Fire Suppression Camp Number 8 at Malibu for their hospitality during the ARIES station deployments; R. Mitchell and J. McMillan, Los Angeles County Survey Division, for geodetic connections to ARIES sites; J. Gummow of N.G.S. for the triangulation of ARIES station intersection of axes to permanent ARIES site geodetic control; W. Mast, N.G.S. Chief of Party, for the laser measurements across Santa Monica Bay; D. Rogstad of J.P.L., M. Ewing and M. Cohen of the Caltech Radio Astronomy Department for the implementation of software and hardware subsystems of the joint Caltech/J.P.L. Mark II cross-correlation processor; to J.P.L. colleagues J. Fanselow for numerous discussions and assistance with ionospheric delay estimations, B. Thomas for valuable discussions and T. Lockhart for data-processing assistance; N. Yamane of J.P.L. and D. Petterson of Ball Brothers Research

Corporation for modifications and operation of the water-vapor radiometers. This paper presents the results of one phase of research carried out at J.P.L., California Institute of Technology, under contract NAS 7-100, sponsored by the National Aeronautics and Space Administration, Office of Space and Terrestrial Applications.

REFERENCES

Berman, A.L., 1976. The prediction of zenith range refraction from surface measurements of meteorological parameters. JPL Tech. Rep. 32-1602. Jet Propulsion Laboratory, Pasadena, California.

Chao, C.C. and Fliegel, H.F., 1970. Polar motion: Doppler determinations using satellites compared to optical results. JPL Space Programs Summary 37-66, Vol. II. Jet Propulsion Laboratory, Pasadena, California.

Clark, T.A., Hutton, L.K., Marandino, G.E., Counselman, C.C., Roberston, D.S., Shapiro, I.I., Wittels, J.J., Hinteregger, H.F., Knight, C.A., Rogers, A.E.E., Whitney, A.R., Niell, A.E., Rönnäng, B.O. and Rydbeck, O.E.H., 1976. Radio source positions from very-long-baseline interferometry observations. Astron. J., 81: 599—603.

Elsmore, B. and Ryle, M., 1976. Further astrometric observations with the 5-km radio telescope. Mon. Not. R. Astron. Soc., 174: 411—423.

Fliegel, H.F. and Lieske, J.H., 1970. Inherent limits of accuracy of Existing UT 1 data. JPL. Space-Programs Summary, 32-62, Vol. II. Jet Propulsion Laboratory, Pasadena, California.

Meade, B.K., 1969. Corrections for refractive index as applied to electro-optical distance measurements. I.A.G. Symp. Electromagn. Dist. Meas. Atmos. Refraction, Boulder, Colorado.

Ong, K.M., MacDoran, P.F., Thomas, J.B., Fliegel, H.F., Skjerve, L.J., Spitzmesser, D.J., Batelaan, P.D. and Paine, S.R., 1976. A demonstration of a transportable radio interferometric surveying system with 3 cm accuracy on a 307 m base line. J. Geophys. Res., 81: 3587—3593.

Thomas, J.B., Fanselow, J.L., MacDoran, P.F., Skjerve, L.J., Spitzmesser, D.J. and Fliegel, H.F., 1976. A demonstration of an independent-station radio interferometry system with 4 cm precision on a 16 km base line. J. Geophys. Res., 81: 995—1005.

Wade, C.M. and Johnston, K.J., 1977. Precise positions of radio sources: V. Positions of 36 sources measured on a baseline of 35 km. Astron. J., 82: 791—795.

Winn, F.B., Wu, S.C., Resch, G.M., Chao, C.C. and von Roos, O.H., 1976. Atmospheric water vapor calibrations: radiometer techniques. The Deep Space Network Prog. Rep. 42-32:38-49. Jet Propulsion Laboratory, Pasadena, California.

Tectonophysics, 52 (1979) 59—67

© Elsevier Scientific Publishing Company, Amsterdam — Printed in The Netherlands

THE MEASUREMENT OF FAULT MOTION BY SATELLITE LASER RANGING

D.E. SMITH [1], R. KOLENKIEWICZ [1], P.J. DUNN [2] and M.H. TORRENCE [2]

[1] *Goddard Space Flight Center, Greenbelt, Md. 20771 (U.S.A.)*
[2] *EG&G, Washington Analytical Services Center, Inc., Riverdale, Md. 20840 (U.S.A.)*

(Revised version accepted for publication March 24, 1978)

ABSTRACT

Smith, D.E., Kolenkiewicz, R., Dunn, P.J. and Torrence, M.H., 1979. The measurement of fault motion by satellite laser ranging. In: C.A. Whitten, R. Green and B.K. Meade (Editors), Recent Crustal Movements, 1977. Tectonophysics, 52: 59—67.

The distance between two points on opposite sides of the San Andreas Fault is being derived from laser tracking of near-earth satellites as part of an experiment to estimate the motion along the plate boundary. The two sites, at Otay Mountain near San Diego and at Quincy in northern California, are nearly 900 km apart and approximately 150 and 270 km, respectively, away from the main strike of the San Andreas Fault. The angle between the fault and the intersite vector is approximately 25°. In the fall of 1972 satellite laser tracking systems occupied these two sites, and from the data collected the relative location of the two sites was determined. The two sites were reoccupied in the fall of 1974 and again in the fall of 1976, and provided two further estimates of the relative positions of the two sites.

The results of these first three measurements indicate a shortening of the intersite baseline between San Diego and Quincy at an average rate of 9 ± 3 cm/year, suggesting a much larger possible present-day motion across the fault system than expected. The main source of error in this analysis is the motion of the spacecraft which is significantly affected by unmodeled anomalies in the earth's gravity field. However, major advances in our knowledge of the gravity field are expected over the next few years and as these occur the accuracy of the present results will improve.

INTRODUCTION

In order to estimate the gross plate motion across the San Andreas Fault in California an experiment was conceived in 1972 to repeatedly measure the intersite distances between several points along, but back from, the fault over an extended period of time. The technique proposed was laser ranging to near-earth satellites. Three sites were selected in the United States: one

60

at Otay Mountain, near San Diego; the second at Quincy in northern California; and the third at Bear Lake, Utah. In addition, two sites were proposed for Mexico near Las Paż and Mazatlan but have not yet been established.

The two prime sites, San Diego and Quincy, were chosen as they are on opposite sides of the fault and far enough back from the fault to represent (hopefully) the motion of the plate. On the western side of the fault, the Pacific Plate, it was difficult to get far away from the fault because of the proximity of the Pacific Ocean, and weather conditions precluded the possible use of offshore islands. The site eventually chosen was Otay Mountain (Fig. 1), about 20 km from San Diego, 1000 m above sea-level and approximately 150 km from the main strike of the fault. The second site, Quincy (Fig. 1), is situated in the Plumas National Forest, about 270 km east of the fault and at an altitude of 1100 m above sea-level. In selecting this site care was taken not to get too close to the active volcanic region near Lassen. The technique of laser ranging to spacecraft for precise geodetic positioning was expected to be strongest in the intersite distance measurement, and the observation of plate motion was to be derived from changes in intersite distance. Thus, the angle between the intersite vector and the direction of

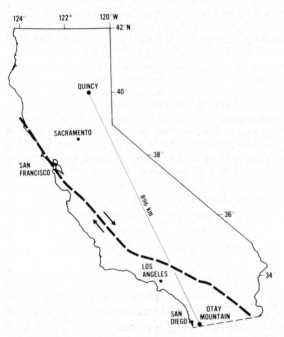

Fig. 1. Laser tracking sites were established at Otay Mountain and Quincy and operated in the falls of 1972, 1974 and 1976.

fault motion should preferably be small. Therefore, the locations of San Diego and Quincy were chosen so that the baseline was nearly along the fault but that the sites were reasonably distant from it. Quincy is 896 km from Otay Mountain and the baseline between the sites makes an angle of approximately 25° with the fault.

The third site at Bear Lake, Utah, was chosen because of its considerable distance from the plate boundary and also because the Quincy—Bear Lake baseline spanned the basin and range province where extension is possibly taking place.

The San Diego and Quincy sites were occupied in the falls of 1972, 1974 and 1976, and Bear Lake was occupied for the first time in 1976. This paper deals exclusively with the preliminary results of the analysis of the three sets of measurements from San Diego and Quincy. This experiment, the San Andreas Fault Experiment (SAFE), is being conducted by NASA in cooperation with scientists from the University of California at San Diego and the Lamont-Doherty Geological Observatory, Columbia University.

METHOD

A satellite laser tracking system determines the distance (range) to a spacecraft by measuring the round-trip travel time of a pulse of light from the system to the spacecraft, where it is reflected from a laser retro-reflector array, back to the ground system. The NASA laser tracking systems make this measurement once per second when the spacecraft is above 20 degrees elevation (weather permitting). In 1972 the quality of the range measurements was only about 70 cm r.m.s. noise for a single laser pulse, but in 1974 and 1976 the systems were upgraded to 10 and 8 cm, respectively. The accuracy of the laser measurements is generally considered to be much better than the noise level because the system is calibrated before and after each track of a satellite by ranging several hundred times to a ground target at a known distance away. If any drift or change occurs in the system during the tracking pass it will show up in a comparison of the two calibration measurements.

If the two laser systems track a spacecraft simultaneously from different sites it is possible to derive the relative locations of the two systems if one has a knowledge of the motion of the spacecraft. Furthermore, if several consecutive tracks of the same spacecraft are obtained from the two systems then it is possible to derive both the relative locations of the tracking system and the motion of the spacecraft (Smith et al., 1973). All that is required in addition to the range measurements is a detailed knowledge of the forces acting on the spacecraft, mainly the earth's gravity field. This is the method that is presently being used in this experiment.

The primary spacecraft used in the experiment so far has been Beacon Explorer C (BE-C) in a near-circular orbit at about 1000-km altitude and inclined to the equator at 41°. Three consecutive passes of this spacecraft

TABLE I

The quality of the laser data (r.m.s. noise) and the orbital fit to three consecutive passes of the spacecraft

Year	Station	Data quality (single range point) (cm)	Orbital fit (three conse-consecutive passes of BE-C)	
			GEM 9 (cm)	GEM 10 (cm)
1972	San Diego	55	80	75
1972	Quincy	85	115	102
1974	San Diego	10	13	15
1974	Quincy	10	16	18
1976	San Diego	8	17	18
1976	Quincy	8	17	18

are used, tracked simultaneously from San Diego and Quincy. Unfortunately, our present knowledge of the earth's gravity field does not enable us to compute accurately the motion of BE-C for three consecutive passes and consequently we are unable to fit the laser tracking data at the noise level for such an orbital arc. Table I shows the quality of the laser data so far obtained in the experiment and our ability to fit the data to three consecutive passes of BE-C for two gravitational-field models, Goddard Earth Models (GEMs) 9 and 10 (Lerch et al., 1977). Errors in the gravity models in effect degrade the data. However, subsequent improvement in our knowledge of the gravity field does allow us to go back and reanalyze the same data to a more precise level. Indeed, our initial attempts to analyze the 1972 and 1974 data were limited to data fits of about 2 m and 1 m, respectively, by our lack of detailed knowledge of the gravity field at that time (Smith et al., 1973, 1977). Additional significant improvements are still expected in gravity-field modeling and it is fully anticipated that the three-pass orbital fits will eventually approach the noise level of the data, enabling the maximum information to be extracted.

RESULTS

In the data analysis the three-dimensional coordinates of the Quincy site have been determined from each data set (1972, 1974 and 1976), together with a single estimate of the height at San Diego for the whole experiment. The latitude and longitude of the Otay Mountain site (San Diego) were assumed to be: latitude 32°36'2.5323"N; longitude 243°9'32.8642"E. The spheroid height of .San Diego that was recovered in the solution was 981.10 m.

Table II shows the preliminary results of the three sets of measurements from San Diego and Quincy for two different gravity fields. The change in

TABLE II

Preliminary results for the location of Quincy, and the baseline distance from San Diego, derived using two different models of the gravity field

Gravity model	Year	San Diego–Quincy baseline (m)	Quincy coordinates		Height (m)
			Latitude (N)	Longitude (E)	
GEM 9	1972	896,275.479±0.086	39°58′ 24.4797″ ± 0.0029	239°03′ 37.6205″ ± 0.0022	1053.52 ± 0.12
	1974	5.279 ± 0.027	24.4808 ± 0.0009	37.6428 ± 0.0006	1053.46 ± 0.03
	1976	5.124 ± 0.026	24.4747 ± 0.0010	37.6436 ± 0.0005	1053.77 ± 0.03
GEM 10	1972	896,275.331±0.082	39°58′ 24.4750″ ± 0.0027	239°03′ 37.6226″ ± 0.0021	1053.54 ± 0.11
	1974	5.184 ± 0.030	24.4773 ± 0.0010	37.6436 ± 0.0007	1053.59 ± 0.03
	1976	4.894 ± 0.029	24.4667 ± 0.0010	37.6440 ± 0.0005	1053.72 ± 0.03

baseline over the four-year period is clearly evident and, if true, is considerably larger than expected, amounting to nearly 9 cm/year for GEM 9, although the standard deviation is admittedly large at ±3 cm/year. Such a rapid motion is surprising since it is 50% larger than that predicted by the Minster et al. (1974) model for this plate boundary, although it is by no means impossible and not the first time that such large rates of motion have been thought to be observed in southern California. Whitten (1956) has reported that triangular measurements extending from Yuma, Arizona, across the Imperial Valley in 1941 and 1954 indicate an equivalent of nearly 9 cm/year of motion in a direction consistent with plate motion. Furthermore the 1975 precise geodimeter traverse across this region confirms the motion with 3.76 m since 1941, equivalent to an average of over 10 cm/year (C.A. Whitten, private communication, 1977).

If the motion we have observed is not entirely plate motion, then the most likely source of systematic error is from the gravity model which perturbs the motion of the spacecraft. For this reason we show the result obtained with another gravity model, GEM 10, which suggests even larger motion. Although we feel the GEM 9 solution is superior, a comparison of the two solutions shows that the effect of gravity-model error could be significant and therefore affecting our interpretation of the GEM 9 solution. Fig. 2 shows the baseline solutions of GEMs 9 and 10. The standard devia-

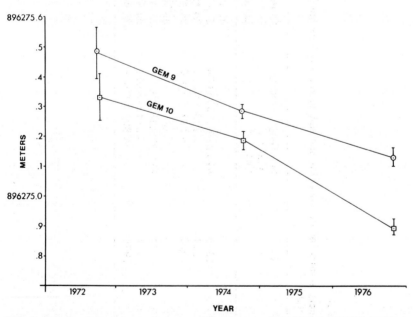

Fig. 2. The baseline distance between San Diego and Quincy for 1972, 1974 and 1976 for two gravity models. The average rate of change of the baseline between 1972 and 1976 was 9 ± 3 cm/year, based on GEM 9.

65

tions shown in Table II and the error bars in Fig. 2 are formal values based on the fit of the tracking data to the orbit.

Figure 3 shows the heights obtained for Quincy and suggests a possible increase since 1972 of about 20 cm. In the solution, however, the height at San Diego was assumed the same in all three years, which may not be true, and consequently an apparent increase in height at Quincy could really be a decrease at San Diego. Considering the error bars in Fig. 3 the differences in height are hardly significant.

Figure 4 shows the three latitudes and longitudes of Quincy obtained in the solutions, together with their error ellipses. The enormous lateral movement between 1972 and 1974 shown by both gravity models is not considered real and could possibly be evidence of a timing error in 1972 at one of the tracking sites. Since the general track of the BE-C spacecraft near the stations is from west to east, a timing error tends to show up more in the recovered longitude than it does in latitude. However, a careful inspection of the original data does not show any evidence of timing problems.

Another possible explanation is that the longitude tends to be more sensitive to errors in the gravity field than either latitude or height. Therefore, the movement from year to year (in latitude and longitude) appears more like a random walk than either the height or the baseline. Gravity-model errors are known to be larger in the direction of motion of the spacecraft and therefore might be expected to show up in the E—W direction.

The possibility of local movements immediately surrounding the San Diego or Quincy site contributing to the change in baseline has been ruled

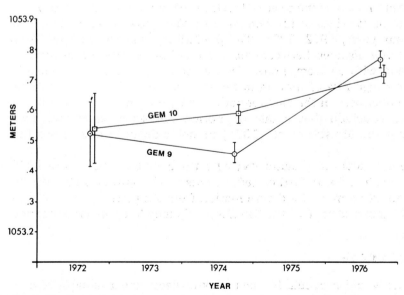

Fig. 3. The height of Quincy above the spheroid for 1972, 1974 and 1976.

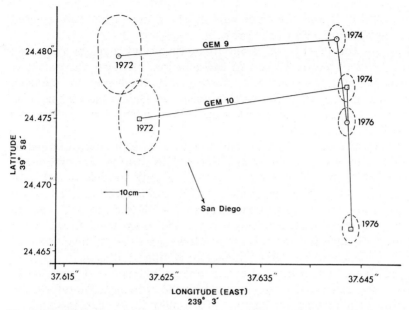

Fig. 4. The latitude and longitude of Quincy in 1972, 1974 and 1976.

out above the 1 cm level. Surveys in 1972 and 1976 (Scholz, 1977) show no significant movement along faults in the vicinity of either site.

CONCLUSIONS

Three measurements of the relative locations of two points, San Diego and Quincy, on opposite sides of the San Andreas Fault have been made over a period of four years, 1972—1976. The preliminary analysis of these data indicate that the distance between the stations has decreased significantly over the period with an average rate of 9 ± 3 cm/year. It seems unreasonable to explain this high rate of motion in terms of errors, either in the data or the gravity model used in the data reduction or by local movements near the sites, and we conclude that a rate of motion somewhat larger than that derivable from the Minster et al. (1974) model is likely to exist along the fault.

The present results are preliminary; over the next few years more 1976 data will be analyzed and further refinements in the gravity field will be made that should improve the determination from the present data. In addition, a fourth measurement of the San Diego—Quincy line is planned for the fall of 1978.

ACKNOWLEDGEMENTS

The authors would particularly like to acknowledge the invaluable advice, assistance and encouragement given by the following colleagues in the planning and execution of the experiment over the past few years: Lynn R. Sykes and Chris Scholz of Lamont-Doherty Geological Observatory, Columbia University; James Savage of the U.S. Geological Survey; Don Tocher of Woodward-Clyde Consultants, San Francisco; Jon Berger of the Institute for Geophysics and Planetary Physics, University of California at San Diego, and Fritz Vonbun of NASA Goddard Space Flight Center.

REFERENCES

Lerch, F.J., Brownd, J.E. and Klosko, S.M., 1977. Gravity model improvement using GEOS-3 (GEM 9 and 10). EOS Trans., Am. Geophys. Union, 58 (6): 371.
Minster, J.B., Jordan, T.H., Molnar, P. and Haines, E., 1974. Numerical modeling of instantaneous plate motions. Geophys. J. 36 (3): 541—576.
Scholz, C.H., 1977. Resurvey of site stability quadrilaterals, Otay Mountain and Quincy, California. Suppl. Rep. to NASA Grant NGR 33-008-146, Lamont-Doherty Geological Observatory, Columbia University.
Smith, D.E., Kolenkiewicz, R., Agreen, R.W. and Dunn, P.J., 1973. Dynamic techniques for studies of secular variations in position from ranging to satellites. In: R.S. Mather and P.V. Angus-Leppan (Editors), Proc. Symp. on Earth's Gravitational Field and Secular Variations in Position. The School of Surveying, The University of New South Wales, Sydney, pp. 291—314.
Smith, D.E., Kolenkiewicz, R., Wyatt, G.H., Dunn, P.J. and Torrence, M.H., 1977. Geodetic applications of laser ranging. Philos. Trans. R. Soc. Lond., 284 (1326): 529—536.
Whitten, C.A., 1956. Crustal movement in California and Nevada. Trans. Am. Geophys. Union, 37 (4) 393—398.

Tectonophysics, 52 (1979) 69—73
© Elsevier Scientific Publishing Company, Amsterdam — Printed in The Netherlands

POSSIBLE HIGH-MOBILITY LAGEOS RANGING STATION

P.L. BENDER [1]*, J.E. FALLER [1]*, J. LEVINE [1]**, S. MOODY [1]**,
M.R. PEARLMAN [2] and E.C. SILVERBERG [3]

[1] *Joint Institute for Laboratory Astrophysics, National Bureau of Standards and University of Colorado, Boulder, Colo. 80309 (U.S.A.)*
[2] *Smithsonian Astrophysical Observatory, Cambridge, Mass. 02138 (U.S.A.)*
[3] *University of Texas McDonald Observatory, Austin, Tex. 78712 (U.S.A.)*

ABSTRACT

Bender, P.L., Faller, J.E., Levine, J., Moody, S., Pearlman, M.R. and Silverberg, E.C. 1979. Possible high-mobility LAGEOS ranging station. In: C.A. Whitten, R. Green and B.K. Meade (Editors), Recent Crustal Movements, 1977. Tectonophysics, 52: 69—73.

A relatively low-cost system for determining both the vertical and horizontal coordinates of several dozen points per year with an accuracy of about 2 cm appears feasible. One approach considered is to use a subnanosecond pulse length laser with a few millijoules per pulse output energy and to employ single photoelectron detection of the returned signal. The single photoelectron approach has been thoroughly tested in the Lunar Laser Ranging Experiment. With a laser average power of about 50 mW, a 30-cm diameter transmit-receive aperture, 10 arc sec pointing accuracy and a beam divergence of 20 arc sec, the expected returned signal level is about 70 pulses in a 3-min interval. If the differences between the observed ranges and those calculated from a reasonably good LAGEOS ephemeris over a 3-min interval are considered, the expected standard deviation of the mean is <0.7 cm.

The calibration procedure used in lunar ranging appears capable of reducing any bias due to the photomultiplier or timing system to 0.5 cm. The other main error source we have considered for the measured optical transit time is a possible difference in arrival time in different parts of the far field pattern because of laser mode structure. This effect needs to be checked experimentally, but we expect it to be 0.5 cm or less for a laser pulse length of about 200 psec. Based on these error estimates, simulations for one week of observations from the high-mobility station have been carried out for us at the National Geodetic Survey. When a refraction model error of 0.15% was used, the uncertainty of the high-mobility station position with respect to a reference station 500 km away was found to be 2.4 cm or less in each coordinate. After the gravity-field uncertainties have been reduced, the station location accuracy will be improved further and the limitation of measuring with respect to a regional reference station can be relaxed.

---

* Staff member, Quantum Physics Division, National Bureau of Standards.
** Staff member, Time and Frequency Division, National Bureau of Standards.

GOALS FOR THE HIGH-MOBILITY STATION

By 1979 or 1980, the main limitation on the use of the LAGEOS satellite (Johnson et al., 1976) for determining crustal movements at many points with the highest possible accuracy is likely to be the lack of suitable ground stations. We believe that there is a strong need to get started quickly on monitoring the motions of several hundred points per year throughout the world, which appears to be possible at moderate cost through the use of high-mobility ground stations. For this reason, we have considered a possible design for a dedicated LAGEOS ranging station to be used in such measurements.

There are five main types of crustal movement measurements where space techniques are likely to play a major role. These are:

(1) Monitoring of anomalous crustal movements in seismic zones for earthquake prediction research purposes.

(2) Verification of the present overall picture of plate tectonics, and determination of whether the present rates of plate motions are equal to the long-term average rates.

(3) Mapping of the strain rates in broad plate convergence zones such as western South America and southern Asia.

(4) Measurement of transient displacements after large earthquakes as they propagate along plate boundaries and into the plates.

(5) Determination of large-scale distortions due to intraplate processes, such as uplift in the Appalachians or spreading in the Basin and Range province of the North American Plate.

For all these types of measurements there are strong arguments in favor of the space techniques being used with high-mobility stations for measurements at intervals of the order of 300 km. Even if the proposed spaceborne ranging system technique is used in the future for monitoring a much higher density network of points in seismic zones, a framework of points at roughly 300-km spacing determined with the highest possible accuracy by LAGEOS ranging or long-baseline microwave interferometry is needed in order to calibrate the spaceborne ranging system results. For all but the last type there is considerable urgency in an early start to the measurements, if they are to contribute in an optimum way to earthquake prediction research and the understanding of earthquake mechanisms.

The objectives of the design for the station were to meet the following objectives:

(1) A range-measurement accuracy capability of 1 cm for a 3-min averaging time, not including atmospheric and earth tide effects.

(2) Short shut-down and start-up times before and after transportation of several hundred kilometers over dirt roads.

(3) Full station operability by a single operator if desired.

(4) Low construction and operating costs and low air-cargo transportation expenses.

We believe that all of these objectives are important to the planned operation of the station, and are fully achievable.

DESIGN CHARACTERISTICS

The error budget corresponding to objective(1) consists of 0.5-cm allowance for the calibration procedures, 0.5 cm for possible differences in arrival time in the far field and 0.7 cm for random errors. The calibration procedure used in the McDonald Observatory lunar ranging operations (Silverberg, 1974) should be capable of 0.5-cm accuracy for 200-psec laser pulse lengths. Achieving this requires making sure that the illumination of the detector photocathode by the calibration stop pulse is the same in both intensity and spatial distribution as for the stop pulses for LAGEOS reflections. The asymmetric distribution in time delays at the satellite will not cause errors in the analysis with the single photoelectron detection method. The difference in arrival times in the far field (Pearlman et al., 1975) is associated with different time profiles for the growth of intensity in different laser modes. This effect can be adjusted to be small, even for laser pulse lengths of 3—5 nsec. However, insuring that it stays adjusted to a small value under field conditions could require frequent additional measurements. If this problem arises, it can be avoided by keeping the total pulse length short, so that substantial fractional differences in time profiles compared with the pulse length can be tolerated. The 0.5-cm error allowance corresponds to 30 psec, or about 30% of the roughly 100-psec halfwidth of the transmitted pulse. This is expected to require little or no adjustment.

The 0.7-cm allowance for random errors for a 3-min averaging time can be achieved even at $20°$ elevation angle for the system which we have considered. We assume the following parameters: a 20-arc-sec beam diameter at the half-power points for a Gaussian beam shape; an offset in pointing of 10 arc sec; a receiver diameter of 30 cm; a 50-mW average laser power; a two-way atmospheric transmission (at $20°$ elevation angle) of $\frac{1}{8}$; an overall optics transmission of $\frac{1}{2}$, an interference filter transmission of $\frac{1}{2}$, a quantum efficiency of 7%; and the average LAGEOS cross-section of $7 \cdot 10^6$ m^2 (Johnson et al., 1976). This gives 73 signal counts in a 3-min interval. The expected shot-to-shot jitter due to the laser pulse width, the distribution of delay times at the satellite, the photomultiplier jitter and other timing system jitter is $\frac{1}{3}$ nsec, or 5 cm in one-way distance. Thus the chosen design parameters give a 0.6-cm random error for a 3-min averaging period at $20°$ elevation angle.

By choosing a low energy per pulse and transmitting through the receiving aperture we can considerably simplify the operation of the ranging system. For example, with 4 mJ/pulse from the laser, 30% loss in the optics before transmission and our 30 cm diameter for the transmitting optics, we find a transmitted flux density of only 4 μJ/cm^2. This flux density does not appear to require safety precautions with respect to aircraft, such as the operation of a radar system or even the use of a visual observer. The suggested 50-mW

72

average power level for the laser appears to us to be achievable in a strengthened unit suitable for extended field use. This can be done with a continuously operating mode-locked oscillator plus pulsed amplifiers. However, we believe that a simpler system based on a pulsed mode-locked oscillator plus amplifier deserves investigation. The requirement of only 10-arc-sec pointing for the 30-cm diameter transmit/receive optics should minimize the cost for this part of the system, and also simplify operations so that the entire ranging process can be controlled by a single person.

POSITION DETERMINATION ACCURACY

As a means of verifying that a LAGEOS ranging station of the type considered would indeed be capable of very useful measurements, computer simulations have been carried out for us by Dr. C.C. Goad of the National Geodetic Survey. These simulations assumed basically the same ranging accuracy as discussed above for one fixed station at Quincy, California, and one mobile station 520 km to the south at Bakersfield. Three other stations — at Mt. Haleakala, Hawaii, at Greenbelt, Maryland, and at Wettzell, Germany — were also assumed to track LAGEOS, but with only one-third of the accuracy. Five additional stations with still lower accuracy were assumed, but they did not significantly affect the results. These weak assumptions about the accuracy of the fixed stations outside California were intended to show that the accuracy of relative measurements is not substantially influenced by the accuracy of the distant stations. A plane-parallel atmospheric model was used, with the assumed uncertainty of 0.15% in the integrated density corresponding to a 1-cm range uncertainty at 20° elevation (Saastamoinen, 1973; Gardner, 1976, 1977; J.P. Hauser and P.L. Bendar, private communication, 1977). A pole-position uncertainty of 4.5 cm was included. Data were assumed to be taken by each station for a 3-min period every 10 min when LAGEOS was above 20°. A 50% data loss due to weather was assumed, with each 3-min observing period being treated independently. A common seven-day observing period for all the stations was used. Rough estimates of the present uncertainties were included for all of the gravity-field coefficients up to degree and order 8.

With the above assumptions, the coordinates for the Bakersfield station were solved for with respect to those of the Quincy station. With only a few of the gravity coefficients solved for simultaneously, the results were poor. However, if about a quarter of the total number were solved for, then the resulting uncertainties in the Bakersfield coordinates with respect to Quincy were roughly 2 cm for each coordinate. As an example, the total uncertainty in the vertical coordinate was 2.4 cm, made up of the following contributions:

Gravity field coefficients	1.4 cm	Quincy refraction	0.8 cm
Bakersfield bias	1.1 cm	Pole position	0.4 cm
Bakersfield refraction	0.9 cm	Random errors	0.3 cm
Quincy bias	0.9 cm		

All other contributions were 0.2 cm or less. Thus, even with roughly the present gravity field uncertainties, very useful measurements could be obtained between two high-accuracy stations located up to at least 500 km apart with one week of data. In view of the small effect of the random errors, a three- or four-day measurement period should be feasible.

The more important question is what will happen in about 1979 or 1980 when the contributions to the uncertainty due to the gravity field should be greatly reduced. Unfortunately, this has not yet been investigated. As far as we know, other simulations that have been done are not directly relevant to our case because of differences in the assumptions which were made. We expect that the results would be similar in this case, even if the high-mobility station were not close to a reference station, as long as about six well-distributed reference stations with accuracies comparable with that of the high-mobility station were in operation. This needs to be checked by additional simulations. Other effects which should be included in the simulations, but which we do not expect greatly to affect the conclusions include the following: fluctuations over periods of one or two days in UT 1 and polar motion; uncertainties in the ocean tide effects on the satellite orbit; uncertainties in the ocean loading effects on the tidal displacements of the observing stations; deviations of the actual atmospheric refraction from the plane-parallel model; and inaccuracies in modeling the earth-albedo radiation pressure effects on the LAGEOS orbit.

ACKNOWLEDGEMENTS

We wish to thank the following for their comments and suggestions concerning the high-mobility LAGEOS ranging station concept and its application to geodynamics: C.C. Goad, B.H. Chovitz, W.E. Carter and W.E. Strange of the National Geodetic Survey; E.M. Gaposchkin of the Smithsonian Astrophysical Observatory; R.O. Castle and R.C. Jachens of the U.S. Geological Survey; E.A. Flinn and D.E. Smith of NASA, and P. Wilson of the Deutsches Geodätisches Forschungsinstitut. We particularly want to thank Dr. Goad for carrying out the computer simulations described in this article.

REFERENCES

Gardner, C.S., 1976. Effects of horizontal refractivity gradients on the accuracy of laser ranging to satellites. Radio Sci., 11: 1037—1044.
Gardner, C.S., 1977. Correction of laser tracking data for the effects of horizontal refractivity gradients. Appl. Opt., 16: 2427—2432.
Johnson, C.W., Lundquist, C.A. and Zurasky, J.L., 1976. The LAGEOS satellite. Paper presented at Int. Astronaut. Fed. 27th Congr., Anaheim, Calif., Oct. 10—16.
Pearlman, M.R., Lehr, C.G., Lanham, N.W. and Wohn, J., 1975. The Smithsonian satellite ranging laser system. Paper presented at the Int. Assoc. Geod. Gen. Assem., Grenoble, France.
Saastamoinen, J., 1973. Contributions to the theory of atmospheric refraction: Part II. Refraction corrections in satellite geodesy. Bull. Géod., 107: 13—34.
Silverberg, E.C., 1974. Operation and performance of a lunar laser ranging station. Appl. Opt., 13: 565—574.

Tectonophysics, 52 (1979) 75—76   75

DETECTION OF CRUSTAL MOTION USING A SPACEBORNE LASERING SYSTEM *

MUNEENDRA KUMAR and IVAN I. MUELLER

National Geodetic Survey, NOAA, Rockville, Md., 20852 (U.S.A.)
The Ohio State University, Columbus, OH., 43210 (U.S.A.)
(Accepted for publication March 24, 1978)

ABSTRACT

The spaceborne laser ranging (or lasering) system provides a method of precise positioning of a large number of points on the earth's surface in a short period of time. That is, a measure of the relative location of geodetic markers from a space platform can maintain horizontal and vertical control to 2—5 cm. At this level of control, small earth surface crustal motions should be detectable. Development of a model for the strain field can be constructed. Furthermore, the spaceborne lasering system can survey an area in a very short period of time (1—2 weeks) and resurvey the area as required.

System design parameters are now being established by NASA for a possible test flight aboard the Shuttle in 1982. These include design specifications of economical corner cubes for ground retroreflectors, coupled with the evolution of engineering model to flight model development. If the experiment of the Shuttle proves successful, it is hoped to put the laser in a free flight satellite. This paper presents the results of a simulated analysis for this contingency.

The system is conceived as an orbiting ranging device with a ground base grid of reflectors or transponders (spacing 1.0—30 km), which are projected to be of low cost (maintenance-free and unattended) and which will permit the saturation of a local area to obtain data useful in monitoring crustal movements. The test network includes 75 stations with roughly half of them situated on either side of the San Andreas fault. Critical study comparatively evaluates various observational schemes and statistically analyzes crustal motion recovery.

The study considers laser radar as the main ranging system, pending final selection from many possible candidates. The satellite orbit is inclined at 110° and slightly eccentric (e = 0.04) with orbital altitudes varying from 370 to 930 km.

---

* Editor's note: The full text has been published in Bulletin Gèodesique, 52 (2): 115—130.

The results indicate that the geometric mode (simultaneous ranging) with a minimum of five grid and three distant (fundamental) stations and mixed ranging to satellite and aircraft seems the most promising. The fundamental stations are distinguished from the grid station in their location, and this location should be sufficiently "distant" from the area of crustal movement so that they can be considered stationary over the time span of the motion involved.

The recovery of motion rate for magnitude is quite straightforward, while for direction each case may require consideration on merit. The study also recognizes the sensitivity of the results/deductions obtained to any associated experimental design. For the specified setup, a time interval between two sets of station recovery for different motion rates or ranging accuracies has also been suggested.

Tectonophysics, 52 (1979) 77—82
© Elsevier Scientific Publishing Company, Amsterdam — Printed in The Netherlands

Measurement of Strain, Tilt and Gravity

DESIGN OF AN EXTENDED-RANGE, THREE-WAVELENGTH DISTANCE-MEASURING INSTRUMENT

S.E. MOODY * and JUDAH LEVINE *

Time and Frequency Division, National Bureau of Standards, Boulder, Colo. 80302 (U.S.A.)

(Accepted for publication March 24, 1978)

ABSTRACT

Moody, S.E. and Levine, J., 1979. Design of an extended-range, three-wavelength distance-measuring instrument. In: C.A. Whitten, R. Green and B.K. Meade (Editors), Recent Crustal Movements 1977. Tectonophysics, 52: 77—82.

We describe an extension of current multiwavelength Electromagnetic Distance Measurement (EDM) techniques which should allow the range of multiwavelength measurements to be extended to approximately 50 km. The basic modification needed is the replacement of the retro-reflector commonly used by an active station containing lasers and a microwave source. Because the system will always be operated as a full three-wavelength instrument, accuracies of about $5 \cdot 10^{-8}$ at 50 km should be obtainable on a routine basis under reasonably clear weather conditions.

EDM TECHNIQUES

Electromagnetic Distance Measurement (EDM) techniques are an essential part of horizontal control networks, as summarized by Wood (1971). Since all such instruments effectively measure a transit time for an electromagnetic signal, it is necessary to know the actual speed of light at the time and place of measurement in order to derive a distance. For geophysical purposes the measurement will invariably be made through the earth's atmosphere which has a refractivity $(n - 1)$ of order $3 \cdot 10^{-4}$. Since it is desirable to measure distances with an accuracy of better than $1 \cdot 10^{-7}$, the atmospheric correction is very large indeed. The first-order refractivity varies with atmospheric temperature, pressure and humidity (Owens, 1967) so that making the corrections directly using meteorological data is difficult and expensive. For optimum results the atmosphere must be sampled along the whole path, which in turn generally requires the use of an aircraft (Savage and

---

* Also with the Joint Institute for Laboratory Astrophysics of the National Bureau of Standards and the University of Colorado.

Prescott, 1973). Methods which allow for correction of the atmospheric refractivity by repetition of the distance measurement at multiple carrier frequencies have been proposed for some years.

The use of two optical wavelengths allows the cancellation of much of the atmospheric uncertainty, but does leave uncorrected some of the error due to water vapor (Bender and Owens, 1965). The addition of a third measurement at radio frequency allows nearly perfect correction of the measurement (Thompson, 1968) except for a term of magnitude $1 \cdot 10^{-8}/K$. With the use of endpoint temperature measurements, this term should be troublesome only in the most extreme meteorological conditions. Analysis by Thayer (1967) of the three-wavelength system suggests that accuracies of a few parts in 10^8 should be attainable for ranges of order 50 km. The main limitation discussed was the difference in the paths traversed by the two optical wavelengths due to the vertical gradient in the refractivity.

DEVELOPMENT IN INSTRUMENTS FOR ELECTROMAGNETIC DISTANCE MEASUREMENT

Instruments which exploit these multiwavelength methods have undergone extensive development by a number of workers and have now reached the point of being productive geophysical tools under field conditions (Slater and Huggett 1976, and references therein). The one drawback of the current multiwavelength techniques is a rather limited range (15 km being typical) which results from the very high signal-to-noise required to allow successful solution of the multiwavelength distance equations. Principal factors which limit the signal to noise are: (1) spreading of the optical beams in the atmosphere which causes much of the light to overspill the retroreflector and receiver apertures; and (2) attenuation of the optical beam in the atmosphere, which is mostly due to scattering. Typical values for this attenuation are 1 dB/km, and for distances greater than a few kilometers the attenuation rather than optical beam divergence limits the optical efficiency of the system.

It is clear that the retro-reflector in such a system cannot return more than all the light which it receives, which in turn will be tens of decibels below the power originally sent by the source. If the retro-reflector is replaced by a second optical source, then a quite dramatic increase in the returned signal will result, which in turn can be used to increase the range of the instrument. Since the total loss between source and detector is roughly a constant, addition of a source at the far end at least doubles the range.

Figure 1 shows the principle of such a two-ended instrument. At each end of the path there is a modulator, a source and a detector. Light from the source is sent through the local modulator, traverses the path and then goes through the far end modulator before detection. A similar process takes place in the opposite direction. After being twice modulated (or modulated and demodulated), the light at each detector will show variation at both the sum

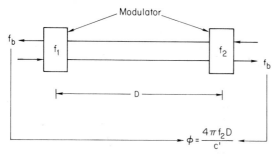

Fig. 1. Principle of two-way optical distance measurement. Note that the measured phase depends only on the frequency of the oscillator at the end where measurements are made.

frequency $(f_1 + f_2)$ and the difference or beat frequency $f_b = (f_1 - f_2)$. In our case the sum frequency is too high to be detected and will henceforth be ignored. Each detector thus has an output at f_b. If one of these signals is sent to the other end, the two signals can be compared in phase. If phases are followed through the system, this measured beat-frequency phase difference is found to be:

$$\phi = \frac{4\pi f D}{c'}$$

where f is the frequency of the *local* modulator, D the path length and c' the actual speed of light for the carrier being used. Just as in the retro-reflecting instrument, the distance is measured in units of the modulation envelope wavelength (c'/f). Similarly, a phase measurement is made at a second optical frequency and at a microwave frequency.

We have chosen operating frequencies for the modulators of approximately 2.7 GHz, or an 11-cm envelope wavelength. The difference frequency $f_b = f_2 - f_1$ is set to be approximately 1.2 kHz, so the necessary phase measurement can be made at audio frequencies. The same microwave frequencies, after multiplication by three for technological convenience, are used to drive a direct phase measurement microwave system which is an exact analog of the one described by Wood and Thompson (1969), although the actual implementation is somewhat different.

Figure 2 shows a block diagram of the complete two-way instrument. An optical carrier at 632.8 nm is used in the configuration outlined in Fig. 1. A second carrier at 441.6 nm is used in the same optical system, but only in one direction since the optical dispersion can be measured as a phase difference between red and blue during a one-way traverse of the path. The microwave system is also bidirectional, and all beat-frequency signals are telemetered to one end of the system for phase measurement. Since all signals are derived from only two oscillators, one at each end, a single set of frequency-control electronics is required to stabilize the frequency difference, while the other oscillator is allowed to run free.

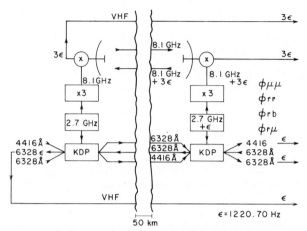

Fig. 2. Block diagram of the prototype system. VHF indicates low-bandwidth telemetry channels whose actual implementation may vary with operating conditions. The various phase combinations measured are indicated. When naming the various beat signals, r refers to red (632.8 nm), b blue (441—446 nm) and μ microwave (8.1 GHz).

Figure 3 shows the calculated range of the instrument, and compares it with a retro-reflecting instrument using similar technology. The three lower curves are for the retro-reflecting instrument, and the upper ones for the two-way design. Assumptions are 10% modulator efficiency and 10% detec-

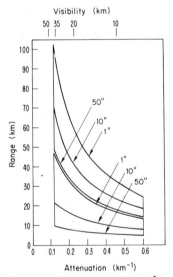

Fig. 3. Calculated range of two-way (upper curves) and retro-reflecting (lower curves) EDM instruments. Different values for the optical beam spread due to the atmosphere are considered.

tion efficiency, using laser powers of 5 mW. Both instruments are assumed to use 10-cm diameter optics throughout. The range is defined as the distance at which shot noise limitations allow a measurement of $1 \cdot 10^{-8}$ precision in 10 sec with full three-wavelength index correction. The calculation is repeated for angular beam spreads of 1, 10 and 50 arcsec, with 10 arcsec being a fairly typical value in realistic atmospheres. One, arcsec corresponds approximately to the diffraction limit of the optics, and cannot be expected to occur except under very special circumstances. About 50 arcsec can be taken as the most extreme case likely to occur. The correspondence between visibility and attenuation is also shown to give the reader some feeling for the attenuation scale.

This rather crude calculation gives the anticipated factor of >2 in range. Since it also corresponds fairly well with experience when checked against operating two-wavelength instruments, there are some grounds for confidence in the range estimates shown. Under conditions which give 15 km for the retro instrument, a 45-km range is predicted for the two-way instrument. Our estimate of a 50-km range results from the slightly lower attenuation expected on longer paths, which will generally lie higher above ground.

An instrument designed in accordance with these principles is currently under construction. When compared with present two-wavelength instruments, there will be a significant loss of portability owing to the need for active systems at each end of the path, as well as the inclusion of the rather bulky microwave antennas as integral components of the instruments. Our prototypes can be broken down into smaller units, the heaviest of which is about 60 kg, so that two people should be able to assemble them at any location accessible by four-wheel-drive vehicle or helicopter. In exchange for this reduced mobility, we anticipate a meaningful improvement in range due to the two-way design, and in accuracy because of the full use of the capabilities of the three-wavelength concept. Therefore, both types of instrument should find complementary areas of application where the advantages of each can be exploited.

ACKNOWLEDGEMENT

This work was supported by the National Aeronautics and Space Administration through Grant No. NSG-7344.

REFERENCES

Bender, P.C. and Owens, J.C., 1965. Correction of optical distance measurements for the fluctuating atmospheric index of refraction J. Geophys. Res., 70: 2461—2462.

Owens, J.C., 1967. Optical refractive index of air: dependence on pressure, temperature, and composition. Appl. Opt., 6: 51—59.

Savage, J.C. and Prescott, W.H., 1973. Precision of geodelite distance measurements for determining fault movements. J. Geophys. Res., 78: 6001—6008.

Slater, C.E. and Huggett, G.R., 1976. A multiwavelength distance measuring instrument for geophysical experiments. J. Geophys. Res., 81: 6299—6305.

Thayer, G.C., 1967. Atmospheric effects of multiple frequency range measurements. Tech. Rep. IER 56-ITSA53. Environ. Sci. Serv. Adm., Rockville, Md.

Thompson, M.C., Jr., 1968. Space averages of air and water vapor densities by dispersion for refractive correction of electromagnetic range measurements. J. Geophys. Res., 73: 3097—3102.

Wood, L.E., 1971. Progress in electronic surveying, U.S. National Report, 1967—1971. Eos. Trans. Am. Geophys. Union, 52: IUGG17—IUGG21.

Wood, L.E. and Thompson Jr., M.C., 1969. Technique for improving accuracy of air-to-ground radio distance measurement systems. Tech. Rep. ERL108-ITS76. Environ. Sci. Serv. Adm., Rockville, Md.

Tectonophyscis, 52 (1979) 83   83

LONG BASE LINE TILT AND STRAIN MEASUREMENT

PAUL DAVIS, KEITH EVANS, JOHN HOSFALL and GEOFFREY KING

Department of Geophysics, Cambridge (Great Britain)

(Accepted for publication March 24, 1978)

ABSTRACT

Two types of noise afflict strain and tilt measurement. They may be categorized as "active" noise, which is due to atmospheric pressure variations, temperature variations, water-table variations and so forth; and "passive" or signal-generated noise which is a consequence of the interaction of the strain field of interest with inhomogeneities of material properties local to the measurement site.

The reason why both types of noise are normally reduced by the use of long base line instruments is explained and a simple, practical long base line tiltmeter is described.

Tectonophysics, 52 (1979) 84
© Elsevier Scientific Publishing Company, Amsterdam — Printed in The Netherlands

A LONG BASE LINE PRECISION TILTMETER

G.R. HUGGETT and L.E. SLATER

Applied Physics Laboratory, University of Washington, Seattle, Wash. (U.S.A.)

(Accepted for publication March 24, 1978)

ABSTRACT

A two-fluid tiltmeter which eliminates the thermally induced errors that have long plagued water-tube tiltmeters has been constructed. While the small tiltmeters currently used tend to be limited because of their great sensitivity to contamination by local nontectonic tilting, this two-fluid instrument has the advantages of a long base line tiltmeter.

Thermally induced errors, dominant in water-tube tiltmeters, are eliminated by putting two fluids, having densities of different temperature coefficients, into a dual tube so that the fluids experience the same thermal environment. By measuring the heights of the fluids at two stations, a correction for thermally induced error can be determined. This correction is then applied to the measured difference in the height of one fluid between stations to yield the true difference in elevation.

Laboratory-induced thermal errors of as much as 45 mm caused by temperature changes of 30°C in a 2-m section were reduced to zero after applying this correction. This test was quite dramatic and clearly indicates the effectiveness of the two-fluid correction technique. The corrected data have a standard deviation from the mean of 0.13 mm. This value could easily be improved by automatic level recording and selection of more optimum fluids.

The application of this method allows precise measurements over long base lines (a resolution of about 10^{-8} to 10^{-9} rad in a 1-km instrument) without expensive level interconnecting tubes or deep burial.

Tectonophysics, 52 (1979) 85—86
© Elsevier Scientific Publishing Company, Amsterdam — Printed in The Netherlands

PRELIMINARY RESULTS FROM COMPARISONS OF REDUNDANT TILTMETERS AT THREE SITES IN CENTRAL CALIFORNIA

C.E. MORTENSEN and M.J.S. JOHNSTON

U.S. Geological Survey, Menlo Park, Calif. (U.S.A.)

(Accepted for publication March 24, 1978)

ABSTRACT

The U.S. Geological Survey has been operating a network of shallow-bore-hole tiltmeters in central California since June 1973. At six sites redundant instruments have been installed as a check on data coherency. These include the Sage Ranch, Tres Pinos, New Idria, Aromas, Bear Valley and San Juan Bautista tiltmeter sites. Preliminary results from the comparison of redundant data from the Aromas, Bear Valley and San Juan Bautista sites for periods of eight, three and seven months respectively, suggest that short-period tilt signals in the range 5 min $< T <$ 3—5 h and ranging in amplitude from $5 \cdot 10^{-8}$ to 10^{-6} rad, but not including step offsets, show excellent agreement on closely spaced instruments. Agreement is not as good in this period range for instruments at San Juan Bautista with a separation of 200 m. Signals of interest observed in this period range include coseismic tilts, teleseisms and tilts associated with creep events. Tilt signals in the period range 3—5 h $< T <$ 2— 5 weeks are not always coherent at all three of the redundant tilt sites studied. Tilt signals in this period range have ampli-tudes up to $5 \cdot 10^{-6}$ rad and wavelengths down to at least the instrument separation at the closely spaced sites (~several meters). Regarding longer-term coherency, the instruments at San Juan Bautista with 200-m spacing, agree within 0.5 μrad for the N—S component and 0.7 μrad for the E—W component for a period of two months. The closely spaced redundant instru-ments at Aromas agree within 2 μrad for the N—S component and 1 μrad for the E—W component for the eight-month period of operation. Data from the three sites have been checked for effects of temperature, atmospheric pres-sure and rainfall. The latter appears to be critically site dependent. The worst case tilts for 1 inch of rainfall can be more than 1 μrad with a duration of a few days to a week. Typical rain-induced tilts are less than 0.3 μrad for 1 inch of rain. The two instruments at the Sage Ranch site have been in operation for the longest period. However, they have shown local site or ground instability, high drift and lack of coherency since installation. Data are not yet available from the Tres Pinos or New Idria instruments. Deeper installation appears necessary for these instruments and two alternative methods of tiltmeter emplacement are currently being tested in an attempt to evaluate the depth, spatial and temporal dependency of surface tilt sources.

Tectonophysics, 52 (1979) 87—96
© Elsevier Scientific Publishing Company, Amsterdam — Printed in The Netherlands

CANADIAN PRECISE GRAVITY NETWORKS FOR CRUSTAL MOVEMENT STUDIES: AN INSTRUMENT EVALUATION *

A. LAMBERT, J. LIARD and H. DRAGERT

Earth Physics Branch, Energy, Mines and Resources, Ottawa, Ont. K1A 0Y3 (Canada)

(Accepted for publication March 24, 1978)

ABSTRACT

Lambert, A., Liard, J. and Dragert, H., 1979. Canadian precise gravity networks for crustal movement studies: an instrument evaluation. On: C.A. Whitten, R. Green and B.K. Meade (Editors), Recent Crustal Movements, 1977. Tectonophysics, 52: 87—96.

Precise gravity networks and profiles have been established in three locations in Canada for the purpose of measuring possible gravity changes associated with seismic events. All measurements were made using two LaCoste and Romberg model D gravimeters. The standard deviation of a single observed gravity difference ranges from 50 to 120 nm/sec². The precision of the measurements appears to depend mainly on the amount of exposure to vibration during transportation. A preliminary comparison between results of different instruments reveals unexplained discrepancies, and calibration tests show that the D meter scale factor varies significantly with either dial reading or reset screw position or both.

INTRODUCTION

The purpose of this paper is to report on the performance of two LaCoste and Romberg (LCR) model D gravimeters used in precise gravity networks in Canada for the purpose of measuring possible gravity changes associated with seismic events. It has been shown (Strange and Carroll, 1974; Whitcomb, 1976, Rundle and Jackson, 1977) that such measurements could play a significant role in the interpretation of crustal deformation when combined with first-order levelling data. To date, the best available gravity data for interpretation in areas of vertical crustal movement come from measurements made with spring gravimeters such as the LCR model G gravimeter. Recent studies in Europe (Brein et al., 1977) have examined the problems encountered with this instrument and indicate that the achievable standard deviation of a tie is 100—150 nm/sec². This performance has now been im-

---

* Contribution from the Earth Physics Branch No. 721.

proved with the introduction of the LCR model D gravimeter. Early tests (McConnell et al., 1975) and small-scale surveys (Lambert and Beaumont, 1977) showed that, with care, a precision of 50 nm/sec^2 was possible with the model D. Although the instrument was not originally intended by the manufacturer for secular gravity studies, it is at present the best available commercial instrument for that purpose. During the last two years, three gravity networks have been established in areas of different tectonic setting and under a variety of operating conditions. This practical experience has revealed some of the more important problems associated with the routine resurveying and synthesis of model D gravity data.

LOCATION AND DESCRIPTION OF THE NETWORKS

The three precise gravity networks described below have been established in areas of known seismic activity and/or where vertical crustal movement is suspected to have occurred. Features common to these precise networks areas follows:
(1) All gravity stations are monumented and consist of three brass plates fixed to bedrock which serve as the instrument pad.
(2) The average distance between stations is roughly 10 km.
(3) Measurements are made in duplicate using two LCR model D gravimeters simultaneously.
(4) An average of eight gravity differences between a given pair of stations are measured before proceeding to the next pair.
(5) The number of connections to each station ranges from two to five.

Manic 3 network

The filling of the reservoir at Manic 3, P.Q., was marked by an increase in local seismicity culminating in a magnitude 4.3 event at a location about 10 km north of the dam site (Leblanc and Anglin, 1976). In order to detect possible vertical crustal movement associated with further seismic activity, a precise gravity network was established around the epicentral area in 1976, together with first-order relevelling to all points accessible by road (Fig. 1a). Precise gravimetry by helicopter was the only practical way of providing some sort of vertical control to otherwise inaccessible locations. The standard deviation of a helicopter tie with respect to the least-squares network solution averaged about 120 nm/sec^2 for the two instruments. However, this result was achieved, only after several cushions were used to decouple the instruments from helicopter vibrations during transit. From the network solution a typical interstation gravity difference is known within 95% confidence limits of 70 nm/sec^2, corresponding to a precision of 2—3 cm in elevation (depending on the mode of deformation). These results are compatible with the 95% confidence limits on the horizontal control for a co-located strain polygon established by the Geodetic Survey of Canada.

Charlevoix network

A precise gravity network was established in 1976 about 130 km northeast of Quebec city in an area marked by a concentration of microseismic activity (Leblanc et al., 1973; Leblanc and Buchbinder, 1977) and a history of large earthquakes occurring roughly every 60—90 years (Fig. 1b). This network replaces an older gravity network and complements first-order levelling done in the area by the Geodetic Survey of Canada from 1926 to 1977. For the 1976 network, the standard deviation of a tie with respect to the network solution was about 110 nm/sec^2, corresponding to 95% confidence limits on an interstation gravity difference of about 60 nm/sec^2. This unexpectedly high deviation was attributable, in part, to vibrations incurred during transit of the instruments over rough roads via automobile, and led to the development of an air-cushioned transit case.

Vancouver Island network

The belt of high seismic activity in coastal and offshore British Columbia includes populated areas subject to the greatest earthquake risk in Canada.

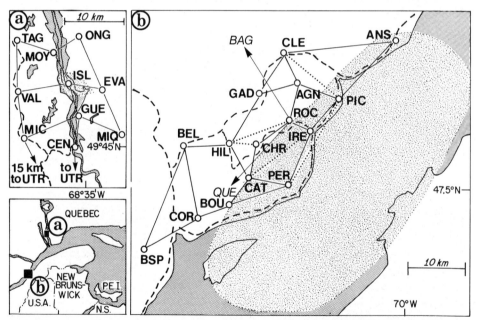

Fig. 1. Precise gravity networks (solid lines) and first-order levels (dashed lines) at (a) Manic 3, P.Q., and (b) Charlevoix, P.Q. Shading denotes areas of seismic activity. Each connection in the gravity networks comprises an average of 8—10 gravity ties. Dotted lines at Charlevoix represent connections added to the network in 1977. Arrows represent 1976 air connections to airport stations at Quebec City (130 km) and Bagotville (100 km).

Although the main concentration of strain release is in the deep ocean in the form of moderate-magnitude earthquakes, there appears to be an almost parallel onshore belt of strain release in the Vancouver Island—Puget Sound area characterized by less frequent but larger-magnitude earthquakes (Milne et al., 1977). In 1946, a magnitude 7.3 event caused damage in the area of Campbell River in central Vancouver Island, and studies of strain release rates suggest that there may be a considerable amount of accumulated strain available for another large event in the region (Milne et al., 1977). In order to detect anomalous vertical crustal movements, extension and relevelling of the first-order level network were carried out, and two precise gravity profiles were established by road across the central part of Vancouver Island in early 1977 (Fig. 2). The northern profile, running between Campbell River and Gold River, comprised a series of interlocking triangles forming what might be called a "single-chain" structure. The southern profile, running west from Qualicum Beach, so far has only a simple structure. The standard deviation of a tie with respect to the network solution was about 55 nm/sec^2 for the two gravimeters, a precision that rivals the results achieved in an earlier network of much smaller scale (Lambert and Beaumont, 1977). The internal consistency of the northern profile for a particular gravimeter suggests that interstation gravity differences are known to better than 50 nm/sec^2 with 95% confidence.

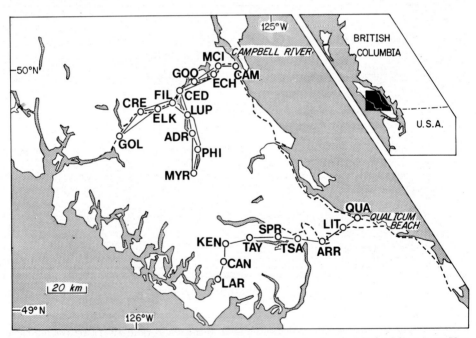

Fig. 2. Precise gravity networks (solid lines) and first-order levels (dashed lines) on Vancouver Island, B.C. Each gravity connection comprises an average of eight ties.

PERFORMANCE OF THE INSTRUMENTS

Investigations on LCR model G gravimeters (Brein et al., 1977) have demonstrated the importance of such influences as levelling and clamping errors, the reading system, transportation vibration, air temperature and pressure, and the earth's magnetic field on the repeatability and biasing of results. Although no intensive investigations were carried out by us to evaluate the significance of these influences, a number of equipment innovations and survey procedures have been adopted to reduce their effects. Survey techniques that have been employed include recoverable instrument orientation, aligned with magnetic north, to reduce possible magnetic effects; an accurately timed 5-min period between unclamping and reading the gravimeter to minimize spring relaxation errors; and the simultaneous deployment of two gravimeters to help reveal systematic errors. To aid in obtaining a more accurate null, 30 sec-per-division instrument levels are used to maintain an accurate vertical alignment, and larger external galvanometers are utilized to allow more consistent nulling even under noisy conditions. Since distances between stations rarely exceed 15 km and times between readings range from 20 to 40 min, the effects of atmospheric pressure changes are not significant when stations are tied in pairs rather than in multi-station loops.

The influence of vibration during transport has been investigated to a greater extent. Fig. 3 shows the standard deviation of gravity ties for LCR-D6 and LCR-D13 for four different "levels" of vibration. Ties between adjacent instrument pads of gravity stations at Charlevoix give the smallest standard deviation (similar tests where the time increment is increased to 60 min give a 50% higher standard deviation). The Vancouver Island survey carried out on good paved roads with an improved air-cushioned transit system had the next higher standard deviation, followed by the earlier Charlevoix survey where the roads were rougher and the vibration insulation was less adequate. Finally, the Manic 3 survey that was carried out by helicopter gave the worst results, even though the time increment between stations was shorter. The contention that vibration plays a major role in degrading results is also supported by preliminary analysis of resurvey data from Charlevoix that shows a significant decrease in the standard deviation of ties when an air-cushion transit system is used.

As pointed out previously, two gravimeters are used simultaneously to detect systematic errors. Fig. 4 shows preliminary differences in station values from Vancouver Island between the two LCR gravimeters D6 and D13 after independent least-squares adjustments of the network for each instrument. The instrumental discrepancies are compared with the differences, denoted by the error bars, that have a one-in-twenty chance of being exceeded. The comparison shows that 5 out of 19 compared gravity differences do not fall within allowable tolerances.

This result suggests that there are probably systematic effects present in the measurements that cannot yet be explained. Since all but one connection

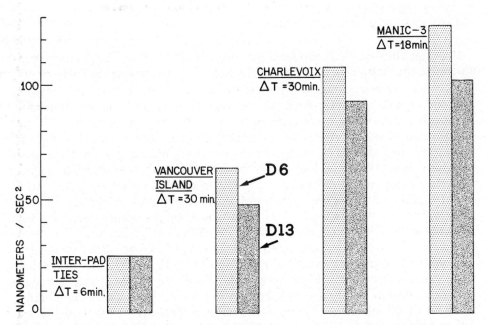

Fig. 3. Variability of standard deviation of gravity ties for LCR-D6 and LCR-D13 under different levels of vibration; ΔT denotes the average time interval for a gravity tie.

were made simultaneously with both meters, earth tide effects could not be the cause. Tides would tend to bias the measurements of both instruments equally. Whether such discrepancies turn out to be repeatable on subsequent

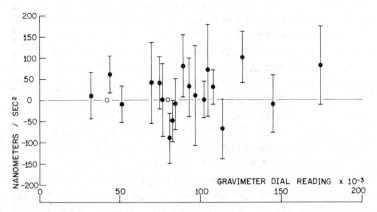

Fig. 4. Differences between LCR-D6 and LCR-D13 station gravity values as a function of gravimeter dial reading for the Campbell River and Qualicum Beach gravity profiles of Fig. 2. The error bars are 95% confidence limits computed by combining the errors in station values for the two instruments. The larger error bars reflect the accumulation of error with increasing distance from the reference stations indicated by open circles.

Vancouver Island surveys will be crucial. Further comparisons between instruments must be made in order to reveal the extent of the problem.

CALIBRATION OF INSTRUMENTS

One of the most important tasks in secular gravity work is to accurately determine the calibration of spring gravimeters and to ensure that the calibration remains constant throughout the survey of a network. These instruments measure gravity differences against the counterbalancing forces of a system of springs and levers. Therefore, it must be assumed that the mechanical properties of such a system could be altered by mechanical shock or change in temperature. A programme of routine recalibration is necessary so that adjustments in data analysis can be made should changes in gravimeter scale factors occur. In the absence of a suitable absolute gravimeter with the required accuracy, it has been necessary to compare the measured gravity differences in a network with gravity differences over certain calibration ranges. Three calibration ranges have now been established for this purpose: Mont Tremblant, P.Q. (633.142 μm/sec^2), Mont Ste. Marie, P.Q. (676.205 μm/sec^2) and Mount Seymour, B.C. (868.432 μm/sec^2). All three ranges are located on solid rock in areas where an abrupt change in elevation gives a large gravity difference over a horizontal distance of less than 2 km. Measurements on these ranges have provided the following observations:

(1) There is a change of parts in 10^3 in the ratio of scale factors for LCR-D6 and LCR-D13 between Texas and Canada.

(2) For a fixed position of the reset screw no significant changes in scale factor have occurred in the two instruments during normal operation.

(3) Both instruments suffered changes in scale factor of parts in 10^4 when they were returned to the manufacturer for internal repairs.

(4) Changes in scale factor of parts in 10^4 occur in both instruments when the reset screw is changed and the calibration is performed over the same gravity range but on a different portion of the dial screw.

In order to investigate in more detail the variation in scale factor, a 207.920 μm/sec^2 calibration range was established in a 22-storey Ottawa office tower where a set of 10 gravity ties could be made in a period of 2 h using a high-speed elevator. Mean gravity differences were observed at about 200 μm/sec^2 intervals across the dial range of the gravimeters by adjustments of the reset screw. Fig. 5 shows that significant variations in gravity difference were observed with both gravimeters. The measured gravity differences were then converted into scale factor curves for each gravimeter (Fig. 6). The variations in scale factor are significant when comparing results from different calibration ranges for a given gravimeter or when relating the results of one gravimeter to those of another in a precise network survey. A typical gravity difference of 200 μm/sec^2, for example, could vary by 140 nm/sec^2 due to such changes in scale factor. In order for these calibration results to be useful, it is necessary to separate the dependence on dial reading from the

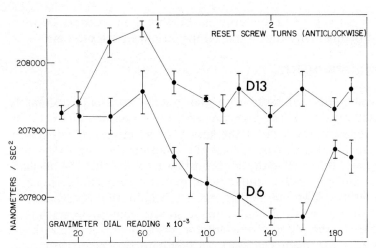

Fig. 5. Apparent change in gravity difference for LCR-D6 and LCR-D13 as a function of dial reading and reset screw position. The error bars are standard error limits.

dependence on reset screw position. Preliminary results show that a dependence on dial reading alone is unlikely. A resolution of the Vancouver Island network was performed using the scale factors of Fig. 6 as interval factors applied to the dial readings, and a significant deterioration in agreement

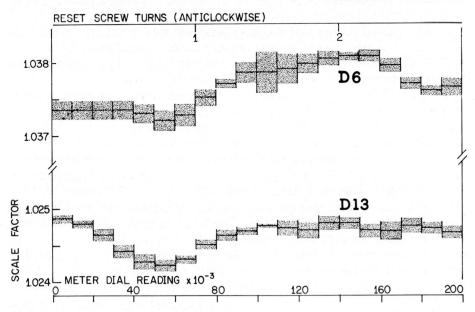

Fig. 6. Scale factor curves for LCR-D6 and LCR-D13 computed by dividing interpolated values from Fig. 5 into a nominal gravity difference for the calibration range. The error bars are standard error limits.

between the two gravimeters was noticed. Since the Vancouver Island survey was done at a single position of the reset screw, an improvement in results would have been expected in the absence of a strong reset screw effect. At present, there are insufficient data to determine whether the scale factor is dependent on reset screw position alone or a combination of dial reading and reset screw position. In order to do this, measurements on a second calibration range shifted by one anticlockwise turn of the reset screw are required. Alternatively, separate relative calibration curves for the dial—screw level system and for the reset-screw system could be determined by the manufacturer using a method that involves adding and removing a small calibrating mass to the gravimeter beam (G. Hamilton, LaCoste and Romberg, Inc., personal communication, 1977).

DISCUSSION

An attempt has been made to give a realistic assessment of the problems facing users of the LCR-D meter. The internal consistency of our networks show that, with care against vibration, relative gravity values between stations can be measured to better than 50 nm/sec^2 with 95% confidence. However, comparisons between instruments suggest that there are systematic effects still unaccounted for that may invalidate the above error estimates. Calibration tests show that even larger discrepancies between instruments are to be expected due to changes in dial reading or reset screw position. Present efforts attempt to overcome this problem by resurveying a network with the same instrument at the same reset screw and dial screw position. However, it may not be possible, owing to instrument drift and reading tares, to guarantee a return to the same position merely by resetting to the same dial reading as before. Moreover, it is difficult, if not impossible, to find an appropriate calibration range covering exactly the same instrument settings as a particular network. At present, therefore, it is necessary to assume that any time changes in scale factor that are revealed on a calibration range accurately reflect the changes in scale factor at the different reset screw position of the survey area. Systematic effects between instruments will continue to receive close examination. Studies on the effect of reset and dial screw positions will be pursued with the objective of developing procedures for accurate and convenient determination of gravimeter scale factors.

REFERENCES

Brein, R., Gerstenecker, C., Kiviniemi, A. and Petterson, L., 1977. Report on high precision gravimetry. Prof. Pap. Natl. Land Surv., Gavle, Sweden.
Lambert, A. and Beaumont, C., 1977. Nano variations in gravity due to seasonal groundwater movements: Implications for the gravitational detection of tectonic movements. J. Geophys. Res., 82: 297—304.
Leblanc, G. and Anglin, F., 1976. Induced seismicity at the Manic 3 reservoir, Province of Quebec (abstract). EOS Trans. Am. Geophys. Union, 57: 759.

Leblanc, G. and Buchbinder, G.G.R., 1977. Second microearthquake survey of the St. Lawrence valley near La Malbaie, Quebec. Can. J. Earth Sci., 14: 2778—2789.

Leblanc, G., Stevens, A.E., Wetmiller, R.J. and DuBerger, R., 1973. A micro-earthquake survey of the St. Lawrence valley near LaMalbaie, Quebec. Can. J. Earth Sci., 10: 42—53.

McConnell, R.K., Hearty, D.B. and Winter, P.J., 1975. An evaluation of the LaCoste—Romberg model D microgravimeter. Bull. Inform., 36: 35—45, Bur. Gravim. Int., Paris.

Milne, W.G., Rogers, G.C., Riddihough, R.P., Hyndman, R.D. and McMechan, G.A., 1977. Seismicity of Western Canada. Can. J. Earth Sci., 15: 1170—1193.

Rundle, J.B. and Jackson, D.D., 1977. Uplift and crustal dilatancy (abstract). EOS Trans. Am. Geophys. Union, 58: 495.

Strange, W.E. and Carroll, D.G., 1974. The relation of gravity change and elevation change in sedimentary basins (abstract). EOS Trans. Am. Geophys. Union, 56: 1105.

Whitcomb, J.H., 1976. New vertical geodesy. J. Geophys. Res., 81: 4937—4944.

Tectonophysics, 52 (1979) 97   97

ESTIMATION AND REMOVAL OF NON-TECTONIC EFFECTS FROM GRAVITY SURVEY DATA

J.H. WHITCOMB [1], W. FRANZEN [1] and W.E. FARRELL [2]

[1] *Seismological Laboratory, California Institute of Technology, Pasadena, Calif. 91125 (U.S.A.)*
[2] *University of California, Berkeley, Calif. 94720, (U.S.A.)*

(Accepted for publication March 24, 1978)

ABSTRACT

When gravity survey accuracies of a few microgals are sought, many correction factors must be accounted for, including meter calibration constants, water-table level fluctuations, solid-earth tides, ocean tides and in some cases rapid atmospheric fluctuations. Calculation of most of these correction factors is relatively straightforward. However, the effects of ocean tide loading are not as easily estimated, partly due to the lack of knowledge of the ocean tides themselves. Amplitude and phase factors for the better-known ocean tide components O_1 and M_2 have been theoretically computed for a grid in southern California in order to correct gravity survey data at arbitrary locations for these ocean tidal-loading components. The gravity data from a three-month period were recorded on a tidal gravimeter at the station PAS and then hand-digitized in order to test the ocean tide estimation program. The O_1 and M_2 ocean tidal components were effectively reduced to less than 0.5 μGal. The remaining high-frequency tidal components appear to be K_1 and S_2. If the ocean tides are not taken into account, as much as 16—20 μGal of error can occur solely due to the effect of ocean loading on the gravitational tides when comparing two surveys near Pasadena. The effect increases towards the coastline and decreases inland. Examples of reduced data from the CIT gravity survey network, which has been observed on an approximately monthly basis since 1974, will be shown.

Tectonophysics, 52 (1979) 99—105   99
© Elsevier Scientific Publishing Company, Amsterdam — Printed in The Netherlands

CONTINUOUS MEASUREMENTS WITH THE SUPERCONDUCTING
GRAVIMETER

JOHN M. GOODKIND

Department of Physics, University of California, San Diego, La Jolla, Calif. 92093
(U.S.A.)

(Accepted for publication March 24, 1978)

ABSTRACT

Goodkind, J.M., 1979. Continuous measurements with the superconducting gravimeter. In:
    C.A. Whitten, R. Green and B.K. Meade (Editors), Recent Crustal Movements, 1977.
    Tectonophysics, 52: 99—105.

    Small drifts in gravity could, in principle, be used as a measure of vertical crustal mo-
tion. However, this could be the case only if other real and apparent causes of gravity
variations can be identified and removed from data. We report here on work with the
superconducting gravimeter. It measures the influence of atmospheric pressure variations
on gravity and indicates that continuous gravity and pressure records are required in
order to account for the effect at the 1 μGal level. In addition, we discuss some preliminary results with two instruments at the same location which suggest that gravity began
to change at the rate of a few microgals per month during the measurements.

INTRODUCTION

The free-air anomaly of gravity is of the order of 3 μGal/cm. Thus, if vertical crustal motion of about 1 cm/year and its variation is to be measured,
gravity variations of the order of 1 μGal must be observable over arbitrarily
long periods of time. Even with a totally drift-free instrument there are a
number of phenomena other than crustal motion which can affect gravity at
this level. The biggest variations in gravity are the tides, but they can be
removed from the data with very high precision. The next largest effect is
from variations in atmospheric pressure. This can also be accounted for with
adequate precision. Finally, a redistribution of mass beneath the surface of
the earth can cause a change in gravity without any vertical displacement.
Groundwater is likely to be the largest effect of this type. Identification of
all of these effects, as well as possible instrumental drift is best accomplished
through the technique of least-squares fits between continuous records of the
relevant variables.

EARTH TIDES

The tidal forces can be computed precisely from the known orbits of the earth and moon and Newton's law of gravitation. However, the gravity variations are enhanced by 16% over these forces as a consequence of their distortion of the elastic earth. The tidal gravity variations are also affected by the tides of the ocean which, however, are not well determined over most of the earth's surface. Fortunately, the gravity tides, including the ocean effect, are stable to better than 0.1 μGal, and the frequencies of all possible tide terms are known precisely. This means that, at any one location, the tidal amplitudes and phases can be determined by least-squares fit of sine functions at each frequency. The sum of all of these sine waves can then be subtracted from the data to yield a tide-free signal. This procedure is described in detail by Warburton and Goodkind (1976a, b; 1977).

ATMOSPHERIC PRESSURE EFFECTS

The effect of atmospheric pressure variations on gravity has also been reported by Warburton and Goodkind (1976b). The dominant effect of the pressure variations on gravity are due to the gravitational attraction of the

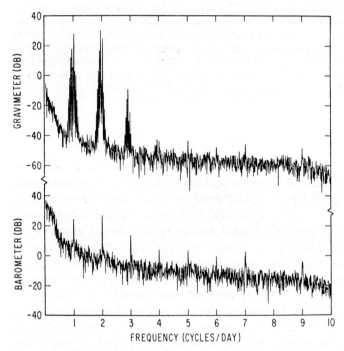

Fig. 1. A comparison of the power spectra of the gravimeter and barometer signals for a one-year record from Piñon Flat, California.

varying mass of air overhead. The spectrum of the pressure fluctuations is shown in Fig. 1 with the spectrum of the gravity signal. This figure shows the atmospheric tides at harmonics of 1 cycle/solar day and shows how they appear on the gravity signal. In addition, the relation between the incoherent background of the two signals is evident. The very strong correlation between the two at very low frequencies is evident in Fig. 2, where the pressure versus time and the gravity residual (tides subtracted) versus time are shown. The ratio of gravity variation to pressure variation as a function of frequency is shown in Fig. 3 with the coherence and phase relation between the two. Tidal terms display a somewhat different frequency dependence than the incoherent background, but the important feature for crustal-motion measurements is the frequency-independent admittance value at low frequencies. The pressure effect can apparently be assumed to be about $0.3\ \mu\mathrm{Gal/mbar}$ with a probable error of less than 15%. This is not a true admittance since its value depends on the size and shape of the region over which the pressure is varying coherently, and this will vary in time. For this reason, the most accurate procedure for removing the pressure effect is to fit the pressure signal to the gravity residual for the specific time period under investigation. The apparent linear drifts of four month-long records were altered by between 1.0 and 1.3 $\mu\mathrm{Gal/month}$ by this procedure.

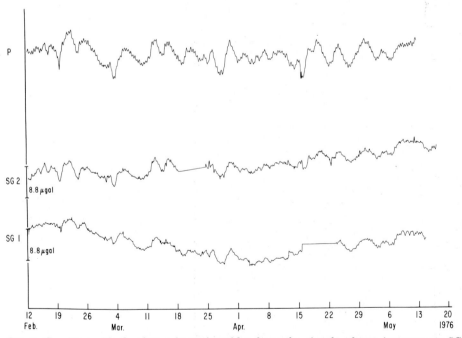

Fig. 2. Gravity residuals after subtracting tides from the signals of two instruments SG1 and SG2. Strong correlation with the atmospheric pressure, P, is evident.

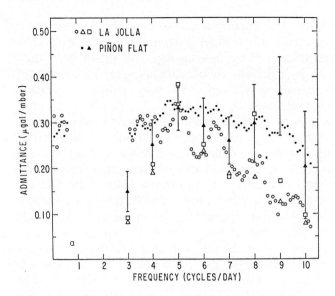

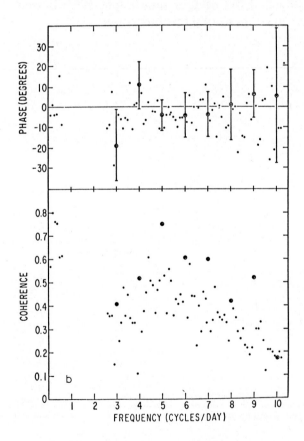

GROUNDWATER

All of our measurements thus far have been in semi-arid southern California. The longest record is from a mountain location with little soil on top of the granite. Thus, we have not measured anything concerning groundwater. However, the magnitude of the problem is indicated by a simple computation. The gravitational attraction of an infinite planar sheet of density σ g/cm is $\Delta g = G2\pi\sigma$. In c.g.s. units this means $\Delta g = 4 \cdot 10^{-7}\,\sigma$. Thus, if a sheet of 1-m thickness is added to or removed from the water table where the ground has 1% porosity $\Delta g/m = 4 \cdot 10^{-7} = 0.4\ \mu\text{Gal/m}$. This means that it is possible to have groundwater effects which are larger than $1\ \mu\text{Gal}$. In order to account for them it will be necessary to have some measure of groundwater variations and to fit it to the gravimeter record for the same time period.

GRAVITY DRIFTS MEASURED BY THE SUPERCONDUCTING GRAVIMETERS

At present, all gravimeters other than the superconducting one have drift rates which are too high to allow continuous measurements to detect crustal uplift. The drift of the present version of the superconducting instrument is so low that it has not yet been possible to determine if it drifts at all. The best information thus far has come from a few months of data for which we had two of the instruments operating in the same place at the same time.

Figure 4 shows the residual signals from the two instruments for the same period after subtraction of both the tides and the atmospheric pressure effects. The lower trace is from our first instrument, SG1, in which we measured a logarithmic decay of the magnetic field which supports the test mass. During the first portion of this record (prior to April 1, 1976) a least-squares fit to a linear drift yielded a slope of $+7\ \mu\text{Gal/month}$. After April 1 a similar fit yields a slope of $-6\ \mu\text{Gal/month}$. This instrument had never previously showed a drift in this direction so that, by itself, it provided some indication of a change in gravity. The upper trace is from an instrument, SG2, in which additional coils had been wound to stabilize the field. These coils reduced the measured effect of current change in the magnet coil on the field by a

Fig. 3. a. Admittance as a function of frequency computed from 53 blocks of 10-day segments of data. Closed circles, ●, and open circles, ○, represent the admittance of the background continuum at Piñon Flat and at La Jolla, smoothed by convolving with a square window of width 1 cycle/day. The continuum cannot be computed in the 1- and 2-cycles/day tide bands so that no points are shown there. The solid triangles, ▲, are the admittances to pressure at Piñon Flat for the tidal frequencies. The error bars represent the 95% confidence limits. The open squares, □, represent the admittance at La Jolla to pressure recorded at Miramar Naval Air Station, and the open triangles, △, are the admittance at La Jolla to pressure recorded at Lindbergh Field.
b. Phase of the admittance and the coherence for Piñon Flat (unsmoothed). The circled data, ⊙, points are the values at the frequencies of the atmospheric tides.

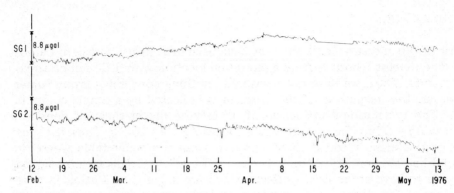

Fig. 4. Gravity residuals of Fig. 2 shown after fitting and subtracting the atmospheric pressure.

factor of 100 so that the drift rate of the field should be reduced by this factor. The difference from zero was not statistically significant for the record shown. After April 1 the drift rate was -7.5 μGal/month. This indicates a change in gravity in the same direction as that indicated by the other instrument, but at about half the rate. The discrepancy could be explained in a number of ways, but we do not yet have adequate information to determine the correct explanation.

One possibility which is currently under investigation is that tilting of the instruments is responsible. At its tilt adjusted position, the instrument has the ideal tilt sensitivity which is proportional to the cosine of the tilt angle. Thus, the apparent change in gravity will be $dg/g \simeq \frac{1}{2} \theta^2$ so that for $dg/g = 10^{-9}$ (1 μGal), $d\theta = 45 \cdot 10^{-6}$ (45 μrad). The gravimeter can be used to measure its own tilt by periodic readjustment to the position of minimum dependence on θ. Over a period of a year SG1 had required a total adjustment of 150 μrad. During the period of the data shown in Fig. 3, SG2 had been mounted on the walls of the underground vault in a corner. In that position it showed anomalous responses at S_1 and S_2 which we have interpreted as a diurnal tilt. The effect was eliminated when the instrument was remounted near the center of the vault. All these observations suggest that the instruments could be tilting in uncorrelated fashion and that there was tilt associated with the gravity change of April 1. Thus, measurement and/or control of the tilt of the instruments at the 10-μrad level will be included in future measurements.

CONCLUSIONS

Tides and atmospheric pressure effects can be removed from continuous gravity measurements to much better than 1 μGal. With our superconducting gravimeter the instrumental drift is at least at the level of known or suspected other influences on gravity and is probably substantially lower. Thus, changes in gravity at rates of a few microgals per month are currently under investigation. If tilt is eliminated and groundwater effects can be accounted for, our results provide evidence that gravity changes of the order of 1 μGal associated with crustal motion may be measured and identified.

ACKNOWLEDGEMENTS

This report is based on published and unpublished work performed by the author and R. Warburton. Analysis of the relative drifts of two instruments was performed by Dr. J.J. Olson and Mr. P. Czipott. We also wish to acknowledge the support of NASA, the National Science Foundation, the United States Geological Survey and the National Bureau of Standards.

REFERENCES

Warburton, R.J., Beaumont, C. and Goodkind, J.M., 1975. The effect of ocean loading on tides of the solid earth observed with the superconducting gravimeter. Geophys. J.R. Astron. Soc., 43: 707—720.
Warburton, R.J. and Goodkind, J.M., 1976a. Search for evidence of a preferred reference frame. Astrophys. J., 208: 881—886.
Warburton, R.J. and Goodkind, J.M., 1976b. The influence of barometric pressure variations on gravity. Geophys. J.R. Astron. Soc., 48: 281—292.
Warburton, R.J. and Goodkind, J.M., 1977. Detailed gravity tide spectrum between 1 and 4 cycles per day. Geophys. J.R. Astron. Soc., 52: 117—136.

Tectonophysics, 52 (1979) 107—116   107

PLANS FOR THE DEVELOPMENT OF A PORTABLE ABSOLUTE GRAVIMETER WITH A FEW PARTS IN 10^9 ACCURACY

JAMES E. FALLER *, ROBERT L. RINKER and MARK A. ZUMBERGE

Joint Institute for Laboratory Astrophysics, National Bureau of Standards and University of Colorado, Boulder, Colo. 80309 (U.S.A.)

(Accepted for publication March 24, 1978)

ABSTRACT

Faller, J.E., Rinker, R.L. and Zumberge, M.A., 1979. Plans for the development of a portable absolute gravimeter with a few parts in 10^9 accuracy. In: C.A. Whitten, R. Green and B.K. Meade (Editors), Recent Crustal Movements, 1977. Tectonophysics, 52: 107—116.

Successful development of a few parts in 10^9 portable *g* apparatus (which corresponds to a height sensitivity of about 1 cm) would have an impact on large areas of geodynamics as well as having possible application to earthquake prediction. Furthermore, the use of such an instrument in combination with classical leveling or extraterrestrially determined height data would yield information on internal mass motions. The plans for the development of such an instrument at JILA using the method of free fall will be given. The proposed interferometric method uses one element of an optical interferometer as the dropped object. Recent work has resulted in substantial progress towards the development of a new type of long-period ($T > 60$ sec) suspension for isolating the reference mirror (corner cube) in the interferometer. Improvements here over the isolation methods previously available, together with state-of-the-art timing and interferometric techniques, are expected to make it possible to achieve a few parts in 10^9 accuracy with a field-type instrument.

DEVELOPMENT IN GRAVITY MEASUREMENTS

The instruments with which I provided myself were, a transit by Dolland, of three feet and a half in length, constructed on the same principle as the transit at the Royal Observatory at Greenwich, so as admirably to combine lightness with strength.

A repeating circle of one foot diameter by Throughton,

A clock and a box chronometer by Arnold, for the loan of which I was indebted to Henry Browne, Esq. F.R.S., and

An invariable pendulum with its support, a description of which will be given hereafter. To these was added, a chest of tools of various kinds.

A small light waggon was constructed at the Royal Arsenal at Woolwich for the

---

* Staff member, Quantum Physics Division, National Bureau of Standards.

conveyance of these instruments, and a party consisting of a non-commissioned officer, two gunners (one a carpenter), and two drivers with four horses of the Royal Artillery, was placed under my orders: a bell tent, and two others of a smaller description, were issued, which I found particularly useful.

> "An account of experiments for determining the variations in the length of the pendulum vibrating seconds, at the principal stations of the Trigonometric Survey of Great Britain." Capt. Henry Kater F.R.S. in Philosophical Transactions, MDCCCXIX

Since Kater's classic pendulum experiments, tremendous progress in the art of gravity measurements at different sites has been made, most markedly during only the last two decades: technology and methods have now developed to the point that the capability exists for the development of portable absolute "g" apparatuses with accuracies of a few parts in 10^9. Though the value of g has had, and will continue to have, considerable significance for the field of standards, at this level of accuracy, a number of effects of geophysical origin can be expected to appear — in an era in which precision measurements of g should prove to be an exciting new tool for furthering our understanding of the world on which we live.

The significance for geophysics of precision g measurements rests in the fact that if a vertical displacement of 1 cm occurs, the corresponding change in the locally measured acceleration of gravity will be roughly three parts in 10^9. Virtually all our present-day understanding of vertical crustal movements rests on long-term geological evidence or on incremental evidence developed from leveling. However, parts in 10^9 gravity measurements will permit "real-time" observations of vertical displacements at the 1 cm/year level.

Only two basic methods have been used in order to measure the absolute acceleration of gravity. They are (1) timing freely-falling bodies, and (2) timing falling bodies whose motion is constrained in some way, as in the case of pendulums. Only in the past 20—30 years has it been possible electronically to realize the necessary accuracy in the measurement of short time intervals to permit effective measurement of g by direct free fall. Since that time the free-fall method has proved to be the most accurate and straightforward way of measuring the acceleration due to gravity. The first free-fall experiments used geometrical optics to define the position of an object as it fell. Since 1963 (Faller, 1963 and 1965), direct interferometric methods using corner-cube type mirrors, one of which was dropped, have led to more accurate distance measurements during the free fall. The advent of laser light sources further simplified the distance measurement in these free-fall interferometric devices.

DEVELOPMENT OF GRAVIMETERS

The first transportable laser-interferometer absolute gravimeter was that of J.A. Hammond and J.E. Faller (Faller, 1967; Hammond and Faller, 1967, 1971a, b). This instrument was developed at the National Bureau of Standards and Wesleyan University, with major support from the Air Force Geophysics Laboratory then A.F.C.R.L.

The basic design of the Hammond—Faller instrument consisted of a laser interferometer system which looked at the motion of a freely falling retro-reflector. Such a system is shown schematically in Fig. 1. Three fringes, formed near the beginning, middle and end of the fall time, were used to determine three positions for the reflector. The time intervals between the formation of the first and second fringes and the first and third fringes were measured accurately. From these measured quantities, the acceleration of the reflector could be calculated in terms of the laser wavelength, the number of fringes counted between the three measured positions, and the time intervals as measured in terms of a frequency standard.

A stabilized laser provided the required brightness and coherence to achieve high-quality fringes over the required $\frac{1}{2}$ m or so dropping distance. The extreme sensitivity of the pattern to the parallelism of the plates was circumvented by using corner cubes rather than plane mirrors for reflectors.

Absolute measurements using the Hammond—Faller apparatus were made in 1968 and 1969 at the eight different sites: Middletown, Conn.; Bedford, Mass.; Washington, D.C.; Denver, Colo.; Fairbanks, Alas.; Teddington, U.K.; Sèvres, France; and Bogotá, Colombia. The typical accuracy achieved at these sites was five parts in 10^8. The results (Hammond and Faller, 1971a, b) obtained during that period represented the first time that free-fall

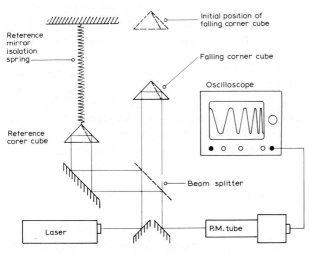

Fig. 1. Laser interferometer system.

absolute measurements had been compared at the same site and this served to corroborate the various measurements. They further provided the first transatlantic absolute transfer, as well as the first series of absolute determinations along the American calibration line. Finally, these measurements played a major role in determining the values used in the International Gravity Standardization Net of 1971 (IGSN, 1971).

These highly successful results, while fully bearing out the promise of this new type of instrument, were nevertheless hard-won in that the apparatus was awkward to transport because of its bulk (approximately 800 kg). As a result of the requirement for a high vacuum and also the complexity of the equipment (involving the assembly and interconnection of a number of components) one week was required to set up and carry out a measurement at each site. The apparatus did, however, represent the best compromise between accuracy and transportability that we could envision at that time (~10 years ago).

A second-generation version of the Hammond—Faller gravity apparatus has been developed by N.B.S. (Faller and Hammond, 1974) and is presently being brought on line by A.F.G.L. as an easily field-portable apparatus. However, the intended uses of this instrument do not require higher accuracy than five parts in 10^8, and as a result (of time and financial limitations) the expected accuracy of this particular new and smaller instrument is not more than a factor of two or three higher than the earlier model.

The only absolute gravimeter which has been operated with substantially higher accuracy than that of the original Hammond—Faller instrument is that constructed by A. Sakuma at the Bureau International des Poids et Mesures in Sèvres, France (Sakuma, 1971, 1973, 1974). In 1970 it was reported to have achieved an accuracy of six parts in 10^9. Further development of the instrument has continued since that time such that its present accuracy now approaches a few parts in 10^9. A second instrument of the same design is being constructed in Japan. However, these instruments are quite complex in design, and are portable only — as someone has quipped — in the sense that the earth goes around the sun.

Recently Sakuma has designed and built a much smaller and transportable version of his rise-and-fall BIPM apparatus. This instrument (the so-called Italian apparatus) is now being used in an extended program of field site measurements by the Italians from the Istituto di Metrologia (Torino). Its achieved accuracy appears to approach one or two parts in 10^8, while the required one week at each site and 600-kg weight are both similar to the Hammond—Faller apparatus.

DEVELOPMENT OF A NEW INSTRUMENT

Recognizing the potential of the laser interferometric technique we are now developing a new instrument of considerably reduced bulk and weight, which is intended to provide higher mechanical reliablity and markedly less set-up time at a given site. The culmination of this effort is expected to result in a highly portable and computer-controlled instrument capable of measuring "*g*" with an accuracy of a few parts in 10^9 or better in a time scale of less than a day at each site. The electronic system to be used in the new instrument has at its heart a time digitizer which, coupled with a mini-computer, will permit us to utilize 100 or more fringes from each drop in solving for the acceleration of the reflector. This will very much improve the signal-to-noise ratio of the measurements and will also permit the analysis to yield the gravity gradient (to better than 1%) as well as the value of *g* itself. The resolution of the instrument is 0.125 nsec, and thus the time measurement will have a resolution of better than one part in 10^9. The time base itself will be derived from a highly stable crystal oscillator.

Besides the smaller size, the design and construction of a new dropping chamber are aimed at correcting the mechanical difficulties associated with the high vacuum (2×10^{-7} mm of Hg) environment required in the initial instrument. These difficulties are (1) a severe restriction on the materials that can be included without severe outgassing, and (2) the mechanical welding of the parts which seems to result after a period of operation. Two alternatives to the high-vacuum requirement have been examined. One of these would be to go to an up-and-down measurement in which the freely falling corner cube is initially launched upward, with timing being done during both the upward and downward parts of the motion. This would permit a less stringent vacuum requirement, since on the upward portion the air drag acts in the same direction as gravity while during the falling part this molecular force is opposite. This cancels out to first order the effects of the residual air pressure on the measured value of *g*.

The second approach considered involves dropping not only the cube but also its container so that air drag results only from the differential velocity between the cube and its container. This differential velocity could then be kept small compared to the falling object's typically 100–300 cm/sec velocity during the measurement portion of its fall. The concept developed for freeing the dropped object from its seat in the vacuum chamber by a reproducible and constant amount during the time of fall involves cocking the corner cube relative to an auxiliary mass by an extended spring. On release this causes the cube and chamber to separate by a fixed amount at the beginning of their fall together.

The question of deciding between the two approaches is not an easy one: it involves the inherent accuracy advantage of the up-and-down approach against the greater simplicity and freedom from systematic errors of the straight-fall approach. And, though we have not completely committed ourselves as yet on the issue, our present inclination is go with the mechanically simple straight-fall method, noting that its accuracy capability per drop still approaches one part in 10^9.

The optical system of the new instrument will be relatively straightforward and similar to what we have used in the earlier instruments. An iodine stabilized laser will most probably be used for the light source. Such lasers have been built by N.B.S. personnel in Washington and have proved to have adequate intensity such that the dynamic length measurement can be made to better than $\lambda/500$. The wavelength uncertainty of such a laser is considerably less than one part in 10^9 (Schweitzer et al. 1973).

The use of standards of length and time which are only one step removed from the primary standards offers a degree of confidence in the long-term stability of the instrument, and the advanced electronics and computing instrumentation should provide a convenience and precision impossible to obtain a few years ago.

Finally, recent work done by one of us (R.L.R.) has resulted in substantial progress toward the development of a new type of long-period ($T > 60$ sec) suspension for isolating the reference mirror (corner cube) in the interferometer from all background disturbances. The need for this isolation stems from the fact that during a measurement, the dropped cube while in free fall is completely isolated from the earth's microseismic and all man-made noise. However, the reference corner cube (in the other arm of the interferometer) is not. In practice, this ground-introduced noise can limit the per drop accuracy to as low as one in 10^6 (Hammond and Faller, 1971b).

Accordingly, if one is to achieve a few parts in 10^9 accuracy in g, an effective method for isolating either the entire system of the reference cube (hung vertically so that vertical motion of the base shortens both arms equally) must be achieved. In the past, use has been made of an astable spring system as employed in commercially available long-period vertical seismographs. However, these systems are somewhat awkward to use and suffer from internal (violin-string) modes in the main system spring.

We noted that a mass (i.e., the reference cube) suspended from a long spring is effectively isolated from vibrations (which enter at the top) for all frequencies greater than the natural resonance of the system. Furthermore, for any given frequency, the lower the mass-spring resonance, the better the isolation. In practice, one would like the system resonance to be 0.05 Hz or lower (20 sec period or longer), and this involves a fairly long spring. For example, it would take a spring 1 km in length to yield a period of about 60 sec.

In practice, we electronically terminate a tractable length of spring (i.e., 30 cm) so that it behaves exactly as if it were, for example, 1 km long, with the mass on the end oscillating up and down with a period of 60 sec ($\nu = 0.017$ Hz) and therefore isolating for all periods shorter than this.

To understand this electronically generated "super spring", imagine you have, say, a 1-kg mass hanging on the end of a spring which extends 1 km vertically. This mass will oscillate up and down with a period of 60 sec, and as it does so the coils of the spring will also oscillate up and down. The coils very near the mass will have an amplitude nearly equal to the amplitude of the mass, and the coils that are far away from the mass will have an amplitude less than that of the mass. In fact the coils near the top will hardly be moving at all. Now if one were to grasp the spring 30 cm above the mass and move that point on the spring just as it moved when the lower mass was in free oscillation, the motion of the mass would remain unchanged. Having done this, one could then cut off the top of the spring and have left a spring 30 cm long that has the same resonance and behaves in all ways exactly like a spring 1 km long. Our "super spring" uses a servo system to generate such a virtual point of suspension.

Figure 2 is a schematic drawing of the system. The two side springs supply the force to support a bracket on which a mass is attached by a central spring; this bracket is free to move in a vertical direction.

The light from the LED is focused by the sapphire ball onto a split photo-diode. The outputs from the two halves of this diode are amplified and dif-ferenced, producing an analog signal that is proportional to this displace-ment of the weight. This signal is processed by a servo-compensating ampli-fier which drives a loudspeaker voice coil. This coil then supplies the needed force on the bracket to cause it to track the motion of the bottom weight. Since the top of the spring is attached to the bracket, the top of the spring moves nearly the same amount as the bottom. The degree of tracking is determined by the gain setting of the servo system, and this in turn sets the effective length of the spring and thereby the achieved period. By changing this gain we have achieved periods from 10 sec to as long as 300 sec (which in the latter instance corresponds to a spring length of 22 km).

In principle the approach would permit one to achieve periods of up to 55 min (corresponding to a spring length equal to half the earth's radius), at

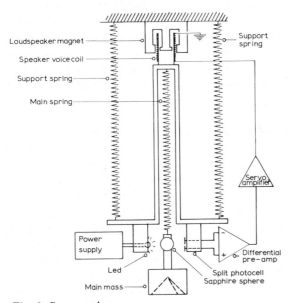

Fig. 2. Super spring.

which point the change in force on the mass for a small displacement result-ing from the earth's gravity gradient exceeds the spring's restoring force and as a result the system becomes unstable.

In practice, other considerations have limited our effective length to only one-tenth of this theoretical limit. Nonetheless, improvements using this type of system are expected to make it practical to achieve the few parts in 10^9 level in g with a field-type instrument by, for all practical purposes, free-ing the measuring instrument from all vibrational problems in the environ-ment.

The scientific and practical reasons for proceeding with this development are based on the fact that an accuracy of three parts in 10^9 for gravity corre-sponds to a height sensitivity of about 1 cm. Instruments of this accuracy could be used to monitor strain fields, and would open up an important new parameter for possible use in the prediction of earthquakes. Absolute gravity measurements also will be important in helping to understand large-scale ver-tical motions when used in conjunction with a distributed net of relative gravity measurements and as a stand-alone instrument at a number of selected sites.

It needs to be recognized that there is some uncertainty associated with the interpretation of gravity-change measurements because of lack of com-plete knowledge of density changes taking place below the earth's surface and because of possible variations in the water-table height: what gravity directly tells us is whether or not the combination of elevation and local mass distribution is stable. Indeed, horizontal motions of material within the aesthenosphere or discharges of deep aquifers do occur over long periods of time, while changes in the local water table can offset gravity on a short time scale. However, it appears probable that these limitations will not seriously degrade the value of absolute-gravity measurements for determining vertical motions, provided that sites are selected on nonporous bedrock or that the effects of superficial geological effects such as changing water tables are sub-tracted on the basis of direct measurement (water height in nearby wells), differential gravitational measurements or differing characteristic frequen-cies.

The main alternative to gravity measurements for the accurate determina-tion of vertical motions appears to be leveling or the use of extraterrestrial methods. These latter techniques include laser range measurements to artifi-cial satellites and to the moon, as well as long-baseline radio interferometry. All three extra-terrestrial methods are capable, with suitable precautions, of achieving accuracies of between 1 and 3 cm in determining crustal move-ments in all three directions. However, the expense is very much higher than it would be for absolute-gravity measurements at the desired sites. In the past, vertical height information was chiefly derived from leveling data, but the cost per kilometer of leveling is high (Castle et al., 1976). Thus, gravity offers an economical tool for initial reconnaissance as well as the high (centi-meter-level) sensitivity required to complement these other techniques.

The primary reason this type of an instrument lends itself to a field-usable version lies in the method of measurement which does not depend on the "tare-free" character of any material-bearing surfaces or mechanically delicate system or systems of springs. Rather the accuracy of the instrument depends on the reproducibility of the basic standards of length and time. These standards are less subject to the ordinary vibration-in-transit, environmental temperature, etc., problems which have proved difficult with traditional gravimeters at the microgal level of sensitivity. It is further recognized that the practical field use of such an instrument will require that the highest degree of portability possible be achieved consistent with the requisite accuracy. Particular attention will be paid to maintain this during the development.

An additional motivation for achieving the utmost in portability is that the use of such an absolute instrument would be desirable as a replacement for relative gravimeters in studying secular changes. Here one needs to note the great care and elaborate precautions taken by the Finnish Group in their study of the Fennoscandian lines with relative gravity meters. Still, in spite of this great care, including limiting observations to seasons of the year with minimum temperature variations, etc., their actual reading accuracy under field conditions (for a single instrument per occupation of a given site) was ±13 μGal (Kiviniemi, 1974). Here the use of a portable absolute instrument — in addition to removing any questions about possible systematic errors — would greatly assist in establishing this type of line and at the same time would remove most of the operating constants that are necessary when using relative meters. This advantage is particularly important for distances between sites of the order of 200 km as was the case in the Finnish measurements.

ACKNOWLEDGEMENTS

This work was supported in part by a grant from the Air Force Geophysics Laboratory (A.F.G.L.).

REFERENCES

Castle, R.O., Church, J.P. and Elliott, M.R., 1976. Aseismic uplift in southern California. Science, 192: 251.
Faller, J.E., 1963. An absolute Interferometric Determination of the Acceleration of Gravity. Ph.D. Thesis, Princeton Univ., 108 pp.
Faller, J.E., 1965. Results of an absolute determination of the acceleration of gravity. J. Geophys. Res., 70: 4035.
Faller, J.E., 1967. Precision measurement of the acceleration of gravity. Science, 158: 60.
Faller, J.E. and Hammond, J.A., 1974. A new portable absolute gravity instrument. In: Bull. Inform., Bur. Gravim. Int., 35: I-43—I-48.
Hammond, J.A. and Faller, J.E., 1967. Laser interferometer system for the determination of the acceleration of gravity. I.E.E.E. J. Quant. Electron. QE-3 (11): 597—602.

Hammond, J.A. and Faller, J.E., 1971a. Results of absolute gravity determinations at a number of different sites. J. Geophys. Res., 76: 7850.

Hammond, J.A. and Faller, J.E., 1971b. A laser-interferometer system for the absolute determination of the accelerators due to gravity. Proc. Int. Conf. on Precision Measurement and Fundamental Constants, Nat. Bur. Std., Spec. Publ., 343, U.S. Gov. Printing Office, Washington, D.C., pp. 457—463.

IGSN, 1971. The International Gravity Standardization Net, 1971. Bur. Cent. Assoc. Int. Geod., 19, rue Auber-75009 Paris, 194 pp.

Kiviniemi, A., 1974. High precision measurements for study in the secular variations in gravity in Finland. Publ. Finn. Geod. Inst. No. 78, Helsinki, 68 pp.

Sakuma, A., 1971. Recent developments in the absolute measurement of gravitational acceleration. In: D.N. Langenberg and B.N. Taylor (Editors), Proc. Int. Conf. on Precision Measurement and Fundamental Constants, Nat. Bur. Std. Spec. Publ. 343. U.S. Gov. Printing Office, Washington, D.C., pp. 447—456.

Sakuma, A., 1973. A permanent station for the absolute determination of gravity approaching one micro-gal accuracy. In: Proc. Symp. on Earth's Gravitational Field and Secular Variations in Position, pp. 674—684, Sydney, 1973.

Sakuma, A., 1974. Report on absolute measurements of gravity. In: Bull. Inform., Bur. Gravim. Int., 35: I-39—I-42.

Schweitzer, Jr., W.G., Kessler, Jr., E.G., Deslattes, R.D., Layer, H.P. and Whetstone, J.R., 1973. Description, performance, and wavelengths of iodine stabilized lasers. Appl. Opt., 12 (12): 2929—2938.

Tectonophysics, 52 (1979) 117

MONITORING OF SECULAR GRAVITY CHANGE IN SOUTHERN CALIFORNIA

WILLIAM E. STRANGE

National Geodetic Survey, NOS, NOAA, Rockville, Md. 20852 (U.S.A.)

(Accepted for publication March 24, 1978)

ABSTRACT

Repeat gravity measurements have been carried out over more than 1000 km of level lines in southern California. The initial surveys over a number of these lines were made in 1962. Repeat gravity survey lines are located primarily in the San Joaquin Valley, in the Los Angeles and Palmdale areas and in the Imperial Valley. Secular changes of gravity have been noted in a number of areas. In areas of subsidence, due to groundwater withdrawal, direct correlation between elevation and gravity change is found. In areas of vertical crustal motion along fault zones, such as in the Palmdale area, gravity changes are also noted. However, these are not, in general, directly related to elevation change. Apparently, subsurface mass redistributions are associated with the vertical motions noted in fault-zone areas before and after earthquakes. Several possible models to explain the gravity observations are being explored.

Tectonophysics, 52 (1979) 118—119
© Elsevier Scientific Publishing Company, Amsterdam — Printed in The Netherlands

IN-SITU STRAIN RELIEF MEASUREMENTS IN ICELAND

KARLHEINZ SCHÄFER

Geologisches Institut der Universität, 75 Karlsruhe 1 (G.F.R.)

(Accepted for publication March 24, 1978)

ABSTRACT

In-situ measurements of rock stress in Iceland have been made by Hast in 1967—1968 at five sites and two sites by Haimson and Voight in 1976. All of the received data revealed compressive stress, although three sites (Keflavik, Burfell and Akureyri) investigated by Hast and the two (Reykjavik) measured by Haimson and Voight are located close to the axial rift zone.

During the summer of 1976 in-situ strain relief was measured at 20 new sites. The localities were concentrated in the axial rift zone in Kelduhverfi/North Iceland. Strain relief was recorded during an episode of intensive seismic activity and of major rifting. Pre-existent fissures and faults were subjected to considerable vertical and horizontal movements a few months earlier. Additional in-situ strain measurements were carried out between the northern axial rift zone and the east coast, and along the south coast from Höfn to Reykjavik.

Most of the measurements have been made at the surfaces of Interglacial and Tertiary flood basalts polished by glacial abrasion. These surfaces were most suitable for in-situ measurements since bedrock outcrops were unaltered. There were no major geological criteria that influenced the stress field after the recession of the glaciers other than stress caused by the driving mechanism. At seven sites within the axial rift zone and at thirteen sites outside, in-situ strain was relieved by overcoring and horizontal strain components were measured. The orientations and magnitudes of maximum and minimum horizontal stress were calculated after determination of Young's modulus and Poisson's ratio.

Outside the axial rift zone the orientations of maximum horizontal stress ($\sigma_{H_{max}}$) are between N30°W and N70°W, and the magnitudes of stress vary from 650 to 10 bar (most values: 50—100 bar). The magnitudes of minimum horizontal stress ($\sigma_{H_{min}}$) vary from 150 to —160 bar (most values: 0—50 bar).

Within the axial rift zone there are two trends of orientation of $\sigma_{H_{max}}$, one of N35°E to N40°E in the central part of the axial rift zone in the south as well in the north, and another of E—W in the marginal areas of the rift zone. The $\sigma_{H_{max}}$ values vary from 120 to 10 bar (most values: 50–100 bar) and $\sigma_{H_{min}}$ was calculated to vary from 100 to −100 bar (most values: 50 to −50 bar).

Absolute tensile stress could be found only in a narrow zone within the northern rift area.

In July 1976, shortly after the episode of intensive recent crustal movements and during increased seismic activity in the northern rift zone only a N—S-striking corridor of about 2.5-km width was subjected to tensile stress, while to the east and west of that zone $\sigma_{H_{max}}$ was measured to be perpendicular to the gaping fissures.

Tectonophysics, 52 (1979) 120

ON RADON EMANATION AS A POSSIBLE INDICATOR OF CRUSTAL DEFORMATION

CHI-YU KING

U.S. Geological Survey, Menlo Park, Calif. 94025 (U.S.A.)

(Accepted for publication March 24, 1978)

ABSTRACT

Radon emanation has been monitored in shallow capped holes by a Track-etch method along several active faults and in the vicinity of some volcanoes and underground nuclear explosions. The measured emanation shows large temporal variations that appear to be partly related to crustal strain changes. This paper proposes a model that may explain the observed tectonic variations in radon emanation, and explores the possibility of using radon emanation as an indicator of crustal deformation. In this model the emanation variation is assumed to be due to the perturbation of near-surface profile of radon concentration in the soil gas caused by a change in the vertical flow rate of the soil gas which, in turn, is caused by the crustal deformation. It is shown that, for a typical soil, a small change in the flow rate ($3 \cdot 10^{-4}$ cm sec^{-1}) can effect a significant change (a factor of 2) in radon emanation detected at a fixed shallow depth (0.7 m). The radon concentration profile has been monitored at several depths at a selected site to test the model. The results appear to be in satisfactory agreement.

Tectonophysics, 52 (1979) 121—138
© Elsevier Scientific Publishing Company, Amsterdam — Printed in The Netherlands

STRAINS AND TILTS ON CRUSTAL BLOCKS *

R.G. BILHAM and R.J. BEAVAN

Lamont-Doherty Geological Observatory of Columbia University, Palisades, N.Y. 10964 (U.S.A.)

(Accepted for publication March 24, 1978)

ABSTRACT

Bilham, R.G. and Beavan, R.J., 1979. Strains and tilts on crustal blocks. In: C.A. Whitten, R. Green and B.K. Meade (Editors), Recent Crustal Movements, 1977. Tectonophysics, 52: 121—138.

We present some of the evidence for continental crustal block structures revealed by geodetic work during the last century. Block dimensions ranging from 5 to 50 km appear to be common in regions of intense tectonic activity. The properties of block structures are discussed in terms of the measurement of tectonic deformation by surface strain-meters and tiltmeters. Block boundaries are often significantly weaker than the contiguous crustal blocks they divide, and this feature results in the concentration of strains at boundaries and the diminution of strains within blocks. Tilts on crustal blocks can vary in phase and magnitude from block to block. In addition, there is some indication that block boundaries may respond viscoelastically or exhibit strain-dependent elastic properties. Such behavior can account for the slow transmission of tectonic deformation reported in Japan and elsewhere. If nonlinear behavior is a characteristic of regions fragmented by crustal blocks it introduces problems for the interpretation of observed surface strain and tilt. In particular, it will generally not be possible to apply a site correction factor based on the observed distortion of seismic or tidal strains to the interpretation of secular strains.

INTRODUCTION

In recent years the problems of measuring strains and tilts near the earth's surface have received the attention of numerous authors. Two principal difficulties have been identified as posing serious threats to the effectiveness of surface deformation measurement. The first of these is the distortion of applied strain fields by local elastic inhomogeneity and the second is the existence of noise generated at the earth's surface by nontectonic effects. It is now established that strain-field distortion arising from geometric and elastic inhomogeneity can be estimated with reasonable precision (Berger and

---

* Lamont-Doherty Geological Observatory Contribution No. 2739.

Beaumont, 1976; King et al., 1976), and some progress has been made in discriminating between certain types of surface noise and signals of tectonic interest (Wood and King, 1977). New instruments with longer baselines may also result in improved data (Davis et al., 1978).

We are concerned about a more fundamental problem associated with measuring strains and tilts near the surface. There is considerable evidence that parts of the earth's crust behave as a mosaic of loosely coupled blocks. Can surface fragmentation result in behavior that cannot be predicted in terms of linear elastic theory? Nonlinear behavior may result in the failure of several assumptions. For example, if block interaction is frequency dependent, it is clearly incorrect to assume that the response of a region to periodic tidal strains is the same as the response to secular strains. If the coupling of blocks is stress dependent, the transmission of strains and tilts across a region may exhibit phase and magnitude anomalies not predicted by present inhomogeneity theory. It is clearly of the utmost importance to clarify the influence that block structures may have on surface strain fields and tilt fields.

GEODETIC EVIDENCE FOR CRUSTAL BLOCKS

Although geodetic survey data of high quality have been obtained in many parts of the world in the last century, it is only in Japan that large numbers of strainmeters and tiltmeters have been operated simultaneously for many years (Harada, 1976; Hosoyama, 1976). The results of geodetic releveling rather than trilateration or triangulation are of interest since the precision obtained by leveling is systematically better than other techniques at any given time and also because the distribution of fixed benchmarks (approximately 2-km spacing in Japan) is far denser than triangulation or trilateration points.

In the late 1920s and early 1930s, a series of papers describing releveling lines in Japan were published by some half-dozen authors. Pivotal to their interpretation of these leveling data was an intense discussion of *block movements* which were clearly identified in central and southern Japan (Yamasaki, 1928; Muto and Atsumi, 1929; Imamura, 1930a; Miyabe, 1933; Tsuboi, 1933). The geodetic blocks that were identified were sometimes, but not always, the structural units into which geologists and geomorphologists had subdivided the crust. The Japanese geodetic data present compelling evidence that at least some of the geological processes that gave rise to observed formations are presently active.

In Figs. 1—4 we present typical data from the papers of Imamura and Tsuboi that illustrate evidence for block movements. Note that the data presented have not been corrected for "spurious" data points nor have they been smoothed by linear or weighted averaging processes. In many cases the data have not been adjusted for the misclosure of long geodetic lines. The intentional display of raw data enables the indentification of discontinui-

123

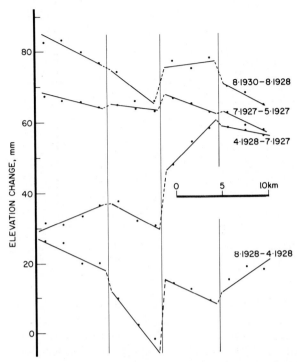

Fig. 1. Block movements in the Tango district (Tsuboi, 1933) during the period May 1927 to September 1930 along a leveling route from Ebara to the Sea of Japan. The Gomura and Yamada faults on which significant postseismic movement was observed, are about 15 km to the east of this N—S line.

ties in the leveling lines if they exist. This is especially easy if several releveling profiles are available. Unfortunately, spurious bench-mark movements can sometimes appear as systematic discontinuities in the data; however, two

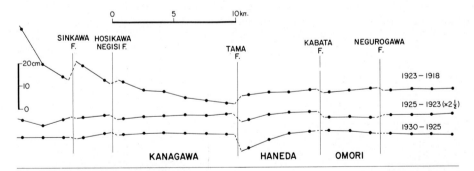

Fig. 2. Block movements west of Suruga Bay (Imamura, 1931). The upper profile shows changes that may have occurred during the 1923 Kwanto earthquake and the lower curve shows changes after the 1926 Haneda earthquake.

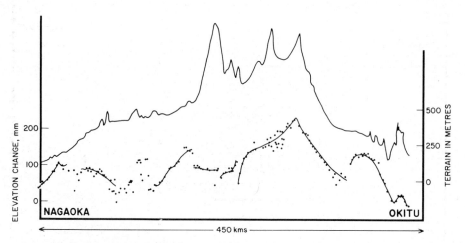

Fig. 3. Major land blocks revealed between 1894 and 1927 in a N—S line from Nagaoka to Okitu in Suruga Bay (Tsuboi, 1933).

remaining tests are available. First, the location of the apparent discontinuity can be related to the known geology. In many cases discontinuities correspond to known active faults, but occasionally they correspond to "dormant" geological faults (Imamura, 1929—1935). Secondly, the discontinuity may coincide with an inflection of the leveling profile that frequent releveling reveals as a consistent hinge-line about which the adjacent surfaces appear to rotate.

Imamura (1933a) generalized the observed behavior of a sequence of block surfaces into two categories: surfaces with "N-shaped faults" in which adjacent block surfaces rotate in the same sense and surfaces with "V-shaped faults" where adjacent block surfaces rotate in an opposing sense (Fig. 5). He remarks further that surfaces with multiple N-shaped faults are associated with compressive horizontal strains and that surfaces with V-shaped faults are generally associated with tension. We adhere to Imamura's original visually descriptive terminology in which he describes the crustal surface

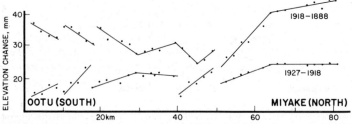

Fig. 4. Block movements between 1888 and 1927 in the region between Ootu and Miyake along the coast of Lake Biwa (Tsuboi, 1933).

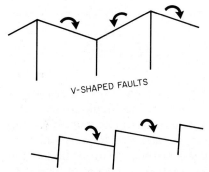

Fig. 5. Imamura defined block boundary faults as V-shaped or N-shaped, depending on whether adjacent block surfaces rotated in opposite or similar directions.

pattern in terms of the discontinuities as it has the advantage of brevity. An important property of V-shaped faults in that the hinge-type motion which occurs need not necessarily involve vertical or horizontal offset and therefore the fault may be dismissed as inactive during surface geological mapping. Moreover, if a V-shaped fault is covered by superficial deposits, its presence may only be detectable by geodetic methods (Imamura, 1931). Activity suggestive of V-shaped fault behavior has been reported in California by Buchanan-Banks et al. (1975), Savage et al. (1975), Bennett et al. (1977) and Hill et al. (1979).

Block activity is most pronounced at times of seismic energy release or volcanic activity. Coseismic block motion forms the bulk of the observed phenomena, but postseismic activity on boundaries has been documented on numerous occasions (Scholz, 1972). For example, earthquakes in 1925 (Tazima) and 1927 (Tango) activated the same block boundaries (Tsuboi, 1933), but in succeeding years, block motions continued (Fig. 1). Imamura (1929) pointed out that block motions associated with the 1923 South Kanto earthquake were bounded by existing geological faults. In the same region, Scholz and Kato (1977), using a 70-year data base with 23 repeat levelings along several leveling lines, showed conclusively that these block boundaries were active in the following 15-year postseismic episode and their presence could still be detected in 1963. Of great significance is their observation that the faults bounding blocks between Tokyo and Atami assume different roles during the various phases of deformation. Thus, the Oiso Hill block mimics the crustal deformation to the NE between 1924 and 1939 and is decoupled from the SW region by the proposed Odawara fault (Sakawa fault, Imamura, 1930a), whereas between 1939 and 1963 inflection occurs at the Isehara fault such that the Oiso Hill block now follows the general deformation of the region to the SW. In a less conspicuous fashion the faults bounding the Kanagara, Haneda and Omori blocks to the west of Suruga Bay are alternately active and passive between 1883 and 1930 (Imamura, 1931 and Fig. 2). Miyabe (1933) notes several examples of adjoining blocks which sometimes move independently and sometimes behave as though locked together.

BLOCK DIMENSIONS

It is of interest to review what is known of the lateral dimensions of observed crustal blocks. Tsuboi (1933) classes *minor* land blocks as generally being of the order of 7—10 km wide and *major* blocks approximately 50 km in extent. In the works of Tsuboi and others, smaller and larger blocks are documented. It is significant, however, that the normal 2-km spacing of the Japanese leveling network provides little positive information on the existence of blocks with a width of less than 6 km. The exception to this occurs when a leveling line runs obliquely across an elongated block, leading to the detection of block motion but to the overestimation of the block size.

The West Momaya 4.5-km block was clearly defined by Imamura (1934) using a 500-m spacing of bench marks near Osaka. Miyabe (1933) argues that secondary block fragmentation down to 1 km may occur, based partly on the observation that horizontal strains exceeding 10^{-4} have been measured within 10-km-wide blocks. In the Tien Shan region blocks with dimensions of 3—5 km have been documented (Pevnev et al., 1975) and in Iceland block dimensions of less than 500 m have been reported by Tryggvason (1970, 1974) and Tomasson (1976).

It is curious that the inferred block structures invoked to explain anomalous tilts at tidal periods (Tomaschek and Groten, 1961; Melchior, 1967, Lambert, 1970) and also to explain anomalous tilting in response to atmospheric loading (Tomaschek, 1953, 1959) often require blocks with dimensions exceeding several hundred kilometers. Block dimensions of this size are seldom apparent on geodetic releveling profiles although oscillatory movements with long wavelengths frequently occur (e.g., Mizoue, 1967). Unfortunately, the tilt data upon which the above-mentioned claims for block structures were based contain systematic errors that were unsuspected by early investigators (King and Bilham, 1973a; Zschau, 1977) and must be regarded as unreliable.

CONSEQUENCES OF BLOCK STRUCTURE

The dangers of measuring surface strain and tilt in the presence of vertical elastic discontinuities have been known for some time (King, 1971, King and Bilham, 1973b). The rigorous correction of tiltmeter data for geometric and elastic inhomogeneities has been achieved recently by Harrison (1976a) and for strainmeter data by Berger and Beaumont (1976) and Levine and Harrison (1976) for a limited number of sites in the U.S.A. The agreement between observed tidal data and predicted values is encouraging (Table I, parentheses) and indicates for these sites, as Berger and Beaumont point out, that the elastic theory is complete and problems of "site" effects can be

TABLE I

Tidal data obtained by strainmeters *

Sources **	Location and azimuth (clockwise in degrees from N)	Latitude (degrees N)	Longitude (degrees E)	O_1		M_2	
				Admittance	Phase (degrees)	Admittance	Phase (degrees)
1	Burdale, 146	54.05	−0.68	0.70	−7	1.0	−1
2	Queensbury, 45 (laser)	53.77	−1.86	0.74	+6	0.82	+1
	(mean of 9 wire)			0.84	+6	0.79	+2
3	Denholme, 128 (mean of 2)	53.78	−1.83	0.77	+6	0.82	0
4	Clayton, 67	53.77	−1.86	0.68	+12	0.75	+17
5	Woodhead, 60 (mean of 3)	53.50	−1.37			0.46	+3
6	Harecastle, 155 (mean of 3)	53.03	−2.25			0.80	−5
7	Haddon, 132	53.19	−1.65			0.89	−7
8	Alfreton, 143	53.09	−1.35			1.25	−4
9	Nottingham, 157	52.95	−1.17			0.75	+5
10	Colwall, 62 (mean of 3)	52.08	−2.33			0.53	−10
11	Mythe, 144	52.00	−2.15			0.65	−16
12	Whittington, 70	51.89	−2.00			1.02	+5
13	Chipping Norton, 140	51.95	−1.55	0.49	−1	0.52	+12
14	Catesby, 162	52.22	−1.23	1.87	−5	1.02	+6
15	Warden, 150	52.69	−0.38			0.18	−16
16	Madingley Rise, 2	52.22	0.09			1.56	+31
17	Schiltach, 2	48.33	8.33	0.78	−6	0.90	+4
	Schiltach, 61			0.63	−5	0.75	+3
	Schiltach, 120			0.60	+5	0.78	+3
18	Camp Elliot, 45	32.88	−117.10	1.02	+2	0.95	+13
	site corrected			(1.00)	(+2)	(0.97)	(+13)
19	Pinion Flat (areal)	33.59	−116.46	0.80	−6	0.96	−1
	site corrected			(0.98)	(−5)	(1.12)	(−3)
20	Mina (areal)	38.44	−118.15	0.72	+5	0.70	−6
	site corrected			(0.97)	(+5)	(0.99)	(−5)
21	Round Mountain (areal)	38.70	−117.08	0.88	−1	0.74	+4
	site corrected			(1.02)	(0)	(0.89)	(+2)

TABLE I (continued)

Sources **	Location and azimuth (clockwise in degrees from N)	Latitude (degrees N)	Longitude (degrees E)	O_1 Admittance	O_1 Phase (degrees)	M_2 Admittance	M_2 Phase (degrees)
22	Flat River (areal)	37.83	−90.48	0.99	−3	1.08	+7
	site corrected			(0.99)	(−3)	(1.08)	(+7)
23	Ogdensburg (areal)	41.09	−74.56	1.01	+5	0.96	+3
	site corrected			(0.99)	(+4)	(0.95)	(+4)
24	Poorman mine, 353	40.03	−105.33	0.60	−6	0.79	−7
	site corrected			(0.81)	(−3)	(1.04)	(−9)
Average of European observations				0.81	+2	0.82	+3
Average of U.S.A. observations				0.86	−1	0.88	+2
Average of site corrected U.S.A. observations				0.97	0	1.01	+1

* Only locations with both diurnal and semidiurnal tidal results are included in the averages. Figures between parentheses refer to data that have been corrected for local elastic inhomogeneity.

** Sources: 1–16 = Evans (1975); Beavan (1976) and Beavan and Goulty (1977); 17 = Emter and Beavan (personal communication); 18–23 = Beaumont and Berger (1975) and Berger and Beaumont (1976); 24 = Levine and Harrison (1976); Levine (1977).

overcome. In practice this means that a site transfer function can be computed for surface instruments in the presence of gross elastic and topographical inhomogeneities by observing any precisely known strain signal, rather than by predicting the effects of the inhomogeneities. King et al. (1976) suggest a method for determining this *site tensor* which requires the observation of at least three strain signals that must be different but whose amplitudes need not be precisely known.

Can block movements or the presence of crustal blocks result in types of inhomogeneity which cannot be defined in terms of the effects of underground cavities, surface topography and the lateral distribution of elastic moduli? We discuss some of the observed behavior of crustal blocks with this question in mind. In the discussion we frequently make use of the term "applied strain field". This we define as the strain that would be present on a laterally homogeneous earth.

1. Incoherent strainfields

Independent motion of adjacent blocks may give rise to the fragmentation of the applied strain field into small-scale strain fields which are related in a complex way to the applied strain field. To an observer within the fragmented region, the strains on adjacent blocks could be described as spatially incoherent with each other. However, strains within a block will appear coherent. In the context of the idealized V- and N-shaped faults of Imamura, a strain array between two faults has internal coherence, but a strain array with one instrument per block shows poor coherence or no coherence. An important corollary is that if coherent strains or tilts are sought in order to reduce the surface noise in strain or tiltmeter arrays, for example, the spacing of instruments should be much smaller than the block dimensions. Kasahara (1973) argued that the direction and amplitude of tilts and strains measured with tiltmeters and strainmeters at Nokogiriyama and Aburatsubo are not in close agreement with each other or with geodetic data in the same region, and attributed this to inhomogeneity and block structure. Scholz and Kato (1977) demonstrate convincingly that the two observatories are on different crustal blocks which move differently from each other and from the general trend of the region. However, Yamada (1973) and Shichi (1973) reveal a curious similarity between the tilt vectors if the vector coordinates of one station are bilaterally inverted, rotated slightly, an isotropic scaling factor applied and the phase of the Aburatsubo Station delayed by about a year. Except for the phase delay, the coordinate transform required can be interpreted as the result of complex block interaction. The phase delay is suggestive of a migrating deformation source (Kasahara, 1977) although it is possible to derive similar delays by invoking nonlinear elastic behavior at block boundaries (see section 5 below).

An important feature of surface strain fields is that although topographic and elastic inhomogeneity can distort applied strain amplitudes considerably, the rotation of the principal strain axes is minimal (Berger and Beaumont, 1976). It is possible, however, that a mosaic of loosely coupled blocks might introduce large rotations of the applied strain axes under certain conditions. For example, if a strain field is applied to one side of a sequence of crustal blocks, the observed strains well into the fragmented region may be largely those arising from random contacts on block boundaries.

2. Strain attenuation

Horizontal strains within a block bounded by free surfaces will tend to zero although the decoupled block will respond to basal tilts. This suggests an argument in favor of using geodetic leveling to search for the existence of active blocks. In practice it is not possible for a fissure to remain open at any great depth (e.g., Reasenberg and Aki, 1974) so that strains will be diminished to a marked degree only near the block boundaries and according to the degree and depth of the weaker boundary zone. Latynina (1957b) reasons that vertical discontinuities may be widespread, resulting in a global attenuation of tidal strain. Recent observations of tidal strain do not favor this hypothesis. In Table I we compile tidal admittance results from several observatories. The admittance is the ratio of the observed tidal strain amplitude to the combined amplitude of the theoretical tide and the estimated ocean-loading strain tide. Figures in parentheses are the tidal admittance after correction for the distorting effects of local cavities, surface topography and known geology. Unity admittance and zero phase are expected where the predicted value agrees with the observations. Although the average tidal admittance is low for data where site corrections have not been applied, it would appear that except in a few cases, site geometry can account for much of the observed attenuation (Berger and Beaumont, 1976). The lack of evidence for a general attenuation of strain amplitudes at tidal frequencies need not necessarily exclude the occurrence of crustal blocks if block boundaries behave viscoelastically.

3. Strain concentration at block boundaries

Strain applied to a region of elastic contrast will concentrate in the relatively compliant zones. That block boundaries are generally zones of low elastic moduli has been substantiated by many investigators. Seismic velocity anomalies associated with major faults have been reported by Alewine and Heaton (1973), Bakun and Bufe (1975), and Healy and Peake (1975). In a finite-element model of the San Andreas fault zone, Wood et al. (1973) show that tidal tilts are particularly sensitive to the effective stiffness of the fault zone. The enhancement of strains across a block boundary was studied by Latynina (1975a, b) who reports tidal strain magnifications of 100% across parts of the Kondarinsky fault and abnormally large secular strains. Tidal strains within the San Andreas fault zone are not reported, although the apparent modulation of the velocities of seismic waves traversing the zone suggests substantial enhancement (McEvilly and Johnson, 1977). Discrepancies in the magnitudes of transient tilts measured by tiltmeters near the San Andreas fault zone are presumably caused by contrasting rock types (McHugh and Johnston, 1976) although these data were not corrected for topographic effects. The effects of varying the shear modulus in the zone are discussed by McHugh and Johnston (1977).

4. Local block boundary strains

Differential movement of adjacent blocks may generate local strains at block boundaries. Local strainfields may be generated near locally rigid asperities on adjoining block surfaces. They are important because they will generally occur at the same times as changes in the applied strains inducing block motion. Consequently, it will be difficult to distinguish between the applied strains and the induced local strain fields, especially if the latter are much larger in magnitude. Whereas strain concentration described in section (3) will be restricted to the weak parts of block boundary zones, local strain fields will be transmitted within the adjacent blocks. The principal axes of these induced local strains need not necessarily be the same as the applied primary strains. A possible manifestation of block behavior which may include the generation of local strains is the observation of strain-steps in certain tectonic regions and to the somewhat uneven way which strain-steps have been observed to decay with distance (Japanese Network of Crustal Movement Observatories, 1970).

The generation of strain fields at block boundaries by nontectonic stresses may occur. Strain measurements across the Yamasaki fault zone (Oike et al., 1976) indicate that strains within the fault occur at times of heavy rainfall. Similar effects are observed on the Lake Cavazzo fault, Italy (Caloi and Migani, 1972). Tomasson (1976) and Tomasson et al. (1976) present data for a fissured region in Iceland which indicates clearly that block movements are related to water-table changes. Herbst (1976) proposes a model for near-surface, hydraulically induced tilt noise that requires a similar mechanism.

5. Non-linear elastic behavior

As has been remarked before, adjoining blocks do not always move independently but sometimes appear to coalesce and behave as a larger unit. Is this to be interpreted as a variation of the properties of a fault zone subject to a uniform applied strain or as the response of weakly coupled blocks to changing applied strain fields?

Evidence of short-term elastic moduli changes may be found in the study of the Kondarinsky fault zone, north of Dushambe, Tadjikistan. Fig. 6 is compiled from Gzovsky et al. (1972), Latynina et al. (1974) and Latynina and Rizaeva (1976). During the first few months of 1967, strains across part

132

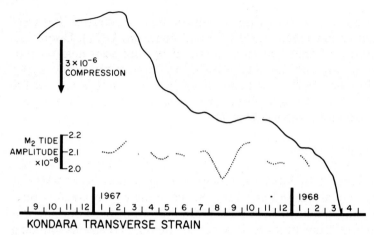

KONDARA TRANSVERSE STRAIN

Fig. 6. Measurements of secular strain and M_2 tidal amplitude across the Kondarinsky fault zone (Latynina et al., 1974 and Gzovsky et al., 1972).

of the fault indicate compression and insignificant tidal amplitude variation. In about July a decrease in the strain rate occurs, and at about the same time the fault-zone stiffness increases as indicated by the decrease in tidal admittance. The compression of the fault zone resumes when the tidal amplitude assumes its former value. The observed elastic modulus change is transitory and does not coincide closely with the strain compression data. Nonuniform compression of the fault zone is to be expected in view of the probable uneven boundaries between the two blocks, and this may account for the lack of closer agreement of the two curves.

Flexing of block boundaries characteristic of V-shaped faults may give rise to time-dependent changes in elastic modulus or viscoelastic behavior. For small variations about a quiescent strain level in the fault zone, the block boundary may behave linearly, but if the strain departs significantly from this level the fault zone may exhibit nonlinear properties.

It is interesting to combine the properties of a viscoelastic material with the additional constraint of nonlinear elasticity within the fault zone. A simple model is shown in Fig. 7. For rapid strains over a small range, the elastic properties of k_1 are dominant. For slowly applied compressive strains, the system has negligible stiffness until k_2 is engaged. If the stiffness of k_2 approaches that of the block material, displacements are transmitted across the system with good fidelity at all frequencies. Experimental evidence for fractured rock behaving in this way is presented by Goodman (1976, p. 172). An important feature of the model is that it enables the transmission of long-period (secular) deformation with a delay. A series of such block boundaries could result in an extremely slow (delay-line) propagation of crustal strain. A second feature of the model is the increase in strain rate that accompanies the eventual transmission of the applied strain. These two char-

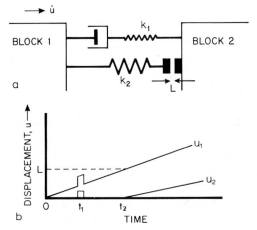

Fig. 7. a. A viscoelastic model for a block boundary. The upper spring (stiffness k_1) and dashpot result in viscoelastic behavior. The lower spring is of stiffness k_2, but is engaged only when the adjoining block surfaces approach and close the gap L. The stiffness of the system increases when this occurs and the viscous contribution to the overall behavior becomes insignificant if $k_2 \gg k_1$. The linear spring and gap L are a simple approximation to a nonlinear spring whose spring constant increases with compression. b. Relationship between applied displacement (u_1) of block 1 and observed displacement (u_2) of block 2 for the above model. Between $t = 0$ and $t = t_2$ slow movements of block 1 are not transmitted to block 2, although rapid movements are (e.g., at time t_1). After $t = t_2$ the stiffness of the block boundary increases and slow displacements are transmitted, such that $\dot{u}_2 \leqslant \dot{u}_1$. An important feature of the model is the time delay involved in the transmission of secular block movements.

acteristics of the model agree with the observed tilt data from Japan where slow-propagating tilts are recognized by rapid inflections in tilt vectors (Kasahara, 1977). The model allows the propagation of crustal deformation without the entity of a deformation front (Scholz, 1977; Kasahara, 1977) or a strain wave (Bott and Dean, 1973; Spence, 1977), however, it still requires the existence of a remote source of strain to initiate the disturbance. A discussion of the physical conditions resulting in behavior similar to the proposed model is outside the scope of the present article, although it is probable that the role of fluids is significant.

CONCLUSIONS

We have reviewed some of the evidence for block motion in areas of intense tectonic deformation. It is important to understand whether active block structures occur in a given area where strainmeters or tiltmeters are to be installed singly or in arrays, since some of the elastic properties of the resulting surface fragmentation can cause interpretive ambiguity in observed data. In general, observed strain fields will be incoherent across block boundaries compared to strains within integral blocks. Strain magnitudes can be reduced significantly within blocks which may exhibit large tilts. The azimuth of observed principal strain axes may differ from the azimuth of applied strain axes in places where local strain fields are generated at block boundaries. Applied strain will normally be intensified at block boundaries if these are represented by relatively weak fault zones.

In many ways the effects of crustal blocks can be described in terms of geometrical and geological inhomogeneities (Harrison, 1976b; Blair, 1977). The theory developed for treating these forms of inhomogeneity assumes linear elasticity. It is possible, however, that some block boundaries must be modelled as nonlinear rheological systems involving frequency-dependent or strain-dependent elastic properties. If these occur, their effects do not appear to modify tidal strains significantly (Levine and Harrison, 1976; Beavan and Goulty, 1977), although they may affect secular strains. A possible strain-delay mechanism based on viscoelastic block-boundary behavior is suggested that can account for the slow transmission of secular tectonic signals.

Blocks with surface dimensions ranging from below 1 km to greater than 50 km have been inferred. Though not of great tectonic significance, blocks with dimensions of a few kilometers potentially create serious difficulties to the interpretation of strainmeter or tiltmeter data. A history of leveling data can reveal the presence of active block motion in a given region better than other geodetic techniques since the density of fixed markers is greater than for triangulation or trilateration data, the precision is consistently good and, in Japan at least, the identification of active block motion is assisted by the recognition of discontinuous surface tilt. For this application, it is permissible to use parallel nearby releveling profiles which may not have a common datum. However, normal geodetic leveling data will not reveal blocks with dimensions less than about 5 km. The density of strainmeters or tiltmeters required in regions in which blocks may be less than a few kilometers is clearly economically prohibitive. A compromise between a dense array of continuously monitoring instruments and frequently repeated geodetic work is indicated in which the spacing of fixed reference markers is some fraction of a kilometer.

An understanding of the behavior of crustal blocks and block boundaries is important if we are to interpret long-period strainmeter and tiltmeter data. The study of block boundaries is a promising area for future research.

ACKNOWLEDGEMENTS

We are indebted to Chris Scholz who drew our attention to the wealth of geodetic data published by Japanese researchers and who provided considerable encouragement for the work. Klaus Jacob and Lynn Sykes have read the manuscript critically and suggested several improvements.

The work has been supported by the N.S.F. (EAR 76 03957) and NASA (NGR 33008-146).

REFERENCES

Alewine, R.W. and Heaton, T.H., 1973. Tilts associated with the Pt. Mugu earthquake. Proc. Conf. Tectonic Probl. San Andreas Fault System, Stanford Univ., XIII: 94—103.

Bakun, W.H. and Bufe, W.H., 1975. Shear wave attenuation along the San Andreas fault zone in central California. Bull. Seismol. Soc. Am., 65: 439—459.

Beaumont, C. and Berger, J., 1975. An analysis of tidal strain observations from the United States of America: I. The homogeneous tide. Bull. Seismol. Soc. Am., 65: 1613—1629.

Beavan, R.J., 1976. Earth Strain: Ocean Loading and Analysis. Ph.D. Thesis, Cambridge Univ.

Beavan, R.J. and Goulty, N.R., 1977. Earth strain observations made with the Cambridge laser strainmeter. Geophys. J.R. Astron. Soc., 48: 293—305.

Bennett, J.H., Taylor, G.C., and Toppozada, T.R., 1977. Crustal movement in the Northern Sierra, Nevada. Calif. Geol., 30: 51—60.

Berger, J. and Beaumont, C., 1976. An analysis of tidal strain observations from the United States of America. II. The inhomogeneous tide. Bull. Seismol. Soc. Am., 66: 1821—1845.

Blair, D., 1977. Topographic, geologic and cavity effects on the harmonic content of tidal strain. Geophys. J.R. Astron. Soc., 48: 393—405.

Bott, M.H.P. and Dean, D.S., 1973. Stress diffusion from plate boundaries. Nature, 243: 339—341.

Buchanan-Banks, J.M., Castle, R.O. and Ziony, J.I., 1975. Elevation changes in the central transverse ranges near Ventura, California. Tectonophysics, 29: 113—125.

Caloi, P. and Migani, M., 1972. Movements of the fault of the Lake of Cavazzo in connection with local rainfalls. Ann. Geofis., 25: 15—20.

Davis, P., Evans, K., Hosfall, J. and King, G., 1979. Long baseline tilt and strain measurement (abstr.). Tectonophysics, 52: 83.

Evans, J.R., 1975. Spatial Inhomogeneity of the Earth Strain Tide. Ph.D. Thesis, Cambridge Univ.

Goodman, R.E., 1976. Methods of Geological Engineering in Discontinuous Rocks. West Publ. Co., St. Paul, Minn., 472 pp.

Gzovsky, M.V., Latynina, L.A., Ostrovsky, A.E. and Pevnev, A.K., 1972. Slow deformations of the earth's crust and their relation to earthquake in the U.S.S.R. Phys. Earth Planet. Int., 6: 235—240.

Harada, T., 1976. Accumulation of data by the Geographical Survey Institute. J. Geod. Soc. Jap., 22: 228—234.

Harrison, J.C., 1976a. Tilt observations in the Poorman Mine near Boulder, Colorado. J. Geophys. Res., 81: 329—336.

Harrison, J.C., 1976b. Cavity and topographic effects on tilt and strain measurements. J. Geophys. Res., 81: 319—328.

Healy, J.H. and Peake, L.G., 1975. Seismic velocity structure along a section of the San Andreas fault near Bear Valley, California. Bull. Seismol. Soc. Am., 65: 1177.

Herbst, K., 1976. Interpretation von Neigungsmessungen im Periodenbereich oberhalb der Gezeite. Ph.D. Dissertation, Technische Univ., Clausthal-Zellerfeld.

Hill, R.L., Sprotte, E.C., Bennett, J.H. and Slade, R.C., 1979. Fault location and fault activity assessment by analysis of historic level line data, oil well data and ground water data, Hollywood area, California (abstr.). Tectonophysics, 52: 303.

Hosoyama, K., 1976. Observational accuracy of crustal movements. J. Geod. Soc. Jap., 22: 235—241.

Imamura, A., 1929. On the multiple source of origin of the Great Kwanto earthquake and its relation to the fault system connected with the earthquake. Proc. Imp. Acad. Jap., 5: 330—333.

136

Imamura, A., 1930a. Topographical changes accompanying earthquakes or volcanic eruptions. Publ. Earthquake Invest. Comm. Foreign Lang., 25: 1—143, Tokyo, Japan.

Imamura, A., 1930b. On the block movement accompanying and following the Great Kwanto earthquake of 1923. Proc. Imp. Acad. Jap., 6: 415—418.

Imamura, A., 1930c. On the chronic block movements in the Kyoto—Osaka district. Jap. J. Astron. Geophys., 7: 93—101.

Imamura, A., 1930d. On changes of topography, both chronic and acute in the southern part of Sikoku. Proc. Imp. Acad. Jap., 7: 101—104.

Imamura, A., 1931. On the northward movement of crustal deformation along the western boundary of the Kwanto plain. Proc. Imp. Acad. Jap., 7: 315—318.

Imamura, A., 1932a. On slow changes of land level, both related and unrelated to earthquakes. Proc. Imp. Acad. Jap., 8: 247—254.

Imamura, A., 1932b. Further studies on the block movement of the Kii Peninsula. Proc. Imp. Acad. Jap., 8: 163—166.

Imamura, A., 1933a. On crustal deformations preceding earthquakes. Jap. J. Astron. Geophys., 10: 82—92.

Imamura, A., 1933b. On the crustal deformations that preceded and accompanied the severe Haneda earthquake of August 3, 1926. Proc. Imp. Acad. Jap., 7: 271—274.

Imamura, A., 1933c. On crustal deformation in west-central Kii Peninsula. Proc. Imp. Acad. Jap., 9: 39—42.

Imamura, A., 1934. Chronic movements of a minor crustal block as revealed by the revision of a levelling line into one with closely-spaced benchmarks. Proc. Imp. Acad. Jap., 10: 69—72.

Imamura, A., 1935. Crustal deformations associated with the Dewa earthquakes of 1804 and 1894 as revealed through revisions of precise levels. Proc. Imp. Acad. Jap., 11: 51—54.

Japanese Network of Crustal Movement Observatories, 1970. Spatial distribution of strain steps associated with the earthquake of the central part of Gifu Prefecture, Sept. 9, 1969. Bull. Earthquake. Res. Inst., 48: 1217—1233.

Kasahara, K., 1973. Earthquake fault studies in Japan. Philos. Trans. R. Soc. Lond., A274: 287—296.

Kasahara, K., 1979. Migration of crustal deformation. Tectonophysics, 52: 329—341.

King, G.C.P., 1971. The siting of strainmeters and tiltmeters for teleseismic and tidal studies. Bull. R. Soc. N.Z. 9: 239—247.

King, G.C.P. and Bilham, R.G., 1973a. Tidal tilt measurement in Europe. Nature, 243: 74—75.

King, G.C.P. and Bilham, R.G., 1973b. Strain measurement instrumentation and technique. Philos. Trans. R. Soc. Lond., A274: 209—217.

King, G., Zurn, W., Evans, R. and Emter, D., 1976. Site correction for long-period seismometers, tiltmeters and strainmeters. Geophys. J.R. Astron. Soc., 44: 405—411.

Lambert, A., 1970. The response of the Earth to the ocean tides around Nova Scotia. Geophys. J.R. Astron. Soc., 19: 449.

Latynina, L.A., 1975a. On horizontal deformations at faults recorded by extensometers. Tectonophysics, 29: 421—427.

Latynina, L.A., 1975b. On the possibility of studying faults in the Earth's crust on the basis of tidal linear strains. Izv. Earth Phys., 3: 16—26.

Latynina, L.A., Karmaleiva, R.M., Rizaeva, S.D., Starkova, E.Y. and Mordonov, B., 1974. Deformation of the earth's surface at Kondava before the earthquake of 3-x-1967. The Search for Earthquake Forerunners at Prediction Test Fields. Nauka, Moskow (in Russian).

Latynina, L.A. and Rizaeva, S.D., 1976. On tidal strain variations before earthquakes. Tectonophysics, 31: 121—127.

Levine, J., 1978. Strain tide spectroscopy. Geophys. J.R. Astron. Soc., 54:27—41.

Levine, J. and Harrison, J.C., 1976. Earth tide strain measurements in the Poorman Mine near Boulder, Colorado. J. Geophys. Res., 81: 2543.

McEvilly, T.V. and Johnson, L.R., 1977. In-situ seismic wave velocity monitoring. National Earthquake Hazards Program, U.S.G.S. Semi-annual Tech. Rep. Summ., 4: 51.

McHugh, S. and Johnston, M.J.S., 1976. Short-period nonseismic tilt perturbations and their relationship to episodic slip on the San Andreas fault in central California. J. Geophys. Res., 81: 6341.

McHugh, S. and Johnston, M.J.S., 1977. Surface shear stress, and shear displacement for screw dislocations in a vertical slab with shear modulus contrast. Geophys. J.R. Astron. Soc., 49:715—722.

Melchior, P.J., 1967. Oceanic tidal loads and regional heterogeneity in western Europe. Geophys. J.R. Astron. Soc., 14: 239—244.

Miyabe, N., 1933. Block movements of the Earth's crust in the Kwanto district. Bull. Earthquake Res. Inst., 11: 639—691.

Mizoue, M., 1967. Modes of secular movements of the Earth's crust, Part 1. Bull. Earthquake Res. Inst., 45: 1019—1090.

Muto, K. and Atsumi, K., 1929. An investigation into the results of the new and old measurements of the levelling net in the Kwanto district. Bull. Earthquake Res. Inst., 7: 495—522.

Oike, K., Kishimoto, Y., Nakamura, K. and Nakahori, Y., 1976. Some characteristic behaviors of the Yamasaki fault zone observed by extensometers. J. Geod. Soc. Jap., 22: 284—285.

Pevnev, A.K., Guseva, T.V., Odinev, N.N. and Saprykin, G.V., 1975. Regularities of the deformations of the Earth's crust at the joint of the Pamirs and Tien-Shan. Tectonophysics, 29: 429—438.

Reasenberg, P. and Aki, K., 1974. A precise, continuous measurement of seismic velocity measurement for monitoring in situ stress, J. Geophys. Res., 79: 399—406.

Savage, J.C., Church, J.P. and Prescott, W.H., 1975. Geodetic measurement of deformation in Owens Valley, California. Bull. Seismol. Soc. Am., 65: 865—874.

Scholz, C.H., 1972. Crustal movements in tectonic areas. Tectonopysics, 14: 201—207.

Scholz, C.H., 1977. A physical interpretation of the Haicheng earthquake prediction. Nature, 267: 121—124.

Scholz, C.H. and Kato, T., 1978. The behavior of a convergent plate boundary: Crustal deformation in the South Kanto District, Japan. J. Geophys. Res., 83: 783—797.

Shichi, R., 1973. J. Geod. Soc. Jap., 19: 4.

Spence, W., 1977. The Aleutian arc: tectonic blocks, episodic subduction, strain diffusion and magma generation. J. Geophys. Res., 82: 213—230.

Tomaschek, R., 1953. Nonelastic tilt of the Earth's crust to meteorological pressure distributions. Geofis. Pura. Appl., 25: 17—25.

Tomaschek, R., 1959. Schwankungen tektonischer Schollen infolge barometrischer Belastungsänderungen. Freiberg. Forschungsh., C60: 35—55.

Tomaschek, R. and Groten, R., 1961. The problem of residual ellipses of tilt measurements. Commun. Obs. Belg. Sér. Géophys., 58: 78—92.

Tomasson, H., 1976. The opening of tectonic fractures at the Langalda Dam. Comm. Int. Grands Barrages, 12th Congr. des Grands Barrages, Mexico.

Tomasson, H., Gunnarson, H. and Ingolfsson, P., 1976. Langulda veita, National Energy Authority of Iceland, Tech. Rep., August 1976 (OS-ROD-7642).

Tryggvason, E., 1970. Surface deformation and fault displacement associated with an earthquake swarm in Iceland. J. Geophys. Res., 75: 4407—4422.

Tryggvason, E., 1974. Vertical crustal movement in Iceland. In: L. Kristjansson (Editor), Geodynamics of Iceland and the North Atlantic Area. Reidel, Dordrecht, pp. 241—262.

Tsuboi, C., 1933. Investigation on the deformation of the earth's crust by precise geodetic means. Jap. J. Astron. Geophys., 10: 93—248.

Wood, M.D., Allen, R.V. and Allen, S.S., 1973. Methods for prediction and evaluation of tidal tilt data from borehole and observatory sites near active faults. Philos. Trans. R. Soc. Lond., A274: 245—252.

Wood, M.D. and King, N.E., 1977. Relation between earthquakes, weather and soil tilt. Science, 197: 154—156.

Yamada, J., 1973. A water-tube tiltmeter and its application to crustal movement studies. Rept. Earthquake Res. Inst., 10: 1—147 (in Japanese).

Yamasaki, N., 1928. Report on the precise levellings in the meizoseismal area of the Tango earthquake. Proc. Imp. Acad. Jap., 4: 60.

Zschau, J., 1977. Air pressure induced tilt in porous media. Proc. 8th Int. Symp. Earth Tides, Bonn, 1977.

Tectonophysics, 52 (1979) 139—155
© Elsevier Scientific Publishing Company, Amsterdam — Printed in The Netherlands

Observed Vertical Crustal Deformation

SOLVABILITY AND MULTIQUADRIC ANALYSIS AS APPLIED TO INVESTIGATIONS OF VERTICAL CRUSTAL MOVEMENTS

SANFORD R. HOLDAHL and ROLLAND L. HARDY *

National Geodetic Survey, National Ocean Survey, National Oceanic and Atmospheric Administration, Rockville, Md. 20852 (U.S.A.)
Iowa State University, Ames, Ia. (U.S.A.)

(Accepted for publication April 4, 1978)

ABSTRACT

Holdahl, S.R. and Hardy, R.L., 1979. Solvability and multiquadric analysis as applied to investigations of vertical crustal movements. In: C.A. Whitten, R. Green and B.K. Meade (Editors), Recent Crustal Movements, 1977. Tectonophysics, 52: 139—155.

A variety of geodetic measurements can be combined, in network fashion, to yield adjusted velocities of elevation change. However, it is not always apparent which network junctions have solvable point velocities. When a velocity surface is desired, it is not always apparent how many coefficients should be used. A solvability algorithm, devised to operate on observation equations, answers these questions, and therefore permits the adjustment process to continue with the assurance that the result will be mathematically justified. Using both the hyperboloid and the reciprocal hyperboloid as quadric forms, multiquadric (MQ) analysis has been applied to leveling and tide gauge data in the vicinity of Puget Sound, to obtain heights corresponding to a selected date, and coefficients which collectively define a velocity surface. The solvability algorithm was used to tell which junctions in the level network had solvable point velocities, and consequently where MQ nodal points should be placed for an optimized solution. Networks of simulated data were also used with the solvability algorithm to help determine data requirements for height—velocity adjustments, and to evaluate the ability of MQ analysis to predict velocities.

INTRODUCTION

Computational maintenance of a level network and the investigation of vertical crustal movements are similar and often inseparable tasks. Maintenance of a network is simple enough if releveling is accomplished frequently and comprehensively. However, the expense is usually not justified. Therefore, the task of updating the heights of bench marks in a level network

---

* Performed during an appointment as Senior Research Scientist, National Research Council, National Academy of Sciences, Washington, D.C., while on leave from Iowa State University, Ames, Iowa.

becomes a mathematical problem of extracting velocities of height change from a collection of time-inhomogeneous observations and scattered relevelings.

There are many computational methods which can be used to determine vertical land movements from levelings, some of which have been described by Holdahl (1975) and Vanicek et al. (1974). Methods which simultaneously produce a velocity surface and adjust heights to a common point in time are preferred because:

(1) Velocity prediction is often required if all heights in an incompletely releveled network are to be reduced to a common point in time.

(2) Error propagation for the velocity surface is more thoroughly and conveniently accomplished.

(3) Full advantage can be taken of all measured height differences.

(4) Automated graphics routines can be used to portray results.

Multiquadric (MQ) analysis and its applications to geodesy have been discussed by Hardy (1976). The use of MQ analysis to adjust heights and detect patterns of crustal movements is one topic of this paper. The other topic is solvability, a computational tool which has been used to gain insight and resolve questions concerning the allowable density and optimal distribution of MQ kernels, and the constraints and data configurations which are required to produce acceptable solutions.

For some applications of surface fitting, there may be concern because it is not always easy to tell how many coefficients could or should be used to describe a surface. Erratic solutions may be obtained when use of an excessive, but seemingly legitimate, number of coefficients is attempted. Insensitive solutions are obtained when too few coefficients are used. For height—velocity adjustments, this concern is unnecessary because the solvability algorithm can determine how many coefficients may be used.

Simulated leveling data and the solvability algorithm have been used to verify ideas concerning the best use of MQ nodal points, and what constitutes a suitable data configuration for a height—velocity adjustment. Real data in the vicinity of Puget Sound have also been used to determine velocities by means of MQ analysis. The example well illustrates the aesthetic benefits of selecting a quadric form which is appropriate for graphics resulting from a height—velocity adjustment.

ADJUSTMENT MODELS

Leveling observations have been made for over 100 years in the United States, and during that time the leveling process has been very precise. The regional level networks of today look quite different from those of 40 years ago. They now have a healthy percentage of releveling and can often yield height-change velocities as well as adjusted heights. However, adjustment models which are now commonly used do not allow for change of height with time. When a level network is composed of surveys made at two or more times, and some network junction points are not common to more than one survey, a more flexible model is needed if the land surface is deforming vertically.

Ideally, a leveling adjustment should yield a set of heights that can be associated with a selected point in time, and a set of velocities that can be used to revise the adjusted heights to reflect earlier or later times. One such model, used by Holdahl (1970), was built arount the following expression for height:

$$h_{a,i} = h_{a,0} + V_a(t_i - t_0) \tag{1}$$

In eq. 1, $h_{a,i}$ is the height of a point A at time t_i; $h_{a,0}$ is its height at the reference time t_0, and V_a is the velocity of height change at A.

The observation equation for a height difference in that model is given by:

$$R_{b-a,i} = h_{b,i} - h_{a,i} - \Delta h_{b-a,i} \tag{2}$$

where $\Delta h_{b-a,i}$ is the height difference observed between points A and B at time t_i, and $R_{b-a,i}$ is its adjustment correction.

In the first applications, the point velocities were unknowns along with the reference-time heights. The data requirements for such adjustments were somewhat demanding in that all junction points were required to be connected by reference-time leveling to a fixed (reference-time) height, or have a solvable velocity. This requirement is too restrictive because such ideal configurations of releveling are the exception rather than the rule.

An adjustment procedure that produces a velocity surface was suggested by Vanicek and Cristodulidis (1974) as a means of obtaining vertical crustal movement patterns from scattered relevelings. Holdahl (1977) has similarly used the idea of a velocity surface in leveling adjustments to take better and more convenient advantage of nonredundant or heavily redundant data configurations. This was accomplished by expressing the velocity of a point A by the expression:

$$V(X_a, Y_a) = c_0 + c_1 X_a + c_2 Y_a + c_3 X_a \cdot Y_a + \dots \tag{3}$$

Substituting eq. 3 into eq. 1 gives a model which has as its unknowns reference-time heights and the coefficients of a polynomial which describes the velocity surface.

The power series used in eq. 3 can fit data reasonably well, but has the disadvantage of quickly taking on extreme values outside the data area. Multiquadric analysis was investigated with the hope that it would produce surfaces which took on reasonable or likely values outside the data area.

In multiquadric analysis a smooth irregular surface is approximated by summation of regular (mathematically defined) surfaces, particularly quadric forms. Each surface in the summation is associated with a "nodal point", a location for which an undetermined coefficient is to be associated. The undetermined coefficients have a geometric interpretation of slope, amplitude or other intrinsic property associated with quadric forms. The general form for a multiquadric velocity surface may be expressed as:

$$V(X, Y) = \sum_{j=1}^{k} c_j Q[X, Y, X_j, Y_j, D] \qquad (4)$$

where c_j are the undetermined coefficients, X_j, Y_j are positions of nodal points, D is a geometric parameter which may or may not be necessary depending on the quadric form, and Q is a quadric kernel function.

QUADRIC FORMS

Examples of quadric forms, or quadric kernel functions as they are sometimes called, are the cone, hyperboloid, reciprocal paraboloid and reciprocal hyperboloid. The choice of quadric form may depend on the nature of the problem and the data configuration, as well as the insights and objectives of the investigator. In selecting a quadric form for height—velocity adjustments, it was considered important that the form produce surfaces which permitted estimation of vertical crustal motion slightly outside the area where survey observations have been obtained. It would also be advantageous to have a quadric form that produced a surface with edge velocities which did not obscure the view or destroy the scale of automated three-dimensional or contour plots.

The reciprocal hyperboloid, shown in Fig. 1, has the advantageous property that it tends to zero outside the region containing the nodal points. This is desirable because we want the velocity features in the data area to determine the scale of the three-dimensional plots rather than the predicted velocities on the edge of the study area. However, the hyperboloid is preferred because it fits the data as well; it fits distant data even better, and also produces a more realistic velocity error surface. The reciprocal hyperboloid produces a velocity error surface on which error decreases with distance from the data. "This is not a reasonable result, intuitively; but mathematically it reflects the fact that absence of data beyond the test area is being treated as data with zero ordinates. It is the nature of the reciprocal hyperboloid function to approach zero as distance increases. Therefore the difference between 'zero data ordinates' and a continuous function approaching zero is a new function (the error function) which also approaches zero with distance."

The expression for velocity, using the hyperboloid as the quadric form, is as follows:

$$V(X_a, Y_a) = \sum_{j=1}^{k} c_j [(X_a - X_j)^2 + (Y_a - Y_j)^2 + D^2]^{1/2} \qquad (5)$$

Substituting eq. 5 into eq. 1 gives the basis for a height—velocity adjustment by MQ analysis with the hyperboloid as the quadric form.

To improve results, the geometric parameter, D, can also be changed.

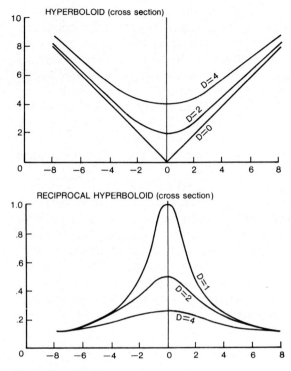

Fig. 1. Quadric forms.

Here, D controls the shape of the hyperboloid and its reciprocal form. Large values permit smooth, broad surface features, while smaller values give a more peaked appearance (see Fig. 1). The choice of a proper D has been studied by Hardy (1976), and should be governed primarily by the spacing of nodal points. In leveling adjustments, D is set to be 0.43258 times the average spacing between nodal points. It is possible to make the D of each quadric kernel function a separate unknown in the adjustment. This may increase the number of unknowns unreasonably. A single D, applying to all kernel functions, might also be treated as an unknown. Unfortunately, when D is an unknown, the adjustment model becomes nonlinear. For height—velocity adjustments, it has been satisfactory to calculate D automatically, based on average nodal-point spacing, before the normal equations are formed.

NODAL POINTS

After selecting a quadric form, the next step is to decide where to place it. These locations are called nodal points or nodes. It is not mandatory that nodal points be located at the junctions of a level network. They may, for example, be evenly distributed throughout the study area. Placement of nodal points in a triangular or rectangular grid within the study area is possible (and has some appeal), but this procedure will generally not take maximum advantage of the information content of the data. For height—velocity adjustments, a good configuration of nodal points would be one in which:

(1) No two nodal points occupy the same location.

(2) A nodal point would be located at or close to any junction with information content.

(3) Nodal points cover the whole area to be mapped/adjusted.

(4) Nodal points are located at velocity maximums and minimums, and along lines where velocity changes significantly.

There is little point in placing nodal points where there are no junctions with velocity information. Thus, the ideal configuration of nodal points can only exist when there is an ideal configuration of junctions with velocity information. A junction with velocity information content may be called "potent" — it is one that has a solvable point velocity. Junctions with highest velocity information content have the greatest number of time-independent connections to a fixed height, and have the greatest time separation to distinguish the connecting links. Best height—velocity solutions are obtained by placing a nodal point at every potent junction.

SOLVABILITY

The solvability algorithm was developed by Allen Pope of the National Geodetic Survey. It uses the coefficients of the observation equations to assess whether the unknowns of a proposed adjustment can, in fact, be determined with the provided observations and constraints. Solvability has been used in conjunction with height—velocity adjustments to gain insight and clarification concerning:

(1) The maximum number of nodal points permissible.

(2) Minimal data requirements for acceptable solutions.

(3) The optimal location of nodal points.

Here we only describe the algorithm. The underlying derivation will be given in a future paper by Allen Pope. The basic solvability algorithm, in its simplest version, proceeds as follows: initially, the solvability matrix, S, is set equal to the unit matrix, $S = I$. Then, each of the n rows of A, the coefficient matrix of the observations, is used to update S as follows:

$$S_i = S_{i-1} - S_{i-1}a[a^tS_{i-1}a]^{-1}a^tS_{i-1} \tag{6}$$

For some observations the update of S, i.e., the evaluation of eq. 6 is bypassed. This happens when the quantity a^tSa is essentially zero. To facilitate testing for zero, a^tSa is divided by a^ta so that it is normalized to a value between 1 and 0. For leveling-adjustment problems, the condition:

$$[a^tS_{i-1}a]/a^ta < 10^{-6} \tag{7}$$

causes the solvability algorithm to skip to the next observation without further computation, i.e., $S_i = S_{i-1}$. The rank, r, of the normal equations is the count of the number of times eq. 7 is not realized. For a leveling adjustment to be successful S_n must have full rank; i.e., the rank must equal the number of unknowns, u. The trace of S_n is the rank deficiency, e. Since $e = u - r$, the trace of S_n provides a check on the rank count. Failure of this identity is an indication of excessive rounding errors in the solvability computations, or of a tolerance in eq. 7 which is too small. Initial applications of the solvability algorithm to height—velocity adjustments by MQ analysis yielded confusing results because of excessive rounding errors. The problem was initially resolved by performing all solvability computations in double precision. A "second-generation" solvability algorithm, devised by Allen Pope, is numerically stable and requires no double precision. It is algebraically equivalent to that described above, but has the computational sequence rearranged, as in the method of the modified Gram—Schmidt orthogonalization, in order to decrease the sensitivity to roundoff.

The solvability algorithm is presently being revamped further to utilize its advantages with a much smaller storage requirement. It is reasonably certain that a newer version will operate with approximately $u \times 25$ storage elements. The present version requires $(u^2 + u)/2$ storage locations.

After the last row of A has been used to update S, the diagonal elements of S_n are reviewed to ascertain which unknowns are solvable with the provided observations and fixed control. Diagonal elements that are zero ($<10^{-6}$, or some other chosen tolerance) correspond to solvable unknowns. Diagonal elements of S_n that equal 1 correspond to unknowns for which there is no information in the adjustment, and consequently are unsolvable. Elements between 1 and 0 correspond to indeterminate unknowns about which the observations contain some information, but not enough to yield a solution.

To gain maximum insight concerning data requirements of height—velocity adjustments, solvability was applied twice for each data set tested. At first, observation equations were formed according to eq. 2, where the velocity at each junction was an unknown. Velocity unknowns were blindly put at every junction to make the diagonal elements of S_n clearly identify which junctions had solvable velocities. This is exactly the information needed to locate MQ nodal points for an optimized MQ solution. Secondly, solvability was applied using observation equations formed with the MQ expression for velocity. In this case, solvability gives a proper count of rank and rank deficiency, but S_n is not so explicit in identifying where to look for trouble when a problem arises. For example, if too many nodal points are used, all diagonal elements of S_n which correspond to MQ coefficients, are identified as unsolvable. By knowing the rank deficiency the investigator can decide how many nodal points to remove, but will not know which to retain for an optimized solution. Therefore, although solvability applied to the direct velocity formulation of the problem does not correspond to the normal equations used in the MQ solution, it was found to be very useful for optimizing the MQ solution.

RESULTS

Test networks, fabricated from fictitious measurements, were used to evaluate the relative merits of different nodal-point arrangements, and to determine the data requirements for a height—velocity adjustment. Sets of height differences, corresponding to epochs (times), were generated using assumed junction velocities. The heights of the junctions in the test networks were considered zero in 1900. Fixed heights and observed velocities were added in various ways until solutions could be obtained. The solvability algorithm was used to isolate the specific junctions where height or velocity information was lacking.

The example networks shown in Fig. 2 are composed of single measurements, rather than relevelings as we normally think of them. This illustrates that relevelings of the usual type are not a necessary ingredient of a height—velocity adjustment. However, it should be understood that when single levelings are used the adjustment should be accomplished using geopotential units. Otherwise, circuit misclosures are theoretically not equal to zero. For small study areas, where variation in the gravity anomaly is small, or where relevelings exist over the entire net, the use of geopotential units is not important if only the velocities are desired. The heights, rather than the velocities, are most sensitive to gravity anomalies. The gravity effect tends to cancel in the velocity determination when relevelings are complete. Networks with a significant percentage of nonreleveled segments should usually be adjusted using geopotential units since this cancellation may be incomplete or nonexistent.

In the following examples (see Fig. 2), velocity surfaces and reference-time heights were obtained successfully. The variance of unit weight in each adjustment was zero; meaning that in each case a velocity surface was found which required no correction to the observations. This was possible because the observations were error free, and the movements used to generate the observations were perfectly constant.

Example 1

The leveling was accomplished at two different times, 1900 and 1975. The reference time, t_0, is 1950. A fixed height (1950), is established at Station 41. A 1950 height difference connects Station 41 to Station 1. A velocity has been observed at Station 1. The optimized MQ solution is derived by placing nodal points at the stations marked by double circles. These have information content. The other stations cannot yield velocities directly because they were not leveled over on more than one occasion. The fit ob-

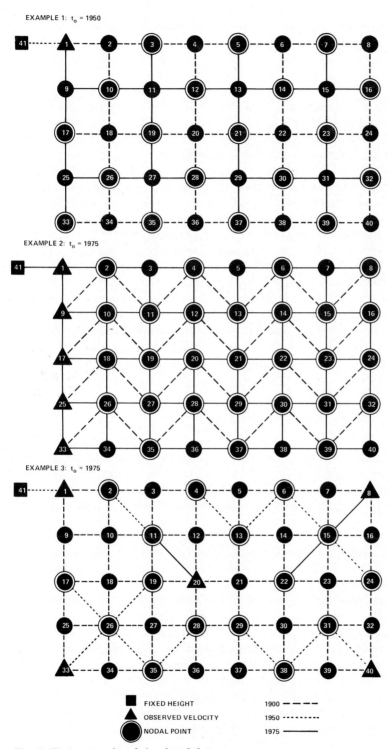

EXAMPLE 1: t_o = 1950

EXAMPLE 2: t_o = 1975

EXAMPLE 3: t_o = 1975

■ FIXED HEIGHT 1900 ― ― ―

▲ OBSERVED VELOCITY 1950 ·············

◉ NODAL POINT 1975 ―――

Fig. 2. Test networks of simulated data.

tained at nodal-point junctions was perfect, and the velocity fit at the remaining junctions was very good [standard deviation (SD) = 0.5 mm/yr].

Example 2

This network was completely leveled in 1975. Prior leveling was done in 1900 as four separate components. A 1950 fixed height is established at Station 41, and a 1975 observed height difference connects it to Station 1. A velocity has been observed at Station 1. It is important to note in this example that only Station 1 would have velocity information content were it not for the four additional observed velocities at Stations 9, 17, 25 and 33. Without the additional observed velocities, or some other acceptable measurements, assumptions or constraints, the problem would be unsolvable. The observed velocities relate the velocity components, which by themselves have only shape information. Related absolutely, a surface can pass through them. Note that height prediction was not needed in this problem because the network was totally leveled at time t_0. Only Stations 3, 5, 7, 34, 36, 38, 40 and 41 did not have velocity information content. Nodal points were placed at all other junctions. The MQ solution produced perfect velocity fit at all junctions having nodal points, and predicted velocities very well (SD = 0.5 mm/yr) at the remaining seven junctions. An MQ fit was also attempted using only 14 nodal points spaced in a triangular grid, but the results were inferior.

Example 3

Like Example 2, this net consists of components which would have only relative information if it were not for initialization of each by an observed velocity. Unlike Example 2, height prediction was required at junctions 1—40. The reference time is 1975. Velocity prediction was required at the 20 junctions which had no velocity information content. A perfect velocity fit was obtained at nodal-point junctions, and good velocity predictions (SD = 0.56 mm/yr) were obtained at the remaining junctions. This data configuration best illustrates the different levels of velocity information content: Stations 2, 4 and 19 have some information; Station 38 has more; and Stations 8, 11, 15, 20 and 26 have even more.

Solvability and MQ analysis were also used on a level network surrounding Puget Sound in Washington State. Four tide stations provided observed velocities. Fig. 3 shows schematically the Puget Sound net, the locations of the four tide gauges, and where 40 nodal points were placed. Relevelings were made at many different times, and some single levelings were used. MQ solutions were tried, using nodal points in a triangular grid covering the entire study area. The results were no better than those previously produced using a power or Fourier series to express velocity. Like the test-net examples, the best solution was obtained by placing nodal points at junc-

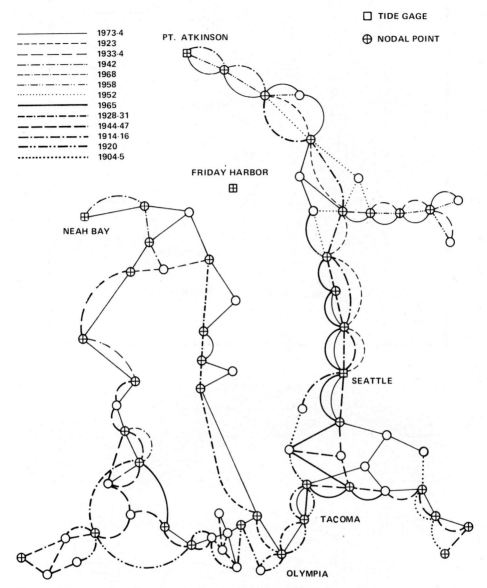

Fig. 3. Diagram of leveling observations — vicinity of Puget Sound.

tions which the solvability algorithm indicated as having velocity information. Before correction, a few of the potent junctions were accidentally collocated because their positions were provided only to the nearest tenth of a
minute. If two or more nodal points have the same location, a singular set of
normal equations will result. To avoid collocated nodal points a programmed
check should be made prior to applying solvability.

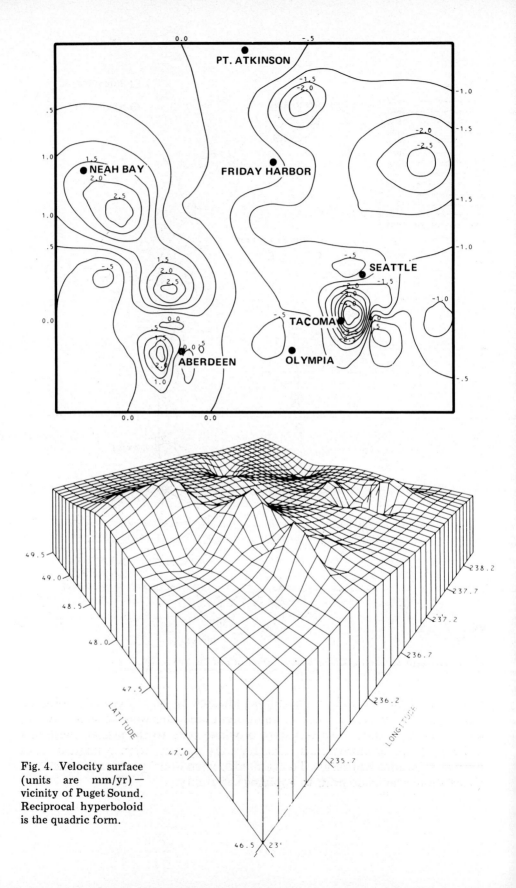

Fig. 4. Velocity surface (units are mm/yr) — vicinity of Puget Sound. Reciprocal hyperboloid is the quadric form.

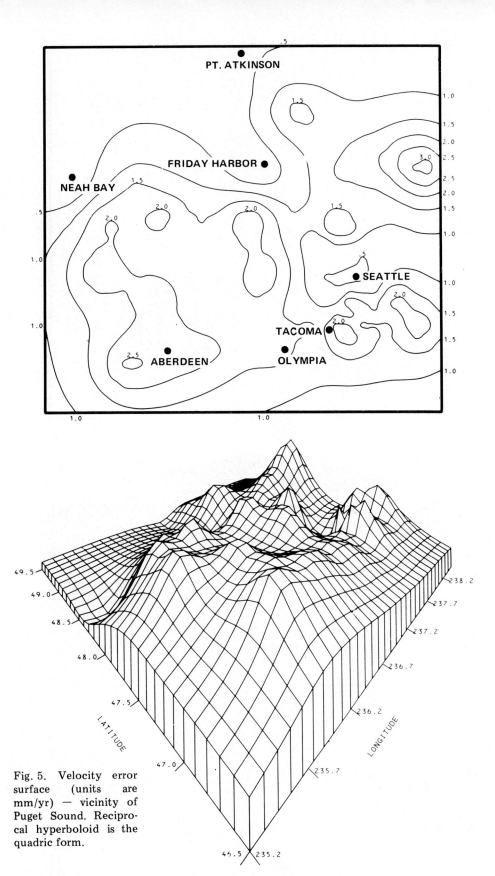

Fig. 5. Velocity error surface (units are mm/yr) — vicinity of Puget Sound. Reciprocal hyperboloid is the quadric form.

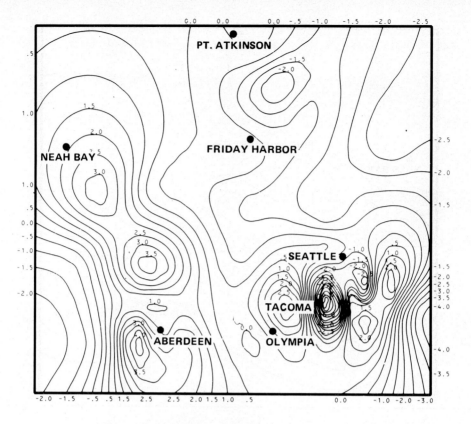

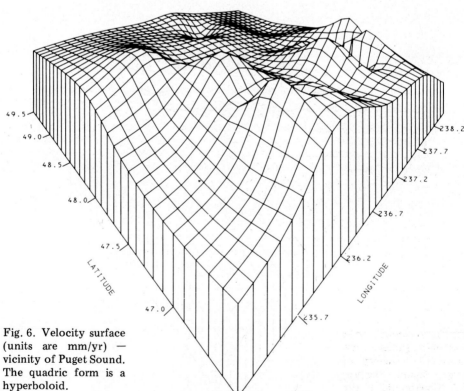

Fig. 6. Velocity surface (units are mm/yr) — vicinity of Puget Sound. The quadric form is a hyperboloid.

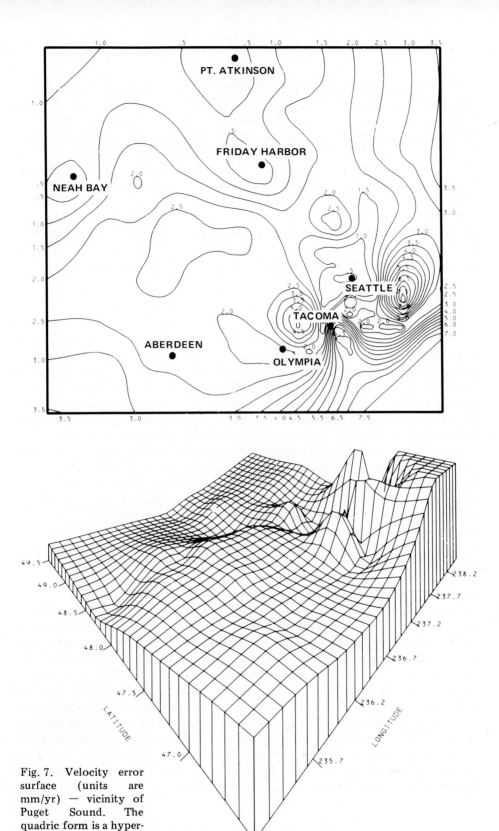

Fig. 7. Velocity error surface (units are mm/yr) — vicinity of Puget Sound. The quadric form is a hyperboloid.

The Puget Sound example (Fig. 4) illustrates the ability of the reciprocal hyperboloid to provide the desired response of the surface at the outside edges of the data. The scale ot the three-dimensional plot is determined by the surface features rather than by the uncontrolled edges. Unfortunately, Fig. 5 also shows that the reciprocal hyperboloid produces a velocity error surface which takes on unreasonably low (but aesthetically pleasing) values beyond the nodal points. Fig. 6 shows the velocity surface derived using the hyperboloid. Fig. 7 gives the corresponding velocity error surface which is more realistic than that in Fig. 5. "In this case the hyperboloid function becomes indefinitely large with distance. Mathematically the error function, i.e., the difference between 'zero data ordinates' outside the test area and the hyperboloid, also becomes indefinitely large with increased distance."

CONCLUSIONS

Multiquadric analysis when applied to leveling networks can produce superior evaluations and displays of vertical crustal movements. Other surface-fitting methods, which used a power series or Fourier series to describe a velocity surface, were less successful. "The geophysical interpretation of MQ coefficients is a subject requiring further study. When the reciprocal hyperboloid is used in MQ analysis to describe the geoid, the MQ coefficients can be interpreted as point mass anomalies. When topography is modeled with hyperboloid kernels, the MQ coefficients may reflect the isostatic condition of the Earth's crust. In crustal-movement studies the coefficients apparently refer to the rate of change of point mass anomalies, or of isostatic conditions, with time. Now lacking in these interpretations are the proper scale factors and refinements. Changes in gravity and orthometric height may also relate to redistributions of mass or changes in density. Therefore, it is expected that physical interpretations can be given to the MQ coefficients resulting from height—velocity adjustments. This would permit the mapping of the coefficients themselves, scaled perhaps to reveal information about mass redistribution in the Earth's interior." (Hardy, 1977.)

There is a need for the solvability algorithm. It should be used to indicate which junctions have velocity information content, and consequently where nodal points should be located. The solvability algorithm is a useful tool which must be utilized more fully by geodesists. While height—velocity adjustments can become complex, the need for the solvability algorithm is even greater when three-dimensional or horizontal position—velocity adjustments are attempted. In the latter adjustments, the variety of measurement types, configurations of repeat measurements, and fixed control can get very complex. For special applications, it may be required that adjustment models be complicated further by parameters that are used to describe nonlinear or episodic movements. In these more difficult problems it is expected that the solvability algorithm will be extremely helpful.

155

REFERENCES

Hardy, R.L., 1971. Multiquadric equations of topography and other irregular surfaces. J. Geophys. Res. 1976.
Hardy, R.L., 1976. Geodetic applications of multiquadric equations. ERI Project 1070-S, Iowa State Univ., Ames, May. (Available only from N.T.I.S., Springfield, Va. 22151. No. PB255296.)
Hardy, R.L., 1977. Least squares prediction. Photogramm. Eng. and Remote Sensing, 43 (4): 475—492.
Hardy, R.L., 1977. The application of multiquadric equations and point mass anomaly models to crustal movement studies. NOAA Tech. Rep. NOS NGS — Rockville, Md., to be published.
Hardy, R.L. and Gopfert, W.M., 1975. Least squares prediction of gravity anomalies, geoidal undulations, and deflections of the vertical with multiquadric harmonic functions. Geophys. Res. Lett., 2 (10).
Holdahl, S.R., 1970. Studies of precise leveling at California Fault Sites. ESSA Tech. Rep., Rockville, Md.
Holdahl, S.R., 1975. Models and strategies for computing vertical crustal movements in the United States. Presented at the International Symposium on Recent Crustal Movements, International Union of Geodesy and Geophysics, Grenoble, France.
Holdahl, S.R., 1977. Recent elevation change in southern California. NOAA Tech. Mem. NOS NGS-7, Rockville, Md., Feb.
Vanicek, P. and Christodulidis, D., 1974. A method for the evaluation of vertical crustal movement from scattered geodetic relevelings. Can. J. Earth Sci., 11: 605—610.

Tectonophysics, 52 (1979) 157—165
157

GEODETIC HIGH-PRECISION MEASUREMENTS IN ACTIVE TECTONIC AREAS; EXAMPLE: THE RHINEGRABEN

ERWIN GROTEN, CARL GERSTENECKER and GUENTER HEIN

Institute of Physical Geodesy, The Technical University at Darmstadt, Darmstadt (G.F.R.)

(Accepted for publication April 4, 1978)

ABSTRACT

Groten, E., Gerstenecker, C. and Hein, G., 1979. Geodetic high-precision measurements in active tectonic areas. In: C.A. Whitten, R. Green and B.K. Meade (Editors), Recent Crustal Movements, 1977. Tectonophysics, 52: 157—165.

Geodetic measurements in the Rhinegraben area are discussed. Repeated levellings, together with horizontal control data, torsion-balance measurements, very dense gravity coverage, tiltmeter observations and a high-precision gravity test net were available in order to test present concepts of the Rhinegraben structure and sources of recent crustal movements. The ambiguities of levelling data are stressed, but investigations of high-precision levelling data clearly indicate subsidence of several centimetres in the graben area, whereas multivariate studies of the data available on the graben shoulders do not indicate any significant movement on the shoulders themselves.

INTRODUCTION

In contrast to large horizontal as well as vertical movements along boundaries of main tectonic plates as in California, Japan, etc., or in active intraplate areas as the southern part of the U.S.S.R., recent movements in the western part of the Eurasian Plate are relatively small. Unfortunately, geodetic information along the active zone in northern Italy is still too scarce for detailed investigation. Therefore, graben areas in Western Europe have attracted attention for several decades. At present, the Rhinegraben zone is the best observed area as far as vertical control is concerned. But owing to well-known ambiguities of levelling data, intricate statistical procedures and independent supplementary geodetic data of great accuracy need to be incorporated in order to obtain unambiguous results. In this study the northern graben zone was especially investigated. The following data types are associated with that area.

(1) A dense gravity coverage yielding a Bouguer field of the accuracy of ±0.5 mGal within and around the northern part of the graben.

(2) Several thousand torsion-balance data within the graben zone.

(3) Horizontal control which, however, in the area of investigation was not able to yield any significant information because of *relatively* low accuracy in view of small recent horizontal movement.

(4) Groundwater table gauges data which, however, were not applied in detail in the present study; even though the gauges net is extremely dense, large-scale information was found to be sufficient as is explained below.

(5) Tiltmeter data obtained at three stations situated along the border of the northern part of the graben zone.

(6) High-precision gravity measurements as obtained from repeated surveys within a net connecting the shoulders, as well as the parts of maximum subsidence within the graben, with an absolute gravity site; owing to the short interval between repetitions these data have not yet been fully exploited and do not yet significantly contribute to the present study.

Present geological and geophysical theories on the Rhinegraben are numerous and partly contradictory; there are a few relevant geodetic studies which, however, are also not free of contradictions; recent geological, geophysical and geodetic aspects are well explained in Neugebauer and Braner (1975), Illies and Greiner (1976) and Schwarz (1976).

METHODS AND TECHNIQUES

Bouguer gravity and gravity gradients in relation to levellings

The first-order levelling is very dense in the graben area so that maximum distances from levelling lines are of the order of 20 km; the rms error of unit weight is found to be 0.6—0.7 mm/$\sqrt{km}$ for the network located along the western part of the Rhinegraben (Rheinland—Pfalz) and about 0.2—0.4 mm/$\sqrt{km}$ for the eastern side (Hessen).

A regression study was made in order to investigate straightforward regression of Bouguer gravity anomalies with areas of subsidence and uplift, respectively. Special care was necessary to take into account artificial (industrial, etc.) interference which led to subsidence not noticeably affecting the gravity field. Significant correlation coefficients $r > 0.7$ were found only in a few areas where the Bouguer field is very smooth.

As small mass variations at shallow depth might affect gravity gradients more than gravity itself, a similar study relating areas of subsidence and uplift, respectively, to positive and negative gravity gradients was undertaken. Because of well-known "topographic" and "cartographic" corrections to be applied with torsion-balance data, reduced free-air gradients do not significantly differ from rigorous Bouguer gradients. In contrast to Bouguer gravity data which give information about mass heterogeneities at greater depth, the present torsion-balance data are directly related to mass variations at depths of a few kilometres. This is realized on inspecting the autocorrelation curves shown in Figs. 1 and 2 where the correlation length is of the order of 2—3 km. By fitting polynomials of the order $n \leqslant 4$ to the data periodicities in the

159

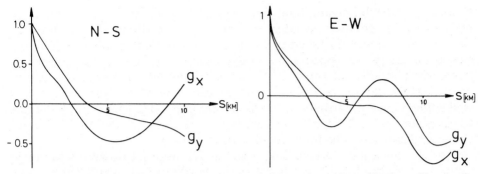

Fig. 1. Autocorrelation functions for torsion-balance data along N—S profiles in the Rhinegraben area.

Fig. 2. Autocorrelation functions for torsion-balance data along E—W profiles.

autocorrelation curves of the torsion-balance data, residuals arise similar to those in some geophysical density anomaly studies, but the significance of those periodicities needs further investigation.

In the area investigated, only positive correlation is found where correlation coefficients are typically high and gradients are strong. Again industrial (waterworks, etc.) interference has been taken into account. In the whole, correlation coefficients vary between $0.03 \leqslant r \leqslant 0.81$.

When paying special attention to the graben boundaries the first general conclusion is drawn: there is no significant relation of deep-seated mass anomalies with subsidence or uplift (there are only very few uplift areas according to repeated levelling data), but there are some local areas where remarkable correlation exists between horizontal gravity gradients and subsidence. Most purely local dislocations are associated with artificial (nontectonic) interference.

Horizontal control data

Shearing (Illies and Greiner, 1976) as well as downgliding graben shoulders (Neugebauer and Braner, 1975) would imply increasing distances between stations on both sides of the graben. But detailed studies of the trigonometric net, including recent geodimeter measurements, reveal accuracies which are too low for relevant investigations.

Tiltmeter data

Four tiltmeter stations around the northern half of the graben (outside the graben itself and more than 20 km from the graben boundaries) were installed in underground mines at depths $d > 50$ m. The results of three of those stations are incorporated in the present study, i.e., Wolfstein,

160

Bruschied and Obrigheim. Full details cannot be given here, but the main features are as follows: at each site at least two tiltmeters (Hughes and Ascania meters) were installed in order to obtain redundant observations and checks on cavity and topographic effects. Because of well-known uncertainties in the periodic (tidal) part of the tiltmeter output due to sea tidal disturbances, which cannot be fully eliminated using present sea tidal models, and well-studied local perturbations in the secular part, which are supposedly due to rainwater-induced temperature gradients, etc. (even at depths $d > 50$ m), tiltmeter data have to be interpreted with great care.

Two of the three stations mentioned above were set up across the main fault (Hunsrücksüdrandverwerfung) at the northern end of the Rhinegraben (left side; the third is on the right side, far away from the main faults).

On subtracting from the results of harmonic analyses (as obtained from records over several months, up to more than one year) the "theoretical" part, corresponding to a yielding earth by allowing for the above-mentioned uncertainties, the residuals indicate a reasonable regression with the directions of "main motion" as discussed by Illies and Greiner (1976). But corresponding correlations are significant only at one site directly situated at the main fault (distance $d < 10$ km). The precise station locations are: Wolfstein ($\phi = 49.60°$N; $\lambda = 7.60°$E); Obrigheim ($\phi = 49.33°$N; $\lambda = 9.12°$E); Bruschied ($\phi = 49.80°$N; $\lambda = 7.40°$E). The conventional Hughes tiltmeters have meanwhile been modified such that they can be installed in deep (depth >30 m) boreholes where the base is now essentially larger than with the original short-base instrument.

Detailed investigations of levellings

Levelling data usually refer to the geoid or, more precisely, to a zero bench mark connected with MSL (mean sea level) at a specific epoch. If repeated levellings are compared over long time intervals (e.g., several decades) the variations of the geoid due to internal mass changes with time as well as to pure subsidence and uplift, including related groundwater table variations, need to be considered. This is often neglected. In practice, repeated levellings are often referred to different bench marks. In those cases the variations of geopotential surfaces at corresponding bench marks should be taken into account.

A second aspect plays a major role, especially in the adjustments of levelling nets in the Rhinegraben area: in general, if levellings are interrupted in areas of recent movement for longer time intervals so that observations made at different epochs are connected within one loop, then *apparent* subsidence and uplift arise whenever those levelling results are compared with repeated levellings where such interruptions of measurements did not occur. It should be noted that whenever the original levelling has been done at different epochs, then in areas of subsidence apparent uplift occurs and vice versa, unless corrections for movements within those epochs are applied.

Another effect is also often overlooked: levelling data are well known to be "inexact" differentials; they need to be converted into "total" differentials using gravity as an integrating factor which, however, is not always available. Whenever repeated levellings do not run along the same line the elevation differences are not directly comparable. Since we do not yet have sufficient information on secular gravity changes, the conversion of path-dependent "inexact" differentials into "total" differentials which are independent of path should be done with great care along very long lines. However, using available geophysical and geological models, as well as groundwater table data, corresponding effects were found to be insignificant.

On the other hand, the above-mentioned concept by Neugebauer and Braner (1975) and further investigations by H.J. Neugebauer (personal communication, 1976) can imply relative vertical movement of the shoulder areas with respect to the graben as well as to those areas north of the above-mentioned E—W-running main fault at the northern graben boundary. Fig. 3

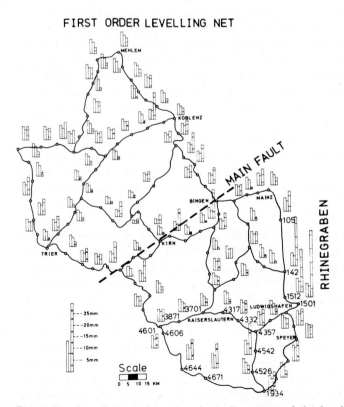

Fig. 3. Results of the multivariate free adjustment of the levelling network on the left of the Rhinegraben area including the Hunsrück fault.

shows the area on the left of the graben where a multivariate analysis was applied to a regional first-order levelling net. A similar study was done on the right of the graben where essentially higher accuracy (random part: ±0.3 mm/$\sqrt{\text{km}}$ and less systematic distortion) is found. However, because of different accuracies the two nets have not yet been combined in one investigation. The left side results of the present study were further considered in connection with previous data given by Schwarz (1976) for the graben area itself and its right side boundary as well as a recent study (Schwarz, personal communication, 1977).

Even though single variate analysis has sometimes been applied in studying time-dependent phenomena in geodetic systems, multivariate analysis which is much more effective has seldom been used in geodesy. The deficiency of most statistical procedures previously applied to repeated levellings is due to the specific accuracies of parts of the system not being fully taken into account, so that the clear separation of real elevation variations with time from simple error modelling is difficult. In other words: the interpretation of systematic distortions as indications of apparent subsidence or uplift rates cannot often be avoided. Advantages and deficiencies of multivariate analysis cannot be treated in detail in this paper; we wish only to mention the Fisher—Behrens problem which still might imply open questions. However, the main feature of our approach is the comparison of standard deviations associated with different (repeated) levellings and the elevation changes within the interval of repetition. In this case one repetition running along the original levelling lines was available. In Fig. 3 the standard deviations for the first levelling at each station (left-hand column), for the second levelling (middle column) and for the elevation changes between both levellings (right-hand column) are shown. Further details are given in the figure. The standard deviations refer to a free adjustment (using the well-known generalized inverse approach). On inspection of Fig. 3 it is realized that no significant elevation changes are found from the repeated levellings in that area except at stations 142, 1512 and 1501 which belong to the graben area itself where measurements of 1938 and 1950 have been tied together within one loop. The systematic rise in the southern part of the left-hand shoulder (points 3701, etc., around Kaiserslautern (see Fig. 3) and points 4357, etc., around Speyer) is only slightly greater than the standard deviation, so that the movement there is not yet considered significant. The possibility of purely local effects like subsidence at Stations 4542 and 4671 cannot be excluded — even though the monuments are older than 40 years — but it is improbable. The strong local subsidence of the graben Stations 142, etc., is corroborated by the results of the network on the eastern side (Hessen) which, because of essentially greater accuracy, clearly indicate a significant downward trend of the Rhinegraben. Some values at identical points of the Rheinland—Pfalz and the Hessen net are given for comparison in Table I. The above-mentioned study by Schwarz (personal communication, 1977) fully corroborates the subsidence in that part of the graben. Schwarz used

TABLE I

Recent vertical crustal movement and elevation accuracies at specific points of relatively strong movement

Station	Period of observations	Levelling network of the Rhinegraben			
		western side (Rheinland—Pfalz)		eastern side (Hessen)	
		rms error of adjusted height (mm)	movement in mm (mm/yr)	rms error of adjusted height (mm)	movement in mm (mm/yr)
142 Worms	1938—55 1968—73	±5.6 ±4.4	−21 (−0.7)	±2.9 ±3.3	−16 (−0.5)
105 Oppenheim	1938—55 1968—73	±5.9 ±4.6	−9 (−0.3)	±3.2 ±3.6	−4 (−0.2)
1512 Ludwigshafen	1938—55 1968—73	±5.9 ±4.6	−18 (−0.6)	no value	

four repeated levellings of the highest accuracy within an interal of 40 years.

Precise gravity observations

The above-mentioned difficulties in dealing with repeated levellings and the high cost, low speed and necessary skill in performing repeated levelling led to the concept of using gravity measurements together with levellings.

Since gravity combined with levellings are able to yield the separation of potential changes with time (conventional "elevations" are basically modified potential data) from true elevation changes, the relation of gravity changes to "elevation" changes can be established for specific mechanisms in uplift or subsidence areas. Repeated gravity can then supplement repeated levellings. In addition, on introducing absolute-gravity observations an absolute reference is obtained.

The above-mentioned effectiveness of gravity measurements was the main reason for the establishment of the Rhinegraben high-precision test net. The repeated gravity observations obtained during the last two years do not yet indicate significant secular changes, but some remarks on the principles are in order. Using the LaCoste model G (modified so that a resolution of ±0.002 mGal and essentially lower drift rates are achieved) and D meters, a test net was set up, rigidly connected to the International Gravity Standardization Net 1971 and to an absolute site at the northern graben boundary (see Fig. 4).

In order to overcome well-known difficulties encountered in precisely

164

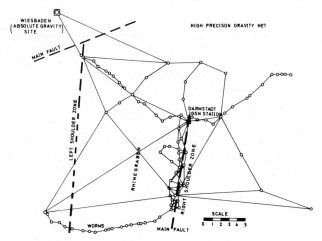

Fig. 4. Gravity test net, including a small test area (Pfungstadt), for studying artificial interferences. (Scale in km.)

measuring gravity in Holocene areas, criss-crossing ties were used in the graben zone, mainly around the primary fault areas. Sufficient connections are made in the graben shoulders in order to monitor the graben as well as the surroundings; single measurements (with two gravimeters) yield an accuracy of about ±3 μGal so that microgal accuracy can be achieved within reasonable time if meteorological effects (amounting to a few microgals) are allowed for. Consequently, subcentimetre accuracy is achieved in monitoring elevation variations.

CONCLUSIONS

Only at small distances ($d < 10$ km) do tiltmeters indicate strong motion across the main fault at the northern border of the Rhinegraben. In view of the different studies, the recent crustal movement within the Rhinegraben can be considered as a subsidence with a mean rate of movement of $\geqslant$0.5 mm/yr.

The levelling data along the western part of the shoulder area do not indicate recent movements of the southern area of the shoulder relative to the area north of the Hunsrück fault.

Unfortunately, only the main results have been given in very concise form, without giving details on the mathematical background. Nevertheless, the consequences of possible distortions in levelling nets in dealing with repeated levellings became clear.

The effect of water-table variations on levelling is small. It has to be taken into account (in the graben area) in dealing with gravity measurements of microgal accuracy; outside the graben its impact is small. The feasibility of highly precise gravity differences and their usefulness in connection with levelling is shown.

In order to avoid simple error modelling, intricate statistical methods are necessary in handling precise repeated data. Heuristic correlation of torsion-balance data with areas of subsidence indicates a relatively stable graben situation where only local subsidence might be explained by local mass heterogeneities. The use of Bouguer anomalies instead of gravity disturbances (which are not related to the geoid, being, in principle, variable with time) does not imply significant errors.

The main feature of the modified borehole tiltmeters is the base, now being of the order of 1 m so that local effects are averaged out.

REFERENCES

Illies, H.J. and Greiner, G., 1976. Regionales stress-Feld und Neotektonik in Mitteleuropa. Oberrhein. Geol. Abh., 25: 1—40, Karlsruhe.

Neugebauer, H.J. and Braner, B., 1975. Zur Dynamik kontinentaler Gräben, 35. Jahrestagung Dtsch. Geophys. Ges., Stuttgart Apr. 2—5, 1975, 5.4. (abstr.).

Schwarz, E., 1976. Präzisionsnivellement und rezente Krustenbewegungen dargestellt am nördlichen Oberrheingraben. Z. Vermess., 1: 14—25, Stuttgart.

Tectonophysics, 52 (1979) 167—176
167

RECENT VERTICAL MOVEMENTS AND THEIR DETERMINATION IN THE RHENISH MASSIF

H. MÄLZER, G. SCHMITT and K. ZIPPELT

Geodetic Institute of Karlsruhe University, Karlsruhe (G.F.R.)

(Accepted for publication April 4, 1978)

ABSTRACT

Mälzer, H., Schmitt, G. and Zippelt, K., 1979. Recent vertical movements and their determination in the Rhenish Massif. In: C.A. Whitten, R. Green and B.K. Meade (Editors),
Recent Crustal Movements, 1977. Tectonophysics, 52: 167—176.

The uplift of the Rhenish Massif started in late Tertiary time. The determination of
the rates of recent vertical movements helps to interpret the whole phenomenon. The
data obtained by relevellings of high precision will be analysed by two different velocity
models based on the concepts of Holdahl: the single-point model and the velocity surface
model. Both models yield adjusted heights corresponding to a selected reference time and
the velocities. The significance of the model parameters will be increased by introducing
free adjustments with generalized inverses.

INTRODUCTION

The first-order levelling network of West Germany extends over a distance
of 900 km from Denmark in the north to Lake Constance and the Alps in
the south. The average range of the levelling network in the E—W direction
is between 250 and 300 km. The first measurements were made over a long
period from 1912 to 1942. Between 1947 and 1962 relevellings were carried
out and the network was generally completed. This new network is called
Levelling Net 1960 and covers the block mosaic of West Germany (ADV,
1975). The network consists of 128 levelling loops with lengths between 90
and 410 km; the whole loop line of the net is 4132 km and the total length
of the lines is 16,193 km. Since 1970 in some areas — especially in the
Rhenish Massif and the Hessian Depression — relevellings were carried out.

The data obtained by the precise levellings needed for the determination
of recent vertical crustal movements in the Rhenish Massif were made available by the Departments of Surveying of the territories concerned — Rheinland-Pfalz, Hessen and Nordrhein-Westfalen. In the area of study a levelling
net has been formed by existing elements consisting of about 100 junction
points and approximately 160 levelling lines (Fig. 1). The length of the

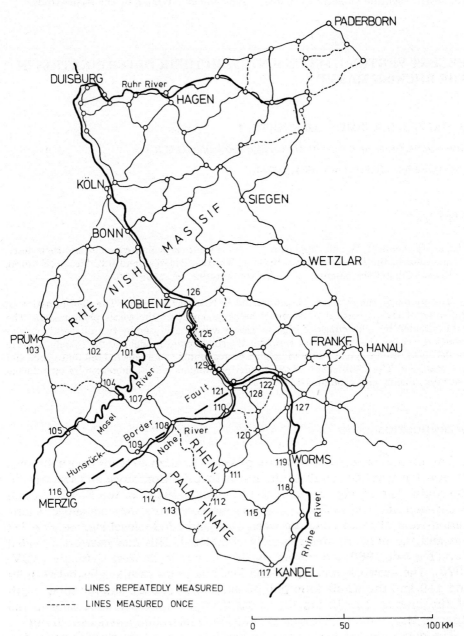

Fig. 1. The levelling network of the Rhenish Massif; 120 point number within the test area (see Figs. 3 and 4).

whole network measured once is about 4000 km, with 3600 km of first-order and 400 km of second-order observations. The time interval between the measurements for the Levelling Net 1960 and the relevellings later than

1970 is 15—20 years. Most levelling lines in the northern part (Nordrhein-Westfalen) and some in the northeast of the test area (Hessen) have been reobserved at least once.

The first measurements were carried out mainly by means of Hildebrand instruments with a nonhorizontal line of sight. Since 1955 exclusively self-adjusting levels with parallel plates have been used in the precise levellings. The accuracy of both observation methods varies between $m_0 = \pm 0.41$ mm/km (old and new measurements) and $m_0 = \pm 0.15$ mm/km (new levellings in Hessen). Large discrepancies between the rms errors m_1, calculated by section differences, and m_2 by line differences point out systematic error influences in the observed data. These may be caused by physical parameters not yet considered, such as temperature effects, incorrectly determined scales and tidal effects. Mass transport and changes of the water table may also influence the datum level and the precision of the measurements. The order of magnitude of these effects is being studied at present.

THE RHENISH MASSIF

The central part of the West German block mosaic is formed by the Rhenish Massif. Its southern border fault crosses transversally the northern end of the Upper Rhinegraben; to the east it is framed by the Hessian Depression (the line Paderborn—Hanau), to the north by the River Ruhr. In the west the massif overlaps parts of France, Luxembourg and Belgium. In a geologically young time — end of Tertiary — the Rhenish Massif began to be rapidly uplifted. The Rhine was forced to cut an antecedent river valley into the rising massif like a canyon. In the present, Pliocene to lower Pleistocene sediments are found as river terraces on the top of the upthrown block more than 200 m higher than the river level today. Near Bonn the river Rhine leaves the massif to enter the tectonic depression of the Lower Rhine Embayment. The determination of rates and sense of recent vertical movements may contribute to a better understanding of the driving mechanism of the regional uplift.

MODELS OF CALCULATION

General remarks

The classical method used in former determinations was to adjust the network for two or more different reference times as related to those observations, which have been carried out in more or less short periods. After these adjustments, movements can be calculated by comparing the sets of adjusted heights and the time differences between the measurement epochs. But it is very difficult to get significant indications for movements in this way because the collection of single levellings to a reference time and the adjustment without a time parameter involves a lot of uncertainties.

Avoiding these disadvantages, Ghitau (1970) and Holdahl (1975) have presented some mathematical models including the time parameter. The origin of these models is the functional relation between the actual height of a point and the time $H_p = f(t)$ (Fig. 2).
At the time $t = t_i$ the height of a point can be calculated by:

$$H_p^i = H_p^0 + \int_{t_0}^{t_i} \frac{df}{dt} dt \qquad (1)$$

with H_p^0 height at the reference time t_0.
Since the relevellings were carried out in periodic intervals, the integral in eq. 1 must be written by differential quotients as:

$$H_p^i = H_p^0 + \sum_{k=t_1-t_0}^{t_i-t_{i-1}} \left(\frac{df}{dt}\right)_k \Delta t_k \qquad (2)$$

If the intervals between the measurement epochs are too large, the differentials in eq. 1 can be expressed in a Taylor's series, a universal equation to determine the actual height values:

$$H_p^i = H_p^0 + \sum_{k=t_1-t_0}^{t_i-t_{i-1}} \left[\left(\frac{df}{dt}\right)_k \cdot \Delta t_k + \frac{1}{2!} \left(\frac{d^2 f}{dt^2}\right)_k \cdot \Delta t_k^2 + ...\right] \qquad (3)$$

Introducing only one reference time equation, eq. 3 will be simplified to:

$$H_p^i = H_p^0 + \frac{df}{dt} \cdot \Delta t + \frac{1}{2!} \frac{d^2 f}{dt^2} \cdot \Delta t^2 + ... \qquad (4)$$

$$\Delta t = t_i - t_0$$

Based on Holdahl (1975), two main velocity models for the determination of vertical movements may be given. These will be described as follows.

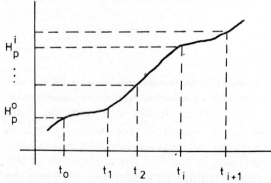

Fig. 2. The actual height of one point H_p^i as a function of the time t_i.

Model 1: the single-point model

Each derivative within eq. 4 will be replaced by a constant term:

$$H_p^i = H_p^0 + p_1 \cdot \Delta t + p_2 \cdot \Delta t^2 + ... \tag{5}$$

The advantage of this model is in the fact that every point of the network gets an individual equation. The network configuration can be utilized in an optimal way, and real movements can be approximated as best possible. This model is applicable both in regional and local areas for arbitrary types of motion. The result for one measuring point is — as the model is concerned — independent from the situation and the velocities of neighbouring points. The first two parts of the Δt polynomial (eq. 5) are velocity and acceleration. The maximum degree of the polynomial is limited by the number of repeated measurements, considering also that besides the coefficients p_i, the initial heights H_p^0 at the reference time t_0, are unknown parameters. Generally, the polynomials can be broken off after the first or second term. Even if the data redundancy is high there will not be much advantage in higher degrees. The model can also be used for linear investigations along single levelling lines.

Model 2: the velocity surface model

The derivatives of eq. 4 are described by two-dimensional polynomials:

$$H_p^i = H_p^0 + f_1(x, y) \cdot \Delta t + f_2(x, y) \cdot \Delta t^2 + ... \tag{6}$$

with a balancing velocity surface:

$$f_1(x, y) = v(x, y) = \sum_{i=0}^{n} \sum_{j=0}^{n} a_{ij} \cdot x^i \cdot y^j \tag{7}$$

and a corresponding acceleration surface:

$$f_2(x, y) = b(x, y) = \sum_{i=0}^{m} \sum_{j=0}^{m} b_{ij} \cdot x^i \cdot y^j \tag{8}$$

The coordinates of the net points form the basal points for the polynomials. The surface model makes the movement of a single point dependent on its situation within the study area; therefore it is suitable for statements concerning regional motion tendencies. However, the determination of relative point motions by regular and comparatively simple surfaces decreases the significance of the results. The degree of a polynomial can be restricted to the postulation of a minimal standard error of unit weight.

Polynomials of low degree characterize mostly regional motion processes such as a tilt for a linear approach or an uparching resp. drop-down for a quadractic approach. First computations carried out have shown that polynomials with an order higher than three will lead to unreasonable results. The main disadvantage of the surface model is that regional fault lines or escarpments are smoothed. Its applicability lies in the determination of regional vertical movements. The redundancy of this model is very high: in addition to the initial heights H_p^0 at the reference time t_0 the coefficients a_{ij} (and b_{ij}) of the polynomial(s) are unknown (at most 18 for a cubic approach). Other advantages of the model are the arbitrary construction of the network in the study area and the possible representation of the results by contour lines.

Problems: numerical and statistical

Both models show singularities concerning the fixing of heights and velocities, which result from the relative character of levellings. The singularities of the problem are removed by the introduction of two constraints:

$$dh_N = 0 , \qquad v_N = 0 \rightarrow H_N = \text{constant} (P_N = \text{reference point})$$

This reference point substantially influences the accordance of the computed relative velocities with the absolute ones. Therefore the following points should be taken into account:

(1) The reference point should be situated in a geologically stable region without individual motion.

(2) In order to obtain absolute height values it should have a connection to a mareograph (this is not possible in the Rhenish Massif).

(3) Its location should be near the centre of the study area. In this case a high stability of the normal equations is guaranteed.

After the introduction of the reference-point constraints, further singularities may appear in both models. This occurs in the single-point model if the number of measurements to one point at different times is not greater than the degree of the velocity polynomial at this point. In such a case no unique solution is possible and the degree of the polynomial has to be reduced. In the surface model the computation of the observational equations by rough Cartesian coordinates leads to diagonal elements in the normal equations increasing rapidly to values of 10^{20} and above. The system of normal equations becomes ill-conditioned and pseudosingularities arise. This numerical problem can be removed by standardizing the coordinate values.

Besides the choice of the reference point the weighting of the observations has a considerable influence on the results and their significance. Usually the weight factor of a levelling line is computed to $p_i = \text{const.}/s_i$, where s_i is the length of the line. This factor does not consider correlations between the observations. To obtain optimal weights, it is possible to calculate the physical correlations after Lucht (1972) or by way of weight estimation, for example by the maximum likelihood method.

Thus, the results of both models and their significance, given by statistical tests (e.g., the t-test), are dependent on the choice of the reference point, the

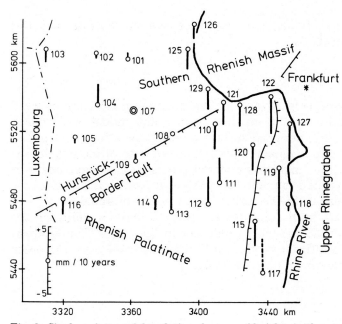

Fig. 3. Single-point model: relative changes of heights in the southern Rhenish Massif and Rhenish Palatinate areas; reference time 1950, reference point 107 (120 point number).

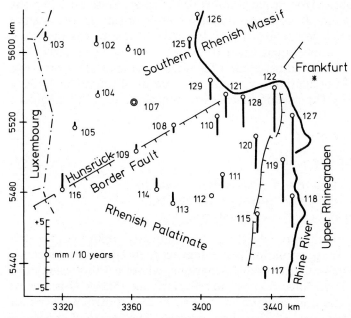

Fig. 4. Surface model (second order); relative changes of heights in the southern Rhenish Massif and Rhenish Palatinate areas; reference time 1950, reference point 107 (120 point number).

weighting of the observations, the degree of the movements and the influences of systematic errors.

First results

For a limited test area (Figs. 3 and 4) first diagnostic calculations were carried out. Hereby the results of the single-point model with linear approach and the quadratic designed surface model show a high agreement. Both models yield the same regional movements, but with different velocity magnitudes for which the single-point model seems to give more realistic values. The results are related to point 107 (Wahlenau) which is situated on a geologically stable plateau of the Hunsrück with a height of 459.6 m.

The characteristics of both solutions are (Figs. 3 and 4): uplift of the Rhenish Massif, up to 0.35 mm/yr, subsidences in the Upper Rhinegraben, up to 0.9 mm/yr; no significant varying motion values at the western border of the test area near Belgium and Luxembourg.

FREE ADJUSTMENT

The interpretation of the results is dependent fundamentally on their significance. This significance, estimated by the inverse Q of the coefficient matrix N of the normal equations is strongly related to the choice of the reference point, because error propagation is based on accurate initial height and velocity at this point. In order to obtain *inner* error estimations for the model parameters, which are free of the *external* influence of the reference point, the singular normal equations can be treated by generalized (g-) inverses. In the case of free adjustments of geodetic triangulation or vertical control networks, the postulation $x'x \rightarrow$ min (x = vector of unknowns) yields an efficient solution by the pseudo-inverse (or minimum-norm least-squares g-inverse) to:

$$x = N(NN)^- A'I = N^* A'I = A^+ I \tag{9}$$

$$Q = N^* N^* N = N^+ \text{ (spur } (Q) \rightarrow \text{min)} \tag{10}$$

where A = coefficient matrix of observation equations,
     I = absolute terms of observation equations,
     $(.)^-$ = g-inverse and $(\ )^+$ = pseudo-inverse.

This solution leads to an optimal fitting of the adjusted unknowns to their approximate values. The postulation $x'x \rightarrow$ min in geodetic networks corresponds to $v'v \rightarrow$ min in similarity transformations, where v = the residuals of the transformation. Therefore, the variance—covariance matrix Q is free of residual errors of positioning, orientation and scale of a control network, or free of small vertical removals of a levelling network. The centre of gravity of the approximate coordinates, or heights, changes over to that of the final values.

In the velocity models, different types of unknowns exist: $x' = (x_1' | x_2')$, where x_1 = initial heights and x_2 = velocity coefficients. The introduction of a reference point with a fixed intitial height value should be interpreted as such a small vertical removal of the levelling network. Therefore, this reference point influences the variances of the unknowns. To obtain an "inner solution" here, the postulation $x'x \rightarrow$ min has to be extended by a diagonal weight matrix M to $x'Mx \rightarrow$ min, with different weighting factors m_1 and m_2 for both unknown groups x_1 and x_2. The solution is written:

$$x = M^{-1}N(NM^{-1}N)^{-}A'I = N_M^* A'I = A_M^+ I \tag{11}$$

$$Q = N_M^* N_M^* N = N_M^+ \quad (\text{spur } (QM) \rightarrow \text{min}) \tag{12}$$

The minimum constraint for $x'Mx$ can be decomposed to $m_1 x_1' x_1 + m_2 x_2' x_2 \rightarrow$ min. An overweighting of the initial heights x_1 against the velocity coefficients x_2, for example by $m_1 = 100 \cdot m_2$, allows a geometrical interpretation in the meaning previously shown. The error influence of small removals of the network is eliminated and the resulting variances of the unknowns are inner ones. Numerical computations of free adjustments will be performed in the near future.

CONCLUSIONS

The determination of recent vertical crustal movements of secular character by repeated levellings of high precision in the area of the Rhenish Massif is based on measurements of the last 20—30 years. After the first test calculations an interpretation of these preliminary results is possible only with a great deal of care. In particular, an extrapolation into past or future times can lead to erroneous statements. A certain periodicity of recent movements can never be absolutely excluded, therefore it is impossible to arrive at conclusions from a short observation period concerning the motion process over thousands or even millions of years. For this reason it is most reasonable to design the mathematical models with mainly simple (even linear) types of movement. The residual errors may consist of observation errors or unidentified local movements of given single points.

ACKNOWLEDGEMENTS

This work was submitted under a grant by the German Research Society. We thank the Departments of Surveying of Hessen, Nordrhein-Westfalen and Rheinland-Pfalz for collaboration.

REFERENCES

Arbeitsgemeinschaft der Vermessungsverwaltungen der Länder der Bundesrepublik Deutschland (ADV) 1975. Nivellementnetz 1960. Bayerisches Landesvermessungsamt, München.

176

Ghitau, D., 1970. Modellbildung und Rechenpraxis bei der nivellitischen Bestimmung
   säkularer Landerhebungen. Dissertation, Bonn.
Holdahl, S.R., 1975. Models and strategies for computing vertical crustal movements in
   the United States. Paper presented at International Symposium on Recent Crustal
   Movements, Grenoble, France.
Lucht, H., 1972. Korrelation im Präzisionsnivellement. Wiss. Arb. Inst. Geod. Photogramm. an der TU Hannover, No. 48.

Tectonophysics, 52 (1979) 177—178
© Elsevier Scientific Publishing Company, Amsterdam — Printed in The Netherlands

ARCHAEOLOGICAL EVIDENCE FOR TECTONIC ACTIVITY IN THE REGION OF THE HAIFA-QISHON GRABEN, ISRAEL

N.C. FLEMMING

Institute of Oceanographic Sciences, Wormley, Godalming, Surrey (Great Britain)

(Accepted for publication April 4, 1978)

ABSTRACT

Thirty-five archaeological sites between Lebanon and Gaza on the Mediterranean shore of Israel were surveyed for evidence of relative changes of sea level; and published and unpublished data were compiled on a further 23 sites. Remains of occupation levels were classified into 13 archaeological periods from Neolithic to Crusader, giving a good dating scale over the last 9000 years. The Bronze Age shoreline (4000 years B.P.) is shown to be very close to the present shore along most of the coast, with vertical changes of less than 1.0 m at most sites. For later periods the accuracy of measurement of relative changes of level was improved to a root sum square error of 28 cm.

The mean and modal sea levels derived from all sites of the same archaeological period were plotted against time to indicate the most probable "eustatic" sea-level curve, though it is accepted that hydro-isostatic adjustment and broad-scale slow earth movements cannot be separated statistically from eustatic changes on a 200-km stretch of coast. The relative sea level is shown to have been between —70 cm and present sea level for the last 4000 years, with no evidence for oscillations of amplitude more than ±30 cm at periodicities longer than 400 years, and no high sea level still stands.

Relative vertical displacement between sites of the same archaeological age is attributed to tectonism. The sites on the alluvium of the Haifa—Qishon graben show no signs of vertical displacement, but there is maximum activity to north and south. Acco, on the north margin, shows submergence of nearly 2.0 m since A.D. 1300, with probable uplift preceding that date. Mount Carmel, south of the graben, shows signs of only very slight relative change, possibly attributable to eustatic change, in the immediate vicinity of the graben. But the southern flank of the mountain, 20—80 km south of the southern fault of the Graben, shows active vertical movements on the whole coast. Dor and Caesarea indicate multiple vertical movements ranging from 1.0 to 5.0 m in the last 2000 years, and a slump or fault has intersected the harbour area of Caesarea about 150 m offshore.

These observations are related briefly to recent geodetic levelling of the coast and graben area; observations on the Dead Sea Rift; and hypotheses of fault control of the linearity of the Israel Mediterranean coastline.

REFERENCE

Flemming, N.C., Raban, A. and Goetschel, C., 1978. Tectonic and eustatic changes on the Mediterranean coast of Israel in the last 9000 years. In: Progress in Underwater Science. Pentech Press, London, 3: 33—93.

Tectonophysics, 52 (1979) 179—180

THE USE OF ASSUMED DISPLACEMENT FIELDS IN FINITE-ELEMENT ANALYSIS TO CALCULATE THE GRAVITY EFFECT OF CRUSTAL DEFORMATION

JACK PELTON

Department of Geology and Geophysics, University of Utah, Salt Lake City, Ut. 84112 (U.S.A.)

(Accepted for publication April 4, 1978)

ABSTRACT

An expression is derived which relates a displacement field u defined on a continuum to the change in gravity (Δg_D) brought about by deformation:

$$\Delta g_D(x, y, z) = G \iiint_{V_R} \rho_R(R_1, R_2, R_3)$$

$$\times \left[\frac{(z - R_3)}{\alpha^3(R_1, R_2, R_3)} - \frac{(z - R_3 - U_3)}{\alpha^3[T_D(R_1, R_2, R_3)]} \right] dR_1 \, dR_2 \, dR_3$$

where G is the gravitational constant; ρ_R is the pre-deformation density function; V_R is the volume occupied by the continuum before deformation; T_D is a transformation which represents the one-to-one mapping of V_R onto the volume V_A occupied by the continuum after deformation:

$$T_D: \begin{aligned} A_1 &= R_1 + U_1(R_1, R_2, R_3) \\ A_2 &= R_2 + U_2(R_1, R_2, R_3) \\ A_3 &= R_3 + U_3(R_1, R_2, R_3) \end{aligned}$$

where A_1, A_2, A_3, are the Cartesian coordinates of a point in V_A; R_1, R_2, R_3 are the Cartesian coordinates of a point in V_R; U_1, U_2, U_3, are the components of u; and α is the distance from the field point (x, y, z) to a general point in V_A or V_R. The derivation is based on the principle of conservation of mass as stated in continuum mechanics. The formulation of the finite-element procedure for problems in linear elasticity typically involves the selection of an assumed function for u within each element. Unknown constants which characterize the function are determined in terms of nodal displacements. The integral in the first equation above may therefore be evaluated for each element, and the results summed to obtain Δg_D. Convergence to the exact solution for Δg_D depends on the convergence of the finite-element solution for u. The procedure is demonstrated for axisymmetric, torsion-free deformation of a solid of revolution in which the elements are tori of triangular cross-section. These results may be useful in the analysis of gravity change associated with tectonic deformation of the earth's crust.

Tectonophysics, 52 (1979) 181

181

TRANSCONTINENTAL PROFILE OF RECENT VERTICAL CRUSTAL MOVEMENTS

M.B. LAWRENCE and L.D. BROWN

Department of Geological Sciences, Cornell University, Ithaca, N.Y. 14853 (U.S.A.)

(Accepted for publication April 4, 1978)

ABSTRACT

A transcontinental profile of vertical crustal movement from San Diego, California, to Meldrim (Savannah), Georgia, has been assembled from precise leveling data of the National Geodetic Survey. Assuming constant movement during the time interval between leveling and releveling surveys, rates of movement have been plotted. The overall rate of movement of the East Coast relative to the West Coast derived from leveling is opposite in sign to the corresponding rate from tide gauges at San Diego and Savannah. Though this discrepancy is within the acceptable limits of normal random error associated with leveling measurements, it may also result in part from unavoidable uncertainties at the connection points between the various segments. In the absence of deterministic evidence on the source of the discrepancy, a least-squares adjustment was performed to distribute this difference in the overall trend through the profile. Apart from this trend, the western end of the profile is dominated by pronounced subsidence (rates 10 cm/yr) owing to water withdrawal and associated consolidation of alluvial sediments. With the exception of these movements, rates of vertical movement of the eastern and western United States are similar, suggesting that if such measurements represent significant tectonic activity, this activity may not be confined to the western United States. Movements are correlated with topography only in the peninsular ranges of California. Although these apparent elevation changes may reflect actual ground movement, this correlation may also result from systematic leveling errors. Comparison with gravity, depth to basement and depth to Moho shows no correlation.

Tectonophysics, 52 (1979) 183—189
© Elsevier Scientific Publishing Company, Amsterdam — Printed in The Netherlands

VERTICAL CRUSTAL MOVEMENTS IN THE CHARLESTON, SOUTH CAROLINA—SAVANNAH, GEORGIA AREA

PETER T. LYTTLE, GREGORY S. GOHN, BRENDA B. HIGGINS and
DAVID S. WRIGHT

U.S. Geological Survey, Reston, Va. (U.S.A.)

(Accepted for publication April 4, 1978)

ABSTRACT

Lyttle, P.T., Gohn, G.S., Higgins, B.B. and Wright, D.S., 1979. Vertical crustal movements in the Charleston, South Carolina—Savannah, Georgia area. In: C.A. Whitten,
R. Green and B.K. Meade (Editors), Recent Crustal Movements, 1977. Tectonophysics, 52: 183—189.

First-order vertical level surveys (National Geodetic Survey) repeated between 1955
and 1975 suggest that modern vertical crustal movements have taken place in the Atlantic
Coastal Plain between Charleston, South Carolina and Savannah, Georgia. The relative
sense of these movements correlates with the sense of displacement of Tertiary strata on
known geologic structures. Whereas regional dip of strata in most of the Atlantic Coastal
Plain is southeasterly, the regional dip of Tertiary strata in this part of the Coastal Plain
averages 2 m/km to the south or southwest. Positive structural features disturb this
regional dip along a poorly defined zone, about 25 km wide, parallel to the coast between
Savannah and Charleston. Structural relief on these features is as much as 20 m. Repeated
level lines that cross the Atlantic Coastal Plain elsewhere generally show an increase in
modern relative subsidence from west to east. However, in the Charleston—Savannah
area, the amount of relative subsidence remains fairly constant or decreases from west to
east across the structural highs. At two localities near Charleston, where Tertiary beds are
offset by faults roughly on strike with one another, an abrupt break in a repeated level
line occurs where the level line crosses the probable extensions of these faults. The
average modern rates of relative uplift and subsidence (assuming they are constant) are
compatible with rates noted throughout the Coastal Plain. Long-term extrapolation of
modern rates appears unreasonable; episodic or oscillatory movements are much more
likely.

INTRODUCTION

The area around Charleston, South Carolina, is a region of anomalously
intense seismicity in the southeastern United States. The 1886 earthquake
(Dutton, 1889), assigned a maximum modified Mercalli intensity of $\bar{x}$
(Bollinger, 1977) in the Charleston—Summerville epicentral region (Fig. 1),
dominates the seismic history of the area. A documented pre-1886 seismic

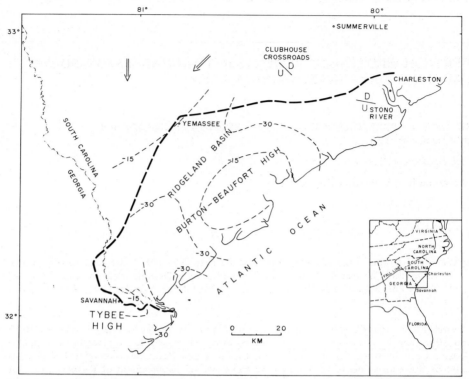

Fig. 1. Map showing generalized geologic structures in the Charleston, S.C.—Savannah, Ga. region. The dashed line represents the location of three level-line profiles. Strike and sense of movement is shown for faults along the Stono River near Charleston and at Clubhouse Crossroads southwest of Summerville. Structure contours in meters below mean sea level on top of Miocene limestone are redrawn from data in Furlow (1969) and Siple (1969). Large arrows show the direction of regional dip of Coastal Plain units in South Carolina.

history exists for the area (Bollinger and Visvanathan, 1977), and earthquakes continue to take place to the present (Tarr, 1977).

The Charleston—Summerville area is near the South Carolina coast on a low-lying part of the Atlantic Coastal Plain (Fig. 1). Approximately 750 m of unconsolidated to weakly consolidated Cretaceous and Cenozoic sedimentary rocks underlie the area (Gohn et al., 1976). The few previous geologic studies of the area have indicated that a poorly defined area of uplifted Tertiary sedimentary rocks is present along the coast between Charleston and Savannah (Fig. 1). The presence of this uplifted area suggests, without providing specific evidence for, a possible correlation between the uplift and the seismicity near Charleston.

The U.S. Geological Survey began investigations into the cause of the Charleston—Summerville seismicity in 1974. Preliminary results of deep drilling and shallow auger drilling indicate that an uplift exists in the Charleston—Savannah coastal area. An examination of existing first-order level-line data for South Carolina and adjacent states shows that the sense of vertical movement of the present land surface corresponds with the sense of vertical movement of Tertiary strata in the area. In addition, the boundaries of the block of uplifted Tertiary strata coincide with the points where offset of the level lines occur.

We have sought correlations between relative vertical movements shown by repeated level lines and geologic structures of local extent in many places in the eastern U.S. Broad correlations can be made over large areas or geologic provinces, such as the Blue Ridge, Piedmont or the Coastal Plain. Within one of these geologic provinces correlations between relative movements discovered by repeated leveling, and local structures discovered by detailed geologic mapping, are difficult to document. One or both sources of data are usually inadequate or missing.

GEOLOGIC SETTING

Figure 1 summarizes the known structural data for the Coastal Plain between Charleston and Savannah. Previous studies based on structure contours on Eocene, Oligocene, and (or) Miocene sedimentary units have defined an area of uplifted Tertiary sediments elongated to the northeast and parallel to the coast that include the Tybee high (Furlow, 1969), the Burton high (Siple, 1969), and Beaufort high (Heron and Johnson, 1966). The data in Siple (1969) and Heron and Johnson (1966) also define a structural trough named the Ridgeland basin that is inland from, and parallel to, the Burton—Beaufort high. The maximum observed structural relief between the crest of the Burton—Beaufort high and the bottom of the Ridgeland basin is about 20 m (Heron and Johnson, 1966).

High-resolution seismic surveys conducted in estuaries near Charleston (Colquhoun and Comer, 1973) define a structure of uncertain nature and geometry along the Stono River just west of Charleston (Fig. 1). According to Colquhoun and Comer, the structure is an asymmetric anticline, the Stono arch, with an axis trending N63°W to N88°W. They suggest that the folding is related to a probable thrust fault dipping at a very low angle to the southwest. Numerous right-angle turns along the line of the seismic profile make detailed interpretation difficult, but divergent angles of dip of various reflecting horizons shown along the line of the profile suggest structural disruption of beds. Preliminary analysis of data from auger holes indicates as much as 10—15 m of relief (vertical displacement?) on the top of the Eocene beds (lower part of the Cooper Formation) across the Stono structure. The sense of movement on the Stono structure (Fig. 1), north side down, complements the sense of movements on the Burton—Beaufort high.

Deep stratigraphic test holes and auger drilling by the U.S. Geological Survey at Clubhouse Crossroads, about 35 km northwest of Charleston (Fig. 1), have defined local vertical relief of as much as 40 m on the Eocene—

186

Oligocene uncomformity. In this area, about 40 m of upper Eocene sedimentary rocks are missing along a probable northwest-trending normal fault
where it is intersected by a deep core hole. The sense of movement on this
probable fault is down to the northeast. This fault and the Stono structure
are areally consistent with, and may form the northeastern boundary of, the
Burton—Beaufort high.

LEVELING DATA

Figure 2 shows relative vertical movements obtained from first-order leveling data in the Charleston—Savannah area; the profiles shown in Fig. 2 can
be compared with the structures in Tertiary strata shown in Fig. 1. The
amount of relative movement exceeds that which can be attributed to

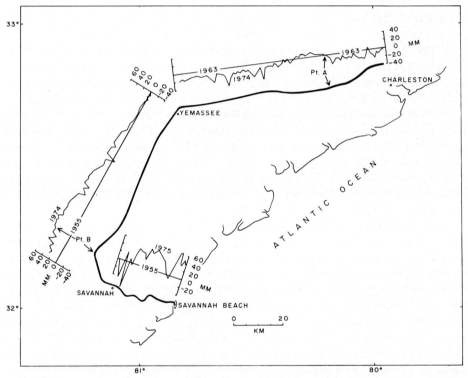

Fig. 2. Three level-line profiles in the Charleston, S.C.—Savannah, Ga. region: (1) Charleston, S.C.—Yemassee, S.C., 1963 and 1974; (2) Yemassee, S.C.—Savannah, Ga., 1955 and
1974; and (3) Savannah, Ga.—Savannah Beach, Ga., 1955 and 1975. The heavy line indicates the location of the level lines. Vertical scale of profiles represents the simple difference (dH) of unadjusted elevations (in millimeters) for the two surveys involved. Point
A represents the location where the level line crosses the probable extensions of both the
Stono and Clubhouse Crossroads structures. Point B represents the sharp bend in the
Yemassee—Savannah profile.

systematic or random leveling errors. Level-line profiles (Lyttle and Wright, unpublished data) in the vicinity of Charleston show that the direction of relative movement has changed between successive relevelings. In addition, as there is no geologic reason to assume constant rates and nonepisodic movement during these time periods, simple differences in the adjusted elevation data rather than rates are used.

For the Charleston—Yemassee profile, a station in Charleston was arbitrarily held constant for the two episodes of leveling in 1963 and 1974. Holdahl and Morrison (1974) have estimated absolute velocities of vertical movements at Charleston. After adjusting the slope of the curve representing mean sea-level variation for the eustatic rise in sea level, they obtained the value of 2.53 ± 0.34 mm/yr for the apparent change in sea level related to vertical movement of the land. That is, the Charleston area shows a subsidence of 2.53 ± 0.34 mm/yr. In addition, holding the Charleston station constant, almost the entire Charleston—Yemassee profile continues to show relative subsidence during this 11-year period. Although leveling data for most of the Atlantic Coastal Plain, not shown here, suggest an increase in relative subsidence to the east, i.e., approaching the coast, in general, the amount of relative subsidence along the Charleston—Yemassee profile increases gradually to the west.

Where the level line crosses the probable extensions of both the Stono and Clubhouse Crossroads structures at a point 35 km west of Charleston (Fig. 2, point A), the profile shows an abrupt offset, and the sense of movement is up to the west. Both the modern and Tertiary displacements have the same sense of movement. Between point A (Fig. 2) and Yemassee the amount of relative subsidence increases slightly again. This relationship possibly corresponds with the structural relief between the Beaufort high and the Ridgeland basin.

Interpretation of the Yemassee—Savannah profile is difficult for several reasons. First, the northern three-quarters of the profile roughly parallels the strike of the geologic structures previously mentioned, and second, the profile makes a fairly sharp bend (Fig. 2, point B) approximately 15 km northwest of Savannah. Nevertheless, the profile shows a gradual relative subsidence in a northeastward direction in the area over the Ridgeland basin.

The profile from Savannah to Savannah Beach continues to show a relative uplift, increasing to the axis of the Tybee high. One complicating factor for both the second and third profiles is local subsidence in the Savannah area related to groundwater withdrawal and sediment compaction. This local subsidence could account for some of the very sharp peaks and valleys in the profiles.

DISCUSSION

Level lines throughout the Atlantic Coastal Plain that have been surveyed more than twice commonly show that the direction of vertical crustal movement has reversed in the period between surveys. The rate of movement (assuming for the moment that the movement is continuous rather than episodic) is in the same range for both uplift and subsidence. Because an absolute net rate and direction is unknown, modern rates interpreted from these short-term studies should not be extrapolated back through geologic time. If one assumes the same tectonic regime for the last 25 m.y., it is clearly impossible for modern rates of movement (approximately 2.5 mm/yr) to be applied throughout that time. It would take only 8000 years to produce 20 m of structural relief (approximately the amount found in the vicinity of the Burton—Beaufort high). This clearly supports the concept of episodic or oscillatory movement.

In South Carolina the relative sense of movement on displaced Tertiary beds and the relative sense of offsets in the present-day level lines appear the same, and highs and lows defined by mapping Tertiary horizons correspond to relative highs and lows on level-line profiles. These relations permit two alternative explanations. In the Charleston—Savannah area, either the same tectonic regime has been in effect since at least the Late Oligocene (for about 25 m.y.) or the displacements shown by Tertiary beds and level-line profiles have been produced solely during the Quaternary. Both these explanations must remain conjectural until such time as faulting is clearly shown to affect or not to affect Quaternary beds. Whatever the timing of the deformation, the Charleston—Savannah area remains an anomalous region where modern crustal movements do not reflect the normal coastal trends of a gradual increase in the amount of relative subsidence towards the coast as a result of sedimentary loading.

ACKNOWLEDGEMENTS

Work by two of us (G.S.G. and B.B.H.) is supported by the U.S. Nuclear Regulatory Commission, Office of Nuclear Research under agreement No. AT (49-25)-1000.

REFERENCES

Bollinger, G.A., 1977. Reinterpretation of the intensity data for the 1886 Charleston, South Carolina, earthquake. In: D.W. Rankin (Editor), Studies Related to the Charleston, South Carolina, Earthquake of 1886: A Preliminary Report. U.S. Geol. Surv. Prof. Pap., 1028: 17—32.
Bollinger, G.A. and Visvanathan, R.T., 1977. The seismicity of South Carolina prior to 1886. In: D.W. Rankin (Editor), Studies Related to the Charleston, South Carolina, Earthquake of 1886: A Preliminary Report. U.S. Geol. Surv. Prof. Pap., 1028: 33—42.
Colquhoun, D.J. and Comer, C.D., 1973. The Stono Arch, a newly discovered breached anticline near Charleston, South Carolina. S.C. State Dev. Board, Div. Geol., Geol. Notes, 17 (4): 98—105.
Dutton, C.E., 1889. The Charleston earthquake of August 31, 1886. U.S. Geol. Surv. Ninth Annu. Rep. 1887—1888, pp. 203—528.

Furlow, J.W., 1969. Stratigraphy and economic geology of the eastern Chatham County phosphate deposit. Ga. Geol. Surv., Bull., 82: 40 pp.

Gchn, G.S., Higgins, B.B., Owens, J.P., Schneider, Ray and Hess, M.M., 1976. Lithostratigraphy of the Clubhouse Crossroads Core; Charleston project, South Carolina [abs.]: Geol. Soc. Am. Abs. with Programs, 8 (2): 181—182.

Heron, S.D., Jr. and Johnson, H.S., Jr., 1966. Clay mineralogy, stratigraphy, and structural setting of the Hawthorn Formation, Coosawhatchie District, South Carolina. Southeast. Geol. 7 (2): 51—63.

Holdahl, S.R., and Morrison, N.L., 1974. Regional investigations of vertical crustal movements in the U.S., using precise relevelings and mareograph data. Tectonophysics, 23: 373—390.

Siple, G.E., 1969. Salt-water encroachment of Tertiary limestones along coastal South Carolina. S.C. State Dev. Board, Div. Geol., Geol. Notes, 137 (2): 51—65.

Tarr, A.C., 1977. Recent seismicity near Charleston, South Carolina, and its relationship to the August 31, 1886, earthquake. In: D.W. Rankin (Editor), Studies Related to the Charleston, South Carolina Earthquake of 1886: A Preliminary Report. U.S. Geol. Surv. Prof. Pap., 1028: 43—57.

Tectonophysics, 52 (1979) 191—192   191

RECENT VERTICAL CRUSTAL MOVEMENTS FROM PRECISE LEVELING DATA IN SOUTHWESTERN MONTANA, WESTERN YELLOWSTONE NATIONAL PARK AND THE SNAKE RIVER PLAIN *

R.E. REILINGER, G.P. CITRON and L.D. BROWN

Department of Geological Sciences, Cornell University, Ithaca, N.Y. 14853 (U.S.A.)

(Accepted for publication April 4, 1978)

ABSTRACT

Repeated levelings in southwestern Montana, the western portion of Yellowstone National Park and the Snake River Plain provide information on the pattern of relative vertical crustal movement throughout this region. Except for the coseismic deformation associated with the 1959 Hebgen Lake earthquake, the most outstanding and best-defined feature of the data is contemporary doming at a rate of 3—5 mm/yr, involving approximately 8000 km^2 including the epicentral area and after-shock zone of the 1959 Hebgen Lake earthquake. Based on observations over different time intervals, doming appears to have continued throughout the time the movements were monitored, beginning at least 25 years prior to the 1959 earthquake and continuing for at least 1 year after the earthquake. The character of the coseismic deformation associated with the 1959 earthquake and the high regional elevation are consistent with the observed doming. It is suggested that doming preceded the earthquake for a considerable time (on the order of hundreds to thousands of years, perhaps longer), giving rise to tensional stresses in the upper crust. When these stresses exceeded some critical value, faulting and collapse in response to gravity occurred, resulting in the 1959 earthquake. The voluminous Tertiary and younger volcanics in the vicinity of the doming region suggest that magma intrusion into the crust is the most likely cause of the observed uplift. The proximity of the doming region to the thermally active Yellowstone area supports this suggestion.
Secondary features of the data include:
    (1) A spatial correlation between tilting and historic seismic activity.
    (2) Uplift within the Norris—Mammoth corridor in Yellowstone National Park relative to nearby bench marks to the north and south.
    (3) Regional subsidence of the eastern Snake River Plain relative to points

---

* Editor's note: Article published in J. Geophys, Res., 82 (33): 5349—5359.

north and west of this physiographic province, including subsidence of the Pleistocene Island Park caldera floor relative to its rim fractures.

(4) Rapid tilting in the vicinity of (a) the Continental fault east of Butte, (b) the intersection of the Gardiner, Mammoth and Reese faults just north of Yellowstone National Park, and (c) the Madison Range fault in eastern Idaho.

Tectonophysics, 52 (1979) 193—201
© Elsevier Scientific Publishing Company, Amsterdam — Printed in The Netherlands

NEW RESULTS ON THE PROPERTIES OF RECENT CRUSTAL MOVEMENTS IN THE BOHEMIAN MASSIF AND ITS BOUNDARY WITH THE WEST CARPATHIANS

PAVEL VYSKOČIL

Research Institute of Geodesy, Topography and Cartography, Obránců míru 35, 170 00 Prague (Czechoslovakia)

(Accepted for publication April 4, 1978)

ABSTRACT

Vyskočil, P. 1979. New results on the properties of Recent crustal movements in the Bohemian Massif and its boundary with the West Carpathians. In: C.A. Whitten, R. Green and B.K. Meade (Editors), Recent Crustal Movements, 1977. Tectonophysics, 52: 193—201.

Preliminary analysis of the new results of repeated levellings within the territory of the Bohemian Massif and its border with the Carpathians is made, and the correlation between horizontal gradients of vertical movements and main fault zones discussed. The results indicated the continuance of the main movement's tendencies, determined by the foregoing measurements. Also discussed are the vertical and horizontal crustal movements between the Bohemian Massif and the Carpathians. The active zones are connected with the main fault systems, and the horizontal movements indicated the spreading tendencies between both geological structures. These results are in accord with the tendencies of horizontal movements in the Soviet Carpathians.

INTRODUCTION

Geodetic measuring results analysed in papers by Vyskočil and Kopecký (1974) and Vyskočil (1975) have recently been extended by additional numerical data (Vyskočil, 1976, 1977). These results may serve as a basis for improving our knowledge of the properties of recent movements in the Bohemian Massif area and its eastern contact with the Carpathian System. While analysing new results at this interim stage, attention is paid especially to the differentiation between active fault lines and single structural blocks. This process of gradual extension of present-day knowledge is aimed at comprehensively rechecking all the information acquired by the end of the 1970's. Such checking will result in a more precise delineation of the individual structures as well as fault lines, in particular their dynamics. This information will not only serve as a means of analysing geological and geophysical phenomena, but will also be an aid in considering still more efficient non-linear movement properties in time, and the like. One important aspect is — and undoubtedly will continue to be — the potential dependence of movement irregularity on the passage of seismic impulses across the area studied. Indications of such relationships were observed when analysing measuring results obtained from the Lišov geodynamic polygon, which lies at a key point of the Alpine earthquake effects on the Bohemian Massif structures. Therefore, this approach will also be retained for the analysis of the results obtained from other parts of the entire area under study.

This paper is largely based on an analysis of the vertical movement (displacement) gradients and their correlation with the corresponding structural elements. Emphasis is placed on the contact between the Bohemian Massif and the Carpathians, where the accumulating data on the horizontal movements of the two main geological units on Czechoslovak territory attain a higher quality than elsewhere. These results of a broader regional importance are believed, as work progresses, to add partly to the knowledge of the dynamic properties of the marginal parts of the Alpine—Carpathian System.

THE BOHEMIAN MASSIF

Dynamics of minor partial structures

The data that follow are based on the repeated measuring results obtained from the central and southern parts of the Bohemian Massif at the end of the 1960's, in addition to present-day levelling data gained from its northern part. New observations provide the possibility, inter alia, to compare movement tendencies as shown on the Map of Vertical Movements in Eastern Europe or as discussed in greater detail in the papers cited above. Furthermore, the results may serve as a means of differentiating more precisely between dynamics for the single structures and lines. In this paper the differentiation is, of course, given in a rather schematic way, but complete repeated measuring results may otherwise be used in more detailed analyses.

From the results previously obtained, here expressed in terms of vertical movement gradients (Fig. 1), there follows the dynamics of certain principal as well as partial structural lines recognized in the Bohemian Massif. The scheme discussed in this paper does not show the single gradients expressed numerically, but includes only gradients of principal tendencies exceeding the value of 1 mm/yr per 10 km. Arrows indicate the direction of subsidence. In addition to the boundary between the Bohemian Massif and the Carpathians, discussed in the next section, there are some other important fault systems produced by recent dynamics. These include systems known as the Blanice and Boskovice furrows, the marginal parts of the Bohemian Cretaceous Basin associated with the Labe (Elbe) lineament and the Krušné hory fault showing a high dynamic differentiation. Second-order lines are represented by the active Mariánské Lázně fault, the fault system in the

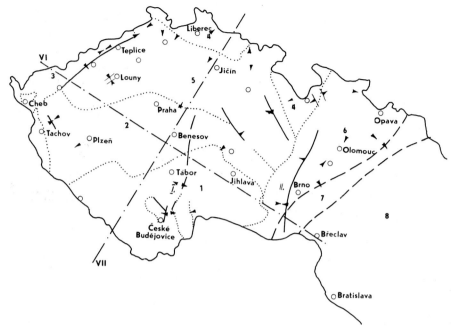

Fig. 1. Gradients of vertical movements in the Bohemian Massif. *1* = Moldanubicum structures; *2* = Assynth structures; *3* = Krušné hory Mtz. structures; *4* = Krkonoše Mts. and Orlické hory Mts. structures; *5* = Cretaceous; *6* = Jeseníky Mts. and Moravicum structures; *7* = Carpathian Foredeep; *8* = Outer Carpathians; *I* = Blanice furrow; *II* = Boskovice furrow.

Louny area and a tectonic subline in the Jeseníky area.

The directions of the horizontal gradients of the vertical movements in the Blanice graben clearly indicate its uplifting nature which has so far been established between the towns of České Budějovice and Tábor. The southern part of this fault system is being studied in detail by geodetic measurements carried out at the Lišov geodynamic polygon. These detailed results have revealed that the dynamics of the Blanice graben fault system is much more complex than previously supposed, and is confined to the single subblocks. The movement-dependent relationship between these smaller blocks is probably determined not only by internal principal forces but also by the effects produced by the passage of seismic impulses. The seismic impulses recorded in early May 1976 are particularly important for a better understanding of the properties of this phenomenon. Its effects, recognized in the results of geodetic observations, are being analysed.

The dynamics of another N—S-trending fault system known as the Boskovice graben is entirely different from that mentioned above. This is a fairly conspicuous graben whose continuation may be traced further to the northwest in the structures of the Bohemian Cretaceous Basin as far as the northwest border of Czechoslovakia. From its rather diagrammatic picture it is apparent that there is a present-day dynamic connection between the two faults. There is no doubt that both fault lines may be internally dynamically subdifferentiated and additionally refined by subsequent analyses. Such internal differentiation may be exemplified by the cross-dynamics known from the northwestern part of the Bohemian Cretaceous Basin as a probable result of the neovolcanics present in the Teplice area.

The northeastern margin of the Bohemian Massif, which has been shown by recent results to attain a higher degree of differentiation, requires separate discussion. What is particularly prominent is the dynamic accentuation of the Bohemian Cretaceous Basin in the northeast, and slightly differentiated phenomena such as the graben at its eastern margin. New geodetic measuring results have also confirmed a systematic subsidence of the Krkonoše Mts., as shown on the previously published Map of Vertical Movements in the Bohemian Massif. This notion has been found in conflict with geological opinion on the Quaternary development of the part of the Bohemian Massif studied, but it may be regarded as proved, based on results so far available. It may be reasonably inferred that systematic movements took place over a long period, as is indicated by geodetic results obtained during the past 50 years.

As yet, the dynamic picture showing the northwestern margin of the Bohemian Massif with the Krušné hory piedmont fault does not seem to be unambiguous. The large and striking differentiation of the movements occurring in the Krušné hory mountain range at the state frontier may also be seen on the Map of Vertical Movements published earlier (Vyskočil and Kopecký, 1974). These and most recent results lead to the conclusion that the mountains and fault line have been subject to recently rejuvenated tectonics, both longitudinal and cross. Because the entire area is characterized by the presence of both neovolcanic structures and fairly intense man-made movements, particular attention is paid to certain selected parts. Geodetic measurements have been used to study both the vertical and horizontal component of the movements. Associated with the area examined is the prominent partial graben running parallel to the Krušné hory piedmont fault in the Louny area. The Mariánské Lázně fault area assumes the shape of transverse graben, due apparently to recent dynamics of the Doupovské vrchy volcanic mountains. The movement activity of the two latter localities has as yet been inferred only from measurements made previously; its duration and further development will be investigated using the results of measurements now being carried out in the study area.

New results of the repeated levelling measurements also make it possible to examine more closely the dynamics of the eastern projection of the Bohemian Massif in the Jeseníky Mts. Again, most recent data have confirmed that the structures continuously tend to rise in this area. In contrast to previous, rather schematic data, it is now possible also to distinguish between the inner active fault lines confined to the core of the mountain massif alone. These structures are characterized by an uplift, not only toward the inner parts of the Bohemian Massif but also toward the marginal Carpathian structures. Such tendencies will be discussed in the next section. Compared with the Krkonoše Mts. area, recent movements tending to occur in the Jeseníky Mts. correspond to those known in Quaternary time.

Present-day dynamics of that part of the Bohemian Massif under study have been confirmed by the partial analytical results obtained from repeated geodetic measurements. This dynamic effect is of a block nature and, if principal movements occurring in Quaternary time are taken into account, its movements are marked by a distinctly variable sense. One of the problems in international collaboration to be resolved in the near future is to determine the dynamic effects of the structures adjacent to the Bohemian Massif on its movement. As far as Carpathian structures are concerned, such problems can be solved, at least in part, on a national scale.

Contact between the Bohemian Massif and the Carpathians: its dynamics

Compared with studies made on the movements of the Bohemian Massif, research into the dynamic properties of the contact between the Bohemian Massif and the Carpathians is of a broader, regional nature. Among others, this may be illustrated by research activities using various geophysical methods to study the structural properties of this area. All these investigations are discussed in a synthesizing paper by Zátopek and Beránek (1975) dealing with these problems in the boundary area related to the Alpine—Carpathian structures of central Europe. This area is also receiving increased attention from the aspect of geodetic measurements, as exemplified in Vyskočil (1977), recently supplemented by additional data. By analogy to the preceding section, the results will also be discussed in terms of vertical movement gradients, supplemented by the data on principal movements tending to occur in the horizontal direction.

The boundary area in general is shown in Fig. 2, the northwestern part of which overlaps Fig. 1. It is based on the schematic map of geological structures with the classification following the pattern used by Zátopek and Beránek (1975). With reference to the preceding section, it can be stated that the marginal structures of the Bohemian Massif related to the Carpathian Foredeep structures are regarded as structures tending to rise. The marginal structures of the Carpathian flysch zone have also been found to rise virtually along the entire length of the boundary between both the principal geological units. In comparing the uplift of the marginal parts of the Bohemian Massif and the Carpathians, it can be seen that the Carpathian structures are noticeably uplifted in the northeastern part of the boundary, whereas the Bohemian Massif structures tend to rise, especially in the northeast. If geographic units are used, it is apparent that in the northeast the Jeseníky Mts. and Moravskoslezské Beskydy Mts. have been uplifted in the Bohemian Massif and the Carpathians, respectively, and in the southwest the Drahanská vrchovina Upland and Ždánické vrchy Mts. respectively, in the

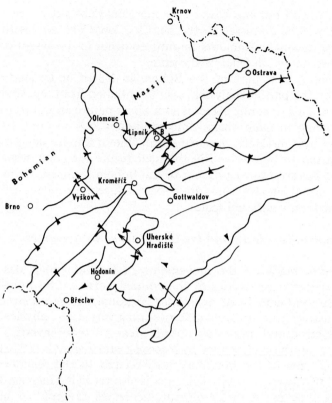

Fig. 2. Gradients of vertical movements and main tendencies of horizontal movements on the boundary between Bohemian Massif and the Carpathians.

same units. In both cases apparently elevated structures of the Outer flysch zone lie within the Carpathians. Regarding vertical movement, the Outer Carpathian flysch zone is more differentiated and is at present active on partial inner fault lines. It is beyond the scope of this paper to discuss such movements in greater detail: consideration is given only to principal tendencies.

The basinal areas of the investigated boundary may be regarded as those showing relative subsidence, in accordance with Fig. 2 and the preceding text. This relates to the entire Carpathian Foredeep as well as to the northern margins of the Vienna basin. Maximum subsidence has been recorded in the Břeclav area. In this connection it seems necessary to emphasize the relativity of the subsidences compared to that of the surrounding structures. At present, the question may arise as to the extent of control of basinal bottom movements by the subsidence of sedimentary basin surface determined by geodetic methods. In discussing this question the possibility cannot be ruled out that the movement on the surface is caused by consolidation processes of the sedimentary beds of great thicknesses and may be opposite to that on the bottom of the particular basin. Attention is now focused on the solution of such problems in the northern projection of the Vienna Basin.

Data on the principal tendencies for vertical movements can also be extended by those on principal horizontal movements. It is interesting to note that significant horizontal movements (Fig. 3) principally tending to expand are associated with three fundamental structural lines in the area under study. One of these lines is the Carpathian Foredeep characterized by the expansion in the Lipník n. Bečvou and Vyškov areas, attaining a value of about 1.5 ± 0.4 cm/yr and 1.1 ± 0.4 cm/yr, respectively. The other significant line is the Lednice zone (Zátopek and Beránek, 1975) cut by horizontal measurements in the northern projection of the Vienna Basin. Here horizontal expansion attains a value of 1 cm ± 0.3 mm/yr. Since the horizontal movement between both lines on the structures of the Outer and Inner flysch zones is below the significant limit, it may be inferred that both structures move as a whole. The third significant line of the Carpathian marginal structures is the Peri—Pieninian lineament (Zátopek and Beránek, 1975), with an evident expansion of about 1.1 cm ± 0.3 mm/yr. Horizontal expansion in the outer flysch and Lednice zones is associated with elevated structures bordering relatively subsiding basinal sectors. The direction of horizontal movement in the Lipník n. Bečvou and Uherské Hradiště areas is marked by an azimuth of about 130°, if we proceed from the Bohemian Massif toward the Carpathians. In the Vyškov and Peri—Pieninian lineament areas the horizontal movement has been determined only from longitudinal measurements. It can only be assumed that the principal direction of the movement is approximately the same as in the cases mentioned earlier.

The values of horizontal movement given above characterize a relative, mutual movement of the structures of the Bohemian Massif and the Carpathians, as well as their parts. Since the continuous longitudinal measurements connect the whole sector between the marginal parts of the Peri—Pieninian lineament and the Bohemian Massif, the horizontal movement can be specified still more closely. The horizontal movement lies below the limit of significance in those sectors cutting flysch zone structures, not only between

Fig. 3. Main tendencies of horizontal movements in the Carpathians.

the Carpathian Foredeep and the Lednice zone but also between the latter and the Peri—Pieninian lineament. It may therefore be reasonably assumed that a fairly intensive horizontal movement is characteristic of only the three significant lines of the Carpathian marginal structures. To confirm this assumption preparations are being made for measurements to be carried out, especially in the northeastern continuation of the Lednice zone and the Peri —Pieninian lineament.

In general, the data on the horizontal movements conflict with the information on the development of the dynamic relationship between the Bohemian Massif and the Carpathians in geological history. In this connection mention must be made of the results obtained from Recent horizontal movement measurements carried out in the Carpathian area on Soviet territory (Sobakin et al., 1975). Here analogous structures tend to move in a uniform manner. As in the contact area between the Carpathians and the Bohemian Massif, the directions of these movements are approximately normal to the basic structural elements of the Carpathian arc. This adds considerably to the reliability of the results thus far obtained and indicates a general movement in the horizontal direction, apparently along the entire arc of the Carpathians. However, the question of the mechanism of Recent movements is left completely open, and can only be resolved by interdepartmental as well as international collaboration. Taking into account the similar features recognized in the Carpathian and Alpine systems, and with regard to basic synthesizing data (Zátopek and Beránek, 1975), it is hoped that such a work will contribute much to a better understanding of the dynamics over much of the European territory. A comparison of the results obtained by the present authors with those gained from the Soviet part of the Carpathian arc suggests that it is sufficient to locate geodetic measurements only at key points of other extensive areas. Essentially, an effort should be made to create a separate system of local geodetic networks permitting determination of both horizontal and vertical relative movement with high accuracy. Principal tendencies of the movements as determined in the single networks are defined and subsequently analysed, and apparently may characterize more general tendencies for the movement of the entire structure as a whole.

CONCLUSION

Some of the recently active structures and fault systems have been distinguished in the present partial analysis of new repeated geodetic measuring results obtained from the Bohemian Massif and its contact with the Carpathians. The new results essentially confirm the principal tendencies previously recognized for the movements and, moreover, permit their differentiation to be made in greater detail. In this way it proved possible to establish the continuous relative subsidence in the Krkonoše Mts. area and the uplift in the Jeseníky Mts. area. Furthermore, the fault system of the Boskovice graben and Labe lineament seems to be a highly continuous fault zone. A more detailed analysis would reveal a still higher differentiation, as is clearly illustrated, e.g., by the northwest flank of the Cretaceous basin in the area containing neovolcanic structures. Recent activity of the Krušné hory piedmont fault, Blanice graben and other fault subzones will be studied in a later paper.

Subsiding and uplifting areas strictly confined to principal and subordinate structural units may be differentiated in the contact area between the Bohemian Massif and the Carpathian System. The Carpathian Foredeep and the northern projection of the Vienna Basin show relative subsidence. Relative uplifts are characteristic of the entire zone of the Carpathian Outer flysch and the opposite structures of the Bohemian Massif. The results so far available indicate that a relative uplift of the Carpathians over the Bohemian Massif prevails in the northeastern part of the boundary area under study. The reverse is true of the southwestern part of the contact, the Bohemian Massif rising relative to the Carpathians. The Carpathian flysch zone alone may be internally subdivided dynamically, with indications of active occasionally local fault zones. Of principal importance is the information on the Recent movement activity recognized on the major lines of the Carpathian Foredeep, the Lednice zone and the Peri—Pieninian lineament in both horizontal and vertical directions. There is a tendency for the structures bordering these lines to expand horizontally, and a relative uplift has been found in the vertical direction.

The results correlate well with separate determinations of movements in the Carpathian area on Soviet territory. On the basis of this comparison it may be tentatively inferred that the Carpathians have recently been detached from the platform structures along the whole Carpathian arc. The results point to the possibility, among others, of determining principal tendencies for the movement of extensive geological structures by comparing and analysing partial movements recognized in measurement-independent geodetic networks. For a complete understanding of the mechanism of movements so far established, interdiscliplinary and international collaboration are required. From the aspect of the Bohemian Massif inner dynamics, importance should be attributed to the extension of such studies so as also to cover some of the Alpine areas.

REFERENCES

Sobakin, G.T., Somov, V.I. and Kuznecova, V.G., 1975. Sovremennaya dinamika i struktura zemnoy kory Karpat i prilegayuschich territoii. Naukova Dumka, Kiev.
Vyskočil, P. and Kopecký, A., 1974. Neotectonics and Recent Crustal Movements of the Earth's Crust in the Bohemian Massif. Monograph of VÚGTK, Prague.
Vyskočil, P., 1975. Recent crustal movements in the Bohemian Massif. Tectonophysics, 29: 349—358.
Vyskočil, P., 1976. To the dynamics of the boundary area between the Bohemian Massif and Carpathians. Res. Works VÚGTK, 11: 29—45.
Vyskočil, P., 1977. Současné poznatky o dynamice styku Karpat a Českého masivu. (Present state of knowledge about dynamics of the joint between the Carpathians and the Bohemian Massif) (in Czech), Geod. Kartogr. Obz., 23 (3): 55—61.
Zátopek, A. and Beránek, B., 1975. Geophysical synthesis and crustal structure in central Europe. Stud. Geophys. Geod., 19 (2): 121—133.

Tectonophysics, 52 (1979) 203—209
© Elsevier Scientific Publishing Company, Amsterdam — Printed in The Netherlands

203

SURFACE DEFORMATION OVER THE GARM GEODYNAMIC POLYGON

A.K. PEVNEV, N.N. ODINEV, T.V. GUSEVA, YA. DAVIDENKO and
V.A. BELOKOPYTOV

Institute of Physics of the Earth, Academy of Science, Moscow (U.S.S.R.)

(Accepted for publication April 4, 1978)

ABSTRACT

Pevnev, A.K., Odinev, N.N., Guseva, T.V., Davidenko, Ya. and Belokopytov, V.A., 1979.
Surface deformation over the Garm geodynamic polygon. In: C.A. Whitten, R. Green
and B.K. Meade (Editors), Recent Crustal Movements, 1977. Tectonophysics, 52:
203—209.

The Garm polygon is confined to the junction between Tien-Shan and the Peter the
Great fault zone separating the Pamirs and Tien-Shan. It is just in the zone of contact
between the fault zone itself and Tien-Shan (the spurs of the Gissar ridge) that a submeridional overthrust of the fault zone on the Gissar block is revealed along the surface
inclined to the horizontal by an angle of 40°, the velocity of the overthrust being approximately 20—25 mm/yr. Recently the area of the polygon has been increased. Investigations over the expanded polygon permit the preliminary conclusion that alongside the
overthrust mentioned above there is a block in the fault zone approaching southwesterly
along the contact with the Gissar block. Attempts are made to enlarge the deformational
network so as to cover the fault zone completely.

INTRODUCTION

Principal results of surface-movement investigations over the Garm geodynamic polygon for the period up to 1973 were reported at the International Symposium on Recent Crustal Movements in Zurich, 1974. Therefore,
the present report is mainly devoted to research carried out during the last
three years. The territory under investigation at the Garm polygon was
enlarged in 1974. The main reason for this was to set up a deformational network completely covering the seismic zone situated between the Gissarkokshaal and Darvaskarakul deep faults (Fig. 1). The ends of the network were
extended into the more rigid blocks of the Gissar (Tien-Shan) and Darvas
(the Pamirs) ridges. According to the age of its rocks, the seismogenic zone is
considerably younger than its framework: Meso—Cenozoic for the zone itself
and Paleozoic and Proterozoic for the Gissar and Darvas ridges. The studies
in such a deformational network are of interest, both from geodynamic posi-

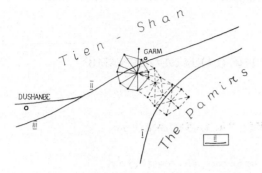

Fig. 1. Scheme of the Garm geodynamic polygon faults. *I* = the Darvas fault; *II* = the Gissar fault; *III* = the Hyak fault.

tions (experimental control of possible rapprochement between the Pamirs and Tien-Shan) and from the viewpoint of studying the mechanics of seismic processes (the search for forerunners of great earthquakes). In Fig. 1 the existing portion of the network is shown by solid lines, and that planned for the near future by dashed lines.

The scheme of the faults of the Afghano-Tadjik depression (seismogenic zone) originated or rejuvenated in Late Pleistocene or Holocene, compiled by V.G. Trifonov, is shown in Fig. 2. The position of the existing deformational network is indicated by a circle, and the net is set up in a tectonically complicated section of the crust. Along with submeridional overthrusts, there are submeridional and sublatitudinal displacements and faults, emphasizing the necessity for increasing the area of investigation at the Garm polygon.

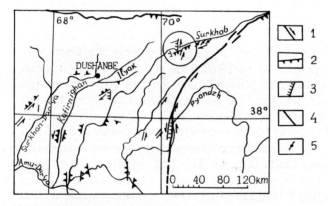

Fig. 2. The faults of the Afghano-Tadjik depression formed or rejuvenated during the Late Pleistocene or Holocene (according to V.G. Trifonov). *1* = strike-slip faults; *2* = thrust faults; *3* = normal faults; *4* = faults with unknown displacement; *5* = volcanic chains.

Only then can one have the opportunity to measure the whole range of existing displacements and obtain an authentic picture of spatial regularities of surface deformations.

PREVIOUS INVESTIGATIONS

Before analyzing measurement results in a large network (at the third level of 3—4-km elevation) let us briefly discuss the results of previous investigations. The first level, 1 km above sea level, of the Garm polygon consists of a very limited geodetic network, confined to the foot of the Gissar and Peter the Great ridges. Both vertical and horizontal movement components are studied here. The total length of the leveling route does not exceed 30 km, while geodimeter lines are approximately 1 km long. The investigations further confirmed the idea of the monolithic nature of the block presented by the foot of the Gissar ridge (Mt. Mindalul, Figs. 3 and 4) and by a considerably fractured character of the Peter the Great ridge foot (the fault zone). Fig. 4 shows displacements of the bench marks installed in the spurs of the Gissar ridge. Bench mark 3040 is used as the reference, which belongs to the same block. The block as a whole undergoes sign-changing tilts, proved by the increase in the amplitude of synchronous displacements for all the bench marks distant from the reference point. The division of the polygon territory

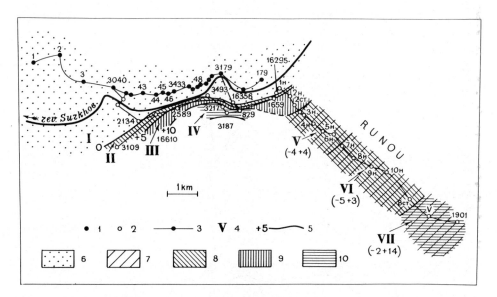

Fig. 3. Scheme of surface vertical displacements over the Garm polygon. *1* = rock bench mark; *2* = ground bench mark; *3* = leveling line; *4* = block number; *5* = isolines of equal velocities; *6* = 0 mm/yr; *7* = from 0 to 5 mm/yr; *8* = from 0 to +5 mm/yr; *9* = from +5 to +10 mm/yr; *10* = from +10 to +15 mm/yr.

206

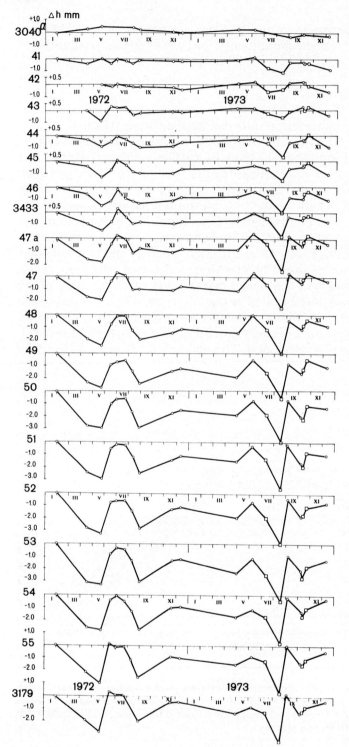

Fig. 4. Displacements of the bench marks located in the spurs of the Gissar ridge (Mt. Mindalul). For location of the bench marks see Fig. 3.

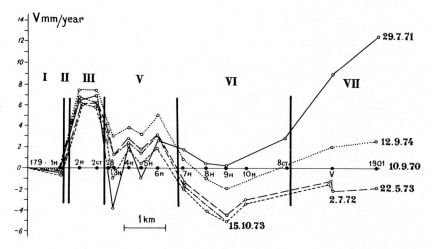

Fig. 5. Velocities of bench-mark displacements over the Garm geodynamic polygon (the Runou profile).

on blocks according to the data of repeated levelings is shown in Fig. 3. Here an interesting peculiarity should be noted — a decrease of average annual rate stability for displacements distant from the line of tectonic contact between the Gissar block and the Peter the Great fault zone (Fig. 5). Blocks *II*, *III*, and *IV*, pressed to the monolithic one, have some constant average annual vertical velocities. This phenomenon is supposed to signify an increase of rock strength characteristics in the fault zone when the latter is pressed against a stronger block. We define this phenomenon as an effect of tectonic contact of rocks of various strengths. In the future we propose to investigate this effect in more detail (experimentally and theoretically).

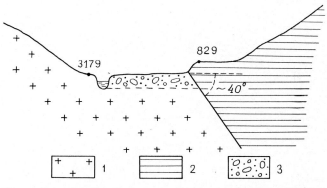

Fig. 6. Scheme of tectonic contact between the Gissar block and the Peter the Great fault zone. *1* = Proterozoic rocks of the Gissar block. *2* = Meso-Cenozoic rocks of the Peter the Great fault zone. *3* = Boulder—pebble—bed alluvium.

208

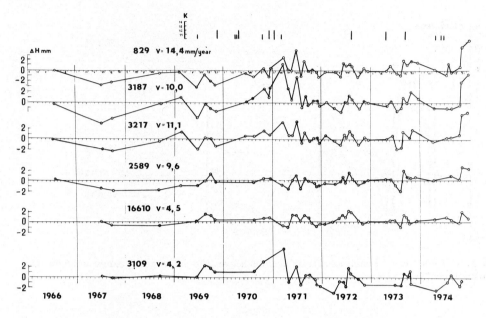

Fig. 7. Bench-mark displacements over the Garm polygon (the main ring).

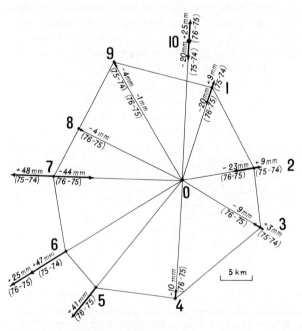

Fig. 8. Scheme of point displacements over the Garm geodynamic polygon.

SEARCH FOR PRECURSORS

Combined analysis of vertical and horizontal shift has shown that within the polygon in question submeridional overthrusting of the Peter the Great zone on the Gissar block took place along the surface, inclined to the horizontal by an angle of approximately 40°, with a velocity of about 20—25 mm/yr (Fig. 6). At the polygon an attempt was made to search for precursors of earthquakes shown by the frequent relevelings between the bench marks placed on opposite sides of the tectonic contact zone. The results of the experiment are shown in Fig. 7. In the upper portion of the figure the moments of earthquakes are shown within a radius of 30 km from the polygon. No connection between bench-mark displacement and seismic events has been established. One possible reason for the negative result seems to be the limited dimensions of the polygon. Hence the latter cannot filter displacements stipulated by various causes.

HORIZONTAL DEFORMATION

Let us consider Figs. 1 and 8, which show the network of the third level of elevation. Points 7, 8, 9, 10 and 1 represent the Gissar ridge; points 0, 2 and 3 the Peter the Great ridge. The net points are located at the peak of the ridges or their spurs. The program specifies the measurement of all the net lines, but so far, due to some difficulties in organization, we have managed to make line measurements only from the central point 0. In 1974 lines 0—4, 0—5, and 0—8 were not measured, but in 1975 and 1976 all the radial lines were measured. So far, it is impossible to estimate the actual measuring accuracy of lines. A theoretical estimation is of little interest. Taking into consideration all pros and cons we may suppose that the measurement error of a line is about ±15 mm or $1 \cdot 10^{-6}$.

The greatest shift was observed at points 5, 6 and 7. Points 5 and 6 are located within the axial zone of the Vakhsh ridge. The large one-sided displacements of point 6 suggest that the axial portion of the Vakhsh ridge is a block being displaced (within the system of the existing network) along the Gissarkokshaal fault. It is quite likely that the boundary between the Vakhsh and Peter the Great blocks is connected with the Obihingou River Valley which in its upper stream follows the main structures, then takes a perpendicular swing and crosses the ridge. This suggests the necessity of verifying the hypothesis of tectonic origin of the Obihingou River Valley. The large sign-changing shift of point 7 may be connected with elastic deformations of the Gissar block, since point 7 is located near the possible intersection of two faults: the Gissarkokshaal fault and the suggested Obihingon fault. Owing to the limited time of the experiment, it is impossible to draw final conclusions. Certainly all of the above considerations provide a source for discussion and the development of plans for future studies.

Tectonophysics, 52 (1979) 211—222
© Elsevier Scientific Publishing Company, Amsterdam — Printed in The Netherlands

HOLOCENE DEFORMATION AND CRUSTAL MOVEMENTS IN SOME TYPE AREAS OF INDIA

L.N. KAILASAM

Geophysics Division, Geological Survey of India, Indian National Committee for Geodynamics, 15 Park Street, Calcutta 700016 (India)

(Accepted for publication April 4, 1978)

ABSTRACT

Kailasam, L.N., 1979. Holocene deformation and crustal movements in some type areas of India. In: C.A. Whitten, R. Green and B.K. Meade (Editors), Recent Crustal Movements, 1977. Tectonophysics, 52: 211—222.

Recent crustal movements have been observed and studied in several parts of India including the Himalayan and sub-Himalayan regions, the Precambrian shield of peninsular India and also the coastal tracts. The results of studies of Holocene deformation and crustal movements in two type areas are presented, one in the extreme southeastern part of the peninsula and the other in northeastern India.

The Precambrian shield in the extreme southeastern part is characterised by a major NE—SW trending fault zone in the Tirupattur—Mattur areas of Tamil Nadu with some major extended faults, one of which apparently cuts through the entire crust and Moho as indicated by gravity data and which is associated with occurrences of alkaline and basic intrusions and carbonatite complex. Evidence of Recent crustal movements in this zone is afforded by geomorphic features and recent and current seismicity of a mild nature which is apparently to be attributed to slow movements along the fault plane.

The Shillong plateau in northeastern India occurs as block-uplifted horst, comprising for the most part Archaean crystalline rocks with plateau basalts and Cretaceous and Tertiary sediments occurring on its southern margin. The plateau is bounded by major faults and is located in a zone of high seismicity lying astride and parallel to the eastern Himalayas intervened by the alluvium of the Brahmaputra Valley. Geomorphic features such as raised terraces, straight-edged scarps, etc., provide evidence for Recent crustal movements with dominant vertical movements along the fault planes which have continued through Tertiary and Recent times. Repeated precision levelling measurements conducted by the Survey of India indicate a rate of uplift of 4—5 cm per 100 years during the period 1910—75.

The gravity data pertaining to this region are also discussed in relation to the crustal movements.

INTRODUCTION

Studies on Holocene crustal deformation and Recent vertical movements in various parts of India have been made during the past several years, and the pace of these studies, especially on a quantitative basis, has gained further momentum during the current International Geodynamics Project. Six areas have been selected for such quantitative studies involving repeated precision levelling and geodetic measurements, gravimetric studies and Deep Seismic Sounding profiles, and work is in progress in these areas (Kailasam, 1975).

The Himalayan orogenic movements are believed to have ended in the Early Pleistocene, but the post-orogene movements in this region are still continuing and earth movements, mainly epeirogenic in nature, have been continuing during the subsequent period through the Recent in the Himalayan region and also in the northern plains and the peninsular shield, as evidenced by sea-level changes, surface geomorphic features and current seismicity.

OBSERVATIONS OF HOLOCENE DEFORMATION AND CRUSTAL MOVEMENTS IN INDIA

In several parts of the vast Himalayan region, where observational data are still rather meagre, evidence of Recent crustal movements has been noticed. For instance, typical cirque-like structures and steep cliff faces are seen in the Poonch and Pirpanjal areas of the western Himalayas. In the latter area in-situ moraine masses have been observed at levels of about 1980 m, while re-sorted marine debris occupy the high reaches of the valley further below to a level of 1220 m (Biswas et al., 1971). In the Kathiawar region of western India which was submerged in Early Eocene, the Tertiary and the Pleistocene sequences deposited therein were raised by more than 300 m in Pleistocene and Sub-Recent times. In the adjoining Kutch area in the north where a prominent E—W fault zone, 5—8 km in width occurs, the northern side of the fault was raised by 3—8 m during the earthquake of 1819. The Saurashtra, Kutch and Cambay regions which were submerged during the Jurassic were subsequently uplifted to the present level during the Quaternary with raised beaches, sand deposits and marine shell fragments indicating coastal uplift (Krishnan, 1966, 1968). Similarly, the west coast further to the south was also affected by earth movements during the Quaternary, as indicated by the high gradient of rivers flowing into the Arabian Sea and the slow rate of deposition of sediments. In the Koyna area of western Maharashtra which witnessed a devastating earthquake of magnitude 6.5 and intensity VIII in December 1967, detailed geological, geophysical, geodetic and seismological studies of the crustal movements have been in progress for the past 10 years. Repeated precision levelling and triangulation conducted in this zone by the Geodetic and Research Branch of the Survey of India indicate a subsidence of 3 cm in the Koyna region subsequent to the earthquake, and a southeastward horizontal movement of 15 cm in the isoseismal region of highest intensity (Chugh, 1973; Kailasam, 1975). Over the Kerala coast in southwestern India there is large-scale manifestation of Holocene deformation and crustal movements as indicated by geomorphic and tectonic features. For instance, the post-Miocene Warkala and Quilon formations occur 30—60 m above the present sea level (Rao, 1976). Evidence of repeated changes in sea level is afforded by Recent and Sub-Recent raised beaches and sand bars along the coast, raised old river terraces and pebble beds in the Palghat gap and Nilamur Valley and submerged laterite in the coastal plains of Trichur. On a regional basis five geomorphological features have been observed in Kerala on the basis of field studies of laterite remnants, viz.: (1) 60—120 m, (2) 150—210 m, (3) 360—420 m, (4) 600—900 m and (5) >1600 m (Murty et al., 1976).

Raised beaches and dead coral reefs are seen at an elevation of 3—5 m above the present high-tide mark, extending up to 2 km inland from the present shore on the east coast between Cape Comorin and Tirunelveli; they are believed to be due to eustatic changes in sea level or tectonic movements during the Pleistocene and Recent times. Similar raised features — terraces, sand bars, etc. — are also noticed further to the north on the Madras coast.

In northeastern India, which is a zone of high seismicity, the observed changes in the course of the Brahmaputra River are attributed to epeirogenic movements during the Quaternary and Recent periods. The Sylhet basin, to the south of the Shillong plateau (see Fig. 5) has subsided to a depth of 10—15 m during the past 100 years. The presence of decayed wood at various depths in the northern parts of the Sylhet basin and the presence of subsiding forest at a depth of 4 m near Port Canning afford further evidence of epeirogenic movements (Biswas et al., 1971). Alluvial terraces are also observed in the Bankura and Puralia districts of West Bengal. The Son Valley of Bihar in the interior of the peninsular shield was also subjected to epeirogenic movements during the Quaternary and Recent periods. For example, two erosional surfaces have been observed in the Cuddapah basin, one at an altitude of 610—915 m and the other at 490—610 m, suggesting crustal uplift during post-Miocene to Recent times (Vadiyanathan, 1964).

The results of studies carried out recently in two of the type areas mentioned above, viz. (a) the Koratti basin and neighbouring areas of Tamil Nadu in the southeast of the peninsula, and (b) the Shillong plateau in northeastern India, are briefly described below.

Koratti basin

The Koratti basin and its neighbouring areas, which are of considerable interest from the viewpoint of Recent vertical movements of epeirogenic type, are located in a region of Lower Precambrian rocks comprising granites, gneisses, schists and charnockites intruded by basic dykes (see Fig. 4) on the western flanks of the Eastern Ghats to the southwest of Madras (see Figs. 1, 2 and 4). This region has been studied intensively by geologists from both structural and geomorphological aspects. These studies indicated the evidence for the changes in river courses here, and also other geomorphological evidence in the form of terraces, raised gravel beds and waterfalls such as Hoganaikal to the west of the Dharmapuri—Koratti—Tirupattur basin. This region has acquired special significance in view of the persistent seismicity, albeit mild with moderate magnitudes of less than 5, particularly along the axis of the deep main fault known as the Eastern Ghat fault which is characterised by the occurrence of syenite, and ultrabasic intrusives such as pyroxenite, dunite and carbonatite.

The main epicentres of the earthquakes which have occurred in the southern Indian peninsula during recent years are indicated in Fig. 1 (after Grady, 1971). The Koratti area and neighbourhood fall in a seismic zone of intensity II and the magnitude of the earthquakes is generally <5 on the Richter scale. In Fig. 1 are also indicated the main faults as postulated by Grady who has described the geological features of this area in detail. Of particular interest among these faults over the crystalline rocks of the peninsular shield

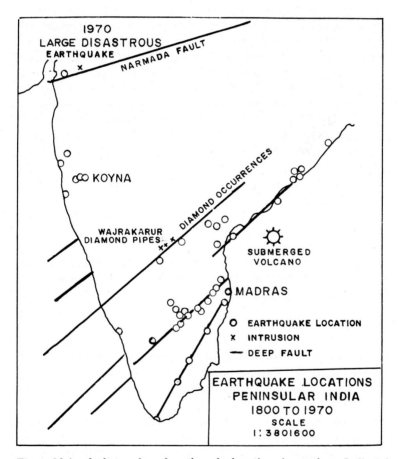

Fig. 1. Major fault trends and earthquake locations in southern India (after Grady, 1971).

is the NE—SW deep main fault traversing the large Elagiri syenite massif, and the carbonatite complex in the Koratti and Samalpatti basins to the south-west along this fault line which, according to Grady, can be traced over a strike length of more than 200 km. It is noted that there are a number of earthquake epicentre locations along this fault which adds significance to this area in the context of the geological and geomorphological features observed. The Elagiri syenite massif rises to 750 m above the plains to an elevation of roughly 1200 m above mean sea level. On the basis of evidence of diorite dykes in contact with the syenite, Grady believes that these hills were formed by replacement of previously folded gneisses. Outcrops of pyroxenite occur scattered on the outer rim of this syenite intrusive to the south, west and east of the Elagiri Hills.

Some 9 km to the southwest occurs the prominent Koratti carbonatite basin-shaped complex with arcuate of syenite in the centre with pyroxenite occurring as scattered outcrops on the outer rim to the west. A carbonatite lens 200 m wide and 2.5 km long occurs as an outcrop on the northwestern side of the basin, distinguished by strong shear zones containing breccia and mylonite on the western and eastern sides of the Koratti basin.

Further to the southwest another structural basin occurs in the Samalpatti area associated with the carbonatite complex and with syenite outcrops in the inner part of the basin in contact with an outer ring of pyroxenite. Carbonatite occurs at the contact between the syenite and pyroxenite at two places, also occurring as dykes outside the basin. Dunite bodies and quartz—barytes veins in pegmatites are also found in this basin.

Further to the northeast along this fault plans there are quartz—barytes—galena vein occurrences in the Alangayam area. To the east of the Javadi Hills, other roughly NE—SW trending faults occur in an echelon pattern in the Eastern Ghats area. The eastern margin of the peninsular shield is characterised by major faults (Fig. 2) with large downthrow to the east (Kailasam, 1968), denoting the western margin of the sedimentary basins of the Madras coast filled with a thick sequence of Cretaceous and Tertiary sediments. It may also be stated that the tectonic map of Eurasia (Semenov, 1968) shows the submerged volcano in the Bay of Bengal to the northeast of Madras roughly along this NW—SE trending main fault.

Detailed gravity-cum-magnetic observations were recently conducted by the Geophysics Division of the Geological Survey of India over six traverses laid roughly in a WNW—ESE direction (profiles I—VI, Fig. 3A) and four N—S traverses (profiles VII—X, Fig. 3B). The locations of these traverses are shown in the sketch map of the Koratti and neighbouring areas (Fig. 4) where the geological features are also indicated. The gravity and magnetic profiles (Fig. 3B) have clearly brought out the deep main fault (known as the Eastern Ghat fault) trending roughly NE—SW across the Koratti—Tirupat-tur—Samalpatti basins, and have enabled a clear delineation of this subsurface fault which is somewhat shifted from that geologically deduced. The Bouguer gravity profiles (Fig. 3A) show an abrupt fall of 15—20 mGal across

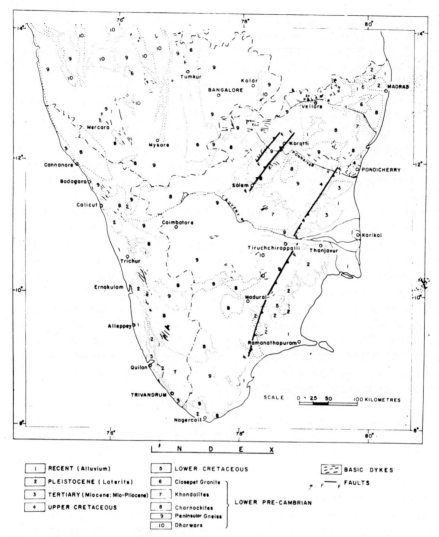

Fig. 2. Sketch map of southern peninsular India showing geological features and the major faults delineated geophysically.

the fault along the traverses with an estimated downthrow varying from 1500 to 2000 m for an assumed density contrast of 0.25 g/cm³ between the charnockites and granite gneisses. The denisty of the charnockite samples determined in the Geophysics Laboratory of the G.S.I. varies from 2.75 to 2.85 g/cm³, while that of the granite gneisses varies from 2.6 to 2.68 g/cm³.

This deep main fault appears to be a crustal fault extending into the man-

217

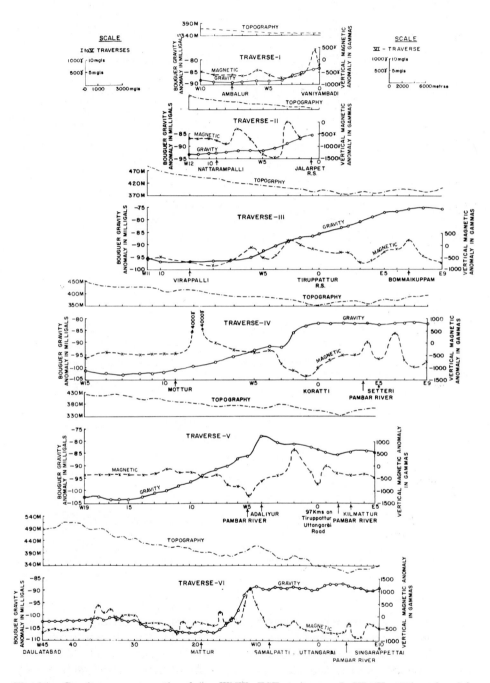

Fig. 3A. Gravity-cum-magnetic along WNW—ESE traverses I—VI, Koratti and neighbouring areas.

218

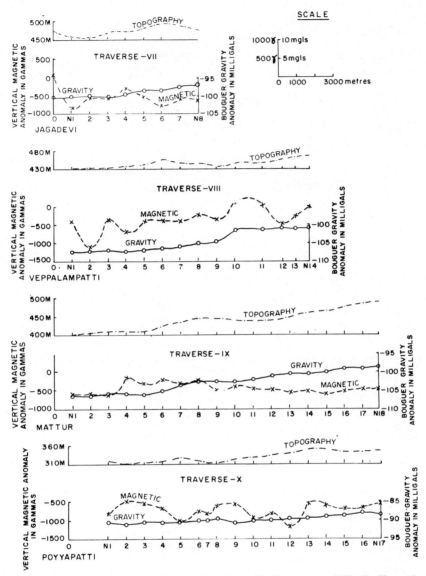

Fig. 3B. Gravity-cum-magnetic profiles along N—S traverses VII—X, Koratti and neighbouring areas.

tle — as can be reasonably inferred from the number of earthquake epicentre locations linearly disposed in this region and the large downthrow of the fault. This aspect can be verified by an E—W Deep Seismic Sounding profile in the region, and if the gravity indications of this crustal fault are confirmed by the DSS results it is not improbable that a major earthquake may occur in this region, even if it is in the distant future. In addition to this deep major fault, another major fault has also been indicated further to the west by the gravity and magnetic profiles roughly trending ENE—WSW with a substantially smaller downthrow to the west (Fig. 4). A N—S shear plane is also indicated along the Pambar River to its west. The gravity and magnetic profiles along the N—S traverses (Fig. 3B) do not indicate any E—W faults. To the east of the Javadi Hills two faults occur in an enechelon pattern as deduced geologically and also indicated in aerial photographs and airborne magnetic data. In this region the Eastern Ghats would thus appear to occur as an uplifted horst in a zone of mild seismicity. Precision levelling and

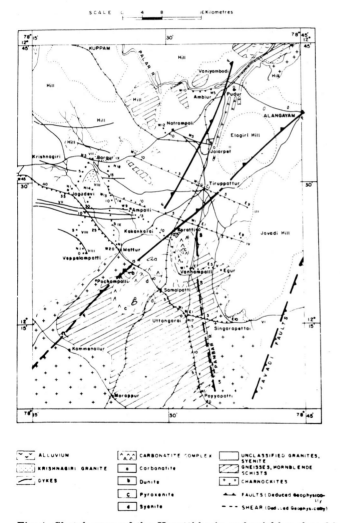

Fig. 4. Sketch map of the Koratti basin and neighbourhood indicating geological features and major faults deduced geophysically.

geodetic triangulation measurments have recently been undertaken by the Geodetic and Research Branch of the Survey of India along geotraverse VI (see Fig. 4) and are to be repeated at regular intervals in the coming years to determine the rate of vertical movement in this region.

Shillong plateau

The Shillong plateau comprising the Garro, Khasi and Jaintia Hills occurs in northeastern India in a zone of high seismicity with frequently occurring earthquakes of magnitude >6. This area witnessed plateau basalt volcanism in the Cretaceous period and the Sylhet traps were erupted along fissures. The plateau, which has a maximum elevation of roughly 1800 m above mean sea level and comprises for the most part Archaean crystalline rocks, occurs as an E—W horst along the southern margin of the Brahmaputra Valley (Fig. 5). Since the Tertiary period, this region has also been subject to epeirogenic movements which have continued through the Recent to the present, as borne out by recent and current seismicity of high intensity. Well-defined

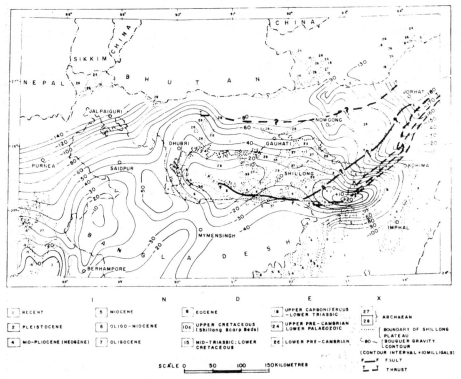

Fig. 5. Bouguer gravity map and geological features of the Shillong plateau and neighbouring areas.

geomorphic features indicative of continued recent crustal movements can be seen in the Brahmaputra alluvium to the north of the plateau in the Goalpara district of Assam where straight-edged scarps of older alluvium within the recent alluvium are seen, apparently as a result of the Recent vertical movements of the basement blocks. The plateau, which is believed to have been contiguous with the Rajamhal area in the west until the Mio—Pliocene, is characterised by a prominent E—W fault known as the Dauki fault over its southern margin (Fig. 5) and the plateau is believed to have been uplifted and dragged eastward over 250 km along this fault relative to the East Bengal plains (Krishnan, 1966). Cretaceous and Tertiary (Eocene and Miocene) sediments, including sandstones and limestones, occur in the southern margins of the plateau. The uplift of the basement blocks to the north of the Dauki fault is believed to have determined the disposition of the Eocene sediments with continued vertical movement of the Khasi block along the Dauki fault system which has apparently remained active, as indicated by continued sinking of the Sylhet basin to the south (Murthy, 1971). East—west faults have also been indicated to the north of the plateau in the Brahmaputra alluvium; thus the relief of the Shillong plateau is that of the horst due to continued uplift of the area along a series of fracture systems.

The Bouguer gravity map of the Shillong plateau and neighbouring areas, based on the latest gravity data along with the major geological features, is shown in Fig. 5. The Bouguer gravity map clearly reflects the horst-like nature plateau. The plateau has the highest positive gravity anomaly values varying from +42 to —50 mGal for elevations ranging from a few metres to roughly 2000 m, while in the Brahmaputra Valley mean negative Bouguer anomalies varying from —180 to —250 mGal are indicated for a mean elevation of 85 m. The nature of the Bouguer anomaly and its variation with elevation appear to be related to the underlying crust which may have a density higher than the normal according to Verma and Mukhopadhyay (1976).

Repeated precision levelling and geodetic measurements by the Survey of India are continuing along a N—S traverse across the plateau from Gauhati through Shillong (Fig. 5). The precision levelling measurements conducted during this period (1910—1975) by the Survey of India indicate a rate of uplift of 4—5 cm per 100 years for this plateau (Chugh, 1977).

ACKNOWLEDGEMENTS

The author expresses his thanks to Messrs S. Subrahmanyam and V.L. Raju, geophysicists, Geological Survey of India, who carried out the gravity and magnetic measurements in the field and the reduction of the field data.

REFERENCES

Biswas, A.B., Das Sarma, D.C. and Roy, A.K., 1971. Quaternary deposits of India. Rec. Geol. Surv. India, 101 (2): 209—228.

Chugh, R.S., 1973. Survey of India. Unpublished status report on work on Geodynamics Project.

Chugh, R.S., 1977. Survey of India. Unpublished annual report on Geodynamics Project.

Grady, C., 1971. Deep main faults in South India. J. Geol. Soc. India, 12: 56—62.

Kailasam, L.N., 1968. Some results of geophysical exploration over the Cretaceous Tertiary formations of South India. Geol. Soc. Mem., 2: 178—195.

Kailasam, L.N., 1975. Epeirogenic studies in India with reference to Recent vertical movements. Tectonophysics, 29: 505—521.

Krishnan, M.S., 1966. Tectonics of India. Bull. Ind. Geophys. Union, 3: 1—35.

Krishnan, M.S., 1968. Geology of India and Burma. Higginbothams, Madras.

Murthy, M.V.N., 1971. Tectonic history of the Deccan and Shillong plateaus in relation to their seismicity. Ind. Power. and Riv. Val. Dev. Spec. Nr., 31—37.

Murty, Y.G.K., Thampi, P.K. and Nair, M.M., 1976. Geology and geomorphology of Kerala. Geol. Surv. India Semin. Geol. Geomorphol. of Kerala (Abstr.): 6—9.

Rao, P.S., 1976. Geology and tectonic history of the Kerala region — a recycling of data. Geol. Surv. India Semin. Geol. Geomorphol. of Kerala (Abstr.): 13—17.

Semenov, E., 1968. Tectonic map of Eurasia. United Nations Report, Moscow.

Vaidyanathan, R., 1964. Geomorphology of Cuddapah Basin. J. Indian. Geosci. Assoc., 4: 29—36.

Verma, R.K. and Mukhopadhyay, M., 1976. Tectonic significance of anomaly-elevation relationships in north-eastern India. Tectonophysics, 34: 117—133.

Tectonophysics, 52 (1979) 223—238
© Elsevier Scientific Publishing Company, Amsterdam — Printed in The Netherlands

RECENT VERTICAL CRUSTAL MOVEMENTS FROM PRECISE LEVELING SURVEYS IN THE BLUE RIDGE AND PIEDMONT PROVINCES, NORTH CAROLINA AND GEORGIA *

GARY P. CITRON and LARRY D. BROWN

Department of Geological Sciences, Cornell University, Ithaca, N.Y. 14853 (U.S.A.)

(Accepted for publication April 4, 1978)

ABSTRACT

Citron, G.P. and Brown, L.D., 1979. Recent vertical crustal movements from precise lev-. eling surveys in the Blue Ridge and Piedmont provinces, North Carolina and Georgia. In: C.A. Whitten, R. Green and B.K. Meade (Editors), Recent Crustal Movements, 1977. Tectonophysics, 52: 223—238.

Examination of two lines of repeated leveling in North Carolina and Georgia reveals (1) apparent uplift at the Blue Ridge—Piedmont physiographic boundary (the Atlantic— Gulf drainage divide) relative to the Atlantic Coastal Plain on the east and the Valley and Ridge province to the west; and (2) large tilts over short baselines superimposed upon the regional pattern in the vicinity of the nearby Blue Ridge—Piedmont geologic boundary (the Brevard fault zone). In the North Carolina profile a very pronounced correlation between topography and movement suggests possible systematic leveling error, but the observed movements appear to be larger than those normally attributed to leveling error. Thus, either refraction or rod errors are larger than expected, or the movement is real and strongly correlates with topography along this portion of the leveling line.

Anomalously high stream-gradients over both resistant and nonresistant lithologies are found around the drainage divide in North Carolina, and may be associated with the relative uplift inferred from releveling. The drainage divide in Georgia, also characterized by relative uplift on the movement profile, approximately separates two different types of stream patterns. In both cases evidence presented here suggests that stream morphology may be responding to contemporary deformation as implied by the observed elevation changes. The relative uplift in North Carolina also correlates with a positive Bouguer gravity anomaly of 30—40 mGal in the midst of the regional Blue Ridge gravity low, although the significance of the correlation is unclear.

The close spatial correspondence between the zone of maximum uplift and the drainage divide suggests that the vertical movements and geomorphic anomalies may result from the same mechanism, although the nature of such is unclear. One possible mechanism could be displacement at depth along the nearby Brevard zone. However, on the basis of dislocation modeling it appears that the geodetic observations cannot be adequately explained by surface deformation associated with any simple models of slip on the Brevard zone.

---

* Cornell University Department of Geological Sciences Contribution 634.

INTRODUCTION

Elevations for the southern Blue Ridge highlands along the Atlantic—Gulf drainage divide approach 2 km (7000 ft). Schumm's (1963) estimate of a reasonable denudation rate (1 mm/yr or 1 km/m.y.) for a newly formed mountain system implies that any orogen subject to erosion would not last long geologically as a pronounced topographic feature. Menard (1961) calculated the erosion rate from the volume of Appalachian-derived sediments over the past 125 m.y. to be a more modest 0.062 mm/yr. Even using the lower rate, assuming a 5—8 km average elevation along the southern Atlantic—Gulf drainage divide at the end of the Triassic, it should have taken only about 80—120 m.y. to level the southern Appalachian topography. That the Blue Ridge Province still consists of highlands suggests that the region has either (1) experienced post-Triassic tectonic uplift (Owens, 1970); or (2) inherited a large-scale erosional disequilibrium from the Triassic (Hack, 1973a; personal communication, 1977).

Brown and Oliver (1976), on the basis of regional profiles of relative rates of apparent vertical movement, concluded that parts of the Appalachian highlands appear to be rising at rates on the order of 6 mm/yr with respect to the Atlantic coastal region. Here we examine in further detail segments from two profiles which traverse the Blue Ridge and Piedmont provinces of North Carolina and Georgia, and their geologic boundary, the Brevard fault zone (Fig. 1). The significant features delineated by the data, and of particular interest here, include: (1) tilts up to the east over long baselines within the Blue Ridge belt, with relative movement peaking at or near the Piedmont boundary; (2) long baseline tilts down toward the central Piedmont; and (3)

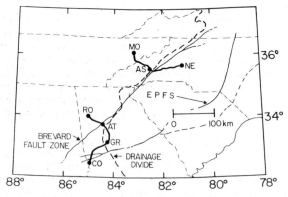

Fig. 1. Location diagram for leveling routes. Brevard fault zone is considered a geologic boundary, and the drainage divide a physiographic boundary between the Blue Ridge and Piedmont provinces. Drainage divide taken from Hack (1973a). MO = Morristown, AS = Asheville, NE = Newton, RO = Rockmart, AT = Atlanta, GR = Griffin, CO = Columbus, EPFS = Eastern Piedmont fault system.

shorter baseline, but larger magnitude tilts spatially associated with the Brevard fault zone. Both of the profiles examined here reveal that the peak of relative upward movement coincides very closely with the drainage divide, which is located a few kilometers west of the Brevard zone in North Carolina and a few kilometers east of the Brevard zone in Georgia. The regional pattern of uplift around the Brevard zone appears to correlate geographically with a band of stream-gradient anomalies in North Carolina, and a stream pattern change in Georgia. The correlation is of particular interest in that it implies a causal link between the short-term Recent crustal movement indicated by leveling and longer-term geomorphic activity.

Most Appalachian geologists recognize the Brevard fault zone as a fundamental structure of the southern Appalachian orogen (Hatcher, 1975). The topographic lineament and zone of cataclasis associated with the Brevard zone can be traced along a remarkably straight trend from the North Carolina—Virginia border to the edge of the Coastal Plain in northwestern Alabama (Fig. 1), a length of 530 km (330 miles). As the Brevard zone splays in a series of faults at its northern terminus, the drainage divide veers sharply to the northwest from its northeast trend. However, the Brevard may not be the only fault structure of great length and significance in the Appalachians. An Eastern Piedmont fault system (Hatcher et al., 1977) has recently been defined by geologic mapping and field checking of aeromagnetic data. This fault system parallels the regional Appalachian trend at the eastern end of the Piedmont, and also extends from Alabama to Virginia (Fig. 1).

Previous interpretations of the Brevard zone have been diverse, although a general consensus has evolved which considers it to be a fault zone with concurrent, but varying (depending on the interpretation) components of thrust and strike-slip motion during its history of faulting (Roper and Justus, 1973). Workers now believe the Brevard zone to have developed in the Middle Paleozoic during an initial phase of collision of the North American and African continents (Hatcher, 1972). If this is correct, the Brevard zone probably extends to great crustal, perhaps even lithospheric depths. Triassic dikes that intrude the Brevard fault zone in several localities have not been offset, and place an upper limit on the time for major faulting recorded on the Brevard (King, 1969). Yet while the stratigraphy records no post-Triassic tectonic activity on the fault zone, the spatial correlation between the Brevard zone and the apparent crustal uplift examined in this paper seems too close to be entirely fortuitous. If the apparent elevation changes discussed here are real, then this correlation could have important implications regarding the role of the Brevard zone, and the neotectonic and geomorphic evolution of the southern Appalachian region.

RECENT CRUSTAL MOVEMENTS FROM LEVELING MEASUREMENTS

The procedure used by the Cornell Recent crustal movements group for deriving vertical motion data from precise leveling surveys is identical to that commonly used by the National Geodetic Survey. A detailed account is described by Brown and Oliver (1976). Briefly, relative elevation changes are determined by subtracting the elevation difference between a given bench mark and a reference bench mark at some time (Δh_0) from the same difference measured at some other time (Δh_1). Relative velocities are determined by dividing the relative elevation changes ($d\Delta h$) by the time interval between surveys (Δt). Therefore, only time-averaged, relative velocities ($\bar{v}$) can be calculated, since the profiles are not connected to any absolute reference. However, there is no assurance that the movements have occurred at constant velocity. Fig. 1 shows the location of the two leveling routes examined in this study, the Atlantic—Gulf drainage divide, and the Brevard and Eastern Piedmont fault zones. Estimates of elevation change rates and their uncertainties along these routes are shown in Figs. 2a and 2b. Relative velocities may be readily converted to total movement ($d\Delta h$) by multiplying the average velocity by the time interval (indicated above the profiles).

The standard deviations of the relative velocity estimates were calculated according to the relation given by Holdahl (1973) and are also given in Fig. 2. Data collected by comparing precise leveling surveys are considered to be potentially significant when relative rates of movement for particular bench marks exceed random measurement error over baselines on the order of several tens of kilometers or more. Restriction of attention to longer baselines helps reduce the possibility of misidentifying bench-mark instability as tectonic activity.

Bench marks for both the North Carolina and Georgia lines are predominantly embedded in concrete structures, but approximately 20% of the bench marks are embedded in bedrock. The overall pattern of movement does not appear to be influenced by different types of bench-mark monumentation.

As can be seen in Figs. 2a and 2b, apparent differential rates on the order of several millimeters per year are commonly observed. Since 1 mm/yr = 1 km/m.y., such rates, if applied to the entire Pleistocene, would imply unreasonable topography unless erosion rates were extremely high. From a study of the Quaternary deposits of the southern and central Atlantic Coastal Plain, Owens (1970) concluded that any uplift that may have occurred during the Quaternary was not of significant intensity or duration compared to the uplift indicated by the Early Cretaceous deposits. Thus, if the movements are real, they are either episodic or oscillatory, probably about some long-term trend (Boulanger and Magnitskiy, 1974; Brown and Oliver, 1976).

The changes in direction of tilt on the profiles in Figs. 2a and 2b correspond to some extent with changes in the direction of profile routes (Fig. 1). Without supporting data, this situation may lead to more than one interpretation of the possible geometry of the relative uplift in the southern Appalachians. For example, Meade (1971) interpreted the releveling results as defining a domical pattern centered near Atlanta, Georgia. On the other hand, Brown and Oliver (1976) interpreted the same data as an elongated up-

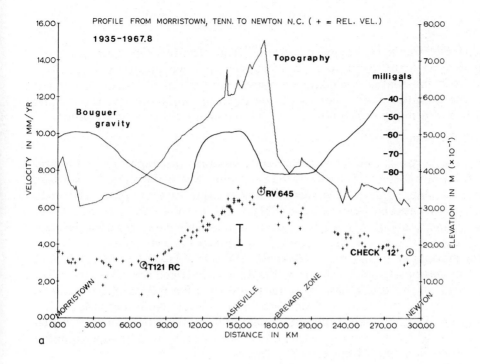

a

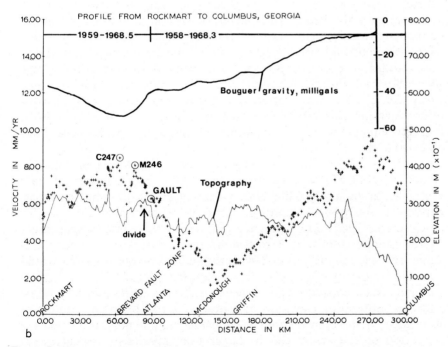

b

Fig. 2. Profile of relative vertical velocity (crosses) and absolute elevation: (a) from Morristown to Newton, (b) from Rockmart to Columbus. Times of level surveys are indicated at top of profiles. Bouguer gravity profiles taken from map by Woollard and Joesting (1964) for North Carolina, and from map by Long et al. (1972) for Georgia. Bar in 2a indicates two standard deviations (one positive and one negative) of the relative velocity ($2m_v$) of BM RV 645 relative to T 121 RC. For 2b, m_v for Rockmart relative to BM GAULT is 1.4 mm/yr, and m_v for BM GAULT relative to Columbus is 1.9 mm/yr.

lift parallel to the Appalachian seismic trend. The measurements presently available are insufficient to decide between the two interpretations. However, it is the spatial correlation between the pattern of crustal movement and the Atlantic—Gulf drainage divide deduced from the leveling profiles which is of primary concern here, not the regional geometry of the uplift.

North Carolina

Figure 2a shows relative vertical velocities and topography along the profile from Morristown, Tennessee, via Asheville, North Carolina, to Newton, North Carolina. This profile reveals an increasing rate of uplift relative to Morristown between bench marks (BM) T 121 and RV 645. The tilt rate for this segment, about 4.4×10^{-8} rad/yr, is relatively large for long leveling route segments crossing the eastern United States (Brown and Oliver, 1976). Total movement ($d\Delta h$) recorded at BM RV 645 relative to BM CHECK at Newton was nearly 13.5 cm during the 32.8 yr between first-order levelings. The peak of the uplift has an average rate of 4.0 ± 0.9 mm/yr relative to bench marks in the vicinity of Morristown or Newton. Note that in the vicinity of the Brevard fault zone the apparent movement of about 10 bench marks east of RV 645 define a pattern of two short, steep tilts superimposed upon the regional uplift pattern (Fig. 2a).

A striking feature of Fig. 2a is the strong correlation between relative velocity and topography. Brown and Oliver (1976) considered the possibility that the apparent movement pattern resulted from elevation-correlated errors that systematically accumulated during the surveying. They determined the amount of correlation between movement and elevation change with distance for sections of the leveling route. The majority of their correlation ratios ($d\Delta h/\Delta h$) for each section were larger than could easily be attributed to an elevation-correlated error such as refraction. Values of $d\Delta h/\Delta h$ were especially large, compared to possible refraction errors, in the vicinity of the drainage divide.

The relation between relative elevation change ($d\Delta h$) and elevation is shown graphically in Fig. 3. As can be seen here, the majority of data points from Fig. 2a lie along a line with a slope of about 38 mm/100 m rise. For comparison, fairly recent studies of refraction effects in Europe (for example, Hytonen, 1967) and California (Holdahl, 1974) cite typical values for this error ranging from only 1 to 6 mm/100 m rise. Also shown on this plot is a line with a slope of 20 mm/100 m rise, which represents what would be severe refraction error in a single survey (Bomford, 1971). However, since refraction error has a constant sign, a net cancellation of its effect could be expected when subtracting elevation measurements to obtain an estimate of elevation change. Without more conclusive evidence, one must infer that (1) either refraction or rod errors can be larger than previously expected; (2) the measurements are contaminated by some other source of error (which seems unreasonably ad hoc); or (3) the movement is real and strongly correlates with topography.

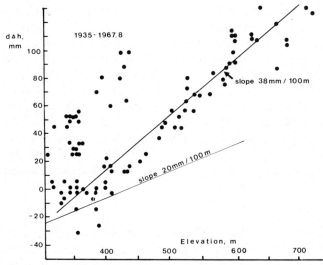

Fig. 3. Plot of relative elevation change (dΔh) versus elevation for bench marks from Morristown to about 17 km before Newton.

Such examinations of possible elevation-correlated errors, although suggestive, are inconclusive. The movement pattern observed on the Georgia profile (Fig. 2b) where no correlation with topography is readily evident, also shows northwest tilting in the Blue Ridge, and two short but steep tilts in the vicinity of the Brevard zone. Such a similarity in profile argues that the North Carolina data are not seriously affected by systematic error. Alternatively, it is possible that there was error in the North Carolina survey but not in the Georgia survey, although this would imply that the similar correlations are merely coincidental.

The relative velocity maximum in Fig. 2a corresponds geographically with the Atlantic—Gulf drainage divide (the topographic peak), about 10 km to the west of the Brevard fault zone. As Hack (1973a) points out, the drainage divide at the escarpment of the Blue Ridge highlands (sometimes referred to as the Blue Ridge front), rather than lithologic differences, defines the Blue Ridge—Piedmont physiographic boundary. However, the southeast boundary of the Blue Ridge geologic belt, is generally considered to be the Brevard fault zone (King, 1955), which is usually found a few kilometers to the east of the divide in this region. Thus, while the Blue Ridge highlands die out in northwestern Georgia, the Precambrian geology of the Blue Ridge belt can be traced southwest across the Piedmont physiographic province to the Coastal Plain onlap (Reed, 1970). The gross pattern of uplift correlates with the geomorphically defined boundary rather than the geologically defined one.

If the large rates of uplift indicated by Fig. 2a are real, the relative velocity pattern suggests that the eastern margin of the Blue Ridge geologic belt

230

may be in dynamic imbalance within the crust or upper mantle. A dynamic imbalance might be expected to give rise to a gravity imbalance. Thus, it seems plausible that modern uplift might be associated with a gravity anomaly. The Blue Ridge geologic belt has the largest Bouguer gravity low in the eastern United States (Figs. 2a, 2b; Woollard and Joesting, 1964). However, accompanying the topographic and relative velocity maxima of the Blue Ridge escarpment in North Carolina and superimposed upon the regional gravity low, is a Bouguer gravity bulge centered near Asheville with an amplitude of 25—30 mGal (Fig. 2a). The same area shows no significant isostatic gravity anomaly (Woollard, 1966). At present, the low resolution of both the available geodetic and gravity data poorly constrain any models of structure related to the inferred uplift for this area.

Stream-gradient anomalies

The pattern of uplift and the topography are both asymmetrical with respect to the drainage divide. Oversteepened Blue Ridge stream profiles (Hack, 1973a) seem to support the idea that an erosional disequilibrium exists between the Atlantic-bound and Gulf-bound systems (Davis, 1903). It is suggested here that streams draining the Blue Ridge highlands in the vicinity of Asheville may be responding to the apparent modern uplift indicated by precise releveling. To examine this possibility more closely, an analysis of stream-gradients in a test area in western North Carolina, very near to a similar study by Hack (1973a), was performed to determine where stream profiles are steep in relation to discharge. This, in effect, determines where anomalous amounts of erosional energy are being expended (Hack, 1973a; Keller, 1977). The geomorphic parameter used to summarize the stream activity of concern is the stream-gradient index, defined as the product SL, where S is the channel slope of the selected reach of stream in feet per mile, and L is the distance from the midpoint of that reach to the source in miles (Hack, 1973b). For all streams mapped on the northeast corner of the Knoxville $1°$ by $2°$ topographic sheet (scale = 1 : 250,000), SL values were calculated from reaches within 3×3 km squares, and the resultant average values from each square were contoured. The values of SL vary considerably, which forced scaling of the contour interval to be proportional to 2^n. The results of these calculations are shown in Fig. 4.

A distinct belt of SL values exceeding 500 trends northwest along the leveling route, which follows the Southern Railroad in the valley of the French Broad River. Here the route covers resistant rocks mapped as the Middle Precambrian Max Patch Granite and Cranberry Gneiss (Hadley and Nelson, 1971; Hack, 1973a) and SL values for some reaches, roughly consistent with lithology, exceed 2000.

Another band of high SL values, although poorly defined, extends along the drainage divide, which strikes slightly east of north in this area. This band is underlain for the most part by resistant members of the Great Smokey Group (Mohr, 1973). However, the Swannanoa Mountains in the southern

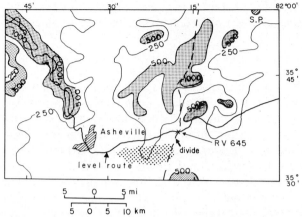

Fig. 4. Stream-gradient index contour map for northern portion of Knoxville 1° by 2° quadrangle. Heavy dashed line marks Atlantic—Gulf drainage divide. Areas above SL = 500 are shaded. SP = Spruce Pine; stipple = Swannanoa Mountains.

part of Fig. 4 are also composed of the Great Smokey Group resistant suite, but do not register such high SL values. Outcrops of hornblende gneiss and amphibolite in the Spruce Pine area underlie the anomaly in the northeast corner of Fig. 4, and the generally well-foliated Henderson Gneiss (Hadley and Nelson, 1971) is also found beneath this anomaly ($SL > 500$) in the southeastern part of the map. In general, the aforementioned gneisses and amphibolites are not as resistant to erosion as the more structurally massive Max Patch Granite and Cranberry Gneiss (Hack, 1973a). Furthermore, where the Henderson Gneiss is cut by the French Broad River, near Asheville, Hack (1973a) obtained the low SL value of 110. These observations suggest that lithologies alone do not control the stream gradients in this area.

The drainage divide, relative velocity maximum, and anomalous belt of SL values around the drainage divide are all nearly coincident. As mentioned earlier, Hack (1973a) has studied stream gradients in the southern Appalachians, and mapped a similar but narrower and more continuous zone of high SL values following the northeast trend of the drainage divide in the Winston—Salem 1° by 2° topographic sheet, directly northeast of the Knoxville quadrangle. As in the Knoxville quadrangle, the zone of high SL values correlates in part, but not entirely, with resistant lithologies. Our study extends these results to the southeast. Hack hypothesized either (1) a tectonic cause for geologically recent uplift of the Blue Ridge escarpment, or (2) an erosional disequilibrium as a feature inherited from the Triassic, because erosion rates may have been slower on the more resistant rocks to the northwest of the Atlantic—Gulf divide. Both theories result in the erosional response of the northwest migration of the divide. Hack preferred hypothesis (2), but the data presented in this study support a neotectonic cause, although the two explanations are not exclusive. We suggest that the eastern band of

232

anomalously high *SL* values in Fig. 4 may represent drainage unadjusted to
the geology in response to the relative uplifted represented by the releveling.

Georgia

The composite profile in Fig. 2b consists of segments leveled between
Rockmart and Atlanta, and Atlanta to Columbus. Each leveling, and releveling was conducted at different times, but all of these surveys are first order,
and share BM GAULT in Atlanta as a common point. Tilts over long baselines are readily noticeable between Rockmart up to the Brevard fault zone,
the Brevard fault zone down to Griffin, and tilting down toward Griffin
from Columbus, which is on the Coastal Plain. The time spans between successive levelings for these segments are 9.5 and 10.3 yr, respectively. Since
random errors are propagated as the inverse of the leveling—releveling time
interval, the standard deviations of the relative velocity for the Rockmart—
Columbus profile exceed 2.0 mm/yr, and caution must be used when interpreting the tectonic significance for these movements. Northwest tilting
(i.e., down toward the northwest at a rate of about 4 to $5 \cdot 10^{-8}$ rad/yr is
characteristic for the Rockmart—Brevard zone segment as a whole, and comparable to the Blue Ridge tilting in North Carolina, although the movement
data in Georgia show more scatter. The Rockmart—Brevard relative velocity
pattern undulates within the trend of northwest tilting at wavelengths of
20—35 km with no obvious relation to geological structure. The maximum
uplift (at BM C247) for this segment closely coincides with the Long Island
fault, which is the northwestern boundary fault of the Brevard zone mapped
by Higgins (1968). Again, a long baseline tilt is evident within the Blue Ridge
belt. Well-defined, short baseline tilts with rates on the order of 10^{-5} rad/yr
spatially correspond to the Brevard zone between BM C247 and M246. Similar tilts are found in the vicinity of the Brevard zone in the North Carolina
profile. Approximately 8 km southwest of M246 (along the profile) the
leveling route intersects the Atlantic—Gulf drainage divide (Figs. 1 and 2b)
as drawn by Hack (1973a). The proximity of the movement maximum to
the divide again constitutes an important correlation, and supports the suggestion that placement of the Appalachian drainage divide is possibly related
to the relative crustal uplift found near the Brevard zone.

Streams northwest of the Brevard fault zone in Georgia have trellis drainage, while those to the southeast (flowing on the Piedmont) have dendritic
drainage. Hence the Brevard fault zone represents a geomorphological discontinuity, which Staheli (1976) attributed to an Oligocene Coastal Plain
transgression extending to the Brevard zone. The observations discussed here
suggest an alternative explanation — that the different drainage patterns are
due in part to asymmetric tilting such as that indicated by the geodetic measurements in Fig. 2b. It is natural to speculate that the steep rates displayed
from Atlanta to Griffin may have an influence in developing the dendritic
drainage pattern. Yet these steep tilt rates (on the order of 10^{-7} rad/yr), as

discussed before, cannot be reasonably extrapolated very far back in time with confidence, and are probably part of some oscillatory or episodic pattern. Therefore, a much more subdued, time-average rate operative over a time span on the order of 10^6 yrs, rather than the presently observed tilt rates may be responsible for the contrasting drainage patterns. LaForge (1925) proposed a similar model of regional uplift at the western end of the Georgia Piedmont that induced northwestward headward erosion to explain the dendritic nature of Atlantic-bound streams.

Along the Georgia profile line, the steep tilts in the vicinity of the Brevard fault zone are followed by a broad V-shaped pattern to produce a relative velocity minimum near Griffin (Figs. 1 and 2b). These two limbs also have a minor undulatory nature, with wavelengths averaging between 25 and 30 km. Large systematic errors unrelated to topography, but possibly resulting from tidal forces or unequal rod lighting (Bomford, 1971) may contribute to the gross pattern, although again, the magnitude of such influences seem inadequate to explain the observations (Holdahl, 1974). Except for a possible interaction between the undulations in the profile from 5 to 20 km before Columbus, and the Eastern Piedmont fault system (Hatcher et al., 1977) (Figs. 1 and 2b), the surface geology bears no apparent correlation with such dramatic and symmetric tilts. However, that there is no apparent geologic correlation should not be construed as evidence against a tectonic origin for these prominent tilts, since they may reflect activity at intermediate or lower lithospheric levels (Artyushkov, 1972).

MECHANISM OF UPLIFT

The inferred uplift around the Atlantic—Gulf drainage divide is intriguing in that consistent tilts extend to the northwest and southeast of the divide on both profiles. In North Carolina, the baseline of this feature is almost 200 km. If such elevation change data represent real vertical crustal movements, as argued, then the long-baseline character suggests a deep causal mechanism. As the Brevard fault zone appears to be in close spatial proximity with the uplift maxima on both profiles, it is a logical candidate to test models of surface deformation associated with downdip slip. A further reason for such modeling is that directly at the Brevard zone a distinct sequence of short, steep tilts superimposed upon the longer baseline uplift is evident on both profiles.

In view of these correlations on two independent profiles separated by about 200 km (along the Brevard strike), it is natural to consider whether or not the deeply rooted Brevard zone may be involved in producing the movement patterns, and possibly the relatively high elevations. To test the validity of this hypothesis, the observed elevation change data was modeled, using dislocation theory, as surface deformation associated with dip-slip movement (Savage and Hastie, 1966) accumulated on the Brevard zone during the leveling time span. Such movements could conceivably be generated from seismic

234

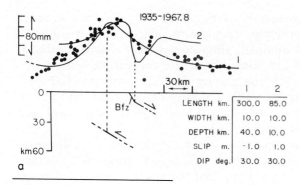

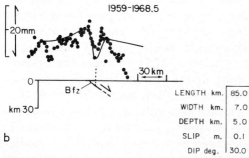

Fig. 5. Relative movement data superimposed upon theoretical surface deformation curves. Positive slip indicates normal fault movement. Dates refer to times of leveling for the (a) North Carolina and (b) Georgia profiles. At bottom is a scale representation of the Brevard fault zone (Bfz) and theoretical fault segments.

faulting or aseismic creep. The results, shown in Fig. 5, indicate that the models which best fit the data require very small fault width to fault length ratios and large amounts of slip. From a consideration of the model results which the best-fitting fault parameters provide, we conclude that a simple creep or faulting model does not accurately depict the physical situation. The model that produces the best match to the North Carolina data (Fig. 5a, model 1), slip on a deep basement fault west of the Brevard, must be considered ad hoc at best. Such modeling, of course, is not unique, and more complex models can be applied. It seems clear, however, that simple dip-slip displacement on the Brevard fault zone, whether produced through seismicity or aseismic creep, cannot satisfactorily explain the surface deformation observed from leveling measurements.

SUMMARY AND CONCLUSIONS

The apparent tilts indicated on relative vertical velocity profiles in North Carolina and Georgia appear to represent crustal uplift at the Blue Ridge—Piedmont physiographic boundary, also the location of the Atlantic—Gulf drainage divide. Steep, short-baseline tilts present in both profiles are spatially associated with the nearby Blue Ridge—Piedmont geologic boundary, the Brevard fault zone.

There is a strong correlation between topography and relative velocity in the North Carolina profile. However, the magnitude of the movements seem to be larger than that which can be reasonably attributed to elevation-correlated error. Therefore, either refraction or rod errors are larger than heretofore realized, the movements are contamined by some other source of leveling error, or the movement represents tectonic deformation which strongly correlates with topography.

In North Carolina, independent evidence in favor of a neotectonic explanation for the apparent vertical movements is found in anomalously high stream-gradients over both resistant and nonresistant lithologies around the drainage divide. The association of the stream-gradient anomalies with the observed Asheville uplift suggests a response of stream activity to the inferred crustal deformation. The similar movement pattern without topographic correlation on the Georgia profile also supports the reality of the movements. The observations suggest that the Asheville area may warrant further investigations of possible neotectonism, since current understanding of the phenomena is not sufficiently advanced to determine if such movement constitutes a geologic hazard.

In Georgia, the presence of uplift across the drainage divide and Brevard fault zone is associated with a drainage pattern discontinuity. Other long-baseline, more enigmatic tilts within the central Piedmont between Atlanta and Columbus are not readily correlatable with known geologic structure or topography.

The reliability of the interpretations presented here are difficult to assess. The temporal and spatial density of the leveling data is presently inadequate. The time differential for segments of the Georgia line are only around 10 yr, while the last levelings of all the routes discussed here were completed over 11 years ago. Additional levelings of the routes would not only aid in establishing (or refuting) a trend of uplift at the southern Appalachian drainage divide where the drainage is unadjusted (Hack, 1973a), but also in testing systematic error accumulation in previous levelings. Better spatial coverage of the leveling will, of course, improve any estimates of the uplift geometry. Another asset of better spatial leveling coverage is that the extent and credibility of the large enigmatic tilts found within the Georgia Piedmont may be examined. The credibility and extent of the intra-Piedmont tilts, which are bounded by two major southern Appalachian fault zones (Brevard and Eastern Piedmont), bears directly upon the value of any tectonic information extracted from the lesser magnitude tilts near or at the Brevard fault zone.

Assuming the movements are real, models applicable to the problem of uplift along the nearby Brevard zone may be chosen from a representative list of mechanisms possibly related to Recent vertical crustal movements in the eastern United States, provided by Brown and Oliver (1976). Such models should be constrained by the observation that the uplift peaks locate at or near the drainage divide. This correspondence suggests not only a genetic relationship between the uplift maxima and position of the divide, but that the parent mechanism may have been operative, in varying intensity, since the drainage divide acquired its unadjusted character along the Atlantic—Gulf divide, sometime after the Triassic. We have shown from simple dislocation theory that the geomorphic and movement anomalies do not result from creep or faulting in any simple manner on the Brevard fault zone itself. Speculatively, viscous, viscoelastic or plastic processes in the mantle interacting with pronounced structural features such as the Brevard and Eastern Piedmont fault zones might account for the movement maxima (Boulanger and Magnitskiy, 1974). The mechanism accountable for the surface deformation pattern around the drainage divide is almost certainly complex, but higher-level creep or minor faulting may be contributing factors, particularly where short-baseline, steep tilts are observed, such as those between the boundary faults of the Brevard and Eastern Piedmont zones.

With the presently available data, it is difficult to infer dynamic mechanisms responsible for modern uplift in the elevated southern Appalachians, as may exist, for example, in the western United States with surface faulting (Savage and Church, 1974) and inferred crustal magma bodies (Reilinger and Oliver, 1976; Smith and Pelton, 1977). Seismic activity for the eastern United States is minor, although of considerable scientific and engineering importance. Epicentral alignment and concentration of earthquakes near the drainage divide in North Carolina generally suggest movement along concealed or unrecognized faults (Hadley and Devine, 1974). Thus, any information releveling data can provide toward understanding the current tectonic regime is important for understanding the mechanisms responsible for crustal deformation, and/or earthquake activity, as well as for more reliably estimating potential geologic hazard in the eastern United States.

ACKNOWLEDGEMENTS

The authors wish to thank Drs. Jack E. Oliver, Bryan L. Isacks, and Mssrs. Robert Reilinger and F. Steve Schilt for much constructive criticism of this manuscript. Steve Schilt kindly allowed use of his surface deformation program. We also acknowledge the generous help of the National Geodetic Survey, and in particular, Mr. Sanford Holdahl. Douglas Waters drafted the figures. Ms. Robin Fisher Cisne provided editorial assistance. This work was supported by the Nuclear Regulatory Commission under Grant AT-(49024)-0367.

REFERENCES

Artyushkov, E.V., 1972. The origin of large stresses in the earth's crust. Izv. Acad. Sci. U.S.S.R. Phys. Solid Earth, 8: 3—25.

Bomford, G., 1971. Geodesy, 3rd ed. Clarendon Press, Oxford, 732 pp.

Boulanger, Y.D. and Magnitskiy, V.A., 1974. Contemporary movements of the earth's crust: state of the problem. Earth Phys., 10: 19—24.

Brown, L.D., and Oliver, J.E., 1976. Vertical crustal movements from leveling data and their relation to geologic structure in the eastern United States. Rev. Geophys. Space Phys., 14: 13—35.

Davis, W.M., 1903. The stream contest along the Blue Ridge. Geogr. Soc. Phila. Bull., 3: 213—244.

Hack, J.T., 1973a. Drainage adjustment in the Appalachians, In: M. Morisawa (Editor), Fluvial Geomorphology, State Univ. New York, Binghamton Publ. in Geomorphology, Binghamton, New York, N.Y., pp. 51—69.

Hack, J.T., 1973b. Stream profile analysis and stream-gradient index. U.S. Geol. Surv. J. Res., 1: 421—429.

Hadley, J.B. and Devine, J.F., 1974. Seismotectonic map of the eastern United States. U.S. Geol. Surv. Misc. Field Studies Map MF-620 (3 sheets) scale 1 : 5,000,000.

Hadley, J.B. and Nelson, A.E., 1971. Geologic map of the Knoxville quadrangle, North Carolina, Tennessee, and South Carolina. U.S. Geol. Surv. Misc. Geol. Inv. Map I-654, scale 1 : 250,000.

Hatcher Jr., R.D., 1972. Developmental model for the southern Appalachians. Geol. Soc. Am. Bull., 83: 2735—2760.

Hatcher Jr., R.D., 1975. Special report: Second Penrose Field Conference, The Brevard zone. Geology, 3: 149—152.

Hatcher Jr., R.D., Howell, D.E. and Talwani, P., 1977. Eastern Piedmont fault system: Speculations on its extent. Geology, 5: 636—640.

Higgins, M.W., 1968. Geologic map of the Brevard fault zone near Atlanta, Georgia. U.S. Geol. Surv. Misc. Geol. Inv. Map I-511, scale 1 : 48,000.

Holdahl, S.R., 1973. Vertical crustal movements — status of NGS investigations. Geop-3 Res. Conf.: Vertical Crustal Movements and Their Causes, Am. Geophys. Union.

Holdahl, S.R., 1974. Times and heights. Paper presented at the Int. Symp. on Problems Related to the Redefinition of North American Geodetic Networks. Int. Assoc. Geod., Paris.

Hytonen, E., 1967. Measuring of the refraction in the second leveling of Finland. Suom. Geodeettisen Laitoksen Julk., 63: 1—22.

Keller, E.A., 1977. Adjustment of drainage to bedrock in regions of contrasting tectonic framework. Geol. Soc. Am. Abstr. with Prog., 9: 1046.

King, P.B., 1955. A geologic section across the southern Appalachians — an outline of the geology in the segment in Tennessee, North Carolina, and South Carolina, In: R.J. Russell (Editor), Guides to Southeastern Geology. Geol. Soc. Am., Boulder, Colo., pp. 332—373.

King, P.B., 1969. Tectonic Map of North America. U.S. Geol. Surv., scale 1 : 5,000,000.

LaForge, L., 1925. The provinces of Appalachian Georgia. Georgia Geol. Surv. Bull., 42: 55—92.

Long, L.T., Bridges, S.R. and Dorman, L.M., 1972. Simpler Bouguer gravity map of Georgia. Geol. Surv. of Georgia, scale 1 : 2,500,000.

Meade, B.K., 1971. Report of the sub-commission on recent crustal movements in North America. Paper presented to the 15th General Assembly of the Int. Union of Geod. and Geophys. Intl. Assoc. Geod., Brussels.

Menard, H.W., 1961. Some rates of regional erosion. J. Geol., 69: 154—161.

Mohr, D.W., 1973. Stratigraphy and structure of part of the Great Smokey and Murphy Belt Groups, western North Carolina. Am. J. Sci., 293-A (Cooper volume): 41—74.

Owens, J.P., 1970. Post-Triassic tectonic movements in the central and southern Appalachians as recorded by sediments of the Atlantic Coastal Plain. In: G.W. Fisher, F.J. Pettijohn, J.C. Reed and K.N. Weaver (Editors), Studies of Appalachian Geology: Central and Southern. Interscience, New York, N.Y., pp. 417—428.

238

Reed Jr., J.C., 1970. The Blue Ridge and the Reading Prong: Introduction. In: G.W. Fisher, F.J. Pettijohn, J.C. Reed and K.N. Weaver (Editors), Studies of Appalachian Geology: Central and Southern. Interscience, New York, N.Y., pp. 195—198.

Reilinger, R.E. and Oliver, J.E., 1976. Modern uplift associated with a proposed magma body. Geology, 4: 583—586.

Roper, P.J. and Justus, P.S., 1973. Polytectonic evolution of the Brevard zone. Am. J. Sci., 293-A (Cooper volume): 105—132.

Savage, J.C. and Church, J.P., 1974. Evidence for post-earthquake slip in the Fairview Peak, Dixie Valley, and Rainbow Mountain fault areas of Nevada. Bull. Seismol. Soc. Am., 64: 687—698.

Savage, J.C. and Hastie, L.H., 1966. Surface deformation associated with dip-slip faulting. J. Geophys. Res., 71: 4897—4904.

Schumm, S.A., 1963. The disparity between present rates of denudation and orogeny. U.S. Geol. Surv. Prof. Pap., 454 H, 1—13.

Smith, R.B. and Pelton, J.R., 1977. Crustal uplift and its relationship to seismicity and heat flow at Yellowstone (abstr.) EOS Trans. Am. Geophys. Union, 58: 495—496.

Staheli, A.C., 1976. Topographic expression of superimposed drainage on the Georgia Piedmont. Bull. Geol. Soc. Am., 87: 450—452.

Woollard, G.P., 1966. Regional isostatic relations in the United States, In: J.S. Steinhart and T.J. Smith (Editors), The Earth Beneath the Continents. Am. Geophys. Union Monogr., 10: 557—594.

Woollard, G.P. and Joesting, H.R., 1964. Bouguer anomaly map of the United States. U.S. Geol. Surv. Spec. Map, scale 1 : 2,500,000.

Tectonophysics, 52 (1979) 239—248
© Elsevier Scientific Publishing Company, Amsterdam — Printed in The Netherlands

ANALYSIS OF RECENT CRUSTAL MOVEMENT IN THE CENTRAL AND NORTHERN SIERRA NEVADA, CALIFORNIA, USING REPEATED GEODETIC LEVELING DATA

D.O. WEST AND J.N. ALT

Woodward-Clyde Consultants, San Francisco, Calif. 94111 (U.S.A.)

(Accepted for publication April 4, 1978)

ABSTRACT

West, D.O. and Alt, J.N., 1979. Analysis of Recent crustal movement in the central and northern Sierra Nevada, California, using repeated geodetic leveling data. In: C.A. Whitten, R. Green and B.K. Meade (Editors), Recent Crustal Movements, 1977. Tectonophysics, 52: 239—248.

A comparative analysis of repeated geodetic leveling data was made along nine subparallel, E—NE-trending leveling lines located in the central to northern Sierra Nevada and the eastern Central Valley. The analysis was made to identify relative changes of elevation and evaluate these changes with respect to the regional geology and tectonics. The analysis used National Geodetic Survey first- and second-order, unadjusted, observed elevations.

The relative changes in elevation indicate that crustal deformation is continuing to occur in the Sierra Nevada along pre-existing zones of crustal weakness and that this deformation is localized along some strands of Late Cenozoic faulting within the Mesozoic Foothills fault system. This deformation is characterized by variable and nonuniform westward tilt of the Sierran block west of the Melones fault zone, and relatively consistent eastward tilt of the Sierran block east of the Melones fault zone. Variable elevation changes occur within the Foothills fault system and are often associated with prominent geological or structural contacts. In addition, subsidence in the Central Valley appears to be of small magnitude and localized in extent, indicating nontectonic changes in elevation problably due to compaction of unconsolidated sediments.

INTRODUCTION

A comparative analysis of repeated geodetic leveling has been made along nine subparallel, E—NE-trending leveling lines located in the central to northern Sierra Nevada and eastern Central Valley of California (Fig. 1). The purpose of this analysis was to identify relative changes in elevation and evaluate these apparent changes in elevation as they may relate to intrablock deformation of the Sierran crustal block.

Of particular interest were the elevation changes in the areas of the Foothills fault system which includes the Bear Mountains and Melones fault zones

240

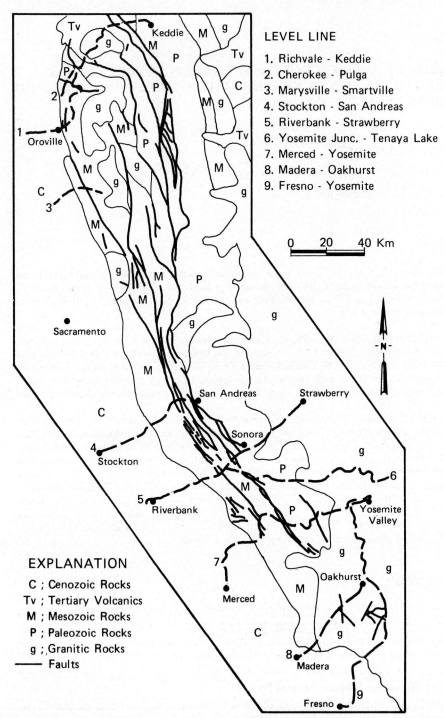

LEVEL LINE

1. Richvale - Keddie
2. Cherokee - Pulga
3. Marysville - Smartville
4. Stockton - San Andreas
5. Riverbank - Strawberry
6. Yosemite Junc. - Tenaya Lake
7. Merced - Yosemite
8. Madera - Oakhurst
9. Fresno - Yosemite

0   20   40 Km

EXPLANATION

C ; Cenozoic Rocks
Tv ; Tertiary Volcanics
M ; Mesozoic Rocks
P ; Paleozoic Rocks
g ; Granitic Rocks
——— Faults

Fig. 1. Location map of leveling lines, Foothills fault system (from Jennings, 1975) and generalized geology.

(Fig. 1). Surface fault rupture associated with the August 1, 1975, magnitude 5.7 Oroville earthquake (Clark et al., 1976) prompted studies to identify other localities of Late Cenozoic deformation in this region of relatively low-level tectonic activity. The present study was part of comprehensive regional geologic and seismologic investigations for the Pacific Gas and Electric Company and the U.S. Bureau of Reclamation to assess Late Cenozoic deformation along the Mesozoic Foothills fault system.

DATA AND TECHNIQUE

The analysis was based on the comparison of National Geodetic Survey (N.G.S.) first- and second-order, unadjusted, observed field elevations. Profiles of relative elevation changes were developed along the nine leveling lines for the lengths shown in Fig. 1.

For all but two of the profiles (Stockton—San Andreas and Madera—Oakhurst), data from only two surveys were available because of the lack of subsequent surveying. Where additional data were available for the Madera—Oakhurst and Stockton—San Andreas lines, they covered only portions of the line. In addition, only the Riverbank—Strawberry line compared first-order data; the remaining profiles were derived from the comparison of first- and second-order or strictly second-order data.

To evaluate the significance of observed relative elevation changes, possible sources of survey errors were evaluated. The three major types of errors considered were blunder errors, random errors, and systematic errors. The N.G.S. leveling data summary sheets and the profiles of elevation change were examined for possible blunder errors.

Random errors for the comparison of two surveys were evaluated using the standard deviation of the computed vertical movement (Holdahl, 1976).

Systematic error, which generally accumulates at a constant rate (<0.1 mm/km) and tends to become random for distances greater than 50 km, also tends to cancel when computing the change in elevation of two surveys over the same route (Holdahl, 1976).

Experience indicates that where conservative estimates of the standard deviation of leveling for 1 km are used, the resulting standard deviation of vertical movement tends to encompass both random and systematic errors. Thus, along the longest profile (comparison of second-order surveys; 140 km), the standard deviation of vertical movement is approximately 33 mm.

The bench mark nearest the boundary between the Sierra Nevada and Central Valley geomorphic provinces (Fig. 1) was held fixed for each profile. This tended to differentiate between relative elevation changes in the bedrock of the Sierra Nevada, which are likely to be tectonic in nature, and those in the Central Valley, where thick unconsolidated Quaternary and older deposits contribute to nontectonic subsidence due to groundwater withdrawal (Poland et al., 1975).

GEOLOGIC SETTING

The Sierra Nevada is a geologically young, westward-sloping, asymmetric mountain range. The axis of the range trends NNW, is approximately 600 km long, and has a core of Mesozoic granitic rock (Fig. 1).

The area under analysis is dominated structurally by the Foothills fault system which includes two major fault zones: the Bear Mountains and Melones fault zones (Fig. 1). Both fault zones displace rocks of Mesozoic and Paleozoic ages. The Melones fault zone generally separates steeply dipping Mesozoic metasedimentary and metavolcanic rocks on the west from Paleozoic metamorphic rocks on the east, whereas the Bear Mountains fault zone is primarily within the Mesozoic metamorphic rocks of the western Sierra Nevada. The relatively undeformed Cenozoic sedimentary rocks in the Central Valley overlap the metasedimentary Mesozoic rocks.

RESULTS

Three profiles have been selected to summarize the results of the evaluation. They typify the location and magnitude of possible Recent crustal deformation along the nine profiles. Each profile consists of the plot of relative elevation change along with the terrain profile and a generalized geologic section.

Richvale—Keddie line

The Richvale—Keddie line (Fig. 2) begins in the Central Valley and follows the north fork of the Feather River to Keddie in the northern Sierra Nevada. The line was surveyed in 1932 and repeated in 1934. Both levelings were second order. Bench mark B205, near Oroville, was held fixed with changes in elevation to the west and northeast, calculated relative to this point.

Because of the short time between surveys and the fact that the leveling was second order, only significant elevation changes would be observed. Except for approximately the last 15 km of the line, all the relative elevation changes are within the standard deviation of vertical movement for the comparison of second-order surveys. However, several significant trends in elevation changes are present along the line.

At the western end of the line, from Richvale to about 30 km northeast of Oroville, there was relative subsidence that did not exceed 20 mm, and there was no significant change in elevation in the Oroville area. Northeast of Oroville, there was generally relative uplift. The survey line crosses the Big Bend fault three times, and there were no significant elevation changes between the two surveys in the vicinity of this fault (Fig. 2).

A portion of the line northeast of Oroville is subparallel to the Cherokee—Pulga line (Fig. 1). The trend in relative elevation changes along this segment

243

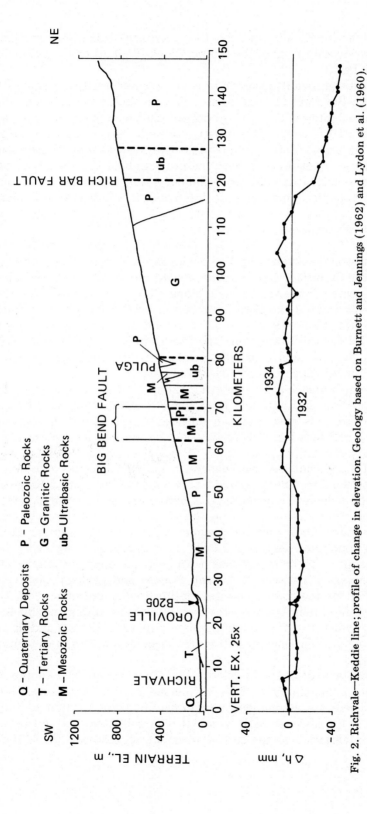

Fig. 2. Richvale—Keddie line; profile of change in elevation. Geology based on Burnett and Jennings (1962) and Lydon et al. (1960).

of the Richvale—Keddie line and the Cherokee—Pulga line are similar and indicate westward tilting of the Sierran block. Bennett et al. (1977) have also noted this trend.

The most significant observation is an apparent relative eastward tilt east of the Rich Bar fault (a trace of the Melones fault zone). Assuming the tilting to be real and constant, this corresponds to an average rate of approximately $6.5 \cdot 10^{-7}$ rad/yr. This observation is in agreement with the geologic evidence of up to 180 m of post-Pliocene, down-to-the-east displacement along a Late Cenozoic fault near the projection of the Rich Bar fault. The geodetic anomaly across this zone indicates that crustal deformation is continuing to take place in a zone associated with this Late Cenozoic fault.

Riverbank—Strawberry line

The Riverbank—Strawberry line begins in the Central Valley and generally parallels the Stanislaus River northeastward into the Sierra Nevada to Strawberry. Repeated first-order leveling along all or parts of this line was done in 1932, 1946—1947, 1956, 1957, 1958, and 1963. However, only the surveys done in 1932, 1946—1947, and 1957 extend far enough east along the line to be useful for this evaluation. The change in elevation for the 1946—1947 and 1957 surveys was calculated relative to the 1932 survey with bench mark Z120 near Knights Ferry held fixed (Fig. 3).

Relative subsidence to the west of Z120 in the Central Valley is minor and substantiates the small amount (less than 30 cm total) reported by Poland et al. (1975) for this area. Relative subsidence near Riverbank was approximately 17 mm between the leveling and amounts to an average rate of 1.1 mm/yr.

Several significant relative changes in elevation occurred to the east of Z120 between the 1932 and 1957 surveys. Directly east of Knights Ferry, the elevation decreases steadily to a minimum of —13 mm at bench mark C121. The profile remains within the standard deviation of vertical movement to about 8 km from Sonora where the elevation decreases steadily eastward to a minimum (relative to Z120) of —108 mm at Strawberry.

There appears to be a correlation between the apparent relative elevation changes observed between Z120 and Sonora and geologic observations of deformation of the Table Mountain latite flow in the same area. The Pliocene Table Mountain latite expresses extensional faulting (within the Foothills fault system) with horst and graben structure. The up and down trend of relative elevation changes in this area suggests that this movement may be continuing.

The constant decrease in relative elevation change east of the Melones fault zone occurs almost entirely within the granitic batholith of the Sierra Nevada. There are no mapped structures along this portion of the profile (other than the Melones fault zone). It is possible that the constant relative decrease in elevation to the east is, in effect, an eastward tilt of the Sierran

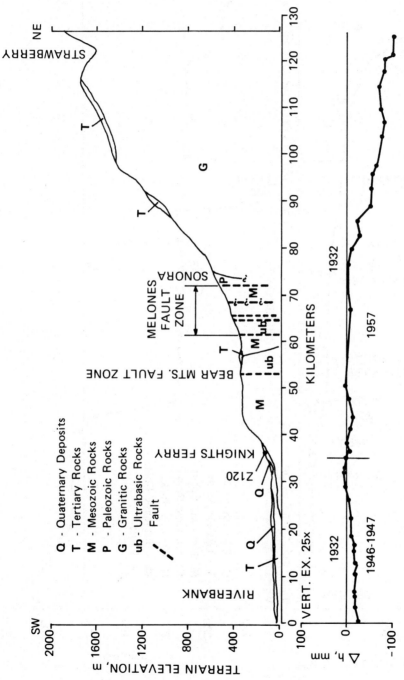

Fig. 3. Riverbank–Strawberry line; profile of change in elevation. Geology based on Rogers (1966) and Strand and Koenig (1965).

246

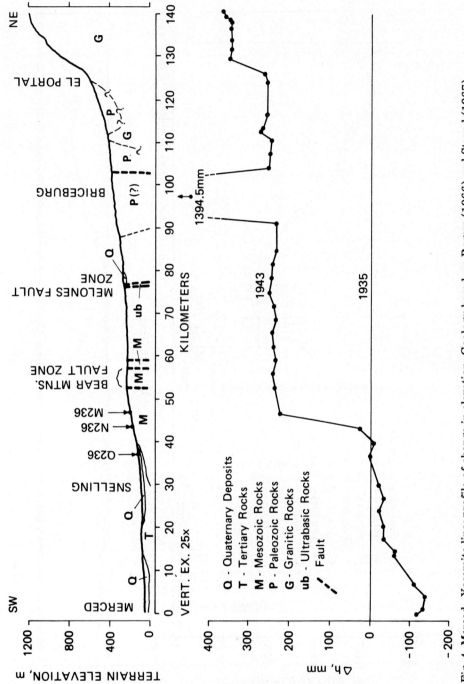

Fig. 4. Merced—Yosemite line; profile of change in elevation. Geology based on Rogers (1966) and Strand (1967).

block with hypothetical hinge near the eastern trace of the Melones fault zone. If an eastward tilt is indeed occurring, then the average rate of tilt would be approximately $8.3 \cdot 10^{-8}$ rad/yr. Apparent eastward tilt was also observed on the parallel Yosemite Junction—Tenaya Lake line (Fig. 1), which had an average rate of tilt $1.22 \cdot 10^{-7}$ rad/yr.

Mersed—Yosemite line

The Merced—Yosemite line generally follows the Merced River from Merced to Yosemite Valley. Repeated second-order leveling along the entire line was done in 1935 and again in 1943. The change in elevation in 1943 was calculated relative to 1935 using bench mark Q236 as the fixed point (Fig. 4).

It is apparent from the profile that relative subsidence in the Central Valley increases toward Merced and reaches a minimum of −117 mm, 5 km north of Merced. The average rate of subsidence at this point was thus approximately 15 mm/yr for the survey period.

The most significant anomaly along this profile occurs directly east of Q236. Approximately 213 mm of apparent uplift has occurred relative to Q236 with approximately 190 mm occurring between bench marks N236 and M236. There are no mapped structures associated with this large magnitude anomaly; however, it occurs within 5 km of the nearest mapped trace of the Bear Mountains fault zone (Rogers, 1966). It occurs near prominent NW-trending topographic lineaments as defined by steep-walled, straight stream segments. In addition, field reconnaissance indicates that the Mesozoic metavolcanic rocks in this area are highly sheared and strongly weathered.

It is possible that the anomaly is due to a blunder error, either from an instrument reading error or subsequent calculation error. However, a field reading error is considered to be unlikely, and the field summary sheets indicated no apparent calculation errors.

The Merced—Yosemite profile does not indicate the graben and horst structure within the Foothills fault system observed on the Riverbank—Strawberry line. In fact, the profile shows little differential change from M236 eastward to El Portal.

At El Portal, there is a 90-mm relative increase in elevation. There are no mapped faults associated with this anomaly; however, it is near the contact of the Sierra Nevada granitic batholith and Paleozoic metasediments. This same trend is also evident just west of El Portal where 30-mm relative increase in elevation occurs. Here, too, it appears to be associated with a small granitic exposure within the Paleozoic rocks.

CONCLUSIONS

Several conclusions may be drawn from the observations of relative elevation changes. Probably the most significant is that minor intrablock deformation is continuing to take place within the Sierran block. This deformation appears to coincide with pre-existing zone of crustal weakness within the Mesozoic Foothills fault system and locations of Late Cenozoic faulting coincident with segments of the Foothills fault system, especially the Melones fault zone, and is expressed by:

(1) Apparent eastward tilt of the Sierran block east of the Melones fault zone.

(2) Less consistent and less evident westward tilt of the Sierran block west of the Melones fault zone.

(3) Variable elevation changes in the vicinity of the Mesozoic Foothills fault system or associated with prominent geologic or structural contacts within the Sierra Nevada.

In addition, subsidence in the Central Valley is variable and localized in magnitude and extent, generally not exceeding 5 mm/yr.

REFERENCES

Bennett, J.H., Taylor, G.C. and Toppozada, T.R., 1977. Crustal movement in the northern Sierra Nevada. Calif. J. Mines Geol., 30 (3): 51—57.
Burnett, J.L. and Jennings, C.W., 1962. Geologic Map of California—Chico sheet. California Division of Mines and Geology, San Francisco, Calif.
Clark, M.M., Sharp, R.V., Castle, R.O. and Harsh, P.W., 1976. Surface faulting near Lake Oroville, California. Bull. Seismol. Soc. Am., 66 (4): 1101—1110.
Holdahl, S.R., 1976. Comment on "new vertical geodesy" by J.H. Whitcomb. J. Geophys. Res., 81 (26): 4945—4946.
Jennings, C.W., 1975. Fault map of California. California Division of Mines and Geology, California Geologic Data Map Series, Map No. 1.
Lydon, P.A., Gay, T.E. and Jennings, C.W., 1960. Geologic Map of California—Westwood sheet. California Division of Mines and Geology, San Francisco, Calif.
Poland, J.F., Lofgren, B.E., Ireland, R.L. and Pugh, R.G., 1975. Land subsidence in the San Joaquin Valley, California, as of 1972. U.S.G.S. Prof. Pap. 436-H, 78 pp.
Rogers, T.H., 1966. Geologic Map of California—San Jose sheet. California Division of Mines and Geology, San Francisco, Calif.
Strand, R.G., 1967. Geologic map of California—Mariposa sheet. California Division of Mines and Geology, San Francisco, Calif.
Strand, R.G. and Koenig, J.B., 1965. Geologic Map of California—Sacramento sheet. California Division of Mines and Geology, San Francisco, Calif.

Tectonophysics, 52 (1979) 249—265
© Elsevier Scientific Publishing Company, Amsterdam — Printed in The Netherlands

EARLY 20TH-CENTURY UPLIFT OF THE NORTHERN PENINSULAR RANGES PROVINCE OF SOUTHERN CALIFORNIA

SPENCER H. WOOD * and MICHAEL R. ELLIOTT

U.S. Geological Survey, 345 Middlefield Road, Menlo Park, Calif. 94025 (U.S.A.)

(Accepted for publication April 4, 1978)

ABSTRACT

Wood, S.H. and Elliott, M.R., 1979. Early 20th-century uplift of the northern Peninsular Ranges province of southern California. In: C.A. Whitten, R. Green and B.K. Meade (Editors), Recent Crustal Movements, 1977. Tectonophysics, 52: 249—265.

Repeated leveling in the northern Peninsular Ranges province identifies an early 20th-century episode of crustal upwarping in southern California. The episodic vertical movement is broadly bracketed between 1897 and 1934, and the main deformation is bracketed within 1906—14 and involved regional up-to-the-northeast tilting of the Santa Ana block of as much as $4 \cdot 10^{-6}$ rad and elevation changes exceeding 0.4 m in the Perris block and parts of the San Jacinto block, Transverse Ranges, and the Mohave block. Primary tide station records containing occasional entries since 1853 at San Pedro and San Diego show no evidence of episodic crustal movement, suggesting that the uplifted area hinged along coastal fault zones forming the west boundary of the Santa Ana block.

Physiographic features and recent studies of Quaternary marine terraces by others show that this episode of regional tilting and uplift is a part of the continuing tectonic process in southern California. A crude, questionable coincidence exists between the uplift episode and a period of increased seismicity (1890—1923) in the northern Peninsular Ranges characterized by a number of moderate-size ($M > 6$) earthquakes on NW-trending strike-slip faults. However, releveling data are too sparse to associate the uplift development clearly with any one event.

INTRODUCTION

The Transverse and Peninsular Ranges of southern California lie immediately south and southwest of the San Andreas fault zone (Fig. 1). Several lines of evidence suggest that the Peninsular Ranges and the western Transverse Ranges are components of the complex strip of continental lithosphere that was rifted from the North American plate in the Late Cenozoic (Larson et al., 1968; Atwater, 1970). This strip of continental crust, including Baja

---

* Present address: Department of Geology and Geophysics, Boise State University, Boise, Ida. 83725 (U.S.A.).

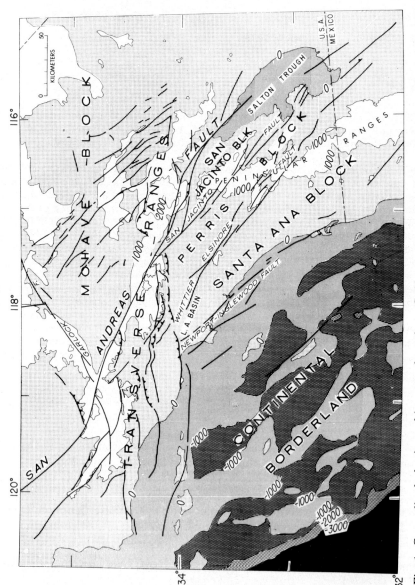

Fig. 1. Generalized physiographic map of southern California showing component tectonic blocks of the Peninsular Ranges and active fault systems. Contour interval is 1000 m.

California (Mexico) and coastal California (U.S.A.), is being translated north-westward with the Pacific plate at a rate of about 5.6 cm/yr (Minster et al., 1974). The relative motion of this crustal unit appears to be accommodated along the many fault zones broadly included in the San Andreas fault system of southern California, largely as slip associated with earthquakes originating on the NW-trending, right-lateral, strike-slip faults in the Peninsular Ranges, the Salton trough, along the main San Andreas zone, and on the E—W-trending high-angle reverse and thrust faults of the Transverse Ranges (Anderson, 1971). The nature and amount of horizontal strain stored elastically or ultimately appearing as either seismic or aseismic slip across active faults along this major plate boundary are a subject of much current research.

Recent discoveries of aseismic, episodic, vertical deformation of regional extent by Castle et al. (1976) have introduced an important new component of the nature of tectonic response in the crust to relative plate motion across southern California. An episode of regional uplift between 1959 and 1963 involved elevation changes of $6 \cdot 10^{-6}$ tilt * (Castle et al., 1976) and horizontal compressive strains of $4 \cdot 10^{-6}$ strain (Thatcher, 1977). Resolution of horizontal strain from older triangulation data is probably $2 \cdot 10^{-6}$, whereas vertical elevation change can be resolved to $1 \cdot 10^{-7}$ tilt, using repeated first-order leveling.

This study concerns relatively large changes in elevation exceeding 10^{-6} tilt in the northern Peninsular Ranges province of southern California that occurred early in the 20th century. These changes probably occurred aseismically and are a part of a more broadly defined uplift episode originally identified in the Transverse Ranges by Castle et al. (1976). This paper supplies an independent confirmation of that early episode of uplift in the Transverse Ranges and focuses on the uplift in the northern Peninsular Ranges province. Lack of early level lines in the eastern Mohave Desert area prevents assessment of its eastern extent, but the early 20th-century uplift appears to have encompassed much of the area of the recent 1957—1974 uplift (Castle et al., 1977) and differs primarily in that it involved the Peninsular Ranges as far south as San Diego.

THE PENINSULAR RANGES PROVINCE

The Peninsular Ranges province can be regarded as a large elongate block uplifted abruptly on its east edge and tilted southwestward (Sharp, 1972). The dominant geologic feature of the province is the great Mesozoic batholith complex that stretches from central Baja California (Mexico) 800 km NNW to the south base of the Transverse Ranges where it appears to be overthrust by the transverse ranges.

The north end of the province is broken into several elongate blocks by

---

* Dimensionless units of tilt are the same as radians (rad) of tilt.

major NW-trending faults of the San Andreas system on which recent right-lateral strike-slip movements have occurred (Fig. 1). Great vertical relief associated with these fault lines may be in part due to differential erosion of fault blocks of different lithology and lateral shifting of topography, rather than to large vertical components of faulting (Jahns, 1954; Sharp, 1972).

The high mountains of the range are bounded from the low-lying Salton trough and Gulf of California by spectacular NE-facing escarpments, 1.8—2.7 km high, disposed in an en echelon pattern (Jahns, 1954). In Baja California, these east-facing, range-front escarpments are active normal faults (Gastil et al., 1975), but the origin of the NE-facing escarpments further north is not well documented.

The crest of the Peninsular Ranges forms a prominent NNW-trending backbone of the range (Fig. 1). Elevations of the highest peaks rise northward from lat 30°31' to lat 34°00', a feature shared by subsea mountains of the offshore borderland province (Emery, 1954). This backbone of elevated terrain is disrupted by San Gorgonio Pass and the San Andreas fault and then seemingly resumes in the Transverse Ranges and the Mohave block, as shown by areas delineated by the 1000-m contour in Fig. 1.

GEODETIC LEVELING DATA

Newly reported data in this paper are elevation changes on two repeated leveling routes from San Diego to the Riverside area, and one repeated route from San Pedro via Riverside to the Salton trough (Fig. 2). Each level survey is referred to the tidal bench mark at San Pedro, which is in turn referred to mean sea level as discussed later. Elevation changes on these three leveling routes provide three independent verifications of the early 20th-century uplift of the northern Peninsular Ranges province because they agree in timing of the episode as well as magnitude of the elevation change where they meet in the Riverside area.

The accuracy to which elevations are determined by first-, second-, and third-order leveling are discussed in Savage and Church (1974), U.S. Department of Commerce (1974), and Bomford (1971). Errors in elevation determinations are composed of random error with a standard deviation proportional to the square root of the distance between points, and of systematic or nonrandom leveling errors. One standard deviation of random error in the elevation change at the end of a 100-km line of compared first-order levels will be less than 50 mm. Compared first- and third-order elevations along a 100-km line will have a standard deviation less than 100 mm associated with the elevation change. Systematic errors are of two types: those that can be assessed through loop closure and those that are independent of loop closure. We have determined loop closures for the surveys reported here. For first-order work the closures are well within one standard deviation of random error. For second- and third-order work we have computed closures on the surveys and have not utilized the few circuits with loop closures

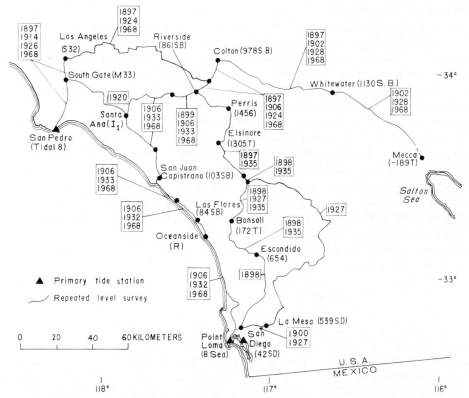

Fig. 2. Routes of profiled elevation changes showing dates of releveling and location of primary tide stations.

above the expected random error. Systematic errors independent of loop closure are chiefly functions of the topographic relief, and although there is some correspondence between topography and elevation changes, the absence of any detailed correlation indicates it is unlikely that these apparent movements are seriously contaminated by systematic error. We also show, where possible, profiles of surveys prior to the datum survey and an additional survey following the survey that detected the elevation change. For example, in Fig. 3, if the 1968—1928 comparison showed an opposite tilt or mirror image of that shown in the 1928—1902 comparison, we would suspect systematic error in the 1928 survey. Since these relations are not observed, systematic errors must be small and unable to account for the observed regional tilting. Furthermore, topographic relief along the coastal route is so slight that topography-related systematic error is very unlikely.

Elevations have been reconstructed from an assumed invariant point (Tidal 8) using only observed elevations; i.e., we have not used adjusted data. Minor uncertainties are associated with starting elevations at Tidal 8 for each survey, but we emphasize that tilts measured by differencing elevations are completely independent of the starting elevations.

A very real problem in reconstructing elevations obtained by leveling surveys results from joining together two surveys at a point when several months to years have elapsed between the two surveys. If there has been vertical movement of the junction bench mark, then the elevation carried forward will be incorrect. These errors are detected when the misclosure on a circuit exceeds the allowable error. It is often difficult to identify whether such movements are tectonic, land-surface subsidence caused by withdrawal of groundwater or oil and gas, or very local instability of the bench mark, unless preceding or succeeding surveys yield information on the history of stability of the junction point. We have, for instance, avoided using the Santa Ana area as a junction point for the 1932 and 1934 surveys because groundwater withdrawal has caused subsidence in the Santa Ana area. We have also avoided using a 1931 survey from Riverside to Mecca because misclosures of several closed loops with the 1928 and 1935 surveys suggest that local tectonic movements up to 100 mm along the Colton to Whitewater route occurred in 1931 or 1932.

San Pedro—Colton—Mecca profile

The level route from San Pedro to Ontario via Los Angeles City Hall (S-32) shows the impressive tilt upward to the north of $4 \cdot 10^{-6}$ rad that defines the south flank of the uplift (Fig. 3). The surveys to bench mark S-32 provide the main constraint on the timing of the uplift, which must have occurred along this line between 1897 and 1914. The 1924 survey simply confirms the elevation change through the Los Angeles basin. Castle et al. (1976) have reported on the continuation of these surveys across the Transverse Ranges to Mohave. Although uplift of the Transverse Ranges may have been as much as 0.5 m by 1914, the 1926 and subsequent surveys show that the uplift collapsed as much as 0.2 m in the northern Transverse Range and the western Mohave block between 1914 and 1926, so that the remaining uplift after 1926 was at most 0.3 m. This profile (Fig. 3) also illustrates upwarping of the northern Perris block centered in the Riverside area and downwarping between Colton and Banning. Although the western part of this downwarped area has more recently been an area of land subsidence owing to groundwater withdrawal, the maximum downwarp is on consolidated sediments that should not subside. Thus, we believe this downwarp to be tectonic warping of the Perris block.

East of Banning is an impressive upwarp centered in the northern Salton trough illustrated by the 1928—1902 comparison in Fig. 3. Tilts on the flanks of this upwarp are $9 \cdot 10^{-6}$ rad. The southeast tilt down the Salton trough is quite uniform to Mecca on the north shore of the Salton Sea. The 1968—1928 comparison shows that most of the uplift remained through 1968, and that parts of the Salton trough had subsided an additional 0.25 m by 1968, suggesting the trough may be undergoing continuing tectonic sub-

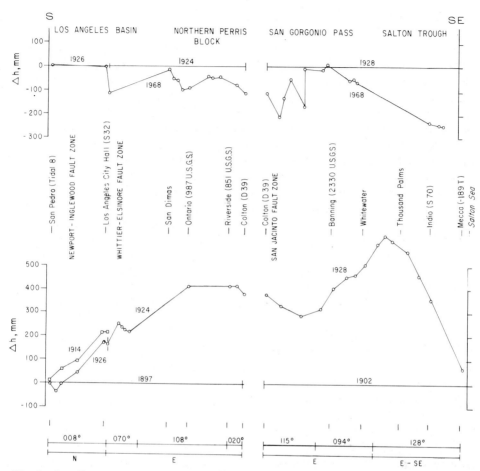

Fig. 3. San Pedro—Mecca profile of elevation changes (1897/1902—1924/26/26) and (1924/26/28—1968): 1897 elevations from U.S. Geol. Survey (1899, p. 381—383), third order, double rodded; 1902 elevations from unpublished U.S. Geol. Survey summary book 9488, third order, double run; 1914 elevations from U.S. Geol. Survey summary book A 9349, second order, class II; 1924 and 1926 elevations from National Geodetic Survey lines "circle 293, 298 and 302", first order; 1928 elevations from National Geodetic Survey line L-11, first order; and 1968 elevations from Southern California Counties Cooperative Level Survey, first order.

sidence. The 1968—1928 local subsidence of Ontario and also the area east of Colton is in part related to groundwater withdrawal in the Chino and Bunker Hill basins. Points that show little or no elevation change correspond to areas of consolidated or crystalline rock and show that most of the earlier uplifted area has not changed significantly in elevation since 1928.

San Diego—Santa Ana—Colton profile

Profiled elevations of these surveys from the San Diego tide station also show an up-to-the-north tilt of the Santa Ana block between 1906 and 1932 (Fig. 4A). We have also shown a comparison of combined 1897/99 third-order surveys with a 1906 first-order survey which demonstrates that an additional 10 cm of uplift may have occurred along this route between 1897 and 1906. Unfortunately, we are unable to connect the 1897/99 segment from Riverside to San Onofre with the 1897/98 inland route (via Escondido and Temecula) because of an unresolvable local surveying error that appears as a 240-mm misclosure in a closed loop of the 1899 and 1897—98 surveys.

The route from San Diego to San Juan Capistrano follows a northwest azimuth along the coast and shows an apparent tilt of 23 cm/100 km $(2 \cdot 10^{-6}$ rad). The tilt steepens to $4 \cdot 10^{-6}$ rad where the route takes a north azimuth 10 km north of San Onofre, indicating that true tilt is up to the NNE direction.

The 1906 survey continues to Barstow, and comparison of a 1924 survey from Riverside to Barstow does not reveal any large tilts. Thus, it appears that the central Mohave block was uplifted with the northern Perris block.

The 1968—1933/34 comparison (Fig. 4A) shows a minor down-to-the-north tilt along the coast and inland toward Riverside. Since there are no subsiding groundwater basins along the coast (except the obvious subsidence bowl at Santa Ana), we consider this apparent downtilt to be tectonic, resulting in as much as 0.07 m subsidence of the previously uplifted area. Otherwise, the essential features of the early uplift remained through 1968.

San Diego—Elsinore—Riverside profile

The inland level route (Fig. 4B) from San Diego to Riverside confirms the uplift observed along the coastal route. The 1897/98 survey, which originated at San Pedro, provides the first geodetic leveling tie between the San Pedro and San Diego primary tide stations. Although this route used third-order instrumentation, 75% of the work utilized second-order methods. Third-order methods are associated with the Riverside to Temecula segment. This 60-km segment is verified by a network of adjacent closed loops. This profiled elevation change between 1897/98 and 1927/35 reflects 1897/98 elevations based on the 1897 San Pedro datum. If a San Diego datum had been used, the differences would decrease by 0.0537 m.

We have used the 1927 first-order inland route to establish elevations at Temecula and Bonsall because the 1935 second-order surveys have poor closure in the area immediately north of San Diego. This reconstruction has remarkably good closure with the 1932/34 coastal route. This profile (Fig. 4b) demonstrates that the Santa Ana block is broadly tilted up to the northeast $4 \cdot 10^{-6}$ rad. The downwarp in the San Luis River Valley centered at Bonsall may be tectonic, for the up-to-the-south tilt is more than twice the expected random error for the second-order comparison from Bonsall to Escondido.

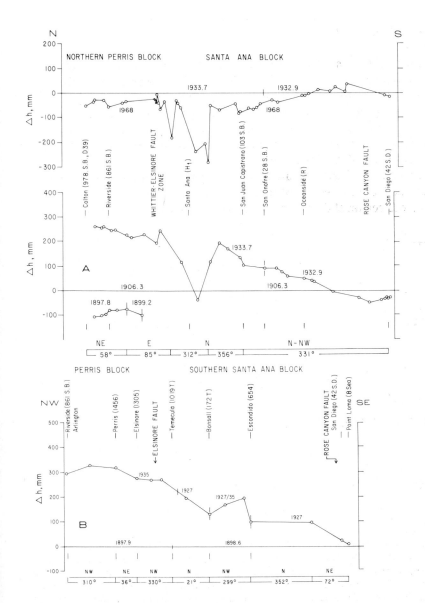

Fig. 4. A. San Diego—Colton profile of elevation changes (1906—1932/34 and 1932/34—1968): 1897 elevations from U.S. Geol. Survey (1899, pp. 381—383), second order; 1906 elevations from U.S. Coast and Geodetic Survey (1914, pp. 32—33), first order; 1932 and 1934 elevations from National Geodetic Survey lines L-570 and L-991, first order; and 1968 elevations from Southern California Counties Cooperative Level Survey, first order. B. San Diego—Riverside profile of elevation changes (1897/98—1927/35): 1897.9 elevations from unpublished U.S. Geol. Survey field books 5831-5835, 5844, third order; 1898.6 elevations from unpublished U.S. Geol. Survey summary book 6396, second order, double rodded; 1927 elevations from National Geodetic Survey line L-305, first order; and 1935 elevations from L-5578, second order.

258

SEA-LEVEL MEASUREMENTS

Sea-level measurements have been made intermittently at San Pedro (Los Angeles outer harbor) and at San Diego since 1853. Annual variations in mean sea level are examined at these two stations for evidence of vertical crustal movement (Fig. 5). Tide-gauge readings provide a continuous record from which the temporal nature of earth movements relative to mean sea level can be determined. However, the records at these two stations show no convincing evidence of regional episodic crustal movement such as that recently identified by Castle et al. (1976) and that reported in this paper from the vertical geodetic record of inland southern California.

Features of the sea-level datum plane in southern California are important.

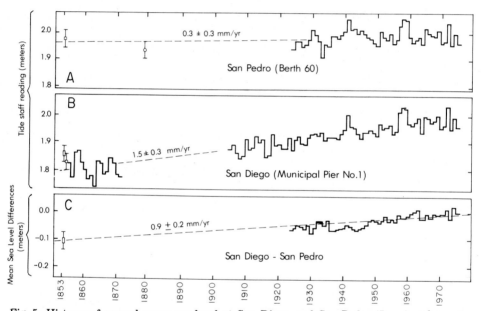

Fig. 5. History of annual mean sea level at San Diego and San Pedro (Los Angeles outer harbor) from records of the National Ocean Survey. The dashed lines are straight-line trends of the sea-level records. The range of possible slopes of these linear fits is written above each line. A. The San Pedro primary station has always been at the San Pedro waterfront on Palos Verdes Peninsula. Points with error bars are the means of 120 days (Nov. 19—March 23, 1854) and 47 days (Sept. 14—Oct. 24, 1878 and Nov. 7—14, 1878). B. The San Diego primary station was at the Quarantine Station on Point Loma for the 1835—1872 and the 1906—1925 series. The station was moved inland 8 km to the San Diego waterfront for the 1925—present series. Corrections obtained from level connections between tidal series have been applied. Points with error bars are averaged monthly means for October—December, 1853 and April—December, 1854. C. The lowermost graph is a plot of the difference in mean sea level between San Diego and San Pedro and can be interpreted as the history of elevation change of San Pedro relative to San Diego. The 1853 difference in mean sea level is based on a partial annual mean (120 days), but should be within ±0.03 m of the annual mean (Marmer, 1951, p. 65).

Castle et al. (1976) determined from data published by Hicks (1972) that the tidal bench marks at San Pedro are undergoing upward crustal movement at about the same rate as the eustatic sea-level rise so that the tide staff readings have not changed significantly (Fig. 5). We show here that the San Diego station has a history of relative vertical stability. However, elevations are more conveniently referred to the San Pedro tidal bench mark (Tidal 8) because level surveys are more frequently tied to San Pedro than to the San Diego tidal bench marks.

It is now recognized that the sea-level reference surface along the Pacific Coast of the United States slopes down to the south with respect to a geodetic level surface (Braaten and McCombs, 1963; Sturges, 1967; Balazs, 1973; and Lisitzin, 1974). The sea-level slope between San Pedro and San Diego has been determined by leveling between tide stations on three separate occasions with remarkably similar results (Table I). The slope does not appear to have changed significantly between 1898 and 1968. The cause of this slope is not known.

Sturges (1967) attributed the southward slope to systematic surveying error associated with leveling in a north or south direction with a magnitude of −35 mm/degree of latitude. Balazs (1973) cites similar results on the east coast of North America and elsewhere in the world. The question of which surface is more closely aligned with the geopotential surface remains unresolved. Balazs (1973) recommends against mixing steric leveling results in adjustments of geodetic leveling networks.

TABLE I

Difference in elevation as determined by geodetic levelings between San Pedro and San Diego, and the discrepancy between the mean sea-level surface and the level surface of geodetic leveling

Date of leveling	Difference in elevation between BM Tidal 8 (San Pedro) and BM Tidal 4 (San Diego) *	Difference in elevation of mean sea-level at San Diego with respect to San Pedro *
1897—1898	+194 ± 78 mm	−125 ± 78 mm **
1932—1934	+191 ± 19 mm	−110 ± 19 mm ***
1968—1969	+206 ± 19 mm	−110 ± 19 mm ***

* Random error depends upon the square root of the distance along the level route which is about 380 km. The 1897—1898 elevations probably meet second-order Class II standards with a standard deviation of ±4 mm $\sqrt{(380)}$. The 1932—1934 and 1968—1969 elevations are first order with a standard deviation of random error of about 1 mm $\sqrt{(380)}$.
** From U.S. Geological Survey (1899, p. 460). We have recalculated this difference from unadjusted levels and derive−150 mm.
*** From Balazs (1973).

Sea-level variations at San Pedro and San Diego

Secular sea-level variations of a non-tectonic origin at an individual station may be attributed to a variety of causes, as discussed by Lisitzin (1974). Much of the oceanographic noise and the effect of eustatic sea-level rise can be removed by differencing with nearby stations. We have differenced the results from two primary stations, 152 km apart, along the southern California coast. What is left should be a record of changes of elevation of one station relative to the other or local long-term variations in salinity, meteorology, coastal currents, or changes in amplification of sea-level fluctuations due to harbor improvements. The individual station records seem to indicate that sea level has steadily risen at San Diego at a rate of 1.5 ± 0.3 mm/yr and at San Pedro at an insignificant rate of 0.3 ± 0.3 mm/yr since 1853 (Fig. 5). The eustatic rise of sea level is reported to be $1.0-1.1$ mm/yr (Lisitzin, 1974, p. 183). Subtracting the eustatic rise, San Diego appears to be subsiding at an average rate of 0.4 ± 0.3 mm/yr since 1853, while San Pedro peninsula appears to be rising at an average rate of 1.2 ± 0.4 mm/yr (Fig. 5). The difference curve (San Diego—San Pedro, Fig. 5) has the least noise and indicates relative upward movement of San Pedro at a rate of 0.9 ± 0.2 mm/yr.

Wehmiller et al. (1977) have obtained geologic estimates of uplift rates from marine terrace studies on the tectonic blocks upon which the San Pedro (Palos Verdes peninsula) and the San Diego (Point Loma) tide stations are situated. They report $0.6-0.9$ mm/yr of uplift for Palos Verdes peninsula marine terraces, and 0.16 mm/yr of uplift for the Point Loma terraces near San Diego. Thus, the geologic studies and the tide-gauge data are in general agreement and emphasize that the San Diego area has relative vertical stability while Palos Verdes is continuously moving upward.

Examination of the difference record shows no evidence of a major episodic or coseismic vertical movement. As noted by Wyss (1975), there is no evidence of coseismic movements associated with the 1933 Long Beach earthquake ($M = 6.2$), although Leypoldt (1943) and Nason (1976) have noted a local oscillation and subsidence event of 0.07 m elevation that occurred from March 1928 to April 1929 on the Los Angeles harbor inner tide gauge 5 km north of the San Pedro gauge. In 1935 annual mean sea level at San Pedro rose about 0.04 m while the sea level at San Diego remained relatively constant. This apparent subsidence in 1935 at San Pedro seems to be a permanent movement and could be tectonic in nature. It cannot be related to Wilmington oil-field subsidence, for production of subsurface fluids did not begin there until 1937. In 1964 annual mean sea level dropped more at San Diego than at San Pedro; however, since sea level underwent a sharp decline at both stations, this drop may be an oceanographic event preferentially amplified on the San Diego record.

Tidal 8 reference mark at San Pedro

Unfortunately, the Tidal 8 reference mark is an upward-moving reference mark. All elevation changes computed with respect to Tidal 8 can be increased by an added factor of +0.9 ± 0.2 mm/yr multiplied by the time interval of the elevation change to refer to a more invariant point such as in the San Diego area. The tidal data (Fig. 5) imply that the difference in elevation between BM Tidal 8 (San Pedro) and BM Tidal 4 (San Diego) has systematically increased +32 ± 14 mm from 1898 to 1934 and +31 ± 14 mm from 1934 to 1968; however, changes of this magnitude are not evident in the elevation difference obtained by three repeated levelings between these points (Table I). Data in Table I shows the mean sea-level datum has consistently sloped down to the south so that the elevation difference has ranged only −125 ± 78 to −110 ± 19 mm between the mean sea-level at San Pedro and San Diego. It is possible that changes in the elevation difference are masked in the random error indicated in Table I that would be associated with leveling over this 380-km route. Thus, it may be only fortuitous that the surveyed elevation difference appears to have changed so very little from 1898 to 1968.

Because elevations have been referred to an invariant Tidal 8 reference, bench marks at San Diego appear to have subsided 0.03 m in the 1932.9–1906.3 profiled comparison of Fig. 4a, but in fact these marks were probably stable and it was Tidal 8 that rose 0.03 m with respect to mean sea level.

DISCUSSION

The early 20th-century uplift of southern California

Vertical crustal movement in the northern Peninsular Ranges is broadly bracketed between 1897 and 1932, but the San Pedro–Riverside profile (Fig. 3) indicates a main deformation episode between 1906 and 1914. It was clearly an episodic uplift for it does not continue into the 1928/34–1968 interval. The data reported here extend the area of tilting considerably further south of the area that was deformed in the recent 1959–74 southern California uplift. A generalized contour map of the uplift in the Peninsular Ranges is shown in Fig. 6.

While the early 20th century and the more recent 1959–74 uplift of the Transverse Ranges and the western Mohave block might be explained as a response to N–S compressive forces (Thatcher, 1977), the 4×10^{-6} rad, up-to-the-northeast tilt of the west slope of the Peninsular Ranges and the Los Angeles basin may relate to whatever processes have uplifted the present NNW-trending crest of the range. It is important to note that the episodic regional tilt and uplift extend into the Los Angeles basin — a feature that has clearly had a Late Cenozoic history of negative elevation changes but that can also be considered a part of the Santa Ana structural block.

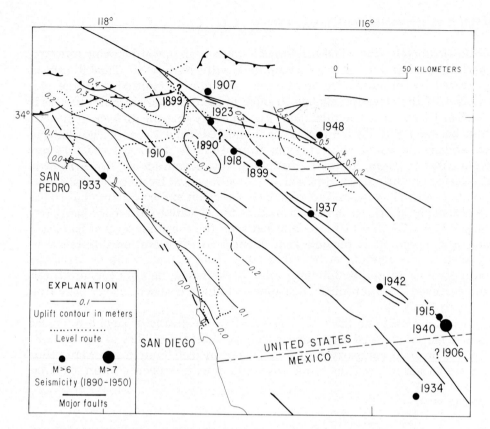

Fig. 6 The early 20th-century uplift in the Peninsular Ranges (1897—1935). Locations of moderate-size earthquakes (1890—1950) from Hanks et al. (1975).

The sea-level data show that neither San Pedro nor San Diego was significantly uplifted in this episode. Therefore, the early 20th-century uplift may have hinged along the coast or perhaps along the Newport—Inglewood fault zone and fault zones to the south that apparently form the western boundary of the Santa Ana block.

Relationship to geology, physiography and seismicity

A remarkably good correlation with the sense of tilting of the Santa Ana block reported here (Figs. 4a and 6) is also found in the deformation of Pleistocene marine and coastal stream terraces (McCrory and Lajoie, 1977). Terraces rise to the northwest from central San Diego County to San Onofre. North of San Onofre, terraces are obliterated by erosion. The higher, and presumably older, terraces at San Onofre slope progressively more steeply toward the ocean. Slopes on upper terraces reach 0.1 rad, suggesting that average rates of tilting during the Quaternary may be about the same order of magnitude as rates derived from the limited history (1906—68) of repeated leveling $4 \cdot 10^{-6}$ rad/62 yr.

The contoured elevation changes in Fig. 6 indicate a NW-trending upwarp of the Perris block. Upwarping along a N—S axis has been suggested by Woodford et al. (1971, p. 3444) to explain diversions of the ancestral Pliocene San Jacinto River, but there are no reports of warping of Quaternary deposits analogous to the upwarp shown in Fig. 6.

The broad upwarp east of San Gorgonio Pass on Fig. 4 lies somewhat east of the NNW-trending backbone of the Peninsular Ranges, but is generally aligned with the north-trending region of higher elevations that continues into the Transverse Ranges and the Mohave block (Fig. 1). This is also a region of maximum elevation change in the recent (1959—74) uplift (Castle, et al., 1977), which suggests that the early uplift may have affected the eastern Transverse Ranges in a similar manner.

The 1968—1928 comparison in Fig. 3 shows continuing subsidence of the Salton trough, which is expected of this large area with elevations below sea level. It is difficult to separate what may be broad uplift of the area from tectonic subsidence of the trough, for these may be separate processes, i.e., continuing subsidence of the trough superposed on episodic broad, crustal uplift. This possibility is suggested because the sea-level-based elevation change 1902—1928 at Mecca is very small, yet the tilt is relatively large. Between 1928 and 1968 Mecca dropped 0.27 m with respect to Tidal 8, but we do not know if this was an episodic or a continuous movement. If subsidence of the trough is a continuous process, the minimum rate suggested by these elevation changes is 6 mm/yr.

Hanks et al. (1975) and Thatcher et al. (1975) have summarized early seismicity in the northern Peninsular Ranges. It is not likely that elevation changes reported here are coseismic because areas of steepest tilt or sharp upwarp do not correspond to the epicentral areas of the moderate-size earthquakes shown in Fig. 6.

While there is no simple correspondence of seismicity with the vertical deformation reported here, it is noteworthy that seismicity in the northern Peninsular Ranges, characterized by a number of moderate-size ($M > 6$) earthquakes on NW-trending, right-slip faults, was higher in a 33-year interval (1890—1923) containing the main uplift episode than in the following 30 years. Seismicity was relatively low in the Transverse Ranges and northern Peninsular Ranges in the interval 1927—52, and indeed, we do not recognize an episode of regional vertical crustal deformation in the interval 1927—1959.

Although the data suggest a crude and questionable temporal relation between episodes of vertical deformation characterized by regional tilts in excess of $4 \cdot 10^{-6}$ rad and intervals of increased seismicity characterized by moderate-size earthquakes ($M > 6$), the repeated leveling data are too sparse to associate the uplift development as clearly precursory to any one of the moderate-size earthquakes.

264

ACKNOWLEDGEMENTS

We thank J.R. Hubbard for providing the tidal data from the National Ocean Survey archives. The paper has benefited from helpful discussions and reviews from R.O. Castle, R. Burford, R.V. Sharp, and E.I. Balazs.

REFERENCES

Anderson, D.L., 1971. The San Andreas fault. Sci. Am., 225: 53—67.
Atwater, T., 1970. Implications of plate tectonics for the Cenozoic evolution of western North America. Geol. Soc. Am. Bull., 81: 3513—3536.
Balazs, E.I., 1973. Local mean sea level in relation to geodetic leveling along the United States coastlines. Unpubl. manuscr., N.G.S., N.O.S., N.O.A.A., Dept. Commerce, 9 pp.
Braaten, N.F., and McCombs, C.E., 1963. Mean sea level variations indicated by a 1963 adjustment of first-order leveling in the United States. Unpubl. tech. rep., U.S. Coast and Geodetic Survey (U.S. Dept. of Commerce), 7 pp.
Bomford, G., 1971. Geodesy, 3rd ed. Oxford Univ. Press, London, 731 pp.
Castle, R.O., Church, J.P. and Elliott, M.R., 1976. Aseismic uplift in southern California. Science, 192: 251—253.
Castle, R.O., Elliott, M.R. and Wood, S.H., 1977. The southern California uplift (abstr.). EOS, Trans. Am. Geophys. Union, 58: 495.
Emery, K.O., 1954. General geology of the offshore area, southern California. In R.H. Jahns, (Editor). Geology of Southern California, Ch II. Calif. Div. Mines Geol., Bull. 170: 107—111.
Gastil, R.G., Phillips, R.P. and Allison, E.C., 1975. Reconnaissance geology of the state of Baja California. Geol. Soc. Am., Mem. 140: 170 pp.
Hanks, T.C., Hileman, J.A. and Thatcher, W., 1975. Seismic moments of the larger earthquakes in the southern California region. Bull. Geol. Soc. Am., 86: 1131—1139.
Hicks, S.D., 1972. Long-period variations in secular sea level trends. Shore and Beach, 40: 32—36.
Jahns, R.H., 1954. Geology of the Peninsular Ranges province, southern California and Baja California (Mexico). In: R.H. Jahns (Editor), Geology of Southern California. California Div. Mines Bull., 170: 29—52.
Larson, R.L., Menard, H.W. and Smith, S.M., 1968. Gulf of California, a result of ocean floor spreading and transform faulting. Science, 161: 68—71.
Leypoldt, H., 1934. Earth-movements in California determined from apparent variation in tidal datum planes. Bull. Seismol. Soc. Am., 24: 63—68.
Lisitzin, E., 1974. Sea-level Changes. Elsevier, Amsterdam, 286 pp.
Marmer, H.A., 1951. Tidal datum planes, Spec. Publ. No. 135, revised (1951) ed., U.S. Coast and Geodetic Survey, Department of Commerce.
McCrory, P. and Lajoie, K.R., 1977. Marine terrace deformation, San Diego County California (abstr.). Int. Symp. Recent Crustal Movements, Stanford Univ.
Minster, J.B., Jordan, T.H., Molnar, P. and Haines, E., 1974. Numerical modelling of instantaneous plate tectonics. Geophys. J.R. Astron. Soc., 36: 541—576.
Nason, R.D., 1976. Vertical movements at Los Angeles harbor before the 1933 Long Beach earthquake. EOS. Trans. Am. Geophys. Union, 57: 1012.
Savage, J.C. and Church, J.P., 1974. Evidence for postearthquake slip in the Fairview Peak, Dixie Valley, and Rainbow Mountain fault areas of Nevada. Bull. Seismol. Soc. Am., 64: 687—698.
Sharp, R.P., 1972. Geology field guide to southern California. Wm. C. Brown, Dubuque, Iowa, 181 pp.
Sturges, W., 1967. Slope of sea level along the Pacific Coast of the United States, J. Geophys. Res., 72: 3627—3637.

Thatcher, W., 1977. Episodic strain accumulation in southern California. Science, 194: 691—695.
Thatcher, W., Hileman, J.A. and Hanks, T.C., 1975. Seismic slip distribution along the San Jacinto fault zone, southern California, and its implications. Bull. Geol. Soc. Am. 86: 1140—1146.
U.S. Coast and Geodetic Survey, 1914. Fourth general adjustment of the precise level net in the United States and the resulting standard elevations. Spec. Publ. 18: 328 pp.
U.S. Dept. Commerce, 1974. Standards of accuracy, and general specifications of geodetic control surveys, Federal Geodetic Control Committee, National Geodetic Survey, N.O.S., N.O.A.A., Rockville, Md., 12 pp.
U.S. Geological Survey, 1899. Twentieth Annual Report of the U.S. Geological Survey, Part 1 — Director's Report, Including Triangulation and Spirit Leveling. Government Printing Office, Washington, D.C., 551 pp.
Wehmiller, J.F., Lajoie, K.R., Kvenvolden, K.A., Peterson, E., Belknap, D.F. Kennedy, F.L., Addicott, W.O., Vedder, J.G. and Wright, R.W., 1977. Correlation and chronology of Pacific coast marine terrace deposits of continental United States by fossil amino acid stereochemistry — technique, evaluation, relative ages, kinetic model ages, and geological implications. U.S. Geol. Surv. Open-file rep. 77-680, 160 pp.
Woodford, A.O., Shelton, J.S., Doehring, D.O. and Morton, R.K., 1971. Pliocene-Pleistocene history of the Perris Block, southern California. Bull. Geol. Soc. Am. 82: 3421—3448.
Wyss, M., 1975. Mean sea level before and after some great strike-slip earthquakes. Pure Appl. Geophys., 113: 107—118.

Tectonophysics, 52 (1979) 267—275
267

RECENT QUATERNARY TECTONICS IN THE HELLENIC ARC: EXAMPLES OF GEOLOGICAL OBSERVATIONS ON LAND

JACQUES ANGELIER

Laboratoire de Tectonique Comparée, Département de Géologie Structurale, Université de Paris VI, 75230 Paris Cedex 05 (France)

(Accepted for publication April 26, 1978)

ABSTRACT

Angelier, J., 1979. Recent Quaternary tectonics in the Hellenic Arc: examples of geological observations on land. In: C.A. Whitten, R. Green and B.K. Meade (Editors), Recent Crustal Movements, 1977. Tectonophysics, 52: 267—275.

Upper Pleistocene and Holocene tectonic movements in the Aegean region are analyzed by geological means (deformation of shorelines, faults in Quaternary deposits, historical seismicity). Examples from Crete, Karpathos, Milos, Chios and Samos are presented. While subduction, indicated by geophysical data, occurs beneath the Hellenic Arc, extensional tectonics (i.e., normal faulting) takes place within and behind the arc, resulting in a slight expansion of the Aegean region towards the Eastern Mediterranean.

INTRODUCTION

Numerous neotectonic studies have been carried out in the Aegean region during the last few years (Mercier et al., 1972, 1976; Angelier, 1973, 1977; Péchoux et al., 1973; Dufaure et al., 1975; Mercier, 1976; Philip, 1976). They are the complement of the geophysical data (not discussed here). The geology of the Aegean region has been summarized by Aubouin (1973).

In this paper we describe examples of Middle—Late Pleistocene and Holocene tectonic features. They are in the Hellenic (=Aegean) Arc (outer arc: Crete, Karpathos; inner, volcanic arc: Milos), or just behind it (Samos, Chios). A more general analysis of Late Miocene and Plio—Quaternary tectonics in the Aegean Arc has been proposed elsewhere (Angelier, 1977).

Table I summarizes the main geological methods applied to neotectonic analysis. On each figure the localization is shown in (a); the stereograms are Schmidt's projections of the lower hemisphere (where fault planes are shown as thin lines, normal striations as black dots, and directions of extension as large arrows).

268

TABLE I

Main geological methods applied to neotectonic analysis

Evidence of recent deformation	Examples	Particular techniques
Altimetric variations of shorelines	Fig. 1 Fig. 2	altimetric measurements paleontological and radiometrical dating, morphology
Tectonic structures (especially faults) in:		
marine terraces	Fig. 3	stratigraphy of quaternary deposits
dunes	Fig. 4	(morphology, paleontology, radio-chronology)
alluvia, fans	Fig. 5	tectonic analysis (especially fault
slope deposits, screes	Fig. 6	mechanisms), including:
	Fig. 7 ··········	separation of successive movements
Morphological data	Fig. 8	search of recent scarps, photogeology study of morphologies related to uplift or subsidence
Historical seismicity	see text	compilation of available descriptions; search and analysis of structures comparison with focal mechanisms of earthquakes

TECTONIC ANALYSIS OF RECENT DEFORMATIONS

Deformation of Quaternary shorelines

(1) Middle—Late Pleistocene shorelines: a comparison of their deformations in the Western Mediterranean ("arc" of Gibraltar and Alboran Sea) and the Eastern Mediterranean (Aegean Arc: Crete) was carried out in a collective study (Angelier et al., 1976). The greatest amplitudes were found in the Aegean Arc: marine terraces are strongly uplifted in Crete (up to 4 cm per century and more, for the last 125,000 years), and submerged in the Cyclads Archipelago (inner arc). Moreover, altimetric changes are sharp (Fig. 1) because of block faulting.

(2) Historical shorelines: a rapid recent uplift of the whole of western Crete is indicated by the elevation of ancient shorelines (Spratt, 1856) 2000 years old (^{14}C: Hafemann, 1965), now up to 9 m at Elafonisi, 15 km west of Paleochora (Fig. 2). The rate of uplift is about 40—60 cm per century (average rate for the last 2000 years).

Tectonic structures in Quaternary deposits (especially Riss, Tyrrhenian, Würm)

In addition to being deformed on a broad scale, marine terraces are cut by faults which are generally normal, as on Karpathos (Fig. 3). Consolidated

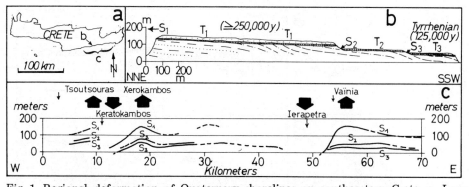

Fig. 1. Regional deformation of Quaternary shorelines on southeastern Crete. a. Location. b. Typical section of the three main marine terraces (T_1, T_2, T_3), with their shorelines (S_1, S_2, S_3), above Neogene deposits: approximate radiometric ages are given after Angelier et al., 1976. c. Altimetric changes of shorelines S_1, S_2, S_3 along the coast.

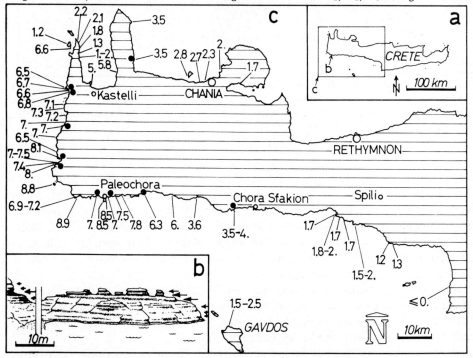

Fig. 2. Regional deformation of ancient historical shorelines, about 2000 years old, on western Crete. a. Location. b. Marine erosional levels near Paleochora: numerous nutches in addition to the main shoreline, complicate the pattern. c. Map of present elevations, in meters, of the main shoreline. It is emphasized that some heights may belong to slightly older or more recent levels — see (b): the ages indicated by Hafemann, at places shown as black dots, range from 1620 to 2375 years. In Gavdos the precise age is unknown, but the terrace contains remains of pottery. After observations of Spratt (1856), Hafemann (1965), Bonnefont (1971), Dermitzakis (1973), and the author. The heights of 1.2—2.2 m indicated in the northwestern peninsula (NW of Kastelli), after Dermitzakis (1973), are probably underestimated (P. Pirazzolli, pers. comm.).

270

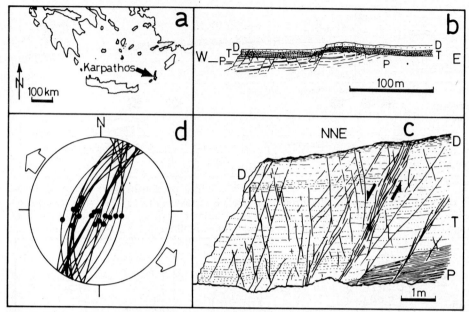

Fig. 3. Normal faults in a Tyrrhenian marine terrace at the southern end of the island of
Karpathos. a. Location. b. Section of a fault scarp: P = marine Pliocene; T = Tyrrhenian
deposits with *Strombus bubonius* LMK; D = dunes. c. Zone of conjugate normal faults.
d. Corresponding stereogram (see text).

Upper Pleistocene dunes are often faulted, for example on Karpathos, above
the marine terrace, or on Milos (Fig. 4), where marine deposits are below sea
level. On Milos (Fig. 4), Karpathos (Angelier, 1973), and in other places, the
main Recent faulting is similar to the older faulting: (1) intense normal faulting during the Late Pliocene or Early Quaternary; (2) unconformable sedimentation during the Middle—Late Pleistocene; (3) reactivation of normal
faults, cutting the latter deposits.

Successive movements closely spaced in time have been observed: for
example, on Samos, two consolidated conglomeratic fans are superposed:
both are cut by normal faults, but a previous fracturation, which is mechanically similar, cuts only the lower fan (Fig. 5).

In the same way the SW-facing scarp of Mt. Kedhros, on Crete, is covered
by screes, the older being consolidated but the younger unconsolidated.
These screes are cut by the NW—SE normal fault on which discrete successive movements have occurred with different mechanisms (Fig. 6). In the
vicinity, near Plakias, three successive tectonic movements are (Fig. 7): (1)
large normal faulting at the end of the Pliocene or during Early Quaternary
times; (2) slight compression generating small reverse faults; (3) extensional
movements again, with probable reactivation of normal faults.

Recent movements are also inferred from morphological data: some fault

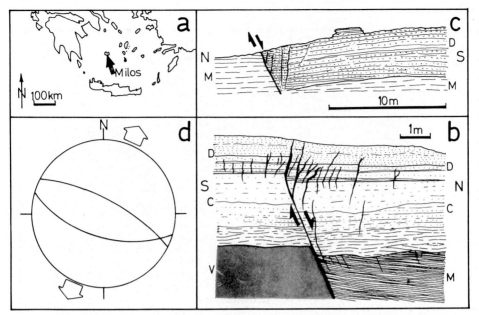

Fig. 4. Recent normal faults northeast of Adamas on the island of Milos. a. Location. b. Section of a fault: V = volcanic Pliocene; M = marine Pliocene; C and D = alluvia and dunes of Upper Pleistocene. Note (1) the curvature of the normal fault near the surface so that it resembles an *open* reverse fault; (2) the evidence of previous normal faulting: the unconformable deposits C and D overlie M on one side of the fault, V (older) on the other side. c. Section of a more important Recent fault. d. Corresponding stereogram, for both faults. Striae of the last movement were not observed, but other observations suggest that the movement is probably purely normal.

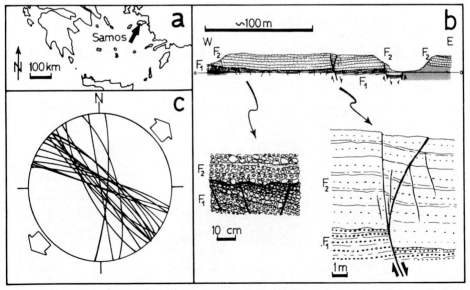

Fig. 5. Fracturation and normal faulting in consolidated alluvial fans west of Marathokambos on the island of Samos. a. Location. b. Schematic section of the two superposed fans F_1 and F_2, successively faulted, and details (note the curvature of one fault near the surface, as on Fig. 4b). c. Stereogram.

272

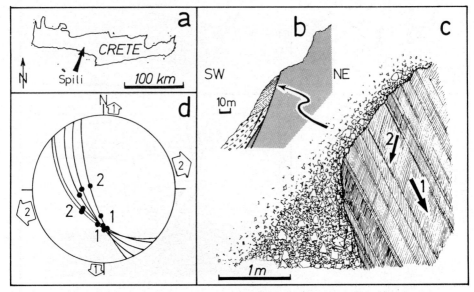

Fig. 6. Recent reactivation of a great normal fault near Spili (Crete). a. Location. b. Schematic section. c. Detailed view: large grooves indicating a normal—sinistral movement *1*, polished by thin recent striations; *2*, purely normal, cutting unconsolidated screes (probably Würm). d. Stereogram: comparison of movements *1* and *2*.

scarps give evidence of recent motion (Fig. 8), and particular morphologies are related to rapid uplift (see Bonnefont, 1971, for Crete).

Relationship with historical seismicity

An example may be described on Chios (Angelier and Tsoflias, 1976): an old N—S fault was reactivated during the Quaternary, since consolidated screes are faulted; on this fault line, near Tholopotami, large open cracks

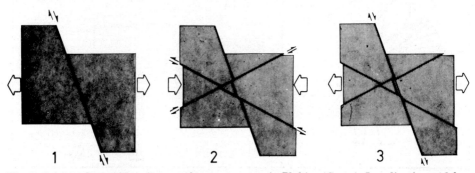

Fig. 7. Schematic section of successive movements in Plakias (Crete). Localization: 13 km WSW of Spili (see Fig. 6a). Explanation in text.

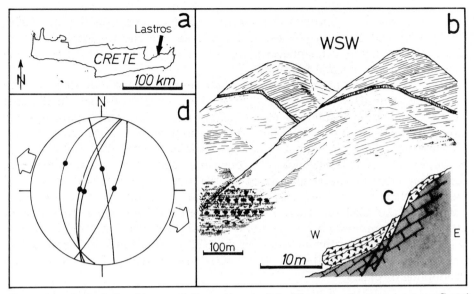

Fig. 8. Morphological evidence of Recent normal faulting near Lastros (eastern Crete). a. Location. b. View of the almost constant height scarp, in homogeneous limestones. c. Detailed section of consolidated slope detritus cut by a similar fault near Mesa Mouliana. d. Stereograms of these faults.

probably originated during a 1546 earthquake, according to contemporaneous documents (note that landslides have also occurred there, some of which were caused by earthquakes, but they have no direct tectonic meaning). The direction of extension is close to WSW—ENE, and seems roughly compatible with focal mechanisms of recent earthquakes near Chios (1949, 1969).

No tectonic structure clearly caused by an earthquake with a known focal mechanism has been found in the islands we have studied. Detailed geophysical studies applied to present surface tectonics are poor or absent (i.e, precise geodetic levelling, strain and stress measurements, precise location of earthquake foci, including microseismicity).

GENERAL PATTERN OF RECENT AEGEAN MOVEMENTS

Middle—Upper Pleistocene and Holocene tectonic mechanisms

(1) Predominance of extensional movements (normal faulting): on the emerging parts of the Aegean Arc and the Southern Aegean, most tectonic mechanisms observed by geological means are extensional. The Africa—Eurasia convergence in the Mediterranean region occurs in two different ways: (i) *westwards, near the Strait of Gibraltar* — strike-slip faulting occurs (continental compression especially, see Bousquet's paper in this volume); (ii) *eastwards, in southern Greece* — subduction of the Eastern Mediterra-

nean crust has been proposed on geophysical criteria (MacKenzie, 1970; Rabinowitz and Ryan, 1970; etc.); the Aegean Arc and the Aegean region, above the plunging lithospheric plate, are mainly affected by extensional tectonics (intense normal faulting).

(2) Traces of compression on the outer margin of the arc: Recent compressional structures (reverse faults) have been clearly observed in the area of the Ionian Islands (Mercier et al., 1972). In addition, on Karpathos, some compressional structures exist but are extremely subdued (Angelier, 1973). The scarcity of compressional structures leads us to make the following hypotheses: (i) *recent compression has occurred near the trenches* — since even the submerged southern margin of the Cretan Arc is affected by normal faulting (Nesteroff et al., 1977), thrusts and transform faults can only be present in the Hellenic trenches and seaward: (ii) *the apparently greater intensity of compression on the Ionian Islands* (Mercier et al., 1972) may be principally related to the closeness of the NW termination of the typical subduction zone beneath the arc (i.e., with the decrease of the mechanical decoupling of plates).

Subduction and the Aegean expansion

Except for a small generalized compressional event during the Early Quaternary s.l. (Mercier et al., 1976), the Plio—Quaternary pattern of normal faults in the Aegean region (Aubouin, 1973) has not changed fundamentally since the Middle Pliocene. A neotectonic analysis including a comparison with available seismic data has been proposed elsewhere (Angelier, 1977). The extensional movements during Upper Pliocene and Early Quaternary times — and probably also during Recent times (with less numerous but roughly analogous observations) — have resulted in an expansion of the Aegean Arc and the Southern Aegean towards the Eastern Mediterranean Sea. This implies that, as Plio—Quaternary subduction processes were occurring beneath the Hellenic Arc, the Crete—Libya rate of convergence was slightly higher than that of the Middle Aegean—Libya.

ACKNOWLEDGEMENTS

The research was supported by the French C.N.R.S./I.N.A.G. (A.T.P. Géodynamique). We must add that without the help of the group "Californie" of this A.T.P., this paper could not have been presented by the author at the 1977 RCM Symposium; special thanks are due to Dr. René Blanchet.

REFERENCES

Angelier, J., 1973. Sur la néotectonique égéenne: failles anté-tyrrhéniennes et post-tyrrhéniennes dans l'île de Karpathos (Dodécanèse, Grèce). C.R. Somm. Soc. Géol. Fr., (7) 15: 105—109.

Angelier, J., 1977. Sur l'évolution tectonique depuis le Miocène supérieur d'un arc insulaire méditerranéen: l'arc égéen. Rev. Géogr. Phys. Géol. Dyn., (2), 19 (3): 271—274.

Angelier, J. and Tsoflias, P., 1976. Sur les mouvements mio-plio-quaternaires et la séismicité historique dans l'île de Chios (Grèce). C.R. Acad. Sci. Paris, D, 283: 1389—1391.

Angelier, J., Cadet, J.P., Delibrias, G., Fourniguet, J., Gigout, M., Guillemin, M., Hogrel, M.T., Lalou, C. and Pierre, G., 1976. Les déformations du Quaternaire marin, indicateurs néotectoniques. Quelques exemples méditerranéens. Rev. Géogr. Phys. Géol. Dyn., (2) 18: 427—448.

Aubouin, J., 1973. Des tectoniques superposées et de leur signification par rapport aux modèles géophysiques: l'exemple des Dinarides: paléotectonique, tectonique, tarditectonique, néotectonique. Bull. Soc. Géol. Fr., 15 (7): 426—460.

Bonnefont, J.C., 1971. La Crète. Etude morphologique. Thèse Géographie Université Paris. University of Lille III, 1972, p. 845.

Dermitzakis, M.D., 1973. Recent tectonic movements and old strandlines along the coasts of Crete. Bull. Geol. Soc. Greece, 10 (1): 48—64.

Dufaure, J.J., Kadjar, M.H., Keraudren, B., Mercier, J., Sauvage, J. and Sebrier, M., 1975. Les déformations plio-pléistocènes authour du golfe de Corinthe. C.R. Somm. Soc. Géol. Fr., suppl. 17 (1): 18—20.

Hafemann, D., 1965. Die Niveauänderungen an den Küsten Kretas seit dem Altertum. Verlag Akad. Wiss. Lit. Mainz, Abh. Math. Naturwiss. Kl., 12: 709—788.

MacKenzie, D.P., 1970. Plate tectonics of the Mediterranean region. Nature, 226: 239.

Mercier, J., 1976. La néotectonique, ses méthodes et ses buts. Un exemple: l'arc égéen (Méditerranée orientale). Rev. Géogr. Phys. Géol. Dyn., (2), 18 (4): 323—346.

Mercier, J., Bousquet, B., Delibasis, N., Drakopoulos, I., Keraudren, B., Lemeille, F. and Sorel, D., 1972. Déformations en compression dans le Quaternaire des rivages ioniens (Céphalonie, Grèce). Données néotectoniques et séismiques. C.R. Acad. Sci. Paris, D, 275: 2307—2310.

Mercier, J., Carey, E., Philip, H. and Sorel, D., 1976. La néotectonique plio-quaternaire de l'arc égéen externe et de la mer Egée et ses relations avec la séismicité. Bull. Soc. Géol. Fr. (7), 18 (2): 355—372.

Nesteroff, W.D., Lort, J., Angelier, J., Bonneau, M. and Poisson, A., 1977. Esquisse structurale en Méditerranée orientale au front de l'arc égéen. In: B. Biju-Duval and L. Montadert (Editors), Int. Symp. Struct. Hist. Mediterr. Basins, Split, Oct. 1976. Editions Technip, Paris, 1977, pp. 241—256.

Péchoix, P.Y., Pégoraro, O., Philip, H. and Mercier, J., 1973. Déformations pliocènes et quaternaires en compression et en extension sur les rivages du golfe Maliaque et le canal d'Atalanti (Mer Egée, Grèce). C.R. Acad. Sci. Paris, D, 76: 1813—1816.

Philip, H., 1976. Un épisode de déformation en compression à la base du Quaternaire en Grèce centrale (Locride et Eubée nord-occidentale). Bull. Soc. Géol. Fr., (7), 18 (2): 287—292.

Rabinowitz, P.D. and Ryan, W.B.F., 1970. Gravity anomalies and crustal shortening in the Eastern Mediterranean. Tectonophysics, 10: 585—608.

Spratt, T.S., 1856. Travels and Researches in Crete, 2 vols. London, 387 pp. and 435 pp.

Tectonophysics, 52 (1979) 277—286
© Elsevier Scientific Publishing Company, Amsterdam — Printed in The Netherlands

QUATERNARY STRIKE-SLIP FAULTS IN SOUTHEASTERN SPAIN

J-C. BOUSQUET

Laboratoire de Géologie Structurale, Université des Sciences et Techniques du Languedoc, Montpellier (France)

(Accepted for publication April 26, 1978)

ABSTRACT

Bousquet, J-C., 1979. Quaternary strike-slip faults in southeastern Spain. In: C.A. Whitten, R. Green and B.K. Meade (Editors), Recent Crustal Movements, 1977. Tectonophysics, 52: 277—286.

The Gibraltar Arc (Western Mediterranean Sea) is traversed by a NE-trending fault system that extends from northern Morocco to southeastern Spain. In this area, three main faults (the Carboneras, Palomares and Alhama de Murcia faults) have been active in Quaternary time. The faults are characterized by left-lateral strike-slip motion. The Quaternary faulting and current seismicity in this part of the Meditterranean area are related to a collision-type tectonics produced by the northwestward relative motion of the African Plate toward the European Plate.

INTRODUCTION

A neotectonic study of the Gibraltar Arc in the Western Mediterranean Sea has recently been under way. A synthesis of the main results of this investigation has already been presented (October 14—15, 1978, in Montpellier (Groupe de recherches néotectoniques de l'Arc de Gibraltar, 1977).

In this paper we will only discuss several strike-slip faults in southeastern Spain which were active in Quaternary time, and which constitute part of a broad, NE-trending fault system that crosses the Gibraltar Arc.

SUMMARY OF REGIONAL TECTONICS HISTORY *

The faults discussed in this report were studied ·in the eastern Betic Cordilleras between the towns of Alicante and Almeria (Figs. 1 and 2). The landscape of this part of southern Spain is composed mainly of wide, low

---

* According to the chronology of Mediterranean areas, we will use the following ages: Serravallian—Tortonian limit 11—10 m.y., Tortonian—Messinian limit 7 m.y., Messinian—Pliocene limit 5.5 m.y., Pliocene—Quaternary limit 1.8 m.y.

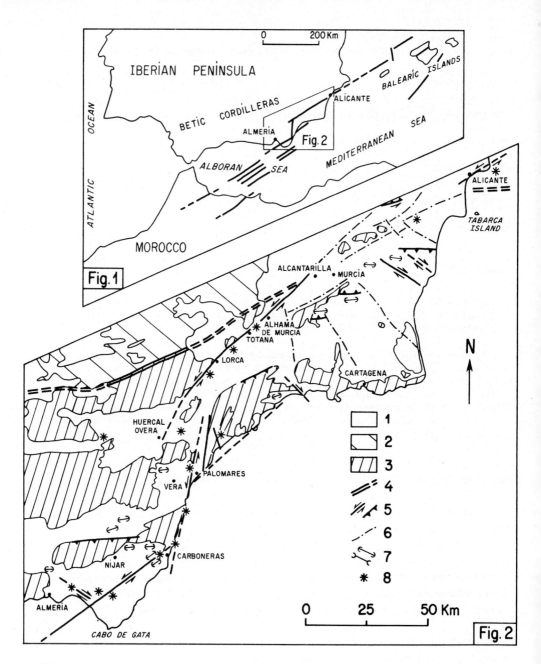

Fig. 1. General location of study area, showing the main faults of the Alboran fault system.

Fig. 2; Quaternary compressive tectonics of the southeastern Betic Cordilleras. *1* = Neogene and Quaternary deposits; *2* = ante-Neogene basement in the external zones; *3* = Ante-Neogene basement in the internal zones; *4* = limit between external and internal zones; *5* = strike-slip and reverse faults active in Quaternary time; *6* = faults according to geophysical data, active or inactive in Quaternary time: *7* = main anticlines or synclines, involving Pliocene and/or Quaternary deposits; *8* = Middle or Recent Quaternary deposits deformed in compressive tectonics.

plains that are separated by elevated sierras. The plains occur in Neogene basins, while sierras are underlain principally by metamorphic rocks (Paleozoic and Triassic with "Alpine" metamorphism). The metamorphic basement belongs to the internal part of the Betic Cordilleras (or Betic zone) and is composed of several thrust sheets (Egeler and Simon, 1969). "Alpine" tectonics related to these great overthrusts occurred before Late Eocene time, according to Paquet (1969). The first Neogene deposits (Early and Middle Miocene) were marine and associated locally with some volcanic rocks in the Cabo de Gata area. These Neogene rocks were strongly deformed in a tectonic episode that occurred before Tortonian time (11 m.y. B.P.). In the internal part of the Betic Cordilleras this tectonic episode was also responsible for the folding and thrusting of the previous overthrusts, and for strike-slip faulting (mainly northeastward left-lateral and northwestward right-lateral). During the same period, great overthrusts, folding, and some strike-slip faulting occurred in the external part of the Betic Cordilleras.

The tectonic episodes that followed the pre-Tortonian deformation, have recently been studied in detail. Here, we will give only a summary of the tectonic changes that occurred in the eastern Betic Cordilleras (cf. Montenat, 1973; Bousquet et al., 1978):

(1) *From Tortonian through Pliocene time* the entire area was subjected to normal-faulting tectonics. Wide, subsiding basins were created between existing faults (trending NE—SW and NW—SE) and new east-trending faults appeared. Sedimentary deposition was directed by downward and upward tectonic movements, with several intervening weak unconformities. In the same period, and only in this part of the Betic Cordilleras, volcanism occurred, first of the "calc-alkaline type" (Tortonian and Messinian) and then of the "alkaline type" (Upper Pliocene, known only near Cartagena, Bellon et al., 1976). Preliminary measurements of minor normal fault trends indicate a N—S extension in Upper Miocene formations, and an E—W extension in Pliocene formations. The same conclusions have been reached in other parts of the Betic Cordilleras. This important change of strain orientation may be related to the change in type of volcanism. Together, the changes indicate that a new geodynamics regime began some time in the Pliocene (see discussion in Bousquet, 1977).

(2) *During the Early Quaternary, compressional tectonics* began with folding, reverse and wrench faulting, which affected Quaternary and older formations. The onset of this tectonic phase is not well dated. It began after some normal faulting visible in Upper Pliocene deposits, some time between 1.5 and 1 m.y. B.P.

The system of mainly normal faults active during the period from the Tortonian through the Pliocene, was reactivated in part during Quaternary time as reverse faults and strike-slip faults (Fig. 2).

The repartition and the importance of tectonic movements related to this tectonic phase are very irregular: in some areas Upper Miocene and Pliocene deposits are relatively undeformed, but in other areas the deposits are strongly folded in E-W-trending folds, except for adjacent folding parallel to the faults. Major E—W folds can be seen east of Orihuela (between Alicante and Murcia). There, Upper Pliocene conglomerates dip vertically in the northern limbs of the folds (cf. map at 1/100,000 in Montenat, 1973). This important folding occurred before Middle Quaternary time. Nevertheless, deposits of the Middle Quaternary and locally those of more recent age are also involved in weaker compressional tectonics, evidence for which can be seen primarily along the principal faults that acted as strike-slip faults during this period. The direction of folding, sense of motion along strike-slip or reverse faults, and microtectonics study (Bousquet and Philip, 1976), indicate a general N—S to NNW—SSE shortening.

Judging from the present elevation of Pliocene marine deposits and Middle to Late Quaternary shorelines in this zone, vertical movement associated with this neotectonic episode was slight when compared to other parts of the Mediterranean area (Angelier et al., 1976; Bousquet, 1977).

PRINCIPAL FAULTS OF SOUTHEASTERN SPAIN ACTIVE IN QUATERNARY TIME

General description

Carboneras fault
Located east of the town of Almeria, this NE—SW fault extends (from the Almeria Gulf coast to the seashore some kilometers north of the village of Carboneras) a distance of 40 km. The fault separates the Cabo de Gata volcanic massif from the Nijar Neogene Basin, and skirts the southern edge of the Sierra Cabrera. For a detailed description of the fault and its history since the Neogene see Bousquet et al., 1978. We present here the general characteristics of the fault related to its activity in Quaternary time.

(1) *Southwestern part.* In this area, the Carboneras fault consists of two main parallel fault breaks with high-angle dips. The fault bounds the low Sierra Serrata, and some NNE—SSW faults are also present in the "horst" between the two main faults, the traces of which are separated by only 1 km. Toward the SW within the horst, Pliocene and Messinian rocks are gently folded locally; toward the NE, more ancient rocks are uplifted (Miocene volcanic rocks of the Sierra Serrata s.s. and Lower Miocene marls overlaying older basement rocks). On both sides of the horst Pliocene deposits are cut and dragged into the faults. Local exposures of the faults show horizontal slickensides. In some places, faulted Quaternary deposits are also visible. For example, in a quarry on the southeastern side of the Sierra Serrata, Recent slope deposits are in fault contact with Miocene volcanic rocks. The fault plane of the SE main fault dips at a high angle toward the northwest, with oblique slickensides indicating a major left-lateral strike-slip movement associated with a reverse vertical movement. It has also been noted that several streams are offset along the NW main fault with a left-lateral separation reaching 100—200 m. On the southeastern side of the Sierra

Serrata some streams are deflected toward the east. This deflection is also in good agreement with left-lateral movement, but it can also be explained by uplift (cf. Wallace, 1975).

In addition, between the Carboneras fault and Almeria, many faults (called the "El Alquian fault net") can be seen cutting thin continental or marine Quaternary deposits, which overlie Pliocene formations. Low, linear escarpments a few meters high along the faults appear to be of recent age. Trending NW—SE, these faults compose an en-echelon fault system corresponding to "Riedel's shears" induced by right-lateral movement along a fault at great depth (Bousquet and Philip, 1977).

(2) *Central part*. In this part, the structure is very simple. The southern fault extends toward the northeast and crosses mainly Pliocene formations. South of the fault, continental Quaternary deposits are locally cut by low-angle (30—50°) reverse faults (Las Piezas) or are folded (Majada del Curica). In this last area, Quaternary pediments are slightly deformed near the fault.

(3) *Northeastern part*. Near the Sierra Cabrera structures are more complicated. Several approximately parallel faults with high-angle dips are present. Between the northeastern end of the Nijar Basin and the Pliocene formations and Miocene volcanic rocks of the Carboneras neighborhood, a "horst-like structure" is composed mainly of Miocene volcanic rocks and schists of the basement. Along the faults, horizontal slickensides can be seen in gently folded Pliocene and Miocene formations. Between two of the main faults, along the Rio Alias, a Quaternary terrace (not dated but strongly cemented: Middle Quaternary?) is folded and faulted.

The NE—SW fault set cuts the southeastern edge of the Sierra Cabrera, and is associated with ENE—WSW and NNE—SSW faults near the coast. Along all these faults (cf. Westra, 1969), Betic rocks (mainly schist and mica-schist) are strongly crushed, and are transformed in an argillaceous mash, in zones 20—50 m wide. These gouge zones are now easily recognizable in the landscape as badlands. Movements older than the Quaternary are largely responsible for this intense deformation, and were certainly produced for a part during the phase preceding the Tortonian period.

Recent activity along these faults has been demonstrated by Cadet, from the study of recent Tyrrhenian shoreline ("Ouljien" or "Tyrrhénien à Strombes", about 70,000—80,000 years, according to Stearns and Thurber, 1965, and 97,000 ± 4,900 years according to Bernat et al., 1978): this shoreline is 3 m above sea level south of Carboneras and is 14 m above sea level south of the main Carboneras fault. At the north, near other associated faults, some remnants of the shoreline are between 8 and 12 m above sea level, and south of Garruchia the shoreline is only 5—6 m above sea level (Bousquet et al., 1975a).

Palomares fault

In the Vera Basin, several fault segments with a NNE—SSW direction (N10° to N20°E) had been previously mapped as crossing Messinian and

Pliocene formations by Völk (1967). It has been proposed that these faults be called the "Palomares strike-slip fault". This fault continues southward to the eastern edge of the Sierra Cabrera and northward into the Pulpi Plain. The fault is mapped for a distance of about 25 km, and seems to extend off-shore southward, at least near Carboneras (Westra, 1969). A general description of its recent activity has been published (Bousquet et al., 1975b) and we will recall only that its fault plane is clearly visible in the Pliocene near Palomares, with horizontal slickensides (photos in above reference and in Bousquet and Philip, 1976); in several places Quaternary pediments are deformed along the fault, and near Garruchia, a marine terrace (Early Tyrrhenian), is also cut by a vertical fault of the same trend.

Alhama de Murcia fault

Mapped by Montenat (1973) between Alcantarilla and Alhama de Murcia, this fault has been observed to extend southward to Huercal-Overa (Bousquet and Montenat, 1974), and has a total lenght of about 100 km. Geophysical studies have traced the fault northward between Alcantarilla and Alicante.

One or two parallel faults, with a direction N40° to N60° E, are present in a narrow zone (1—2 km) and cross the Betic basement (near Sierra Tercia and Sierra de Las Estancias), Upper Miocene formations, or Quaternary continental deposits. As along the Carboneras fault, Betic schists are crushed in a zone up to 10 m wide. Upper Miocene continental or marine deposits are locally near-vertical along the faults, or are folded. Exposures in a few gypsum quarries (northeast of Totana) show metric or plurimetric folds with near-vertical axes, indicating left-lateral movement. This agrees with other microtectonic observations along the Alhama de Murcia fault (Bousquet and Philip, 1976). Pliocene rocks are absent, but Quaternary continental deposits are faulted and folded.

Fan deposits (Middle Quaternary to Würm?) heading in the Sierra Tercia are faulted between Lorca and Totana.

A geophysical study (gravimetry and deep electric well logging — Gauyau et al., 1977) has been made between Alhama de Murcia and Alicante, to define the continuation of the Alhama de Murcia fault under Recent alluvial deposits. A system of several NE—SW faults, crossed by minor NW—SE faults, has been found and is in good agreement with superficial data (Betic Sierras disposition, Neogene basins paleogeography, and flexure or fault segment visible in Neogene formations). In the same area, a subsurface study in Holocene and Recent Quaternary deposits (Echallier et al., 1977), with core drilling and electric well logging, prove Recent activity of some of these faults. In conclusion, the Alhama de Murcia fault certainly exists between Murcia and Alicante, and near Alicante Recent Tyrrhenian deposits are gently folded (Montenat, 1973) along a left-lateral strike-slip fault of the same NE—SW trend.

Horizontal displacement

At present the amount of slip during the Quaternary along the greatest faults is not known. Along the Carboneras and Alhama de Murcia faults, facies of Upper Miocene deposits are locally different on both sides of the faults. However, it is clear that facies changes are very rapid in zones adjacent to faults, and may be related to their vertical movements during Upper Miocene normal faulting tectonics. On the other hand, along the Palomares fault, the outcrop pattern of Upper Miocene volcanic rocks (Sierracica and Cerro Colorados) seem to indicate a horizontal displacement of about 8 km.

Judging from more ancient reference marks, the general distribution of the Sierra Cabrera and the Sierra Almagrera has previously led to the hypotheses of a N—S left-lateral wrench fault (Fernex, 1964; Rondeel, 1965; Völk, 1966), and the Palomares fault is the result of more recent activity of the N—S left-lateral wrench fault. The occurrence of Betic units in these two sierras suggest 20—30 km (or more) of displacement of which 10—20 km occurred before Mio—Pliocene formations of the Vera Basin were deposited.

If we consider the Alhama de Murcia fault strike-slip movement after Upper Eocene or Oligocene, the main tectonic contact between internal and external zones of the Betic Cordilleras can be employed as an indicator of total offset (Fig. 2): west of the fault, this contact crosses the Sierra Espuna (Paquet, 1969) and the La Mula Upper Miocene Basin to join the Alhama de Murcia fault near Alcantarilla; on the other side of the fault, Betic rocks (internal zone) are known to the north, between Murcia and Orihuela, and south of Alicante in the little island of Tabarca. This indicates a left-lateral horizontal displacement of the order of 60 km or more.

Relation with seismicity

In the Betic Cordilleras, instrumentally recorded seismicity is shallow if we except some (two instrumental and one historic) deep earthquakes known in the central part (cf. Beuzart, 1972; Hatzfeld, 1976; Udias et al., 1976). Compared to other Mediterranean regions such as Italy, Yugoslavia, and the Aegean Arc, seismic activity is relatively low (few events with magnitude $\geqslant 5$ occurred between 1900 and 1971, cf. Beuzart, 1972). From old journals and records, we can state, nevertheless, that earthquakes of rather elevated intensity have damaged or destroyed several towns or villages in the past. In the eastern Betic Cordilleras, Almeria was damaged several times (in 1487, 1522, 1659, and 1804) and Vera was destroyed in 1518. More recently, in 1829, near Murcia, an IX—XI intensity earthquake caused several thousand deaths and damaged several thousand houses (Karnik, 1971).

The study of the relationship between seismicity and faults that were active during Quaternary time is only in its infancy. Generally, the accuracy of epicenter location and the number of earthquakes is too low to establish a precise relationship between the two. However, Rey-Pastor (1951), using mainly historical records (for Alicante and Murcia provinces only), defined several earthquakes epicenter patterns that are in good agreement with faults. This is the case for the Alhama de Murcia fault between Lorca and Murcia.

Recently, from March to July 1977, a net of portable seismometers operated in the Alhama de Murcia fault area, south of Lorca (Lopez-Lago et al., 1977; J. Mezcua, personal communication, 1977). A continuous activity at a low level of magnitude was registered, and also some 20 shocks with magnitudes between 3 and 5. The epicenter of the major shock (June 6, 1977) was located southwest of Lorca (37°38'N—1°48'W), 3 km west of the fault, which dips to the northwest in this region.

Some damage produced by this seismic activity has been observed along the fault (Bousquet et al., 1978). New cracks were generated in Lorca's cathedral and a wall was affected by reverse faults with either a left-lateral or a right-lateral component, indicating a NE—SW shortening. Near Totana, a little bridge on a buried aqueduct was also affected by reverse faults, producing a shortening of 1—2 cm along the main fault direction (N40°E).

In both cases the strain direction is in good agreement with that deduced in the same area by microtectonic studies (Bousquet et al., 1976).

According to historical and present data, the Carboneras fault seems to be less active than the Alhama de Murcia fault. Nevertheless, an aqueduct which crosses the fault near El Baranquete village has been slightly damaged (P. Snavely, personal communication, 1977), a pillar has been destroyed and rebuilt, and the aqueduct has been deformed, indicating a left-lateral displacement along the fault.

CONCLUSIONS

The great faults of southeastern Spain are part of a broad fault system, with a NE—SW trend. This fault system begins in north Morocco (Jebha fault) and crosses the Alboran Sea (cf. Olivet et al., 1973). The Carboneras fault is known offshore in the Almeria Gulf (in Rios, 1975) and there is a straight canyon all along the fault from the coast to a depth of 1000 m. Toward the northeast, this fault seems to be offset by the north-trending Palomares fault, which serves as a connection with the Alhama de Murcia fault. It has been hypothesized that this last fault continues offshore to the northern edge of the Balearic Islands (Bousquet and Montenat, 1974).

The presence of this NE-trending fault system crossing the Gibraltar Arc is not in agreement with an E—W boundary between the European and African Plates, as is usually drawn in this area (MacKenzie, 1970; Dewey et al., 1973) or with the existence of an Alboran plate (Udias et al., 1979). As emphasized in conclusions of the recent general study of the Gibraltar Arc (Groupe de recherches néotectoniques de l'Arc de Gibraltar, 1977; Bousquet, 1977), neotectonics, as well as the diffuse seismicity of this area, is related to a *collision-type tectonics* (Molnar and Tapponnier, 1975; Tapponnier, 1977), produced by northwestward relative motion of the African plate toward the European plate (Minster et al., 1974). The strike-slip faults of southeastern Spain are good examples of the expression of such tectonics.

ACKNOWLEDGEMENTS

The author wishes to thank C. Montenat and H. Philip for their collaboration in the field; J.M. Fontbote, C. Sanz de Galdeano, A. Estevez, J. Benkhelil and R. Guiraud for numerous and profitable discussions, J. Mezcua for relevant information; D.G. Herd for translating the manuscript; and R.O. Burford for a critical review. This work was supported by the A.T.P. Géodynamique de la Méditerranée occidentale of C.N.R.S. and the Laboratoire de Géologie Structurale of Montpellier.

REFERENCES

Angelier, J., Cadet, J.P., Delibrias, G., Fourniquet, J., Gigout, M., Guillemin, M., Hogrel, M.T., Lalou, Cl. and Pierre, G., 1976. Les déformations du Quaternaire marin, indicateurs néotectoniques. Quelques exemples méditerranéens. Rev. Geógr. Phys. Geol. Dyn., 13: 427—448.

Bellon, H., Bordet, P., Bousquet, J-C. and Montenat, C., 1976. Principaux résultats d'une étude chronométrique du volcanisme néogène des Cordillères bétiques (Espagne méridionale). 4ème Réunion Annuelle des Sciences de la Terre, Paris, p. 43.

Bernat, M., Bousquet, J.C. and Dars, R., 1978. I_0-U dating of the Ouljian stage from Torre Garcia (southern Spain). Nature, 275 (5678): 302—303.

Beuzart, P., 1972. La séismicité de la région méditerranéenne et de ses bordures. Diplôme d'Ingénieur géophysicien de l'Université de Strasbourg.

Bousquet, J.C., 1977. Contribution à l'étude de la tectonique récente en Méditerranée occidentale: les données de la néotectonique dans l'Arc de Gibraltar et dans l'Arc Tyrrhénien. In: B. Biju-Duval and L. Montadert (Editors), Structural History of the Mediterranean Basins. Editions Technip, Paris: 199—214.

Bousquet, J-C. and Montenat, C., 1974. Présence de décrochements Nord-Est plioquaternaires dans les Cordillères bétiques orientales (Espagne). Extension et signification. C.R. Acad. Sci., Paris, 278: 2617—2620.

Bousquet, J-C. and Philip, H., 1976. Observations microtectoniques sur la compression Nord-Sud Quaternaire des Cordillères bétiques orientales (Espagne méridionale, Arc de Gibraltar). Bull. Soc. Géol. Fr., 3: 711—724.

Bousquet, J-C. and Philip, H., 1977. Observations tectoniques et microtectoniques sur la distension plio-pléistocène ancien dans l'Est des Cordillères bétiques (Espagne méridionale). Cuad. Geol. Granada, to be published.

Bousquet, J-C., Cadet, J-P. and Montenat, C., 1975a. Quelques observations sur le jeu quaternaire de l'accident de Carboneras (décrochement majeur sénestre NE—SW des Cordillères bétiques orientales). 3ème Réunion Annuelle des Sciences de la Terre, Montpellier, p. 73.

Bousquet, J-C., Dumas, B. and Montenat, C., 1975b. L'accident de Palomares: décrochement sénestre du bassin de Vera (Cordillères bétiques orientales, Espagne). Cuad. Geol. Granada, 6: 113—119.

Bousquet, J-C., Montenat, C. and Philip, H., 1978. La evolution tectonica reciente de las Cordilleras beticas orientales. Reunion sobre la Geodinamica de la Cordillera betica y mar de Alboran, 1976, Granada: 59—78.

Bousquet, J.C., Echallier, J.C. and Montenat, C., 1978a. Ruptures dans des constructions situées sur des failles actives du Sud de la péninsule ibérique. 6ème Réunion annuelle des Sciences de la Terre, Orsay, 1978. Résumé des communications en dépôt. Soc. Géol. Fr. Paris.

Bousquet, J-C., Cadet, J.P. and Montenat, C., 1978b. L'Histoire néogène et Quaternaire de l'accident de Carboneras (Cordillères bétiques orientales), in preparation.

Dewey, J.F., Pitman III, W.C., Ryan, W. and Bonnin, J., 1973. Plate tectonics and the evolution of the Alpine system. Geol. Soc. Am. Bull., 84: 3137—3180.

Echallier, J-C., Gauyau, F., Lachaud, J-C. and Talon, B., 1977. Mise en évidence par sondages électriques d'accidents affectant les terrains quaternaires dans la basse vallée du Rio Segura (Province d'Alicante). C.R. Acad. Sci., Paris, 286: 1129—1131.

Egeler, C.F. and Simon, O.J., 1969. Sur la tectonique de la zone bétique (Cordillères bétiques, Espagne). Verh. K. Ned. Akad. Wet. Afd. Nat., XXV, 82 pp.

Fernex, F., 1964. Sur le jeu de la tectonique postérieure aux nappes dans l'Est des zones bétiques. Arch. Sci. Genève, 17 (1): 39—46.

Gauyau, F., Bayer, R., Bousquet, J-C., Lachaud, J-C., Lesquer, A. and Montenat, C., 1977. Le prolongement de l'accident d'Alhama de Murcia entre Murcia et Alicante (Espagne méridionale). Résultats d'une étude géophysique. Bull. Soc. Géol. Fr., (7), 19, (3): 623—629.

Groupe de recherches néotectoniques de l'Arc de Gibraltar, 1977. L'histoire tectonique récente (Tortonien à Quaternaire) de l'Arc de Gibraltar et des bordures de la mer d'Alboran. Bull. Soc. Géol. Fr., (7), 19, (3): 575—614.

Hatzfeld, D., 1976. Etude de sismicité dans la région de l'arc de Gibraltar. Ann. Gèophys., 32: 71—85.

Karnik, V., 1971. Seismicity of the European Area, Part. II. Reidel, Dordrecht.

Lopez-Lago, A., Munoz, D. and Mezcua J., 1977. A seismic active fault region in southeast Spain. E.G.S., Munich 1977. Abstr., p. 911.

MacKenzie, D.P., 1970. Plate tectonics of the Mediterranean region. Nature, 226: 239—243.

Minster, J.B., Jordan, T.H., Molnar, P. and Haines, E. 1974. Numerical modeling of instantaneous plate tectonics. Geophys. J.R. Astron. Soc., 36: 541—576.

Molnar, P. and Tapponnier, P., 1975. Cenozoic tectonics of Asia: effects of continental collision. Science, 4201: 419—426.

Montenat, C., 1970. Sur l'importance des mouvements orogéniques récents dans le Sud-Est de l'Espagne. C.R. Acad. Sci., Paris, 270: 3194—3197.

Montenat, C., 1973. Les formations néogènes et quaternaires du Levant Espagnol. Thèse Sci. Orsay, Paris Sud, 1170 pp.

Montenat, C., 1974. Tectonique et sédimentation pliocène dans les Cordillères bétiques (Espagne méridionale). G. Geol. Bologna, 39: 469—480.

Olivet, J.L., Auzende, J.M. and Bonnin, J., 1973. Structure et évolution tectonique du bassin d'Alboran. Bull. Soc. Géol. Fr., 15: 108—112.

Paquet, J., 1969. Etude géologique de l'Ouest de la province de Murcia (Espagne). Mém. Soc. Geol. Fr., 111: 270 pp.

Rey-Pastor, A., 1951. Estudio sismotectonico de la region Sureste de Espana. Inst. Geog. y Catastral, Madrid, 52 pp.

Rios, J.M., 1975. El mar Mediterraneo occidental y sus costas Ibericas. Rev. Real. Acad. Cienc., 69: 473 pp.

Rondeel, H.E., 1965. Geological Investigations in the Western Sierra Cabrera and Adjoining Areas (South Eastern Spain). Thèse, Amsterdam, 161 pp.

Stearns, C.E., and Thurber, D.L., 1965. T^{230}—U^{234} dates of late Pleistocene marine fossils from the Mediterranean and Moroccan littorals. Quaternaria, 7: 29—42.

Tapponnier, P., 1977. Evolution tectonique du système alpin en Méditerranée: poinçonnement et écrasement rigide plastique. Bull. Soc. Géol. Fr., (7), 19, (3): 437—460.

Udias, A. and Lopez Arroyo, A., 1972. Plate tectonics and the Azores—Gibraltar region. Nature, 237: 67—69.

Völk, H., 1966. Zur Geologie und Stratigraphie des Neogenbeckens von Vera. Thesis, Amsterdam, 160 pp.

Wallace, R.E., 1975. The San Andreas fault in the Carrizo plain-Tremblor range region, California. In: J.C. Crowell (Editor), San Andreas Fault in Southern California. Calif. Div. Mines Geol., Spec. Rep. 118, pp. 241—250.

Westra, G., 1969. Petrogenesis of a composite metamorphic facies series in an intricate fault-zone in the southeastern Sierra Cabrera, S.E. Spain. Thèse, Amsterdam, 166 pp.

Tectonophysics, 52 (1979) 287—300
© Elsevier Scientific Publishing Company, Amsterdam — Printed in The Netherlands

287

FOUR-DIMENSIONAL MODELING OF RECENT VERTICAL MOVEMENTS IN THE AREA OF THE SOUTHERN CALIFORNIA UPLIFT

PETR VANÍČEK, [1] MICHAEL R. ELLIOTT [2] and ROBERT O. CASTLE [2]

[1] *Department of Surveying Engineering, University of New Brunswick, Fredericton, N.B. (Canada)*
[2] *U.S. Geological Survey, Menlo Park, Calif. (U.S.A.)*

(Accepted for publication April 26, 1978)

ABSTRACT

Vaníček, P., Elliott, M.R. and Castle, R.O., 1979. Four-dimensional modeling of Recent vertical movements in the area of the southern California uplift. In: C.A. Whitten, R. Green and B.K. Meade (Editors), Recent Crustal Movements, 1977. Tectonophysics, 52: 287—300.

This paper describes an analytical technique that utilizes scattered geodetic relevelings and tide-gauge records to portray Recent vertical crustal movements that may have been characterized by spasmodic changes in velocity. The technique is based on the fitting of a time-varying algebraic surface of prescribed degree to the geodetic data treated as tilt elements and to tide-gauge readings treated as point movements. Desired variations in time can be selected as any combination of powers of vertical movement velocity and episodic events. The state of the modeled vertical displacement can be shown for any number of dates for visual display. Statistical confidence limits of the modeled displacements, derived from the density of measurements in both space and time, line length, and accuracy of input data, are also provided. The capabilities of the technique are demonstrated on selected data from the region of the southern California uplift.

INTRODUCTION

Vertical crustal movements defined by the results of repeated levelings can be calculated and represented in a variety of ways. The conceptually simplest and least equivocal technique consists of differencing observed elevations obtained through successive surveys over the same route. Because the extension of this technique beyond single-line comparisons generally requires systematic releveling over a closed network, it is usually impossible to relate area-wide groups of such comparisons to a single control point (such as a primary tide station) whose stability can be independently assessed. Alternatively, because releveled segments are commonly scattered in both space and time, areally distributed vertical velocities or displacements can be calculated from geodetically measured tilts through the fitting of these data to a mathematically defined surface or set of surfaces, such as an algebraic polynomial of specified degree. The most obvious advantages afforded by this technique are that it permits the use of otherwise unusable data and provides an objective basis for characterizing variations in the velocity configuration through time. The chief disadvantage of any surface-fitting technique is inherent in the smoothing process that discards or subdues short-wavelength features of potential significance.

Earlier adaptations of surface-fitting techniques to the reconstruction of vertical crustal movements have been developed around a simplifying assumption of constant vertical velocity at every point, i.e., zero acceleration. The resultant velocity surfaces are probably most valid in areas such as Chesapeake Bay (Vaníček and Christodulidis, 1974) or the maritime provinces of Canada (Vaníček, 1975; 1976) where the assumption of constant velocity can be tested through velocity misclosures or a knowledge of the Holocene deformation. Spasmodic or episodic, aseismic crustal movements of the sort recognized in southern California (Castle et al., 1976) have provoked the development of the modeling technique described here, a technique that permits the fitting of a time-varying surface of a prescribed degree. By specifying the power in time as equal to one and by omitting episodic movements, the described model automatically reduces to one characterized by constant velocity.

The data set for testing this newly conceived four-dimensional model has been selected from repeated levelings carried out within the area of the southern California uplift during the period 1897—1934. We have chosen this data set because:

(1) It is large enough that it can be used to assess the logic of the mathematical formulation, yet sufficiently small that it could be put in computer-readable form in something less than a lifetime.

(2) The prescribed period includes independently identified episodes of dramatically changing vertical velocity.

(3) The results can be compared with those obtained through a classic reconstruction involving a network of closed loops.

MATHEMATICAL FORMULATION

The uplift surface, u, is sought in the following form:

$$u(x, y, t) = \sum_{k=1}^{n_t} c_{0k} T_k(t) + \sum_{\substack{i=0 \\ i+j \neq 0}}^{n_x} \sum_{j=0}^{n_y} \sum_{k=1}^{n_t} c_{ijk} x^i y^j T_k(t)$$

$$= T(t)c_0 + X(x, y, t)c \tag{1}$$

where:

$$T_k(t) = t^k , \qquad\qquad\qquad k = 1, n_p$$

$$T_k(t) = \begin{cases} 0, & t < b_k \\ (t - b_k)/(e_k - b_k), & b_k \leqslant t \leqslant e_k \\ 1, & t > e_k \end{cases} \qquad k = n_p + 1, n_p + n_e$$

and c is the n-vector of unknown coefficients c_{ijk} for the previously chosen values n_x, n_y of maximum power in x and y.

In these formulas, x and y are local horizontal Cartesian coordinates calculated from latitude and longitude through the following transformation equations:

$$x = (\phi - \phi_0)R ; \qquad y = (\lambda - \lambda_0)R \cos \phi_0 \qquad\qquad (2)$$

where (ϕ_0, λ_0) is the centroid of all bench marks and R is the mean radius of the earth. Time, t, is reckoned from a stipulated date, t_0, for which $u(x, y, t_0)$ is everywhere equal to zero. In addition, b_k, e_k, for $k = n_p + 1$, $n_p + n_e$, are the dates of the beginning and end of n_e movement episodes so that $n_t = n_p + n_e$. We note that the episodic movements are treated as linear within the duration of the episode.

Observation equations for m releveled segments can now be written:

$$\Delta h(x_1, y_1, x_2, y_2, t_2) - \Delta h(x_1, y_1, x_2, y_2, t_1) = d(x_1, y_1, x_2, y_2, t_1, t_2)$$

$$= \sum_{\substack{i=0 \\ i+j\neq 0}}^{n_x} \sum_{j=0}^{n_y} \sum_{k=1}^{n_t} c_{ijk}(x_2^i y_2^j - x_1^i y_1^j)[T_k(t_2) - T_k(t_1)] + r(x_1, y_1, x_2, y_2, t_1, t_2)$$

$$= B(x_1, y_1, x_2, y_2, t_1, t_2)c + r \qquad\qquad (3)$$

where d denotes the m-vector of the differences of leveled height differences, Δh, and $\mathbf{r}$ is the residual vector. If

$$m > n = (n_x \cdot n_y + n_x + n_y)n_t \qquad\qquad (4)$$

we can find the solution, c, through the method of least squares which yields the following normal equations:

$$(B^T C_d^{-1} B)c = B^T C_d^{-1} d \qquad\qquad (5)$$

where C_d is a diagonal covariance matrix of the releveled segments. The diagonal elements equal the variances of individual height difference differences. In the computer program the normal equations are solved through orthogonalization.

Clearly, the shift coefficients, c_0, cannot be determined from the releveled segments alone. Uplift $u^*(x, y, t)$ of at least one, but generally n_g tide-gauges (x, y), must be determined from sea-level records at n_d dates to allow us to evaluate the shift coefficients. The following $n_g \cdot n_d$ observation equations can be then formulated:

$$u^*(x_i, y_i, t_j) = T(t_j)c_0 + X(x_i, y_i, t_j)c + r^*, \qquad i = 1, n_g; j = 1, n_d \tag{6}$$

If $n_g \cdot n_d > n_t$, then the above equations may be solved for c_0, again using the method of least squares. The minimum condition imposed here is:

$$\min_{c_0} r^{*T}(C_{u^*} + C_{Xc})^{-1} r^* \tag{7}$$

where C_{u^*} and C_{Xc} are covariance matrices of u^* and Xc respectively. Otherwise, everything in eq. 6 is known.

The covariance matrix, C_c, of the coefficients c can be evaluated from the well-known formula:

$$C_c = \frac{r^T C_d r}{m - n} (B^T C_d B)^{-1} \tag{8}$$

Hence the covariance matrix, C_{Xc}, is given as:

$$C_{Xc} = X C_c X^T \tag{9}$$

The covariance matrix, C_0, of the shift coefficients c_0 is obtained from:

$$C_0 = \frac{r^{*T}(C_{u^*} + C_{Xc})^{-1} r^*}{n_g \cdot n_d - n_t} (T^T(C_{u^*} + C_{Xc})^{-1} T)^{-1} \tag{10}$$

We note that for each tide gauge the uplift u^* must be determined so as to satisfy the following condition:

$$u^*(x, y, t_0) = 0 \tag{11}$$

Finally, the standard deviation, $\sigma(x, y, t)$, of the computed vertical displacement, $u(x, y, t)$, may be evaluated from the following relation:

$$\sigma(x, y, t) = \sqrt{\operatorname{diag}[T(t)C_0 T^T(t) + X(x, y, t)C_c X^T(x, y, t)]} \tag{12}$$

COMPUTER PROGRAM

The computer program was designed to solve the problem described in the preceding section. It has the following additional features:

(1) Coefficients in orthogonal solution space that are below the prescribed noise level are removed through filtering. The noise level is prescribed as k — times the coefficient's own standard deviation, where k is input.

(2) Episodes may be specified at the user's option; if they are not, then $n_t = n_p$.

(3) The vertical displacement (eq. 1) and associated standard deviation (eq. 12) surfaces are computed within a minimal rectangle covering all the releveled segments, on a regular grid with specified steps in ϕ and λ. The surfaces are produced for n_s specified dates, $t = \tau_i$, where $i = 1, n_s$. If there is no τ_i specified, a default value, $\tau_1 = t_0 + 100$ years, is chosen automatically.

(4) The surfaces may be plotted at the user's discretion. The scale of the plot is an input parameter.

(5) A plot of the location of the releveled segments may also be requested.

(6) If the vertical displacement and standard deviation surfaces are plotted, then both the contour interval and maximum and minimum values may also be specified. The scale in both x and y directions is determined by the program in relation to the specified steps in ϕ and λ ($\Delta\phi$ and $\Delta\lambda$).

(7) Vertical displacements of the tide stations, u^*, are treated as optional information. If they are not supplied, both the shift coefficients (eq. 6) and

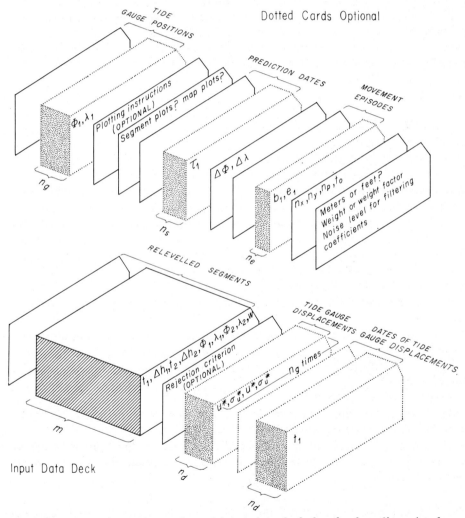

Fig. 1. Diagrammatic representation of input data deck for the four-dimensional computer program.

their covariance matrix (eq. 10) are all set to zero. The uplift surfaces are then held to zero at the centroid (ϕ_0, λ_0) of the bench marks utilized.

(8) Weights ($\sigma_{d_i}^{-2}$) or weight factors (W_i) are also to be supplied for releveled segments. Weights are computed from the standard formulas (see, e.g., Vaníček, 1976). If weight factors are used, they should be introduced as: 2 for two first-order, 5 for one first-order and one second-order, 8 for two second-order, 17 for one first-order and one third-order, 20 for one second-order and one third-order, and 32 for two third-order levelings. These weight factors are automatically converted into the proper weights for each releveled segment, where it is assumed in the conversion that first-order accuracy is equal to 1 mm $\times$ $\sqrt{}$(distance in kilometers).

(9) Segments with residuals larger than a prescribed rejection criterion are automatically eliminated. A default value for the rejection criterion may be chosen (as has been done in this example), so that no segment is rejected. Assembly of the input data deck is shown in Fig. 1.

INPUT DATA

The data set used to test the computer program is based on a broad distribution of repeated surveys that incorporates every combination of leveling precision stipulated in the preceding section. While we have not included all of the data collected within this area (Fig. 2) during the period 1897—1934, incorporation of these additional, relatively recently unearthed segments probably would contribute little additional strength to the solutions, for their distribution closely matches much of the data already included. Examination of the data distribution shows that the chief deficiencies are associated with the concentration of these data in two general time frames (around the turn of the century and the early to middle 1920s) and the poor spatial control west of Ventura.

Input commands and features common to each series of fitted surfaces are:

(1) Computation of vertical displacements at 5' spacings in ϕ and 6' spacings in λ.

(2) Use of a plotting projection in which the angular unit in ϕ equals the angular unit in λ.*

(3) Adopting of $t_0 = 1897$.

(4) Plotting of cumulative vertical displacements from 1897 to 1902, 1908, 1914, 1924, 1928, and 1934.

(5) Inclusion of sea-level changes at the San Pedro Tide Station equal to zero through the period 1897—1934, to which a standard deviation of 3 cm has been assigned.

---

* The horizontal distortion implicit in this projection produces a lengthening in λ proportional to sec ϕ and derives from a deficiency in the plotting subroutines; nevertheless, this distortion has virtually no effect on our chief goal, which is to test utility of the model.

293

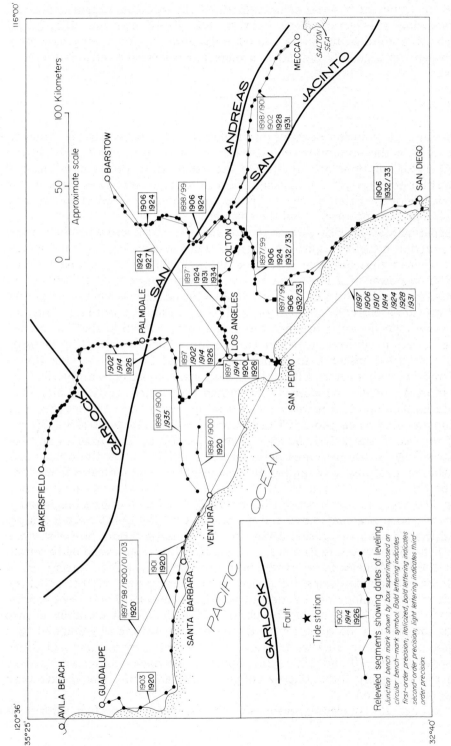

Fig. 2. Generalized map of southern California showing distribution of releveled segments and San Pedro tide station. Only those segments actually used in testing the four-dimensional computer program are shown here.

(6) Inclusion of a series of manufactured tilt segments obtained through differencing sea-level changes between San Pedro and San Diego, and weighted according to an estimated noise level of 3 cm (approximately equivalent to segments obtained through two successive second-order surveys).

RESULTS

Because the solution depends on the degree of the fitted surface, which in turn affects the computer time needed, we have been constrained in the numbers of solutions attempted. On the other hand, preliminary classical reconstructions (as yet unpublished) provide a reasonable basis for selecting the degree of the fitted surface in both ϕ and λ, and a somewhat poorer basis for selecting the power in t and the episodes.

Experimentation with various powers in ϕ, λ, and t, and with the a-priori introduction of several episodes, suggests that the changing vertical displacement field within our area is reasonably approximated by a model with a maximum power of 2 in ϕ, 3 in λ, 3 in t, and no episodes (Fig. 3). The four frames shown here (out of the six actually plotted) are clearly representative and provide a basis for comparison with our continuous profiling. Thus, the modeled solution indicates that the area contained within the 10-cm standard deviation contour sustained virtually no change in elevation between 1897 and 1902, other than modest uplift east of Los Angeles (Fig. 3A), a generalization that agrees with our conventional reconstruction. By 1914 uplift in excess of one standard deviation had propagated westward almost to Santa Barbara and had increased significantly east of Los Angeles (Fig. 3B). Comparison between the 1897—1924 and 1897—1914 plots (Figs. 3B and 3C) suggests a substantial increase in uplift between Los Angeles and the western edge of the map area between 1914 and 1924. On the other hand, continuous profiling between San Pedro and Bakersfield indicates that nearly all of this uplift occurred by 1914. The 1897—1934 displacement surface (Fig. 3D) suggests that most of the previously developed uplift collapsed and that the steep gradient along the north flank of the uplift shifted sharply southward during the period 1924—34. This interpretation, however, again conflicts with our conventional reconstruction, for we are reasonably certain that the Los Angeles Basin experienced no more than a few centimeters of tectonic subsidence during this interval. The breakdown of the model implicit in the seeming continuation of uplift between 1914 and 1924 and the almost total collapse during the period 1924—34 almost certainly derives from the smooth interpolation of the movement in time as stipulated by a degree of 3 in time. Had we increased the power in t by one and added two episodes (1906—10 and 1924—28), uplift would have culminated by 1914 and the seemingly total collapse between 1924 and 1934 would have been sharply diminished.

Although the preceding evaluation of this clearly imperfect solution is obviously subjective, the general validity of the solution is supported by the a-posteriori variance factor of 1.942 (compared with an assumed value of 1.000). Also only 12 coefficients out of 33 differ significantly from zero at the one standard deviation level. The a-posteriori variance factor $r^T C_d r / (m - n)$ assesses three effects that are inextricably entwined: (1) the selection of the model; (2) weighting of the data; (3) the data density and distribution in both space and time; hence it is impossible to say with certainty which of these three contributed most to the value of this quantity. Nevertheless, we strongly suspect that had the data been more evenly distributed both in space and time, the a-posteriori variance factor would have dropped substantially. In any event, to the extent that this factor is a measure of "goodness of fit", the cited value supports the general validity of the 2, 3, 3 solution (Vaníček, 1975, 1976).

For comparison with the selected solution, we show a single frame (1897– 1924) for the solution based on a maximum power of 2 in ϕ, 4 in λ, 3 in t, and no episodes (Fig. 4). Differencing the two solutions through the subtraction of the vertical displacement field of Fig. 4 from that of Fig. 3C shows that, whereas the two displacement surfaces differ very little within the 10-cm standard deviation contour of either solution, they differ significantly away from the central part of the map area (Fig. 5). If it is accepted that the difference, within the region reasonably covered with data, between the two solutions is a variate with normal probability distribution, we might expect those areas in which the differences are greater than one standard deviation to cover about one-third of the map. This appears to be approximately the case here. Parenthetically, those areas in which the difference between the two solutions is greater than one standard deviation are generally coincident with the areas in which the 2, 3, 3 solution least closely accords with our conventional reconstruction. This seems to support our choice of the 2, 3, 3 solution as relatively representative of the changing vertical displacement field within the area of our investigation.

CONCLUSION

The results of the described tests of the four-dimensional modeling approach cannot be considered definitive. Given the restrictions imposed by our decision to limit the degree of the uplift surface, particularly in time, and the use of only 352 segments, these results are, nonetheless, encouraging. Had we been able to obtain a few tens of additional segments randomly scattered through the western and northeastern parts of the area investigated, many and perhaps all of these solutions would have probably been improved significantly. Similarly, even a fragmentary knowledge of sea-level changes at Avila Beach (cf. Fig. 2) during the period 1897–1934 could have sharply constrained all of the attempted solutions. This tide gauge did not become a primary station until 1945.

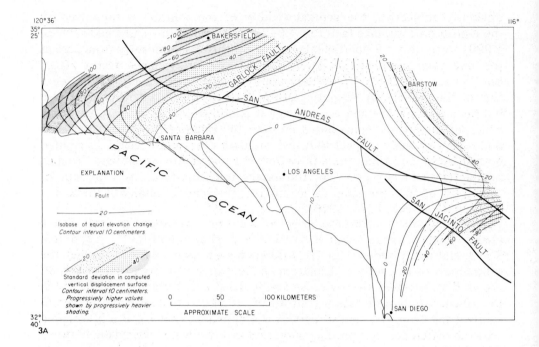

3A

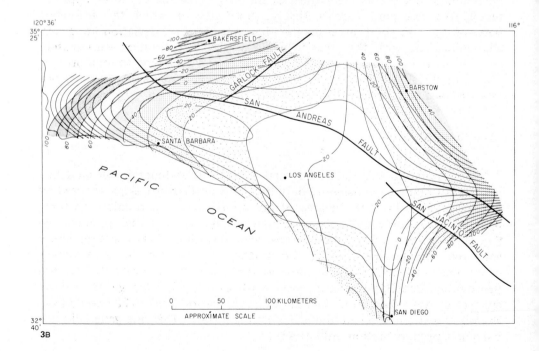

3B

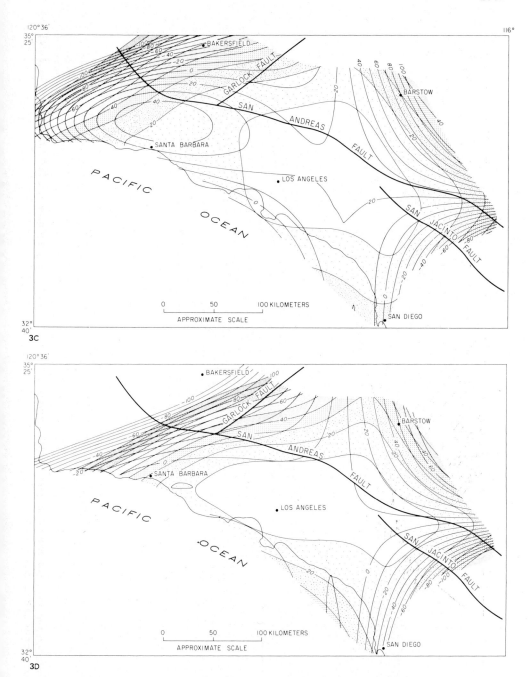

Fig. 3. Computed vertical displacement surfaces within the area of the southern California uplift incorporating a power of 2 in latitude, a power of 3 in longitude, and a power of 3 in time. A. 1897—1902. B. 1897—1914. C. 1897—1924. D. 1897—1934. Note: explanation applies to all four frames.

298

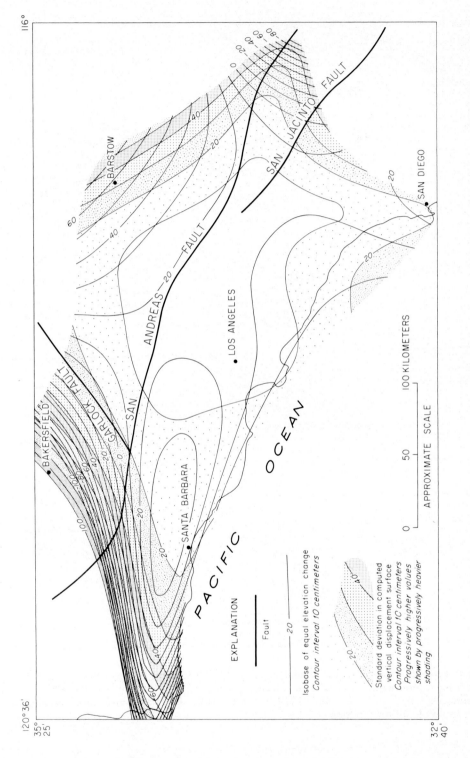

Fig. 4. Computed vertical displacement surface for the period 1897—1924 incorporating a power of 4 in longitude, and a power of 3 in time.

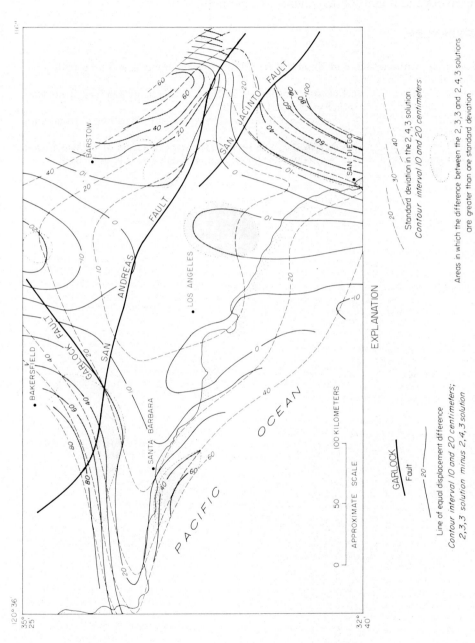

Fig. 5. Differences in the vertical displacement fields and associated standard deviations based on a comparison between the 2, 3, 3 and 2, 4, 3 solutions for the period 1897—1924.

ACKNOWLEDGEMENTS

The development of the four-dimensional model described on this report was supported in part by U.S. Geological Survey Grant No. 14-08-0001-G-408. We thank Robert C. Jachens and M. Darroll Wood for suggestions and comments and a meticulous review of the manuscript.

REFERENCES

Castle, R.O., Church, J.P. and Elliott, M.R., 1976. Aseismic uplift in southern California. Science, 192: 251—253.
Vaníček, P., 1975. Vertical crustal movements in Nova Scotia as determined from scattered relevellings. Tectonophysics, 29: 183—189.
Vaníček, P., 1976. Pattern of recent vertical crustal movements in Maritime Canada. Can. J. Earth Sci., 13: 661—667.
Vaníček, P., and Christodulidis, D., 1974. A method for the evaluation of vertical crustal movement from scattered geodetic relevellings. Can. J. Earth Sci., 11: 605—610.

Tectonophysics, 52 (1979) 301—302

INITIATION AND DEVELOPMENT OF THE SOUTHERN CALIFORNIA UPLIFT ALONG ITS NORTHERN MARGIN

R.S. STEIN [1], W. THATCHER [2] and R.O. CASTLE [2]

[1] *U.S. Geological Survey, Menlo Park, and Department of Geology, Stanford University, Calif. 94305 (U.S.A.)*
[2] *U.S. Geological Survey, Menlo Park, Calif. 94025 (U.S.A.)*

(Accepted for publication April 26, 1978)

ABSTRACT

Analysis of three first-order leveling lines that traverse the White Wolf fault (site of the 1952 $M = 7.7$ earthquake), each resurveyed nine times between 1926 and 1974, reveals probable preseismic tilting, major coseismic movements, and a spatial association between these movements and the subsequently recognized southern California uplift. In examining the vertical control record, we have both searched for evidence of systematic errors and excluded from consideration portions of the lines contaminated by subsurface fluid and gas extraction. Movements have been referred to an invariant datum based on the 1926 position of tidal BM 8 in San Pedro, corrected for subsequent eustatic sea-level change.

An 8 μrad up-to-the-north preseismic tilt (6 cm/7.5 km) was apparently recorded on two adjacent line segments within 10 km of the 1952 epicenter between 1942 and 1947. It is possible, however, that this tilt was in part caused by extraction-induced subsidence at one of the six releveled benchmarks. Data also show evidence of episodic tilts that are not earthquake related. At the junction of the Garlock and San Andreas faults, for example, an $\geqslant 5$ μrad up-to-the-north tilt (7.2 cm/$\leqslant 16$ km) took place between Lebec and Grapevine within three months during 1964.

Comparison of the 1947 and 1953 surveys, which includes the coseismic interval, shows that the SW-fault end (nearest the epicenter) and the central fault reach sustained four times the uplift recorded at the NE end of the fault (+72 cm SW, +53 cm Central, +16 cm NE). A regional postseismic uplift of 4 cm extended $\geqslant 25$ km to either side of the fault after the main event, from 1953 to 1956. An interval of relative quiescence followed at least through 1959, in which the elevation change did not exceed ±3 cm.

The detailed pattern of aseismic uplift demonstrates that movement proceeded in space—time pulses: one half of the uplift at the SW-fault end and extending southward occurred between 1959 and 1961, one half of the uplift at the NE-fault end and extending eastward occurred between 1961 and 1965, while the central fault reach sustained successive pulses of subsidence, uplift, and collapse (−4 cm, 1953—60; +7 cm, 1960—65; −2 cm, 1965—70). In addition, the number of aftershocks concentrated near the fault ends increased in the NE relative to the SW from 1952 to 1974. These observations suggest that the aseismic uplift may have migrated northeastward from 1959 to 1965 at an approximate rate of 7—16 km/yr.

Evidence for a mechanical coupling between the earthquake and the subsequent aseismic uplift is equivocal. At both fault ends, the major NW-bounding flexure or tilted front of the southern California uplift is spatially coincident with the coseismic flexure that preceded it. In addition, the postulated migration of vertical deformation is similar to the 1952 seismic event in which the rupture initiated at the SW end of the fault and then propagated to the NE-fault end. However, the spatial distribution of aseismic uplift, nearly identical at both fault ends and to the south and east, and near zero in the central fault reach, is distinctly different from the nonuniform and localized coseismic deformation.

Tectonophysics, 52 (1979) 303
303
© Elsevier Scientific Publishing Company Amsterdam — Printed in The Netherlands

FAULT LOCATION AND FAULT ACTIVITY ASSESSMENT BY ANALYSIS OF HISTORIC LEVEL LINE DATA, OIL-WELL DATA AND GROUNDWATER DATA, HOLLYWOOD AREA, CALIFORNIA *

R.L. HILL [1], E.C. SPROTTE [1], J.H. BENNETT [1] and R.C. SLADE [2]

[1] *California Division of Mines and Geology, Los Angeles, Calif. 90012 (U.S.A.)*
[2] *Geotechnical Consultants Inc., Burbank, Calif. 91502 (U.S.A.)*

(Accepted for publication April 26, 1978)

ABSTRACT

Transecting the Los Angeles metropolitan area in a general E—W direction are major north-dipping reverse faults comprising the Santa Monica—Raymond Hill fault zone, a segment of the frontal fault system separating the Transverse Ranges from the Peninsular Ranges geomorphic provinces of southern California. Pleistocene or Holocene movement is evident along some segments of these faults, but urban development precludes accurate location and assessment of Quaternary movement by conventional mapping techniques. At present no conclusive evidence of Holocene surface rupture has been found onshore west of the Raymond Hill segment of the fault zone, but the geologic conditions and urban development in the area are such that the possibility of Holocene movement cannot be excluded at this time. Groundwater barriers in Pleistocene sediments are indicative of Quaternary faulting on the Santa Monica fault segment west of the Newport—Inglewood fault zone. Most literature indicates that movement along the Beverly Hills—Hollywood segment east of the Newport—Inglewood fault zone terminated in Late Miocene or Pliocene time, and there is no general agreement on the location of faults in this segment. However, recent work by the Division of Mines and Geology, by Geotechnical Consultants, Inc., and others suggests that the Santa Monica fault transecting the Hollywood area is associated with a zone of differential subsidence that varies from 100 to 400 m wide, depending on the resolution of repeated leveling survey data and with a groundwater barrier determined from analysis of oil-well and water-well data. Additional exploration is essential to test our present geologic model and to evaluate the earthquake hazard and seismic risk of faults in the area.

* Based on research by the Division of Mines and Geology sponsored in part by the U.S. Geological Survey and on an independent study by Geotechnical Consultants, Inc.

Tectonophysics, 52 (1979) 304
© Elsevier Scientific Publishing Company Amsterdam — Printed in The Netherlands

ELASTIC EXPANSION OF THE LITHOSPHERE CAUSED BY GROUNDWATER WITHDRAWAL IN SOUTH-CENTRAL ARIZONA

T.L. HOLZER

U.S. Geological Survey, Menlo Park, Calif. 94025 (U.S.A.)

(Accepted for publication April 26, 1978)

ABSTRACT

Relative crustal uplift observed from 1948—49 to 1967 in the Lower Santa Cruz River Basin in south-central Arizona is attributed at least in part to elastic expansion of the lithosphere induced by the removal, and subsequent loss by evapotranspiration, of 4.35×10^{13} kg of groundwater from alluvium. The area of unloading is approximately 8070 km². Uplift, relative to an apparently stable area west of the unloaded area, was observed in two areas near Casa Grande and Florence where crystalline bedrock is either close to the land surface or crops out through alluvium from which groundwater was withdrawn. The magnitudes of uplift were approximately 6.3 and 7.5 cm respectively. The observations are based on first-order leveling. The observations are significant at three standard deviations for random surveying errors, and are not believed to be affected by systematic errors. However, the 7.5-cm uplift observed at Florence may be from 1 to 2 cm in excess of the actual uplift because of the possibility of subsidence of a tie point due to groundwater pumping during the leveling in 1948—49.

Uplift is attributed to groundwater withdrawal on three bases. First, the observed uplift is consistent with a theoretical evaluation of elastic expansion based on linear elasticity theory. For the observed distribution of unloading and uplift and a Poisson's ratio of 0.25, a Young's modulus for the lithosphere of approximately 0.68 Mbar is implied. This value is comparable to values of the lithosphere reported elsewhere. Second, the magnitude of uplift compares favorably with the magnitude of elastic depression caused by the formation of Lake Mead, Arizona—Nevada, 430 km northwest of the study area, when allowance is made for the different magnitudes and areal distributions of surface (un)loading. And third, in the area near Casa Grande, a reversal in the sense of bedrock displacement form subsidence of tectonic origin to uplift approximately coincided with the beginning of large groundwater overdraft. The uplift from 1948 to 1967 near Casa Grande was preceded from 1905 to 1948 by 7—8 cm of tectonic subsidence; no precise data for the area near Florence are available before 1948.

Tectonophysics, 52 (1979) 305—316

305

© Elsevier Scientific Publishing Company, Amsterdam — Printed in The Netherlands

THE INTERPRETATION OF VERTICAL CRUSTAL MOVEMENTS IN THE TIME—SPACE DOMAIN

TERUYUKI KATO

Earthquake Research Institute, University of Tokyo, (Japan)

(Accepted for publication April 26, 1978)

ABSTRACT

Kato, T., 1979. The interpretation of vertical crustal movements in the time-space domain. In: C.A. Whitten, R. Green and B.K. Meade (Editors), Recent Crustal Movements, 1977. Tectonophysics, 52: 305—316.

In order efficiently to process a file of leveling data which has been built up over the past century, a new computer-generated time—space domain presentation is developed. A critical condition for successful results is the proper interpolation of the original data to avoid undesirable distortion of the time—space pictures. Local procedures proposed by Akima seem most suitable for this purpose.

After solving several technical problems related to numerical work, this technique has been applied to leveling data in several Japanese areas of tectonic interest. Further study of the charts thus prepared has distinguished four elementary components in the patterns, which are:
(1) Local and superficial movements having artificial causes.
(2) Regional crustal movements.
(3) Migrating movements.
(4) Seismic crustal movements.
Discrimination of components of type (4) from the other three may be important for application to earthquake prediction research.

INTRODUCTION

The recent rapid accumulation of geodetic data, especially leveling data, makes it possible to examine in detail the features of vertical crustal movement of an area. Leveling work in Japan began at the end of the 19th century, and the first survey was completed around the turn of the century. Since then, complete revision surveys have been conducted three or four times. In addition, in many areas that suffered large earthquakes, supplementary surveys were carried out to detect coseismic or postseismic deformation. In particular, the route from the Tokyo datum bench mark to Aburatsubo Tidal Station has been surveyed over 20 times in order to determine the vertical movement of the Tokyo datum point relative to sea level.

Since about 1960, leveling surveys with other kinds of geodetic measurements have been accelerated in response to the increasing concern for the study of earthquake prediction, as well as to the improvement of survey techniques. This acceleration has brought a rapid accumulation of geodetic data. Thus, an urgent need has arisen to develop a new way of processing these data by computer. Fortunately, the recent development of computer techniques has enabled us to carry out this task. Kato and Kasahara (1977) recently proposed a new method of presenting leveling data in the time—space domain. This technique facilitates an intuitive view of crustal dynamics in a comprehensive two-dimensional pattern. This paper briefly reviews this technique and outlines recent developments of the computer technique. The accumulation of time—space domain charts by this method indicates that vertical crustal movements can be classified into several kinds of events.

TECHNIQUE AND DATA

Kato and Kasahara (1977) proposed a technique of time—space domain presentation of leveling data in detail, therefore this technique is only briefly reviewed here.

Suppose we consider a portion of a leveling route where a time—space domain contour chart will be designed. We must interpolate the original data when arrayed in the time—space domain, although there is generally a considerable amount of missing data on both the time axis and on the space axis. Therefore, a critical condition for good results is the use of an interpolation technique. Several kinds of techniques for interpolation have been designed (e.g., Vaníček and Christodulidis, 1974; Holdahl and Hardy, 1977). Kato and Kasahara used a cubic spline function, which is characterized by a locally determined third-order polynomial (Greville, 1967). This function satisfies the condition that the integral of the square of the curvature of the fitted curve becomes a minimum, so that this function interpolates the data most "smoothly". We find, however, that this function causes unrealistic undulation if the data include abnormally irregular portions. In order to overcome this difficulty, Akima (1970) proposed a new method of interpolation. His idea is basically the same as the spline function. However, a slight difference exists in the connecting condition at the nodes.

Let us take five successive data points as a set. Then we obtain four line segments that connect the adjoining points. The slope gradient at the center point is determined by taking the average of these four slopes. This gradient and the coordinates of the node successively define a cubic polynomial in each segment. This local procedure never produces unnatural curves, even if the original data include irregular portions. This is advantageous in our case as leveling data sometimes contains abnormal subsidence or uplift due to known or unknown causes.

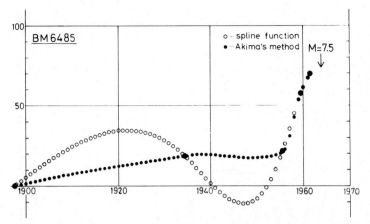

Fig. 1. Interpolation of leveling data by means of spline functions and Akima's local procedure. Original data are taken from Niigata district (see Fig. 2 for reference), and are shown by solid circles. The Niigata earthquake (M = 7.5) occurred in 1964. Only data before that earthquake are used.

Figure 1 compares the results of the two methods. Original data are taken from the Niigata district, Japan, where an abnormal uplift preceding the Niigata earthquake (1964, M = 7.5) was observed, and are plotted as large closed circles. Open and closed circles are the interpolated results using a cubic spline function and Akima's local procedure, respectively. The spline result seems quite "smooth" but is not realistic. For this reason we prefer the result obtained by Akima's method, and use his method here. Of course, unless the original data include abrupt changes of the crustal deformation mode, the two functions show no essential difference.

Using this new method, the author compiled a new computer program which produces contour plots in the time—space domain. First, a chart cumulative from the origin time is obtained. The origin time is set at the time of the first survey or a survey just after a big earthquake to avoid a big coseismic effect. In addition, a chart of annual rate is also drawn by differentiating the previous result with respect to time.

Original data are taken from the Annual Report of the First Order Leveling Measurements in Japan (vols. 1—20, and four supplemental volumes), issued by the Geographical Survey Institute (G.S.I.). References in detail are given in Kato et al. (1977). Regions in which the time—space domain technique is applied, are shown in Fig. 2.

EVENT TYPES OF VERTICAL CRUSTAL MOVEMENT

The time—space domain presentation technique proposed by Kato and Kasahara (see also previous section) is applicable to any leveling route. The accumulation of contour charts using this technique enables us to relate spatial and temporal variations in elevation. Accumulated experience has

308

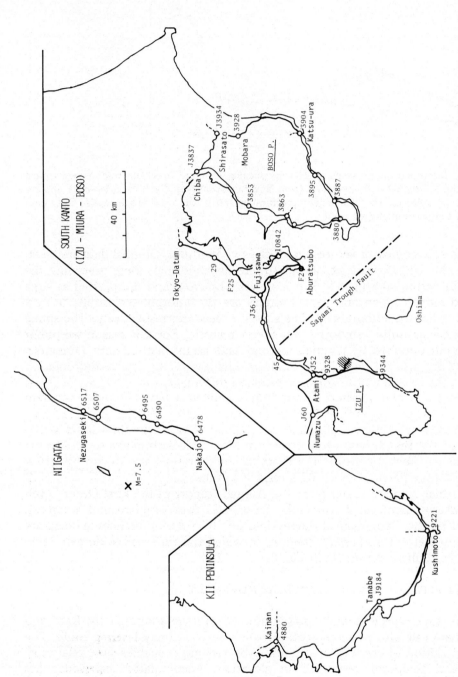

Fig. 2. Local maps of regions in which the time–space domain technique is applied. Numerals are the numbers of bench marks. The shaded area represents the Ito earthquake swarm of 1930. The time–space domain charts derived from the data are shown in Figs. 3–8.

revealed that these patterns include four fundamental types of events which can be readily distinguished.

Local and superficial movements having artificial (or nontectonic) causes

This type is characterized by its narrow area, usually several kilometers or 10 km at most, and its high rate of subsidence or uplift. This is easily distinguished from the other types, which show wider but more gradual movement. The subsidence in the Kawasaki area, south Kanto, shown in Fig. 3, is an example. Another example from Mobara, Boso Peninsula, which is the site of natural gas extraction, is shown in Fig. 4. This kind of event, which is caused by water or gas extraction, hot springs and so on, should be subtracted from the pattern for the analysis of crustal dynamics due to tectonic causes.

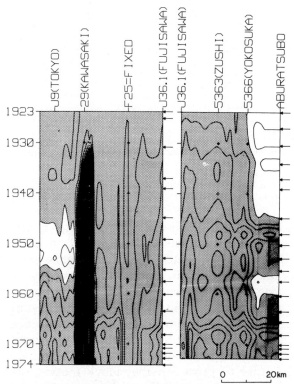

Fig. 3 Cumulative vertical land displacement from Tokyo datum to Aburatsubo Tidal Station, since 1923. Routes are divided into two parts as the epochs of surveys are different between Tokyo—Fujisawa and Fujisawa—Aburatsubo. BM F25 is fixed as the reference point. Meshed portions show subsidence, whereas other areas show uplift. Contour unit is 25 mm. Arrows indicate the dates of leveling surveys. Horizontal axis is taken along the leveling route.

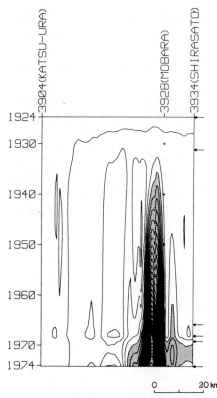

Fig. 4. Cumulative vertical land displacement from Katsu-ura to Shirasato, Boso Peninsula (see Fig. 3 for symbol explanations).

Regionality of crustal movements

We sometimes notice a sharp boundary line on a chart, across which modes of crustal movement differ. Cumulative vertical land displacement from Tokyo to Numazu, in Fig. 5, is an example. In this case, the reference point which is assumed stationary is taken at BM F25, Yokohama. Nevertheless, two sharp boundaries are seen, one near the Kozu—Matsuda fault which is a landward extension of the Sagami trough (see Fig. 2), and the other east of Numazu. This kind of movement is likely to have some relation to block structures in the crust. In the Tokyo—Atami route, the uplifted area corresponds to the neck of the Izu Peninsula, so it might be that this peninsula moves as a block. However, in order to prove whether block motion observed in leveling data is real or not, further study is needed, especially by making comparisons with geological or geomorphological evidence.

Migration of crustal movement

This pattern type is characterized by a trend oblique to the coordinate axes. A well-known example is the land movement during the Ito earthquake

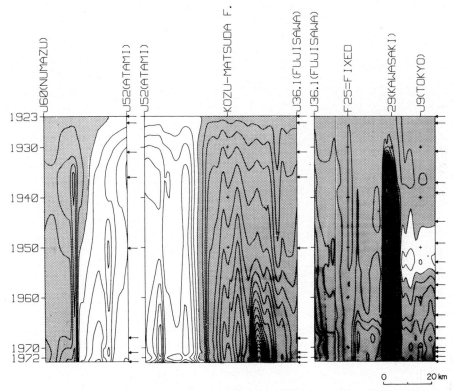

Fig. 5. Cumulative vertical land displacement from Tokyo to Numazu, south Kanto. Routes are divided into three parts as the epochs of the surveys are different (see Fig. 3 for symbol explanations).

swarm in 1930, for which a time—space domain contour chart of annual rate was drawn manually by Tsuboi (1933). Our chart for the same event is drawn in Fig. 6, in which we recognize a migration pattern compatible to that pointed out by Tsuboi. The bold line shown in the chart is the axis of maximum uplift rate. The apparent velocity of migration is about 15 km/yr, from south to north along this route.

As Kasahara (1977) pointed out, the detection of this type of pattern could be a useful method of earthquake prediction. Judging from its propagation rate, 10—100 km/yr, frequent leveling surveys, say, every few years are needed for this purpose.

Seismic crustal movement

The detection of this type of pattern, especially that *preceding* earthquakes, may be most important for earthquake prediction. The best-known example is the case in Niigata district where time—space domain charts are

312

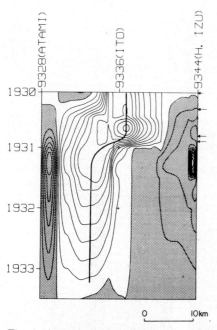

Fig. 6. Annual rate of vertical displacement in the vicinity of Ito. Here the origin time is set at 1923 and only a portion of the original data set — which includes the Ito earthquake swarm in 1930 — is used. Meshed portions show subsiding areas. Contour unit is 2.5 mm/yr.

shown in Figs. 7a and 7b. Crustal uplift preceding the Niigata earthquake is evident in the figure. Unfortunately, as the number of examples is not yet large enough, we cannot conclude that such preseismic crustal movements will always occur. However, if they always occurred, the spatial distribution and the duration of the concentric pattern would predict, to some extent, the magnitude and time of the coming earthquake.

For successful prediction by detecting this kind of pattern, frequent leveling surveys are needed. As we can see in the case of Niigata, the surveys made in 1955, 1959 and 1961 fully define the pattern. So we might say that surveys in other areas should be repeated every few years for the purpose of long-term prediction. If we succeed in detecting such crustal movement before an earthquake, we will also be able to shed further light on the mechanism of both crustal deformation and earthquakes.

Effect of coseismic displacement

It is obvious that our technique cannot be directly applied if the original data contain displacements due to large earthquakes. In other words, direct interpolation over the coseismic step would produce a spurious type of pattern on the time–space domain chart. This pattern distortion may be large, especially preceding an earthquake, since leveling surveys are usually not as

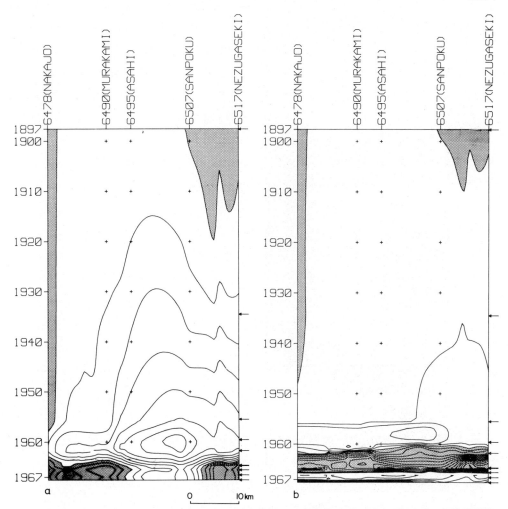

Fig. 7a. Cumulative land displacement in the north Niigata district. Kashiwazaki Tidal Station, fixed during the period 1897—1967, is located to the southwest of Niigata city (not seen in this figure). (See Fig. 3 for symbol explanations). b. Annual rate of vertical land displacement in the north Niigata district. Contour unit is 5 mm/yr and two contours of ±2.5 mm/yr are added.

frequent in this period as after an earthquake.

In order to overcome this difficulty, we should make a few assumptions about coseismic effects. One provisional assumption is, for example, that the coseismic displacement is equal to the displacement between successive revision surveys that span an event. On this assumption, a contour chart of cumulative displacement in the Kii Peninsula, which experienced the Nankaido earthquake (1946, M = 8.1), is shown in Fig. 8. In this case, the periods are separated into two parts — one before and one after the earth-

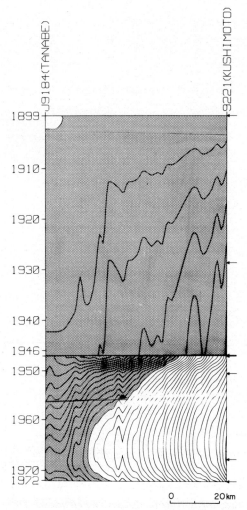

Fig. 8. Cumulative vertical land displacement from Tanabe to Kushimoto, Kii Peninsula, since 1899. This area experienced the Nankaido earthquake in 1946 (M = 8.1). The chart is divided into two parts — one before and one after the earthquake — and are manually connected (see Fig. 3. for symbol explanations).

quake — the two parts being connected manually. Automation of the last process is left as a future problem at this stage.

Eventually, unless we make some kind of assumption about the coseismic step, distortion of the pattern cannot be avoided, except for the case of Niigata because leveling surveys were frequent in this case.

CONCLUSIONS

In this study a computer technique for the time—space domain presentation of leveling data has been proposed, and different types of crustal movement events as seen on the resulting contour charts have been discussed. Several points concerning the further development of the study will be briefly discussed in the following.

Further computerization of technique

The foregoing study has been concerned mainly with the primary and introductory problems of data processing. In other words, refinement of the technique or the completion of the whole program system has been left for the future. In particular, an efficient data filing system is necessary for rapid processing. Such a system should permit easy access to necessary information, together with easy editing of new data.

Comprehensive data processing system

In order to reveal the characteristics of crustal dynamics, many kinds of information such as triangulation data, gravity data, seismicity data, geomagnetic data, etc., should be combined and synthesized. For this purpose, a comprehensive data processing system including the technique proposed here should be developed. If this type of processing system is implemented, it will be possible to derive more meaningful information of crustal movement from the data; furthermore, such a comprehensive study of crustal movement is also very important for earthquake prediction.

Approaches to tectonic implications

An overall view of crustal movement through rapid mass processing of geodetic data in Japan or throughout the world could lead us to a general understanding of crustal dynamics or tectonics. In Japan, especially where the tectonic activity is unusually high, the technique proposed here would reveal dynamic structures related to the effect of plate convergence. One study now in progress by the author (Kato, in preparation) examines whether a marked subsidence along the Pacific coast can be interpreted as crustal drag caused by a sinking slab.

ACKNOWLEDGEMENTS

The author wishes to express his sincere thanks to Professor Keichi Kasahara of the Earthquake Research Institute, University of Tokyo, for his advice during the course of the study and critical reading of the manuscript. The author would like to express his gratitude to Dr. Paul Somerville for a critical reading of the manuscript, and to Ross Stain, Stanford University for helpful discussions.

Numerical computations were carried out with the Earthquake Research Institute's IBM system 370, model 125.

REFERENCES

Akima, H., 1970. A new method of interpolation and smooth curve fitting based on local procedures. J. Assoc. Comput. Mach., 17: 589—602.

Geographical Survey Institute, The Annual Report of the First Order Leveling Measurements in Japan, vols. 1—20, and four supplemental volumes (in Japanese).

Greville, T.N.E., 1967. Spline functions, interpolation, and numerical quadrature. In: A. Ralston and H.S. Wilf (Editors), Mathematical Methods for Digital Computers, vol. 2. Wiley, New York, N.Y., pp. 156—168.

Holdahl, S.R. and Hardy, R.L., 1979. Solvability and multiquadric analyses as applied to investigations of vertical crustal movements. Tectonophysics, 52: 139—155.

Kasahara, K., 1979. Migration of crustal deformation. Tectonophysics, 52: 329—341.

Kato, T., 1979. Recent crustal movement in the Tohoku district for the period 1900—1975. Tectonophysics, in press.

Kato, T. and Kasahara, K., 1977. The time—space domain presentation of leveling data. J. Phys. Earth, 25: 303—320.

Tsuboi, C., 1933. Investigation on the deformation of the earth's crust found by precise geodetic means. Jap. J. Astron. Geophys., 10: 93—248.

Vaníček, P. and Christodulidis, D., 1974. A method for the evaluation of vertical crustal movement from scattered geodetic relevellings. Can. J. Earth Sci., 11: 605—610.

Tectonophysics, 52 (1979) 317

Geologic Studies of Holocene Deformation

POSTGLACIAL DISPLACEMENTS ALONG THE ALPINE FAULT, AND ITS SIGNIFICANCE TO THE NEW ZEALAND TECTONIC FRAMEWORK

G.J. LENSEN

New Zealand Geological Survey, P.O. Box 30368, Lower Hutt (New Zealand)

(Accepted for publication April 26, 1978)

ABSTRACT

Recent field studies demonstrate the southern and northern parts of the Alpine fault to be dominantly under right-lateral shear. The central portion of this fault is dominantly under compression.

The Marlborough—North Island dextral shear zone, together with the Fiordland and NW Nelson sinistral shear zones, demonstrate these shears to result from lateral drag within these zones and is only partially transmitted to the central section of the Alpine fault which is dominantly reverse in character.

Regional extension in the North Island west of the shear belt and regional shortening in the South Island indicate clockwise rotation at the east side of the Alpine fault and its extension in the North Island relative to the west side about a "pole" on the Alpine fault in the north of the South Island.

Tectonophysics, 52 (1979) 319—327
© Elsevier Scientific Publishing Company, Amsterdam — Printed in The Netherlands

EVIDENCE FOR RECENT COASTAL UPLIFT NEAR AL JUBAIL, SAUDI ARABIA

ALBERT P. RIDLEY and MARC W. SEELEY *

Woodward-Clyde Consultants, 3 Embarcadero Center, Suite 700, San Francisco, Calif. 98125 (U.S.A.)

(Accepted for publication April 26, 1978)

ABSTRACT

Ridley, A.P. and Seeley, M.W. 1979. Evidence for recent coastal uplift near Al Jubail, Saudi Arabia. In: C.A. Whitten, R. Green and B.K. Meade (Editors), Recent Crustal Movements, 1977. Tectonophysics, 52: 319—327.

Geologic mapping along the Persian Gulf coast about 15 km north of Al Jubail, Saudi Arabia, reveals a series of elevated beach deposits up to 2.8 m above sea level. Carbon-14 age-dating of shell materials from those beach deposits at an elevation of +2.8 m gives an age of 3812 ± 145 years B.P. Archeological evidence about 15 km south of Al Jubail shows that about 5000—6000 years ago relative sea level was about 4 m higher than present sea level. Uplift of the coast near Al Jubail of 5.0 m in the last 3812 ± 145 years and 9.3 m in the last 5000—6000 years is indicated considering Holocene sea-level data from Flint (1971). Considering the higher Holocene sea level of Clark (1977) uplift is estimated to have been at least 0.8 m in the last 3812 ± 145 years and 1.5 m in the last 5000—6000 years. The mechanism for uplift may be the uplift and warping along the extension of a NW-trending anticline near Bahrein that Kassler (1973) suggests has been rising in the last few thousand years.

INTRODUCTION

Elevated beach deposits were reported along the eastern coast of Saudi Arabia as early as 1958 (Steineke et al., 1958) and later in 1960 (Holm, 1960). Beach deposits identified by the authors (1—2 m above present sea level) were mapped during a geologic study for a planned industrial complex about 15 km north of the town of Al Jubail (Fig. 1). The eastern coastal area of Saudi Arabia is typified by low relief and extensive "sabkhas" (broad saline flats that are at or near sea level). It is generally considered that these sabkhas were formed by accreting beach deposits closing off lagoons.

In contrast to the eastern coast of the Persian Gulf, the western margin of the Gulf is considered to have been comparatively stable throughout the Tertiary and Quaternary periods (Kassler, 1973). If the western margin of

---

* Presently with H M S Associates, Consulting Civil Engineers and Geologists, P.O. Box 177, Hayward, Calif. 94543, U.S.A.

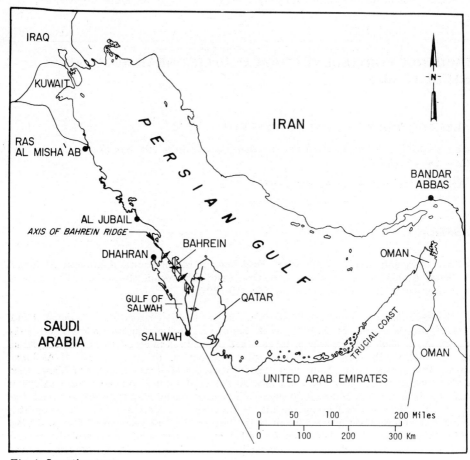

Fig. 1. Location map.

Saudi Arabia has been relatively stable throughout the Quaternary period, the raised beach deposits indicate a higher stand of sea level along the coast near Al Jubail. However, sea-level data (Clark, 1977) for Late Quaternary time in the Persian Gulf do not indicate levels high enough to account for the elevated beach deposits. Therefore, the elevated beach deposits observed near Al Jubail are interpreted to have resulted from Quaternary uplift of the eastern Arabian coast in the vicinity of Al Jubail.

GEOLOGIC SETTING

The low topography of the coastal area in the vicinity of Al Jubail is dominated by marine- and wind-deposited Quaternary sediments. Sabkhas extend 10 km inland from the shallow Persian Gulf containing numerous offshore islands and bars. Interspersed among the coastal sabkhas are dune fields formed by accumulation of sand blown southward from the coastal

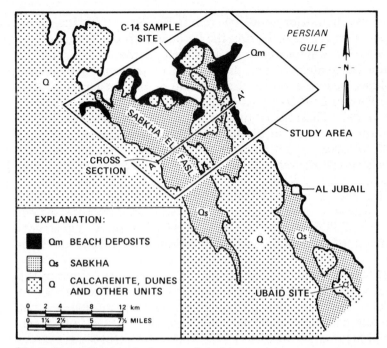

Fig. 2. Generalized geologic map of study area.

area. Partially covered by dune deposits is a flat-lying calcarenite, locally known as the Bahr formation, that occurs along the coast in the vicinity of Al Jubail (Fig. 2). Fossil assemblages sampled for this investigation suggest a Late Pleistocene age for the calcarenite unit, (Durham, 1977). Miocene and Pliocene sedimentary bedrock of the Hadrukh and Dam formations crops out about 20 km inland. The Late Tertiary strata crop out in a broad zone about 100 km wide, extending north and south of the Al Jubail area (Steineke et al., 1958). The Miocene and Pliocene strata in the vicinity of Al Jubail dip 5—10°N beneath the younger coastal sediments and the Persian Gulf. Kassler (1973) shows, by geophysical surveys, that the island of Bahrein, located about 150 km to the south, and the Qatar Peninsula about 200 km to the south (Fig. 1) represent N—S-trending anticlinal structures that are oblique to the trend of the axis of the Persian Gulf.

TECTONIC AND STRUCTURAL SETTING

The structural development of the Persian Gulf is described in detail by Kassler (1973). The island of Bahrein and the Qatar Peninsula, are aligned along anticlinal growth structures that have a N—S or NE—SW trend informally called the "Arabian" trend. These "Arabian" growth structures have two striking characteristics, according to Kassler (1973): prolonged upward movement during sedimentation, and low relief. Kassler (1973) has documented prolonged structural growth from the Permian into the Tertiary.

The Plio—Pleistocene Zagros orogeny is related to the collision of the Arabian and Eurasian plates. During that period of orogenic activity the most intense folding occurred on the Iranian side of the Persian Gulf. The main axis of the Persian Gulf is shown by Kassler (1973) to be a "Zagros" feature formed in Plio—Pleistocene time. He interprets seismic reflection survey data as showing folding in the Salwah—Bahrein—Qatar area (Fig. 1) along a NW—SE "Zagros" trend. Kassler also suggests that NW-trending topographic shoals along Bahrein Ridge indicate Quaternary uplift along those trends.

Seismic profiles and a photogeologic study in Qatar and Abu Dhabi (Kassler, 1973) suggest a NW—NNW-trending fault and fracture system affecting Miocene and Pliocene rocks. On LANDSAT imagery, lineaments are visible northwest of Al Jubail which may represent that fault and fracture system.

Kassler (1973) believes that Bahrein Ridge (Fig. 1) has been rising in the last few thousand years. A northwest extension of Bahrein Ridge through Al Jubail and parallel to the coastline is postulated as a tectonic mechanism that accounts for the raised beaches observed near Al Jubail. The Gulf of Salwah to the southeast is believed by Kassler (1973) to be a Late Quaternary structural feature and an area of subsidence in the Late Quaternary. However, raised beaches reported by Holm (1960) as far south as Salwah are not in agreement with the interpretation of the Gulf of Salwah being an area of subsidence. Kassler (1973) shows a NW-trending syncline along the axis of the Gulf of Salwah (Fig. 1) and a possible fault along the west side of the Gulf of Salwah.

ELEVATED BEACH DEPOSITS AND RELATED FEATURES

Elevated beach deposits are mapped by Steineke et al. (1958) from just north of Al Jubail to as far south as Salwah. Holm (1960) describes beach deposits 1—2 m above sea level extending along the Arabian coast from Ras al Misha'ab, about 200 km north of Al Jubail, southward to Salwah, about 300 km to the south (Fig. 1). Holm (1960) describes the beach deposits as being composed mostly of an accumulation of small turritella-like gastropods that are locally quarried for light aggregate near Dhahran.

Beach deposits, between 1 and 2 m above present sea level and 200—300 m wide, consisting of turritella-like gastropods (*Ceritium* sp., Hughes-Clarke and Keij, 1973) and calcareous sand, occur along the coast about 20 km north of Al Jubail. Field mapping and aerial photograph interpretation show that those beach deposits occur on both sides of a N—S-trending peninsula (Fig. 2) underlain by calcarenite bedrock and separates the Persian Gulf on the east from the Sabkha El Fasl to the west. Recessional strand lines are clearly visible both on the ground and on aerial photographs of the

323

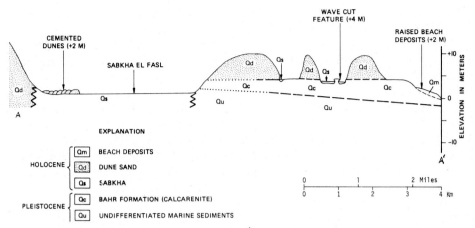

Fig. 3. Generalized geologic cross-section A-A'.

area north of Al Jubail. These strand lines indicate a higher sea level in the recent geologic past.

The beach deposits near Al Jubail rest unconformably upon a flat-lying calcarenite that extends to the west, forming a generally flat surface across the peninsula at an elevation of about 4—5 m (Fig. 2). Fossil assemblage from the calcarenite sampled for this study indicates a Late Pleistocene age (Durham, 1977) for that rock unit. Holm (1960) reports that the calcarenite forms low cliffs along the shore as far north as Ras al Misha'ab and remnants of the formation are found as far south as the Trucial coast (350 km to the southeast of Al Jubail). The calcarenite has been given the informal name of the Bahr formation, from its first observed occurrence in Jabal Bahr, offshore from the village of Al Jubail. The Bahr formation forms stacks and islands offshore and may dip beneath Recent sediments of the Persian Gulf (Holm, 1960).

A leveling survey across the beach deposit east of the peninsula shows it is elevated from 1 to 2 m above present sea level (Fig. 3). Related features that indicate a relatively higher stand of sea level are approximately 2-m high wave cut benches at about the 2—4-m elevation near the center of the peninsula (Figs. 2 and 3). Remnants of cemented dunes at an elevation of about 2 m on the west side of the Sabkha El Fasl (Figs. 2 and 3) were identified. The calcite cemented dunes may have formed during a Holocene or earlier period of relatively higher sea level.

AGE OF RAISED BEACHES

Samples of the shell material from the raised beach deposit (Fig. 2) were taken at an elevation of about +2.8 m by digging approximately 10 cm below the surface. The shells were dated using Carbon-14 methods. The sam-

324

ple has an age of 3812 ± 145 B.P. (Krueger, 1977) which is consistent with the archeological data discussed in the following paragraph.

Bibby (1973) reports evidence of an early Ubaid settlement about 15 km south of Al Jubail (Fig. 2). The Ubaid people were the first agricultural settlers to move into the lower valley of the Tigris and Euphrates 5000 B.C. Bibby (1973) estimates that the Ubaid people were in Saudi Arabia between 3000 and 4000 B.C. (5000 and 6000 years B.P.). The archeological site occurs on a series of low sandy hills that separate the sea from a large sabkha. Bibby (1973) suggests that the hills may have been islands during Ubaid times. There, barnacle-covered remnants of plaster-covered reed walls, occur 4 m above high tide level as surveyed by Bibby (1973). It is unlikely that plaster-covered reed walls would be constructed at sea level. Therefore, remnants of those walls were submerged by a relative sea-level rise and then exposed by uplift or lowering of sea level. However, whether or not the ^{14}C and archeological data confirm relative uplift of from 0.8 to 5 m of the coast of Saudi Arabia over the past 3812 ± 145 years depends greatly upon the interpretation of relative sea-level variations during that period.

SEA-LEVEL VARIATIONS

There is wide disagreement in the literature on variations of relative sea level during the Holocene. Flint's (1971) compilations of data on world-wide sea levels suggest sea level was −125 m about 15,000 years ago and rose rapidly to about −5 m by about 5000 years ago (Fig. 4). The rate of rise decreased rapidly between 5000 years B.P. and the present time (Fig. 4). Using computer modeling techniques and published sea-level data from various parts of the world, Clark (1977) shows that relative sea level varies from place to place in the world, and for the west end of the Makran coast of Iran (Fig. 1), located east of Bandar Abbas, computes a sea-level curve (Fig. 4) showing a relative rise to a high of about +2.5 m about 5000 years ago, followed by a gradual decrease in sea level to the present. Because there is no agreement on the relative sea-level variations during the Holocene, this paper presents the possibilities using the two curves representing the best of the two hypotheses. Holocene tectonic uplift of the coast at Al Jubail is estimated (Table I) using the Holocene relative sea-level curves of Flint (1971) and Clark (1977). The greater amount of uplift is indicated using the curve of Flint (1971). However, even considering the postulated higher relative sea level of Clark (1977), an uplift of 0.8 m in the last 3812 ± 145 years to 1.5 m in the last 5000—6000 years is indicated for the coast near Al Jubail. An uplift of 5.0 m in the last 3812 ± 145 years and 9.3 m in the last 5000—6000 years is indicated using data from Flint (1971). The average rate of uplift in the last 3812 ± 145 years ranges from 0.2 m/1000 years using Clark (1977), to 1.3 m/1000 years using Flint (1971). The average rate of uplift for the last 5000—6000 years ranges from 0.2—0.3 m/1000 years using Clark (1977) to 1.5—1.9 m/1000 years using Flint (1977). Based on what is

TABLE I

Uplift rates

Location [a]	Present elevation (m)	Sea-level position at time of deposition (m) [b]		Amount of uplift (m)		Age (years B.P.)	Average uplift rate (m/1000 yr)	
		Flint (1971)	Clark (1977)	Flint (1971)	Clark (1977)		Flint (1971)	Clark (1977)
Ubaid site	+4 [b]	−5.3	+2.5	+9.3	+1.5	5000—6000 [c]	1.5—1.9	0.2—0.3
Beach deposits near Al Jubail	+2.8	−2.2	+2.0	+5.0	+0.8	3812± 145 [d]	1.3	0.2

[a] From Fig. 2.
[b] From Fig. 4.
[c] From Bibby (1973).
[d] Radiometric age (Krueger, 1977).

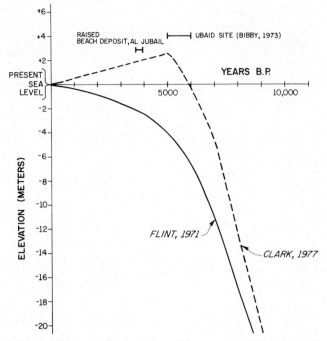

Fig. 4. Relative sea-level curves.

known from the literature (Kassler, 1973) and from observation about the tectonic environment of the coast near Al Jubail, the amount of uplift indicated by Clark's (1977) sea-level curve is most reasonable. This conclusion is supported by the close agreement of uplift rates between the Ubaid site and the ^{14}C site using Clark's (1977) curve (Table I).

SUMMARY AND CONCLUSION

Elevated beach deposits as much as 2.8 m above present sea level occur along the Saudi Arabian coast about 15 km north of Al Jubail. Archeological evidence about 15 km south of Al Jubail shows that relative sea level was about 4 m higher than present sea level 5000—6000 years ago (Bibby, 1973). Carbon-14 age-dating of shell materials from the 2.8 m elevated beach deposits north of Al Jubail give an age of 3812 ± 145 years B.P. There is no general agreement as to the variations in worldwide relative sea level during the Holocene. Clark (1977) suggests a 2.5 m higher stand of sea level about 5000 years ago on the west end of the Makran coast of Iran (Fig. 1), located east of Bandar Abbas. However, the worldwide data suggests a relative rise in sea level over the past 10,000—15,000 years to reach the present level. An uplift of the coast near Al Jubail of 5.0 m in the last 3812 ± 145 years and 9.3 m in the past 5000—6000 years is indicated using data from Flint (1971). These large uplift values may result from the generalized, worldwide nature of Flint's (1971) data. The higher postulated relative sea level of Clark (1977) indicated uplift of 0.8 m in the last 3812 ± 145 years and 1.5 m in the last 5000—6000 years which appears more reasonable. The relationships of the raised beach deposits near Al Jubail to the raised beaches reported to the north and south was not investigated for this study. Elevated beaches should be further studied relative to the regional and local structure to understand better the mechanism of uplift. A northwestward continuation of Bahrein Ridge along the coastline to Al Jubail gives a structural explanation of the raised beaches in Al Jubail. It does not, however, provide an explanation of raised beaches along the Gulf of Salwah, which is reported by Kassler (1973) to be a Recent Quaternary depression. Additional studies along the coastline of Saudi Arabia will provide significant new data regarding local tectonics.

ACKNOWLEDGEMENTS

The authors wish to thank John N. Alt and William D. Page for critically reading the manuscript and giving valuable advice. However, the conclusions presented in this paper are solely the responsibility of the authors.

REFERENCES

Bibby, G., 1973. Looking for Dilmun, proof edition. Export Library Catalog No. 17-184062, 10-6-1973.
Clark, J.A., 1977. Global Sea Level Changes Since the Last Glacial Maximum and Sea Level Constraints on the Ice Sheet Disintegration History. Unpublished Ph. D. Thesis, University of Colorado, 105 pp.
Durham, J.W., 1977. Examination of fossiliferous samples from Al Jubail, Saudi Arabia. Unpublished consultants report to Woodward-Clyde Consultants, San Francisco, Calif., dated April 28, 1977.
Flint, R.F., 1971. Glacial and Quaternary Geology. Wiley, New York, N.Y., 892 pp.
Holm, D.A., 1960. Desert geomorphology in the Arabian Peninsula. Science, 132 (3473): 1369—1379, Nov. 11.
Hughes-Clark, M.W. and Keij, A.J., 1973. Organisms as producers of carbonate sediment and indicators of environment in the southern Persian Gulf. In: B.H. Purser (Editor), The Persian Gulf. Springer, New York, N.Y.
Kassler, P., 1973. The structural and geomorphic evolution of the Persian Gulf. In: Purser (Editor), The Persian Gulf. Springer, New York, N.Y., pp. 11—32.
Krueger, H.W. 1977. C-14 age, Geochron Laboratories sample no. GX-5037, of 3785 ± 145 years B.P., reference year A.D. 1950, based upon Libby half-life (5570 years) for C-14.
Steineke, M., Harris, T.F., Parsons, K.K. and Berg, E.L., 1958. Geologic map of the western Persian Gulf Quadrangle, Kingdom of Saudi Arabia. U.S. Geol. Surv. Map I-208A.

Tectonophysics, 52 (1979) 329—341   329

MIGRATION OF CRUSTAL DEFORMATION

KEICHI KASAHARA

Earthquake Research Institute, University of Tokyo, Tokyo (Japan)

(Accepted for publication April 26, 1978)

ABSTRACT

Kasahara, K., 1979. Migration of crustal deformation. In: C.A. Whitten, R. Green and
B.K. Meade (Editors), Recent Crustal Movements, 1977. Tectonophysics, 52: 329—
341.

Observations on the migration rates of crustal deformation, as recently discovered in
several tectonic areas, such as the south Kanto and central Tohoku districts, Japan and
the West Cordillera Mts., Peru, has opened up a new opportunity for the study of crustal
dynamics. Briefly, these examples from coastal areas are characterized by migration landwards with a velocity of about 10—100 km/yr. This agrees well with the velocity of migration of seismicity as previously known. Dispersion and dissipation of the deformation
waveform are also noted as characteristics.
    Simple extrapolation of the migration path back toward the ocean may locate a pos-
sible origin of the event. In the case of the south Kanto district, for example, the deformation front seems to have originated in the early 1950s from the vicinity of the junction of
the Japan and Izu—Mariana trenches. The deformation front in the central Tohoku district, on the other hand, is thought to have originated in the northern part of the Japan
Trench in the late 1960s. One may suppose that either a repeated irregular aseismic plate
motion generates the deformation events, or that it results from a periodic seismic slip at
a plate boundary. In the latter case, the 1953 Boso-oki and the 1968 Tokachi-oki earthquakes might be suspected of generating the deformation fronts in the south Kanto and
central Tohoku districts respectively.
    As Scholz speculated, the migration of a deformation front might trigger earthquakes,
if it hits areas of high seismic potential. Studies of migration events can contribute significantly to earthquake prediction studies.

INTRODUCTION

Earthquake foci in a seismic zone sometimes appear to migrate systematically in one direction. The most convincing evidence for this is perhaps the
last sequence along the Anatolia fault, Turkey. During the two decades following 1939, a series of major earthquakes occurred which migrated at a rate
of about 80 km/yr from the eastern to the western extremity of the fault
(Richter, 1958; Mogi, 1968a). Further examples of a similar kind have been
recognized in many other areas by Mogi (1968a, 1968b, 1969, 1973),

Kelleher (1972) and Whitcomb et al. (1973). The upper half of Table I is taken from Ida (1974) and compiles earthquake migrations so far reported.

Migration of land deformation as discovered by tiltmeter observations (Yamada, 1973) and by creep measurements (King et al., 1973) have shown us new aspects of the phenomena. In these cases, we observe the reality of migration more directly than in the previous cases, where the earthquake sequence was no more than an indicator of some unknown deformation that migrated in the crust.

Excluding a small number of exceptions, the migration velocity of both

TABLE I

Velocity of migration

Type of event	Locality	Period	Velocity	Author(s)
Major (shallow) earthquakes	Anatolia (Turkey)	1939—57	80 km/yr	Richter (1958), Mogi (1968a), Savage (1971)
	Philippines	1930—60	50 km/yr	Mogi (1973)
	Kita-Izu (Japan)	1930—62	12 km/yr	Mogi (1969)
	Sanriku (Japan)	1926—65	150 km/yr	Mogi (1968a)
	West coast of U.S.	1830—1970	60 km/yr	Savage (1971)
	Chile	1880—1960	10 km/yr	Kelleher (1972)
Deep earthquakes	Mariana Arc	1930—65	50 km/yr	Mogi (1973)
	Tonga Arc	1900—69	45 km/yr	Mogi (1973)
Aftershock sequences	Aleutian	1957.3.9	400 km/yr	Mogi (1968b)
	Alaska	1964.3.27	60 km/yr	Mogi (1968b)
	San Fernando	1971	4—15 km/day	Whitcomb et al. (1973)
Fault creep	Central California	1971	1—10 km/day	King et al. (1973), Nason and Weertman (1973)
Major earthquakes, seismicity increase, ground tilts, etc.	Pohai (NE China)	1966—75	110 km/yr	Scholz (1977)
Ground tilts	South Kanto (Japan)	1950—70	20 km/yr (W) [*]	Yamada (1973), Kasahara (1973a,b)
	Peru	1966—70	60—70 km/yr (NNE)	Tanaka et al. (1977)
Ground strains	Tohoku (Japan)	1968—75	38 km/yr (N50°W)	R.G.C.M. [**] (1976)
	Tohoku (Japan)	1968—75	19 km/yr (W)	R.G.C.M. (1976)
Postseismic fault slip	White Wolf (Calif.)	1957—74	7—16 km/yr	Stein et al. (1979)

[*] Direction of migration.
[**] Research Group for Crustal Movement.

seismicity and land deformation appears generally to be in the range of 10—1000 km/yr (Table I). This may not be an accidental agreement. Several theoretical models have been proposed to explain how crustal activity (seismic activity and/or land deformation) can migrate at rates far below elastic velocities (Savage, 1971; Bott and Dean, 1973; Ida, 1974; Anderson, 1975).

The accumulation of observational data has provided us with several examples of migrating crustal movements, together with new findings on the nature of this interesting phenomenon. This paper will compile the recent data of migrating crustal deformation in order to study its basic characteristics, with special reference to its tectonic implications. A short note has also been added to emphasize its possible role in earthquake prediction research.

PATTERNS OF MIGRATING CRUSTAL DEFORMATION

Ground tilt in the south Kanto district, Japan

Yamada (1973) compared secular tilts at two adjacent stations in the south Kanto district, Aburatsubo and Nokogiriyama, and found evidence for migration. Fig. 1 is taken from his paper. It shows the tilt vector for a recent 10-year interval at Aburatsubo compared with that at Nokogiriyama. We have applied a simple coordinate transformation to the Nokogiriyama data for convenience in comparison. Several irregular events which appear in pairs on the two curves may be noticed. As distinguished by symbols *A*, *B*, *C*, etc., the phases of these events at Nokogiriyama lead those at Aburatsubo by one year or more, suggesting a migration from east (Nokogiriyama) to west (Aburatsubo) at an apparent velocity of about 20 km/yr (separation of the stations is 20 km).

A migration of this kind may be recognized, even at a single station, if the vertical land displacement is known in addition to tilts. In fact, the phase of the tilt at Aburatsubo leads that of the vertical displacement (measured by leveling) by approximately 90°, indicating that the deformation is indeed propagating (Kasahara, 1973b).

Ground strain in the central Tohoku district, Japan

The Research Group for Crustal Movement, Tohoku University (1977) has recently obtained more precise data about migration from their strainmeter array (see also Ishii, 1976). Figs. 2 and 3 are taken from these papers. Fig. 2 illustrates their stations: Sanriku (SNR), Miyako (MYK), Himekami (HMK), Nibetsu (NIB) and Oga (OGA), and Fig. 3 provides a summary of the observed strain accumulation (maximum shear). The symbols in the figure represent dominant events referred to.

The correlation is excellent, especially among the first three stations on the Pacific side, where the events arrive progressively at SNR, MYK and HMK, in that order. This suggests a northwestwards migration. They calcu-

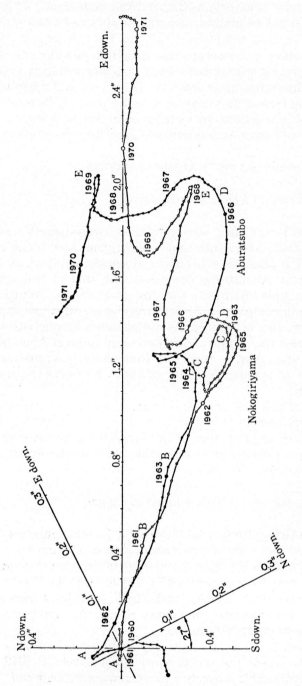

Fig. 1. Locus of the tilt vector at Aburatsubo as compared with the data at Nokogiriyama (Yamada, 1973).

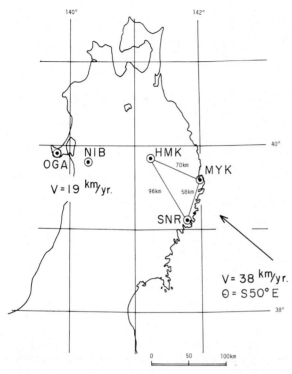

Fig. 2. Array of crustal movement observatories and migration of ground strain (maximum shear) in Tohoku district, Japan (Research Group for Crustal Movement, Tohoku University, 1977).

lated the cross-covariance of the three sets of data and determined a migration velocity at 38 km/yr in the direction N50°W from their phase relationships. As for the other two stations on the Japan Sea side, a velocity of 19 km/yr (westward) was obtained, although the direction was uncertain because a third station was unavailable. The phase relation between the Pacific and the Japan Sea groups is also uncertain, but we may provisionally accept a general trend of westward migration across the Tohoku district, in agreement with the graphical interpretation as given in Fig. 3.

In particular, the waveform is similar at adjacent stations (cf. SNR and MYK), but it becomes less similar for remote stations (cf. SNR and OGA). In other words, migration seems dispersive and dissipative with distance. Ishii (1976) assumed that the waveform change between MYK and HMK represents the propagation of a phase and derived Fig. 4 by conventional seismological techniques. From it we learn that the phase velocity is about 35 km/yr for a period of 5.5 years, and that it tends to decrease with time (Fig. 4, top). The group velocity, as derived by numerical differentiation of

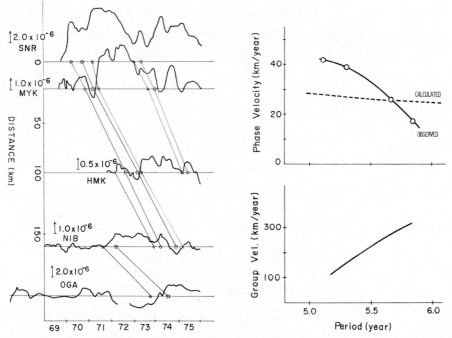

Fig. 3. Changes in maximum shear strain at five stations in Tohoku district. Vertical separation of the horizontal reference lines represents relative location of the respective station in the direction of migration (Ishii, 1976).

Fig. 4. Dispersion of migrating ground strain. Phase velocity (top) and group velocity calculated from phase velocity (bottom), both plotted against period (Ishii, 1976). The broken line in the top figure is added to show the dispersion by the Bott—Dean model (1973), (Ishii, 1976).

the phase velocity, is significantly high, being 100 km/yr for a period of about 5.1 years.

Ground tilt in the West Cordillera Mts., Peru

Long-term observations of crustal movement have been continued in Peru under the cooperative program of Kyoto University, Japan, and San Agustin University, Peru, since the 1960s. Tanaka et al. (1977) studied tiltmeter records from their stations and also found migration in Peru. Fig. 5 is a map of their stations, and Fig. 6 illustrates a summary of data, where they compare episodes of tilt accumulation with respect to temporal azimuth of tilting at the four stations: Arequipa (221), Ayanquera (222), Ongoro (223), and San Gregorio (224). Symbols A, B, C, etc. are assigned to critical portions of the traces to correlate principal events at the respective station.

Good correlation is recognized between the data from the two adjacent

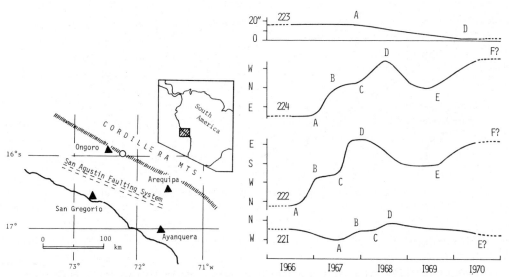

Fig. 5. Array of crustal movement observatories in the West Cordillera Mts., Peru (Tanaka et al., 1977).

Fig. 6. Temporal changes in tilting direction at Ongoro (*223*), San Gregorio (*224*), Ayanquera (*222*), and Arequipa (*221*), Peru. Critical portions of the traces are assigned with symbols *A*, *B*, *C*, etc., for correlation (Tanaka et al., 1977).

stations, 222 and 224, with the former leading the latter for one year as far as the events, *A*, *B*, *C*, and *D* are concerned. Since their separation is 110 km, we obtain an apparent velocity of 110 km/yr to the northwest. Similarly, we obtain 60—70 km/yr northwards, from the pair of stations 221 and 222 (separation: 65 km), although the trace for 221 is of low amplitude in comparison with the former two (high attenuation of the migrating disturbance might be the cause of this). The vertical axis for 221 is scaled in the opposite sense to the former two, for convenience in comparison. This reversal does not necessarily imply that the polarity of the migrating event is inconsistent with those at 222 and 224. The trend of secular tilt at 221 (NW) is opposite to those at 222 and 224 (SE or E), so that the tilt vector may register a counterclockwise turn at 221, and a clockwise turn at 222 or 224, when a deformation front of a southwards tilt approaches.

Combination of the above two velocity components yield a migration velocity of about 60 km/yr to the north. Symbols *A* and *D* at the top of Fig. 6 indicate the theoretical event time at 223, as calculated by this velocity. Evidence for the events is almost nonexistent here, perhaps due to high attenuation. However, we notice significant gradient changes at about the predicted times, which might be slight evidence for the migrated events (Tanaka et al., 1977).

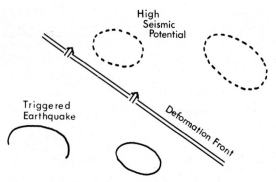

Fig. 7. Hypothetical relation of a deformation front to earthquake occurrence.

Other examples

Scholz (1977) compiled the Chinese data related to the 1975 Haicheng earthquake (NE China) and found that the recent sequence of crustal activity (see below) in and around the Pohai Bay is well explained by a migration effect with a velocity of 110 km/yr to the northeast, which resulted in this earthquake. Here the crustal activity includes major earthquakes and a significant increase of local seismicity, as well as crustal deformations of various types. This example is especially notable since the various kinds of crustal events as stated above seem to have been controlled by a common mechanism. To explain this, Scholz hypothesized a deformation front, which migrates in the crust and reaches areas of high seismic potential to trigger earthquakes there. This idea may be schematized as illustrated in Fig. 7.

Stein et al. (1979) found another example of migration in the postseismic land movement (vertical) in the White Wolf fault which was active in 1952. The migration velocity in this case was 7—16 km/yr (for details, see their paper in this volume).

MECHANISM AND TECTONIC IMPLICATIONS OF MIGRATION

The above examples enable us to compile the basic characteristics of migration and to list several significant points:

(1) Migration of crustal deformations has a characteristic velocity of about 10—100 km/yr, in close agreement with that of seismic activity. This suggests that migration represents some general and fundamental aspects of crustal dynamics.

(2) Migration tends to appear landward from the ocean, so far as the three examples from coastal areas are concerned.

(3) In these cases, the waveform of the migrating deformation attenuates fairly rapidly as it travels inland.

(4) Also, it may be dispersive, with phase velocity higher for short-period

components than for long-period ones, as the events in the Tohoku district suggest.

(5) Seismicity appears to migrate in association with the migration of crustal deformation, perhaps by the mechanism suggested by Scholz.

Origin

The landward trend of migration (2) makes us suspect that the land deformation which we observe might have originated from the trench, as a result of irregular motion of a plate. If this is the case, and if the deformation has migrated at a uniform velocity as observed inland, then we could locate when and where it originated. Fig. 8 schematizes this idea with respect to the events in the south Kanto and central Tohoku districts. It suggests the vicinity of the trench junction as the hypothetical origin during the

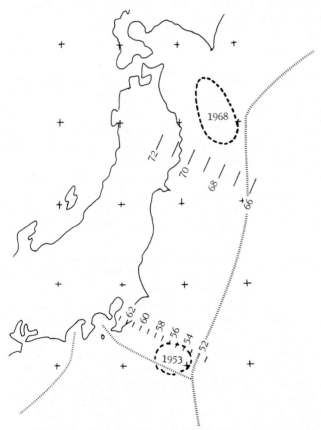

Fig. 8. Hypothetical picture of deformation fronts migrating landward from trench areas. Broken lines illustrate the Boso-oki (1953) and the Tokachi-oki (1968) earthquakes which might have caused the respective observed deformations.

early 1950s for the south Kanto events and the northern part of the Japan Trench as the origin during the late 1960s for the Tohoku events.

The likely mechanism is that the local irregular motion of a plate generates migrating deformation. It is then of interest to inquire whether seismic or aseismic plate motions are the cause. One may suppose, either that plate motion is sufficiently irregular and intermittent to be the origin, even in the interseismic stage, or that only a seismic rebound in a major earthquake can generate a migrating deformation of recognizable amplitude. The spatial and temporal conditions of the two major earthquakes in Fig. 8 seem to suggest the latter possibility as more likely (see next section).

Discrimination between the two possibilities will eventually be achieved by continuing our observations for several more years. If the first possibility (aseismic irregular motion of the plate) is the case, then we shall continue to observe a train of migrating deformation. If the second possibility applies, we are more likely not to observe a further dominant event. In the future we shall come to need ocean-bottom tiltmeters and/or strainmeters, which will provide us with precise data of the plate's behavior.

Modes of slow migration

A more fundamental mechanical problem associated with the events is the mechanism which accounts for the velocity of migration. It is extremely low and can in no way be explained by simple elasticity theory. Several models have been proposed, each hypothesizing anelastic properties of the medium (Savage, 1971; Bott and Dean, 1973; Ida, 1974; Anderson, 1975). Of these models, that by Bott and Dean appears to apply.

Their model is basically a two-dimensional layered system with an elastic plate underlain by a viscous layer, representing, respectively, the lithosphere and asthenosphere between a trench and a fixed continental block. Starting from the unidirectional equation of diffusion, they obtained the solution for the response of an elastic plate of semi-infinite length to a sinusoidal boundary pressure $P = P_0 \sin \omega t$ of period $T = 2\pi/\omega$ applied at the end $x = 0$. Briefly, the stress and strain wave travel along the x-direction with velocity $v = \omega/k = 2 \sqrt{[(\pi\ Ebd)/(\eta T)]}$, and with amplitude falling off with distance as $\exp(-kx)$, where E, d, η, b, and k denote the Young's modulus and thickness of the elastic plate, viscosity and thickness of the viscous layer, and the wave number, respectively.

Bott and Dean took the parameters as: $d = 80$, $b = 250$ km, $\eta = 2 \cdot 10^{21}$ P, $E = 10^{12}$ dyn cm^{-2}, and calculated the velocity v and the penetration distance X, i.e., the distance to the point where the amplitude falls off to e^{-1} of its value at $x = 0$. If we take T as 4, 5 and 6 years, for example, we obtain $v = 31$, 28 and 25 km/yr, and $X = 20$, 22, and 24 km, respectively. The migration velocity thus calculated agrees with the observations (see previous section).

Note that the velocity tends to decrease with period T, and also, that relatively low values of X suggest considerable decay of amplitude with distance. These characteristics of the model are extremely interesting in view of the observational trends as stated in (3) and (4) at the beginning of this section. The amplitude of the vertical land displacement in the present event is about 5 cm at Aburatsubo, based upon leveling by the Geographical Survey Institute (Kasahara, 1973b). We may assume a similar amplitude for falling off during migration from the nearest source boundary (approximate distance 150 km in the south Kanto district, and 70 km in the Tohoku district), then the amplitude at the origin of the 1953 and 1968 events are 90 and 1.6 m, respectively, assuming $X = 20$ km. The latter might be acceptable as an order-of-magnitude approximation, but the former looks one order too high as a likely amplitude at the source.

The broken line in Fig. 4 (top) illustrates the theoretical dispersion of phase velocity by the present model. It agrees with observation at $T = 5.65$ years, and with respect to the trend of velocity decreasing with period. In detail, however, the steep gradient and the convex aspect of the observational curve indicate significant disagreement with the theory. In summary, the Bott and Dean model may explain the modes of slow migration qualitatively, but further improvements are needed, both in theory and observation, for better quantitative agreement.

Migration and earthquake prediction

Studies of migrating crustal deformation may play a critical role in earthquake prediction in two aspects — passive and active. The above examples prove that anomalous crustal deformation may often be migratory. If we disregard its anelastic nature and treat the surface deformation as being purely elastic, we may overestimate the associated strain energy, or the seismic potential in a region. Sufficient knowledge of the migration mechanism will be useful in avoiding this mistake.

The successful prediction of the Haicheng earthquake, China, seems to present a possible active contribution of the migration studies to prediction. As Scholz (1977) reported, the sequence of major earthquakes, increasing local seismicity and crustal deformations in northeast China seem to have migrated toward the future epicenter. To explain this, he hypothesized a deformation front, which migrates in the crust and reaches areas of high seismic potential to trigger earthquakes there (cf. Fig. 7). If this is often the case, it will provide us with a new approach to prediction. Further studies of migration are needed for a better understanding of its nature, as well as for a precise determination of the deformation fronts in the crust.

CONCLUSIONS

Recent progess in crustal-movement studies has yielded convincing evidence for migration of crustal deformation in several areas of tectonic inter-

340

est. The author has compiled the following data (cf. Table I), from which several basic characteristics of migrating crustal deformation may be summarized:

(1) Migration of crustal deformation is dominated by a velocity of about 10—100 km/yr, in close agreement with that of seismic activity.

(2) Migration tends to propagate landwards from the oceans, so far as the three examples from coastal areas are concerned.

(3) In these cases the amplitude of the migrating deformation attenuates fairly rapidly as it travels inland.

(4) Also, it may be dispersive with phase velocity (higher for short-period components than for long-period ones, (i.e., normal dispersion).

(5) In some cases seismicity may appear to migrate in association with the migration of crustal deformation.

On the basis of these findings, the tectonic and mechanical implications have been discussed to yield conclusions as follows:

(6) The origin of migrating crustal deformation as discovered in Japan and Peru, may be located in trench areas. In other words, irregular motions of a plate may generate deformation pulses, either seismically or aseismically. If the former, the 1953 and 1968 earthquakes off the coast of Honshu are likely sources for the events in the south Kanto and Tohoku districts, respectively.

(7) The Bott—Dean model explains the order of magnitude of migration velocities. It may also explain dispersion and attenuation of migrating deformation in a qualitive sense. Quantitatively, however, there is significant disagreement between the theory and observations.

(8) Migration of crustal deformation may provide us with a new approach to earthquake prediction, as the case of the Haicheng earthquake indicates.

As a whole, the above data are the findings from only a few examples. They are to be understood as working hypotheses rather than conclusions. Further studies are needed for a better understanding of the nature of migration of crustal deformation.

ACKNOWLEDGEMENTS

Drs. Hiroshi Ishii and Yutaka Tanaka offered their materials for this paper. Dr. Paul Somerville read the manuscript critically and gave valuable advice.

REFERENCES

Anderson, D.L., 1975. Accelerated plate tectonics. Science, 187: 1077—1079.
Bott, M.H.P. and Dean, D.S., 1973. Stress diffusion from plate boundaries. Nature, 243: 339—341.
Ida, Y., 1974. Slow-moving deformation pulses along tectonic faults. Phys. Earth Planet. Inter., 9: 328—337.

Ishii, H., 1976. Characteristics of crustal movement observed at wide area. In: Z. Suzuki and S. Omote (Editors), Symp. Earthquake Predict. Proc. pp. 116—126 (in Japanese with English abstract).

Kasahara, K., 1973a. Earthquake fault studies in Japan. Philos. Trans. R. Soc. London, Ser. A, 274: 287—296.

Kasahara, K., 1973b. Tiltmeter observation in complement with precise levellings. J. Geod. Soc. Japan. 19: 93—99 (in Japanese with English abstract).

Kelleher, J.A., 1972. Rupture zones of large South American earthquakes and some predictions. J. Geophys. Res., 77: 2087—2103.

King, C.Y., Nason, R.D. and Tocher, D., 1973. Kinematics of fault creep. Philos. Trans. R. Soc. Lond., Ser. A, 274: 355—360.

Mogi, K., 1968a. Migration of seismic activity. Bull. Earthquake Res. Inst., 46: 53—74.

Mogi, K., 1968b. Development of aftershock areas of great earthquakes. Bull. Earthquake Res. Inst., 46: 175—203.

Mogi, K., 1969. Some features of recent seismic activity in and near Japan, 2. Activity before and after great earthquakes. Bull. Earthquake Res. Inst., 47: 395—417.

Mogi, K., 1973. Relationship between shallow and deep seismicity in the western Pacific region. Tectonophysics, 17: 1—22.

Nason, R. and Weertman, J., 1973. A dislocation theory analysis of fault creep events. J. Geophys. Res., 78: 7745—7751.

Research Group for Crustal Movement, Tohoku University, 1977. Analysis of crustal movement in the Tohoku district observed by array system. In: I. Nakagawa (Editor), Proc., Symp. Recent Crustal Movements, J. Geod. Soc. Jap., 22: 225—318.

Richter, C.F., 1958. Elementary Seismology. Freeman and Cooper, San Francisco, Calif., pp. 611—616.

Savage, J.C., 1971. A theory of creep waves propagating along a transform fault. J. Geophys. Res., 76: 1954—1966.

Scholz, C.H., 1977. A physical interpretation of the Haicheng earthquake prediction. Nature, 267: 121—124.

Stein, R.S., Thatcher, W. and Castle, R.O., 1979. Initiation and development of the southern California uplift along its northern margin. Tectonophysics, 52: 301—302.

Tanaka, Y., Otsuka, S. and Lazo, L., 1977. Migrating crustal deformations in Peru. Abstr. Annu. Meet., Seismol. Soc. Jap., 1977, No. 2 (In Japanese).

Whitcomb, J.H., Allen, C.R., Garmany, J.D. and Hileman, J.A., 1973. Focal mechanisms and tectonics. Rev. Geophys. Space Phys., 11: 693—730.

Yamada, J., 1973. A water-tube tiltmeter and its application to crustal movement studies. Rep. Earthquake Res. Inst., 10: 1—147 (in Japanese with English abstract).

Tectonophysics, 52 (1979) 343   343
© Elsevier Scientific Publishing Company, Amsterdam — Printed in The Netherlands

YOUNG MAGMATISM IN AFGHANISTAN AS THE CONSEQUENCE OF THE NEOGENE—QUATERNARY TECTONIC MOVEMENTS

V.M. CHMYRIOV, V.I. DRONOV and SH. ABDULLAH

Afghan Geological Survey, Darlaman, Kabul (Afghanistan)

(Accepted for publication May 22, 1978)

ABSTRACT

A considerable portion of the territory of Afghanistan, having structures of the Mediterranean folded belt, has been subjected to a general tectono-magmatic activization over the Miocene through to the present, resulting in different (predominantly oscillating) tectonic movements, intrusive magmatism, terrestrial volcanism, mineral occurrences, and springs of carbonated and nitrous thermal water.

Three types of young magmatism and volcanism products have been recognized in Afghanistan:

(1) Miocene alkaline granite intrusions, described as the Share—Arman Complex, resulted from the early orogenic stage of the Late Alpine geosynclinal troughs development and were restricted to transversal uplifts, in both the geosynclinal structures and on their extension, in the surrounding median masses. These transversal uplifts also play the role of mineralization-controlling structures.

(2) Late orogenic—Early Quaternary volcanics (the Dash-i-Nawar Complex) cropping out by the periphery of median masses and at the marginal uplifts of the Late Alpine folded area and also restricted to the transversal uplifts with the confined fault zones to them.

(3) Alkaline carbonatitic (the Khanneshin Complex) and trachybasaltic (the SarLogh Complex) Early—Middle Quaternary volcanics in the inner parts of the Central Afghanistan Median Mass and in the southeastern segment of the Turan Plateau.

Areas with products of Middle Quaternary volcanism are restricted to knot areas of the major subcrustal faults which are currently active.

344   *Tectonophysics*, 52 (1979) 344
© Elsevier Scientific Publishing Company, Amsterdam — Printed in The Netherlands

THE NEOGENE—QUATERNARY TECTONICS AND REGULARITIES OF THE MINERAL-WATER SPRINGS ARRANGEMENT IN AFGHANISTAN

B.A. KOLOTOV, V.M. CHMYRIOV and SH. ABDULLAH

Afghan Geolocial Survey, Darlaman, Kabul (Afghanistan)

(Accepted for publication May 22, 1978)

ABSTRACT

There are more than 150 carbonated and nitrous thermal-water springs in Afghanistan, restricted to the major faults and confined to those areas where deep-focus earthquakes are quite common. There are belts and zones with the mineral-water springs coinciding with the major fault zones.

The major structures of Afghanistan, and consequently the major belts with the mineral-water springs, converge in the Shekary—Ghorband area. There is a single intercommunicated system of mineral-water springs discharge, due to the present-day movements in deep-seated and open fault zones. The distribution of the carbonated and nitrous thermal-water types in this system is predetermined by the depth of these open faults, by the presence of heat hearths, as well as by neotectonic movements which form the discharge channelways. The deepest portion of this system, restricted to its major carbonated water resources, is located in the knot area of the Hari-Rod, Qarghanaw, Chaman-Moqur, Central Badakhshan, and Helmand major faults. This area is currently the most seismic in the country and the composition of the carbonated-water springs is unique, having very high content of Li, Rb, Cs, B, Ge, Be, As, Ba and Sr. The high content of Li, Rb, Cs and B in this water classifies it as an industrial-type water.

The isotopic composition (and in particular the O^{18} isotope content of -10.9—11.9%) points to a meteoric origin of the water. The functioning of this artesian-water system suggests present-day tectonic activity in the area. The main orographic point where the major three watersheds of the Amu-Darya, Indus and Seistan drainage areas intersect is located in this area. Away from the center of this deep-seated fault system, the carbonic acid content and the total mineralization of the groundwater are diminished with the increased temperature and discharge of the water springs; there are nitrous thermal-water springs some distance from the center of the system, emphasizing the presence of the currently active movements along the faults.

Tectonophysics, 52 (1979) 345—346

THE CHAMAN—MOQUR FAULT

SH. ABDULLAH

Afghan Geological Survey, Darlaman, Kabul (Afghanistan)

(Accepted for publication May 22, 1978)

ABSTRACT

The Chaman—Moqur fault, the largest in Afghanistan, is located in the southeastern segment of the country and separates the Katawaz zone from the Tarnak zone and from the Central Afghanistan Median Mass. The fault runs through the town of Chaman which is close to the southern frontier of Afghanistan. Southeast of Chaman the fault has been traced mainly in Baluchistan for some 220 km, up to the Kawan Kala township, after which it is concealed under Quaternary formations. North of the Chaman, the fault, appearing as a gently curved arch, is traced for some 140 km up to the town of Moqur. The Chaman—Moqur fault is some 690 km long in Afghanistan where it is well distinguished in aerial and space photographs.

The Chaman—Moqur fault has a northeastern strike with the fault plane dipping steeply southeastwards. In the Katawaz zone, close to the fault, the rocks dip east and southeastwards at 75—90°. The fault zone, according to Sborshchikov et al. (1974), in many places consists the wedge-shaped fault blocks, shattered and altered rocks and has fault wedges consisting of the sandstone, siltstone, and clay slate of the Katawaz zone and limestone of the Tarnak zone.

East of the fault, in the Katawaz zone, and west of it, in the Tarnak zone, are minor (feather) faults having an ENE strike (Sborshchikov et al., 1974).

Many geologists have described the Chaman—Moqur fault as a left-slip wrench fault (Wellman, 1966; Weippert and Wittekindt, 1972).

Between Moqur and Ghazni the Chaman—Moqur fault is concealed under Quaternary formations. North of Ghazni, to the Jabalusaradj area, the fault is well distinguished in relief, in a number of places the fault zone has Paleogene ultrabasic intrusions arranged parallel to the fault.

In the Map of Earthquake Epicenters in Afghanistan, there are numerous epicenters with magnitudes ⋗5.8 along the Chaman—Moqur fault. The majority of the epicenters are on the line of the fault or very close to it. Earthquake hypocenters along the line of the fault and close to it are at depths of up to 60 km. East of the fault, in the Katawaz zone, and west of it, in the Central Afghanistan Median Mass, the density of the earthquake epicenters is greatly reduced.

On the northeastern extension of the fault, between Ghazni, Kabul, and Jabalusaradj, no earthquake epicenters were recorded. However, there are numerous earthquake epicenters along the Ghazni—Altimour—Sarobi and Konar faults, suggesting seismic activity along these faults.

Tectonophysics, 52 (1979) 347
© Elsevier Scientific Publishing Company, Amsterdam — Printed in The Netherlands

ACTIVE FAULTING, RECENT DEFORMATION AND DISPLACEMENT OF EARTH SURFACE OF LARGE SEDIMENTARY BASINS OF THE EARTH'S CRUST

A.T. DONABEDOV [1], V.A. SIDOROV [2], A.S. GRIGORIEV [2], A.V. MICHAILOVA [2] and Z.E. SHACHMURADOVA [2]

[1] *Institute of Geology and Combustible Fuel, Moscow (U.S.S.R.)*
[2] *Institute of Physics of the Earth, Academy of Sciences of the U.S.S.R., Moscow (U.S.S.R.)*

(Accepted for publication May 22, 1978)

ABSTRACT

Further elaboration of the problem concerning the dynamics of the earth's crust is connected with the overall practical problem solution. First, it concerns the prediction of seismic and volcanic events and resource location in the crust.

Following the example of large sedimentary basins — Dneprovsko—Donetskaya depression (introplatform foredeep) and sub-Caucasian foredeep — the relationship of Recent vertical crustal movement to the solution of the problem concerning the location and formation of oil and gas fields has been studied since 1972. The main objects of the research are the faults of different type and order determining the block structure of sedimentary basins and controlling the location of oil and gas fields.

The peculiarities of Recent vertical crustal movement distribution are determined with the existence of narrow (1—3 km), extend (to 150 km), zones of high-amplitude (4—6 to 10—15 mm) movement variations. The relation of these zones to the faults in the crust was studied on the following grounds:

(1) Complex analysis of the existing geological—geophysical data; study of the gravitation field value at the time of rerunning of leveling.

(2) Tectonophysical interpretation by modeling and theoretical calculations.

The degree of recent activity of the faults changes according to their extension and reaches the greatest magnitude at the nodes of intersection of longitudinal and transverse faults. The nodes reflect the zones of greatest strain in the crust, resulting in a high degree of rock jointing.

Tectonophysics, 52 (1979) 349—359
© Elsevier Scientific Publishing Company, Amsterdam — Printed in The Netherlands

HOLOCENE MOVEMENTS AND STATE OF STRESS IN THE RHINEGRABEN RIFT SYSTEM

J. HENNING ILLIES and GERHARD GREINER

Geologisches Institut, Universität Karlsruhe, D 75 Karlsruhe (G.F.R.)

(Accepted for publication May 22, 1978)

ABSTRACT

Illies, J.H. and Greiner, G., 1979. Holocene movements and state of stress in the Rhine-graben rift system. In: C.A. Whitten, R. Green and B.K. Meade (Editors), Recent Crustal Movements, 1977. Tectonophysics, 52: 349—359.

A belt of seismotectonic activity and Holocene crustal deformations traverses Western Europe and forms a 800-km-long subplate boundary. The main segments are the Rhine-graben, the seismic zone straight through the Rhenish massif, the Lower Rhine embayment, and the Zuider Zee depression (The Netherlands). Seismicity and Holocene fault action of the Rhinegraben are controlled by a sinistral shear motion parallel to the graben axis. Accompanying this simple shear motion there are also extension shear, compression shear, and Riedel shear. Extensional tectonics characterize the faulting in the Rhenish massif and the rifting in the Lower Rhine embayment.

In-situ stress data, obtained by using the strain relief technique, confirm principal stress directions, approximately equal to those obtained by fault plane solutions of earthquakes. The calculation of excess stress revealed very high stresses in the Central Alps, whereas minimal to negative values were found in the Rhinegraben and other zones of seismic activity. Stress generation in the area of the Central Alps corresponds with strain release along the rift system which traverses the foreland.

The Upper Cretaceous to end-Miocene process of Alpine plate convergence and folding has been replaced since Pliocene times by epeirogenic uplift and consequent denudation of the mountain range. We assume that the observed state of stress is mainly caused by a sideways extension of the mountain body due to the effects of unloading and topography.

INTRODUCTION

A nearly uninterrupted chain of seismic epicenters traverses Western Europe connecting the Alps and the North Sea basin (Fig. 1). Physiographically, this belt consists of a series of segments of different structure and independent geological case histories. The main segment is the Rhinegraben, in the precise sense of the word. It starts near Basel, where graben structures are inferred from the anticlines of the Jura mountains. About 300 km northward, near Frankfurt, the Holocene rifting is transcurrently interrupted by

350

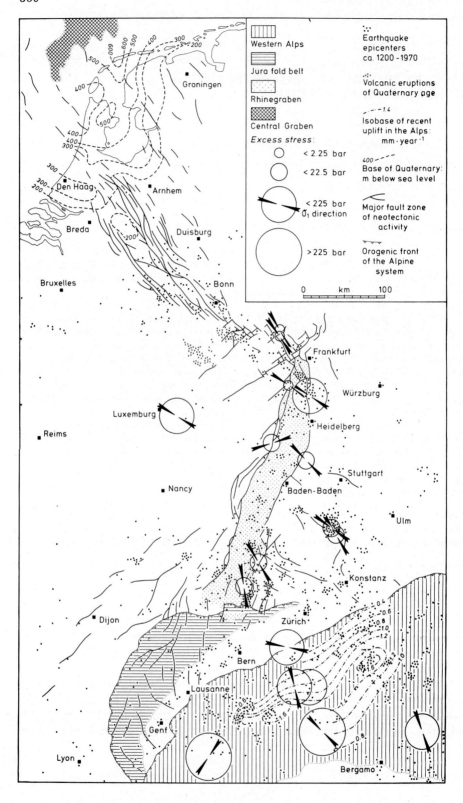

the Hercynian block of the Rhenish massif. Beyond there, the seismicity follows a more northwestward trend, and parallel to it are the Pleistocene to Holocene volcanic chains of the Eifel area, the Neuwied basin, and the Rodderberg. Between Bonn and Arnhem in The Netherlands is the rift segment of the Lower Rhine embayment which is under active block faulting and seismicity. With an eastward offset of about 60 km the neotectonic activity continues in the aseismic furrow of the Zuider Zee depression.

The mechanics of faulting in the individual rift segments vary considerably. This is mainly conditioned by the mutual interrelation between pre-existent basement fractures and the controlling regional stress regime.

THE RHINEGRABEN

Looking for such preconditions in the Rhinegraben we have to consider that its rifting evolved as a two-stage process (Illies, 1978). The primarily extensional graben has been formed in the Middle Eocene—Lower Miocene period. The wedging of the downthrown crustal segment, as well as the antithetic rotations of the internal tilt blocks, reveal about 4.8 km of sideways spreading (Illies, 1974). This kind of extensional rifting became extinct in mid-Miocene times. Subsidence ceased, and the graben floor was affected by fluvial erosion. With an unconsolidated sediment fill up to 4000 m thick, a dense fault pattern in the basement, a mantle bulge or hot spot under its southern part, and geothermal anomalies with respect to this, the Rhinegraben remained a zone of weakness in the lithosphere.

Sinistral shear motion is the second and presently active stage of rifting; it started in mid-Pliocene times (Illies, 1974). At many places it is observed that planes of primary dip-slip faulting have been overprinted by horizontal slickensides. Fault-plane solutions of earthquakes indicate sinistral shear motion parallel to the graben axis (Ahorner, 1975). Locally, high rates of Upper Pliocene to Recent subsidence are shear controlled. By seismotectonic observations Ahorner (1975) calculated the seismic slip rate of the graben to be about 0.05 mm/yr. Geologically, the cumulative amount of axial sinistral shear attains about 500 m. At conjugate dextral strike-slip faults, horizontal displacements up to 400 m are presumed.

The average trend of the Rhinegraben is parallel to a potential sinistral shear plane of the regional stress field. In detail, the Paleogene extensional rifting had caused a crooked course of the rift valley which complicated the

Fig. 1. An intraplate belt of seismotectonic activity, horizontal and vertical block motions, as well as volcanism, all of Quaternary age, traverses Western Europe. In situ stress determinations reveal the Central Alps as the area of stress generation, whereas strain release is indicated along the rift belt of the foreland.

shear motion during the second and Holocene generation of rifting (Knopoff, 1970). Where a local segment of the graben is parallel to the shear plane, only shear motion is found. However, if a segment trends more to the north, extension shear prevails. In a more easterly trend, compression shear is observed (Fig. 2B).

In the northern or Heidelberg segment its N—S trend supports lateral extension and subsidence. Owing to this, the depocenter of Upper Pliocene to Quaternary sedimentation is found in this segment, reaching a maximum thickness of more than 1000 m (Bartz, 1974). As proved geodetically, the actual rates of subsidence there are about 1 mm/yr (Schwarz, 1976). In the central or Baden-Baden segment shear progression had produced an increasing compression. Hereby, the eastern master fault, originally a dip-slip feature, has been overprinted by thrusting (Illies and Greiner, 1976), and the fault-plane solutions of earthquakes show a combination of sinistral shear and thrust (Ahorner, 1975). Uplift due to this ramp mechanism has been measured geodetically (Nivellementnetze, 1977) and is additionally evidenced by erosional landforms on the graben floor. Moreover, it is the only segment where excessive heat-flow values up to about 4.0 HFU are observed (Haenel, 1976). Presumably, increasing pore pressure with depth excited hydrothermal circulation, as well as convective heat transport, along such fracture zones which were opened under local stress conditions. In the southern or Freiburg segment extension shear prevails again. In its center the rigid block of the Miocene-formed composite volcano, the Kaiserstuhl, prevents shear motion, since it acts as a hard inclusion within the weakened shear belt. A zone of Pleistocene to Recent subsidence, combined with microseismic activity (Bonjer and Fuchs, 1974) swings westward around this structure (Fig. 2A). The epicenters of some stronger historical earthquakes are concentrated upon the western and northern front of the Kaiserstuhl (Lais, 1912). Opposite, on the eastern and southern side of the volcanic massif, very low stresses were determined (Illies and Greiner, 1976). This region, i.e., the tilt block area of the Freiburg embayment, is situated on the lee side of active shear motion.

Most of the fracture zones which separate the graben fill, were formed as antithetic dip-slip faults during the extensional stage of rifting. Some of them have been reactivated by shear in Pliocene to Quaternary times, especially those oblique in the counterclockwise sense to the plane of first-order shear motion. Such striking faults of about 160—180° now acted as second-order shears or Riedel fractures in all segments of the graben (Fig. 2A). Morphological scarplets, clusters of seismic epicenters, and hydrothermal activity, illustrate how this mechanism proceeds. Moreover, fault-plane solutions of local earthquakes show some clockwise rotation of the principal stress directions with depth, especially at a hypocenter depth of about 12 km (Bonjer, 1977). This might be influenced by a stress deflection along second-order shear planes which should evidently be restricted to the rigid layer of the crust.

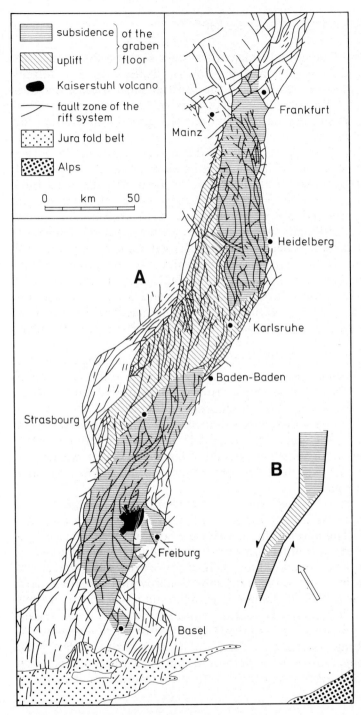

Fig. 2. The neotectonic activity of the Rhinegraben is that of a sinistral shear zone (A). The crooked course of the rift valley causes compression shear motion and uplift in its central segment, but extension shear and subsidence, respectively, in the final parts (B).

THE RHINEGRABEN AND THE RHENISH MASSIF

Near Frankfurt the Paleogene-formed rift system shows a 20-km rift/rift offset between the Rhinegraben and the Hessen depression. The former transform element (Illies, 1972) blocked the sinistral shear mechanism during the second stage of Rhinegraben rifting. Hereby, the northward shift of the block unit east of the graben was transmitted directly to the framing block of the Rhenish massif.

Although the regional stress conditions did not vary substantially during this second and last stage of rifting, it may be shown that the advance of shear motion caused an eastward migration of fault activity with time (Fig. 3). During an Upper Pliocene—Lower Pleistocene interval, shear motion and subsidence had affected the whole breadth of the graben, and related strike-slip faults are additionally spread over the western shoulder (Buchner, 1977). The Rhenish massif, which cuts the shear belt transversally, had still reacted as a monolith. Its pre-existent southern rim, i.e., the Taunus—Hunsrück lineament, split anew and became a mobile fault scarp, in the same way as some parallel fracture zones further south (Illies and Greiner, 1976). This implied a slight anticlockwise rotation of the whole massif which had provoked dextral shear along its eastern rim in the Vogelsberg area (Schenk, 1974).

During the Upper Pleistocene—Recent time episode, shear progression and consequent block separation became increasingly localized in the eastern half of the Rhinegraben, where the seismicity is now concentrated. In the same way the thickness distribution of Pleistocene sediments and the rates of recent subsidence obtained a distinctly monoclinal feature with a maximum near the eastern rim of the graben (Bartz, 1974). Perhaps the finite horizontal displacements along pre-existent second-order shear planes within the graben progressively gave way to a first-order shear motion of less resistance parallel to its eastern frame.

The advance of first-order shear motion towards the block margin east of the Rhinegraben displaced the transmission of horizontal shift to the adjacent Rhenish massif. The upset of the massif now became localized to a relatively short segment at its southern rim west of Frankfurt (Fig. 3). Starting from this kick point, a nearly continuous belt of historical seismic epicenters traverses the massif in a northwestward direction towards the Bonn area (Ahorner, 1975). Regardless of numerous neotectonic extension faults and widespread Upper Pleistocene—Holocene volcanic activity, real graben structures remained absent at this segment. The Rhenish "Schiefergebirge", affected by neotectonic stressing, consists of a thick series of Devonian shales and slates. They show a dense pattern of joints, cleavage, and bedding planes; these act as discontinuities which easily allow internal movements on a different scale.

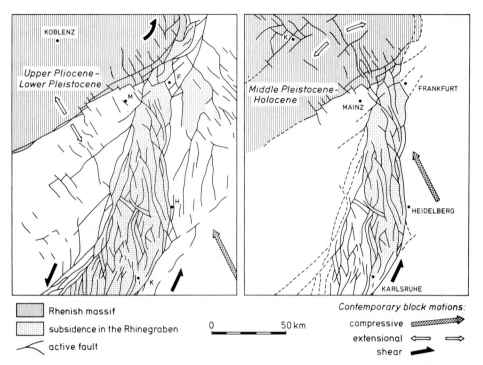

Fig. 3. In Upper Pliocene—Lower Pleistocene times the Rhinegraben was active as a sinistral shear zone for its whole breadth, some shoulder fractures included. The adjacent Rhenish massif had reacted by a slight anticlockwise rotation. Later, the shear became concentrated upon the eastern margin of the graben. Consequent local stress accumulation near the end of its eastern frame forced the Rhenish massif to yield by extensional splitting, parallel to the direction of axial compression.

THE LOWER RHINE EMBAYMENT

The mobile belt across the Rhenish massif joins another pre-existent zone of weakness near Bonn, destined for neotectonic reactivation. It is the fault bundle of the Lower Rhine embayment, first active as cross-fractures during the Hercynian orogeny. Later, in the Upper Cretaceous and Miocene, it became rejuvenated by tilt-block motions. Later the fault bundle underwent extensional rifting and high rates of vertical movement in Quaternary times (Ahorner, 1962). The extensional character of the widespread seismicity, and especially the northwestward increase of crustal spreading, illustrate a funnel-like widening of the whole rift segment. This is in agreement with the observed outward rotation of its tectonic frame, i.e., the splitting halves of the Rhenish massif.

The neotectonic activity and seismicity of the Lower Rhine embayment goes astray near Arnhem. At the same latitude, but with a 60-km eastward offset, the furrow of the Zuider Zee depression sets in, where the Quaternary base is downdropped to a depth of about 600 m, under the island of Ter-schelling. The depression leads up to the southern end of the Central Graben of the North Sea basin. The latter has remained an extinct rift valley since Late Mesozoic times (Ziegler, 1977). When active rift-valley propagation reached the North Sea coast and there was a fit between the controlling stress system and the structural anisotropy of the paleo-rift, the southern appendage of the buried rift was rejuvenated. Storm tides during the 13th century, mainly in 1287, submerged the marshes by creating the Zuider Zee. The actual coastline of this lagoon approximately traces the contours of the tectonic furrow.

IN-SITU STRESS DETERMINATIONS

The active subplate boundary parallel to the River Rhine is about 800 km long. To learn more about the driving mechanism of the Holocene move-ments and the seismicity, we started a program of in-situ stress determina-tions, using the overcoring technique (Greiner and Illies, 1977). The strain relief measurements were made with resistance strain gauges in 2—20-m deep boreholes drilled in quarries, mines, and tunnels at 12 different test sites, with a maximum overburden of about 260 m. The stresses calculated, and some additional data obtained from other papers, were used to obtain a detailed idea of the regional stress conditions in Central Europe.

The area investigated is characterized by a relatively uniform distribution of the σ_1 directions (Fig. 1). A general NW—SE trend of the directions of maximum horizontal compression, found in the Western Alps, as well as in the foreland area, roughly confirm data obtained by fault-plane solutions of earthquakes (Ahorner, 1975; Pavoni, 1977). The Rhinegraben was generally found to be oriented parallel to a plane of maximum shear of the recent stress field. The Lower Rhine embayment, on the other hand, is parallel to the component of maximum horizontal compression of the same regional stress field. Some local variations of σ_1 in and around the Rhinegraben illus-trate its role in the regional stress field and the strain relief.

The determination of excess stress was found to have special tectonic relevance. Excessive horizontal stresses, given in bars, were obtained by the difference between the measured horizontal stresses and the stresses calcu-lated from the gravitational loading at the test sites. The resultant amounts of horizontal stress show considerable but systematic regional variations (Fig. 1). In the whole Alpine foreland the excess stresses are relatively low. The lowest — and sometimes negative — amounts of compressive stress are observed in the Rhinegraben and other related seismically active zones. This is probably due to strain release by seismicity and Recent crustal move-ments. An abrupt increase of stress is observed along the northern rim of the Alps. Excess stresses culminate in the Central Alps of Switzerland and north-ern Italy, where amounts up to nearly 360 bar were determined. The pattern of excess stress reveals the Central Alps as the stress generator responsible for the strain relief in the block mosaic of the foreland. Correspondingly, the epicenter density indicates how in this area stress consumption starts to spread out over the fault pattern of the adjacent part of the European plate (Fig. 1).

To explain the high amounts of horizontal stresses in the Central Alps, several models or a combination of them are considered:

(1) The stresses are of plate tectonic origin. It cannot be denied that an effective part of plate tectonic stresses remained active in this unit. In estimating the geological influence of those stresses, we have to consider that, in this part of the Alpine system, folding and thrusting definitively ended in Late Miocene—Early Pliocene times. Moreover, it seems an open question whether Alpine-type plate convergence is a location of stress generation or of strain release.

(2) The stresses are residual, conserved by former stages of Alpine compressional deformations. However, the stress regime necessary for creating the nappe and fold structures does not fit the stress field as measured in situ.

(3) The stresses are caused by topographic effects of the high mountain range. It is well known that topographic effects of nearby mountain massifs increase horizontal stresses in a considerable and calculable way (Sturgul et al., 1976). But the stresses determined are too high to be explained exclusively by this phenomenon.

(4) The stresses are caused by unloading effects. Over large areas of the mountain range high rates of denudation are observed. On the basis of these data, Jäckli (1958) calculated that within 3—4 m.y. the actual height of the Alps in Switzerland will be reduced by half. Denudation causes unloading and subsequent exfoliation and destressing of near-surface rock materials, as well as voluminous extension in deeper layers of the crust and the upper mantle due to phase transformations of minerals (Neugebauer et al., 1976). In the rising mountain body which is clamped sideways by stable blocks, the extension causes a state of horizontal compression which superimposes the stresses of plate tectonic and topographic origin.

In the Alps a general uplift has increasingly replaced the previous stages of compressive deformations. The Recent tectonic behavior is more of an epeirogenic updoming and isostatic readjustment than a plate tectonic compression. By geodetic observations it has been shown that this uparching in the Central Alps of Switzerland attains 1.7 mm/yr (Gubler, 1976). The extraordinarily high stresses are mainly found in the areas of rapid uplift and denudation (Fig. 1). The stress trajectories in the same region appear dependent on the course of the isobases of Recent uplift. These observations favor the concept that the stresses determined are a mainly self-induced characteristic of the above-mentioned groups (3) and (4).

CONCLUSIONS

In any event, be the cause of the stresses as it may, the rapidly rising and geologically extending mountain body meets a northern frame of foreland

358

blocks, underlain by a strongly consolidated Hercynian basement. Within this stable abutment the Rhinegraben is a weak zone destined to force the frame asunder and to act as an "open door" for attracting the strain relief of Alpine stresses. Since the pre-existent trend of the graben fitted about the orientation of the maximum shear direction of the new regional stress regime, graben tectonics have been revived by shear since mid-Pliocene times. East of the graben, between about Konstanz and Baden-Baden (Fig. 1), some NW—SE-trending lineaments have been reactivated by dextral shear, sometimes combined with rotational effects of sandwiched minor block splinters. By these shear motions, conjugate to the sinistral shear belt of the Rhinegraben, the crust north of the Alps has been considerably weakened. A fan-shaped pattern of seismotectonic activity was formed which links together the Rhinegraben and the Western Alps.

Active rift-valley accretion tracing such elements of weakness in the basement which fitted the recent regional stress conditions, reached North Sea coast. Here, so it seems, today's influence of peri-Alpine stresses disappears.

REFERENCES

Ahorner, L., 1962. Untersuchungen zur quartären Bruchtektonik der Niederrheinischen Bucht. Eiszeitalter Ggw., 13: 24—105.
Ahorner, L., 1975. Present-day stress field and seismotectonic block movements along major fault zones in Central Europe. Tectonophysics, 29: 233—249.
Bartz, J., 1974. Die Mächtigkeit des Quartärs im Oberrheingraben. In: J.H. Illies and K. Fuchs (Editors), Approaches to Taphrogenesis. Schweizerbart, Stuttgart, pp. 78—87.
Bonjer, K.-P., 1977. Azimuth-depth dependence of the continental stress-field in Southern Germany derived from faultplane studies. I.A.S.P.E.I./I.A.V.C.E.I. Assembly, Durham 1977, Abstr. of papers, p. 213.
Bonjer, K.-P. and Fuchs, K., 1974. Microearthquake-activity observed by a seismic network in the Rhinegraben region. In: J.H. Illies and K. Fuchs (Editors), Approaches to Taphrogenesis. Schweizerbart, Stuttgart, pp. 99—104.
Buchner, F., 1977. Tektonische Studien am Rande des Oberrhein-grabens bei Wasselonne (Westl. Strasbourg). Oberrheinische Geol. Abh., 26: 11—21.
Greiner, G. and Illies, J.H., 1977. Central Europe: active or residual stresses. Pure Appl. Geophys., 115: 11—26.
Gubler, E., 1976. Beitrag des Landesnivellements zur Bestimmung vertikaler Krustenbewegungen in der Gotthard-Region. Schweiz. Mineral. Petrogr. Mitt., 56: 675—678.
Haenel, R., 1976. Die Bedeutung der terrestrischen Wärmestromdichte für die Geodynamik. Geol. Rundsch., 65: 797—809.
Illies, J.H., 1972. The Rhine graben rift system — plate tectonics and transform faulting. Geophys. Surv., 1: 27—60.
Illies, J.H., 1974. Intra-Plattentektonik in Mitteleuropa und der Rheingraben. Oberrheinische Geol. Abh., 23: 1—24.
Illies, J.H., 1977. Two stages Rhinegraben rifting. In: I.B. Ramberg and E.-R. Neumann (Editors), Tectonics and Geophysics of Continental Rifts. Nato Advanced Study Institutes Series. Reidel, Dordrecht/Boston/London, pp. 63—71.
Illies, J.H. and Greiner, G., 1976. Regionales stress-Feld und Neotektonik in Mitteleuropa. Oberrheinische Geol. Abh., 25: 1—40.
Jäckli, H., 1958. Der rezente Abtrag der Alpen im Spiegel der Vorlandsedimentation. Eclogae Geol. Helv., 51: 354—365.

Knopoff, L., 1970. Problems of continental rift structures. In: J.H. Illies and St. Mueller (Editors), Graben Problems. Schweizerbart, Stuttgart, pp. 1—4.

Lais, R., 1912. Die Erdbeben des Kaiserstuhls. Gerlands Beitr. Geophys., 12: 45—88.

Neugebauer, H.J., Brötz, R. and Rybach, L., 1976. On the dynamics of the Swiss Alps along the geotraverse Basel—Chiasso. Schweiz. Mineral. Petrogr. Mitt., 56: 703—706.

Nivellementnetze in Rheinland-Pfalz (map), 1977. Landesvermessungsamt Rheinland-Pfalz.

Pavoni, N., 1977. Erdbeben im Gebiet der Schweiz. Eclogae Geol. Helv., 70: 351—370.

Schenk, E., 1974. Die Fortsetzung des Rheingrabens durch Hessen. Ein Beitrag zur tektonischen Analyse der Riftsysteme. In: J.H. Illies and K. Fuchs (Editors), Approaches to Taphrogenesis. Schweizerbart, Stuttgart, pp. 286—302.

Schwarz, E., 1976. Präzisionsnivellement und rezente Krustenbewegung dargestellt am nördlichen Oberrheingraben. Z. Vermess., 101: 14—25.

Sturgul, J.R., Scheidegger, A.E. and Grinshpan, Z., 1976. Finite-element model of a mountain massif. Geology, 4: 439—442.

Ziegler, P.A., 1977. Geology and hydrocarbon provinces of the North Sea. GeoJournal, 1: 7—32.

Tectonophysics, 52 (1979) 361
© Elsevier Scientific Publishing Company, Amsterdam — Printed in The Netherlands

EVIDENCE OF DISLOCATIONS ON THE SOUTHERN SLOPE OF THE NEBRODI-MADONIE MOUNTAINS (NORTHERN SICILY, ITALY): THEIR NEOTECTONICS IMPLICATIONS

F. GHISETTI and L. VEZZANI

Istituto di Geologia della Universita di Catania, Corso Italia 55, Catania (Italy)

(Accepted for publication May 22, 1978)

ABSTRACT

The investigated area is located on the southern slope of the Nebrodi-Madonie Mts. (northern Sicily, Italy). It was studied in order to obtain neotectonic evidence by means of fracture measurements, morphotectonic and hydrographical patterns, aerial photographs and satellite analysis. The results were compared with the well-known geological structures.

This area could be considered an outstanding crustal part of Sicily, as here we find alignments of earthquake foci and a deep gradient of gravity anomalies, testifying to an important isostatic imbalance. This analysis confirms the observed structural picture of surface tectonics. Rose diagrams elaborated from fieldwork and aerial lineations show distinctive E—W trends which conform to the directions of the main faults.

From a chronological viewpoint faults breaks, aerial photographs and satellite lineations seem to indicate that the E—W trend is the less dislocated and, therefore, the most recent.

Tectonophysics, 52 (1979) 363—371
© Elsevier Scientific Publishing Company, Amsterdam — Printed in The Netherlands

A STUDY OF EARTHQUAKE FOCAL MECHANISMS IN THE HOLLISTER AREA, SOUTHERN COAST RANGES, CALIFORNIA *

N. PAVONI

Geophysical Institute, Swiss Federal Institute of Technology, Zurich (Switzerland)

(Accepted for publication May 22, 1978)

ABSTRACT

Pavoni, N., 1979. A study of earthquake focal mechanisms in the Hollister area, southern Coast Ranges, California. In: C.A. Whitten, R. Green and B.K. Meade (Editors), Recent Crustal Movements, 1977. Tectonophysics, 52: 363—371.

The study is based on some 1200 fault-plane solutions derived from local, low-magnitude earthquakes which occurred during 1969—71 in the Hollister area (Fig. 1). The procedure of analysis was as follows: (1) to use a solution with two orthogonal vertical nodal planes (VNP solution), i.e. a pure strike-slip solution, if it would satisfy the distribution of compressions and dilatations; (2) if not, to use a more general solution with inclined nodal planes (INP solution). Emphasis is given to the orientation of P-axes. According to the orientation of P-axes the events are classified into nine classes (Fig. 2). The general reliability of the solutions is demonstrated independently in events associated directly with the Calaveras and the San Andreas fault zones (Fig. 3). Some remarkable regularities in the regional orientation of P-axes are observed: NNE—SSW to NE—SW orientation of P-axes is predominant in the Diablo block, east of the Calaveras—Paicines fault zone, as well as in the San Benito wedge, located between the San Andreas and Paicines fault zones. NE—SW orientation of P-axes is observed within the Gabilan block SW of the San Andreas fault zone. On the other hand, a regime of N—S orientation of P-axes characterizes the Sargent wedge, located between the Calaveras and San Andreas fault zones.

INTRODUCTION

The results of an investigation of fault-plane solutions for local earthquakes of the Hollister area are reported and discussed. The area of investigations extends from 36°30'N lat. to 37°00'N lat. and from 121°00'W long. to 121°30'W long. It includes the tectonically complicated region between the southern part of the Calaveras fault zone, the Paicines fault zone and the San Andreas fault zone (see Fig. 1). The San Andreas and Calaveras—Paicines fault zones divide the region into major structural units:

(1) The Gabilan block, located west of the San Andreas fault zone.

(2) The Diablo block, located east of the Calaveras—Paicines fault zone.

* Contribution No. 227, Institut für Geophysik ETHZ, CH-8093 Zurich, Switzerland.

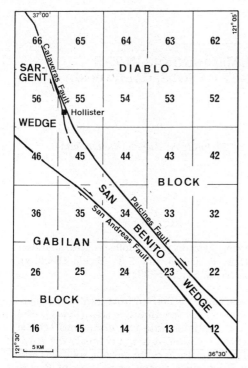

Fig. 1. Map showing the area of investigation, major faults and structural units, as well as the subdivision into quadrangles and numbering of quadrangles.

(3) The Sargent wedge and San Benito wedge, located between the San Andreas and Calaveras—Paicines fault zones.

In order to simplify the description and discussion the area was divided up into 36 5' × 5' quadrangles. Indices were assigned to each quadrangle from 11 in the SE corner to 66 in the NW corner (see Fig. 1).

The investigation is mainly based on local earthquakes that occurred during the years 1969, 1970 and 1971, recorded by the station network of the National Center for Earthquake Research (N.C.E.R.) of the U.S. Geological Survey in Menlo Park. The routine readings at N.C.E.R. and the locations of events with the HYPO 71 computer program (Lee and Lahr, 1972; Lee et al.; 1972a, b, c) were used.

PROCEDURE OF ANALYSIS

A first-motion plot (lower hemisphere, equal-area projection) was produced for an event if five or more station reports contained information about the polarity of first P-wave motion (compression or dilatation). A total of 1275 first-motion plots were analyzed. The total includes a great number of first-motion plots of weak events with local magnitudes $0.4 \leqslant$

$M_L \leqslant 2.0$ which are not generally considered for focal mechanism analysis.

The results of the investigation show that it is indeed worthwhile to consider the weak earthquakes in regional focal mechanism studies. The uncertainties of the single solution are compensated by the great number of individual solutions.

The procedure of analysis was as follows:

(1) First a solution with two orthogonal vertical nodal planes (VNP), i.e. a pure strike-slip mechanism, was used if it would satisfy the distribution of compressions and dilatations.

(2) If a VNP solution was not possible, a general solution with one or two inclined nodal planes (INP) was used.

In 727 cases it was possible to explain the distribution of compression and dilatation by a pure strike-slip mechanism (VNP solution). In many cases of VNP solutions the relatively small number of data points allowed a certain variation in the orientation of the possible nodal planes. In these cases the extreme possible positions of nodal planes, as well as their mean position, were determined. For each solution a corresponding mean orientation of the P-axis and its variation was determined. In the case of VNP solutions the whole solution is determined by orientation of the P-axis. The P-axis bisecting the dilatational quadrant is considered here purely geometrically in order to describe the position of the nodal planes. The fault plane with dextral strike-slip movement, as seen from above, strikes at an angle of 45° counterclockwise, and the fault plane with sinistral strike-slip movement at an angle of 45° clockwise to the azimuth of the P-axis. Given the orientation of the P-axis, the strike of both respective fault planes is easily determined.

In the case of the solutions with one or both inclined (nonvertical) nodal planes (INP solutions) the orientations of the P- and T-axes were determined graphically from a nonextreme position of the nodal planes.

According to the orientation of the P-axis the events were grouped into nine classes. The 180° from west over north to east were divided into

TABLE I

Classification of azimuthal sectors

Sector	Azimuth over E (°)	
1	270—290	90—110
2	290—310	110—130
3	310—330	130—150
4	330—350	150—170
5	350— 10	170—190
6	10— 30	190—210
7	30— 50	210—230
8	50— 70	230—250
9	70— 90	250—270

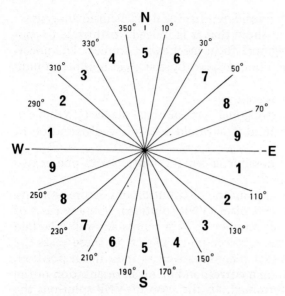

Fig. 2. Classification of azimuthal sectors used for a classification of events. An event receives the class number of the sector in which its *P*-axis is included.

nine equal sectors (see Table I and Fig. 2).

According to the azimuth of the *P*-axis and its corresponding sector the events were grouped into nine classes. An event receives the number of the sector in which its *P*-axis is contained, e.g., an event with a *P*-axis oriented N354°E is classified into class 5.

In the present paper the orientations of *P*-axes derived from VNP-solutions of earthquakes located in the northwest part of the area of investigation are described and discussed. A full description of the fault-plane solutions, including INP solutions, and a discussion of the relationships between focal mechanisms and Pliocene—Quaternary tectonic deformation are given by Pavoni (1977).

THE ORIENTATION OF P-AXES

The following discussion refers mainly to the region between 36°40′N lat.—37°00′N lat. and 121°15′W long.—121°30′W long., i.e., to the quadrangles Hollister *H34, H35, H36, H44, H45, H46, H54, H55, H56, H64, H65* and *H66* (see Figs. 1 and 3).

The orientation of P-axes derived from VNP solutions, 1969—71

Figure 3 shows the orientation of *P*-axes as derived from fault-plane solutions of pure strike-slip type. The numbers shown in the figure indicate: (1)

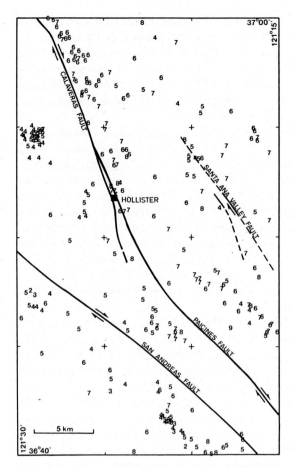

Fig. 3. Orientation of *P*-axes as derived from fault-plane solutions of pure strike-slip type (VNP solutions) of 1969—1971 earthquakes. Numbers indicate the orientation of *P*-axes according to the classification as listed in Table I and shown in Fig. 2.

the orientation of the *P*-axis according to the classification given in Table I and Fig. 2; and (2) according to their position, the location of the event. The numbers refer to the mean orientation of *P*-axes. Within the 12 quadrangles 272 events could be interpreted as pure strike-slip mechanisms. In 205 cases, i.e., in 75%, the orientation of the P-axis shows an azimuthal variation of less than 10°. In 54 cases the azimuthal variation is between 10 and 20° and in 13 cases it is greater than 20°. A regional grouping of events of the same class or two neighboring classes can be recognized.

Within quadrangle *H66* a cluster of class 6 and class 7 events is observed 1—4 km east of the Calaveras fault zone. The class 6 and class 7 events are also found in quadrangle *H55* east of the Calaveras fault zone. The events apparently located in a strip 1—4 km to the east of the Calaveras fault zone most certainly occurred on the fault itself. They are mislocated due to a too simple velocity model used in the location of hypocenters (Eaton et al., 1970; Mayer-Rosa, 1973). In quadrangles *H66* and *H55* the Calaveras fault strikes N20°W. The *P*-axes at an angle of 45° would be expected to be oriented N25°E. The observed orientation of *P*-axes derived from VNP solutions, classes 6 and 7, are indeed in excellent agreement with the observed strike of the fault and a dextral strike-slip movement.

A similar situation is observed along the San Andreas fault zone. The events located 2.5—3.5 km southwest of the San Andreas fault most likely originated on the San Andreas fault itself. Within the segment shown in Fig. 3 the San Andreas fault strikes N48°W—N53°W. The strike of the San Andreas fault is about 30° counterclockwise from the strike of the Calaveras fault. The events directly associated with the San Andreas fault show a clear N—S orientation of *P*-axes (class 5 and class 4). They occur very typically in the quadrangles *H46*, *H35* and *H34* in an apparent distance of 2.5—3.5 km from the surface trace of the San Andreas fault. Again, a clear relationship between the strike of the fault and the orientation of *P*-axes is shown. Classes 4 and 5 are in good agreement with a dextral strike-slip displacement on a N50°W-striking vertical fault plane.

It can be shown that the simplified velocity model used for the calculation of hypocenters does not substantially influence the determination of the azimuth of *P*-axes (Pavoni, 1977). In general, the deviation in the azimuth of *P*-axes caused by the mislocation of hypocenters is expected to be of the order of a few degrees in the case of VNP and INP solutions.

The observed close relationship between *P*-axis orientation and fault strike as shown independently in the Calaveras earthquakes as well as in the San Andreas earthquakes, confirms the general reliability and accuracy of results derived from fault-plane solutions, in spite of the shortcomings of the applied velocity model and the relatively simple procedure of analysis, as described in the preceding section.

TABLE II

Mean orientation of *P*-axes from VNP solutions of 1969—71 earthquakes

Quadrangle (Fig. 1)	Mean orientation of *P*-axes	Stand. dev. (°)	Number of earthquakes
66	N23°E	10	40
65	N17°E	25	13
56	N348°E	18	45
55	N27°E	21	29
54	N21°E	23	18
46	N343°E	20	13
45	N17°E	24	28
44	N30°E	26	32
35	N357°E	21	33
34	N17°E	26	9

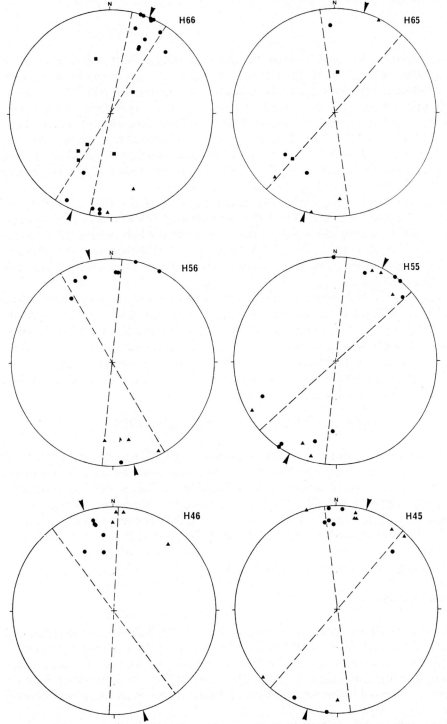

Fig. 4. Orientations of P-axes from INP solutions for 1969—1971 earthquakes of the quadrangles *H45*, *H46*, *H55*, *H56*, *H65* and *H66*. The symbols indicate the orientation of the P-axes in equal-area projection, lower hemisphere, and type of solution. Dots = strike-slip faulting; triangles = reverse faulting; squares = normal faulting. The P-axes are grouped according to the quadrangle in which the events were located. In addition the mean orientation of P-axes from pure strike-slip solutions (arrows) and the sectors defined by their azimuthal standard deviations (dashed lines) are indicated.

The earthquakes not related to the two major fault zones also show remarkable regularities in their P-axis orientation: class 4 and class 5 events are predominantly and typically observed in quadrangle H56. A regime of N—S orientation of P-axes seems to characterize the Sargent wedge. This observation is confirmed by an analysis of earthquakes which occurred after 1971 (Pavoni, 1977). The Thanksgiving Day earthquake of November 28, 1974, along the Busch fault (Rogers, 1977) and its aftershocks all show N—S orientation of the P-axis (Lee, 1974). They are all class 4 and class 5 events and fit excellently in the picture shown in Fig. 3.

In quadrangles H65, H55, H54 and H44, east of the Calaveras—Paicines fault in the Diablo Block, NE—SW orientation of the P-axis is predominant A NE—SW P-axis regime is also characteristic for the San Benito River wedge between the Paicines and San Andreas faults (Ellsworth, 1975; Pavoni, 1977). The relatively few events within the Gabilan block reveal a NE—SW orientation of P-axes.

Based on the individual values of azimuths of P-axes, a mean orientation of P-axes can be calculated for each quadrangle (see Table II). For a really detailed evaluation the subdivision into $5' \times 5'$ quadrangles is still too coarse. Data from differing regimes are obviously mixed together. The mean values are listed if nine or more events were observed in a quadrangle.

The results presented in this section are uniquely based on VNP solutions, i.e., assuming pure strike-slip mechanisms.

The orientation of P-axes derived from INP solutions 1969—71

The regional regularities in the orientation of P-axes as derived from VNP solutions are generally confirmed by additional analyses of INP solutions (Pavoni, 1977). Fig. 4 shows the orientation of P-axes derived from INP solutions 1969—71 for six $5' \times 5'$ quadrangles (equal-area projection, lower hemisphere). The projections are grouped together as the quadrangles shown in Fig. 1. Each symbol indicates the orientation of the P-axis for a single INP solution. Dots represent strike-slip fault solutions, triangles reverse-fault solutions, squares normal fault solutions.

CONCLUSIONS

A close relationship between the orientation of P-axes and the strike and sense of displacement is independently observed in earthquakes directly related to movements along the Calaveras fault, as well as in events associated with the San Andreas fault. This observation confirms the reliability of results derived from fault-plane solutions of low-magnitude earthquakes $(0.4 \leqslant M_L \leqslant 2.0)$.

The earthquakes not related to the two major fault zones also show remarkable regularities in their P-axis orientation: a regime of N—S orientation of P-axes characterizes the Sargent wedge; NNE—SSW to NE—SW orientation of *P*-axes is predominant in the Diablo Block, east of the Calaveras—Paicines fault, as well as in the San Benito wedge, located between the Paicines and the San Andreas faults; NE—SW orientation of *P*-axes is also observed within the Gabilan block. The observed regularities in the orientation of *P*-axes reveal a regional stress field with NNE—SSW to NE—SW orientation of maximum horizontal compressive stress in the upper crust. Along the San Andreas fault zone, as well as within the Sargent wedge, the orientation of *P*-axes undergoes a systematic deflection of 30—40° counterclockwise as compared to the regional orientation of *P*-axes.

ACKNOWLEDGEMENTS

The study was initiated during a stay as visiting scientist at the National Center for Earthquake Research of the U.S. Geological Survey in Menlo Park in 1972/73. I thank the U.S. Geological Survey, the Swiss Federal Institute of Technology and Prof. Stephan Mueller who made the stay in California possible.

REFERENCES

Eaton, J.P., Lee, W.H.K. and Pakiser, L.C., 1970. Use of microearthquakes in the study of the mechanics of earthquake generation along the San Andreas fault in central California. Tectonophysics, 9: 259—282.

Ellsworth, W.L., 1975. Bear Valley, California, earthquake sequence of February—March 1972. Bull. Seismol. Soc. Am., 65: 483—506.

Lee, W.H.K., 1974. A preliminary Study of the Hollister earthquake of November 28, 1974, and its major aftershocks. National Center for Earthquake Research, U.S. Geol. Surv., Menlo Park. Unpubl. Rep.

Lee, W.H.K. and Lahr, J.C., 1972. HYPO 71: A computer program for determining hypocenter, magnitude, and first motion pattern of local earthquakes. U.S. Geol. Surv., Open-file Rep., 100 pp.

Lee, W.H.K., Roller, J.C., Bauer, P.G. and Johnson, J.D., 1972a. Catalog of earthquakes along the San Andreas fault system in central California for the year 1969. U.S. Geol. Surv., Open-file Rep., 48 pp.

Lee, W.H.K., Roller, J.C., Meagher, K.L. and Bennett, R.E., 1972b. Catalog of earthquakes along the San Andreas fault system in central California for the year 1970. U.S. Geol. Surv., Open-file Rep., 73 pp.

Lee, W.H.K., Meagher, K.L., Bennett, R.E. and Matamoros, E.E., 1972c. Catalog of earthquakes along the San Andreas fault system in central California for the year 1971. U.S. Geol. Surv., Open-file Rep., 67 pp.

Mayer-Rosa, D., 1973. Travel time anomalies and distribution of earthquakes along the Calaveras fault zone, California. Bull. Seismol. Soc. Am., 63: 713—729.

Pavoni, N., 1977. Contemporary crustal movements and Pliocene—Quaternary tectonics of the Coast Ranges south of Hollister, California, in preparation.

Rogers, T.H., 1977. Geology and seismicity at the convergence of the San Andreas and Calaveras fault zones near Hollister, San Benito County, California. Unpubl. manuscript.

Tectonophysics, 52 (1979) 373—375
© Elsevier Scientific Publishing Company, Amsterdam — Printed in The Netherlands

SURFACE FAULTS IN THE GULF COASTAL PLAIN BETWEEN VICTORIA AND BEAUMONT, TEXAS

EARL R. VERBEEK

U.S. Geological Survey, Box 25046, Denver Federal Center, Denver, Colo. (U.S.A.)

(Accepted for publication May 22, 1978)

ABSTRACT

Displacement of the land surface by faulting is widespread in the Houston—Galveston region, an area which has undergone moderate to severe land subsidence associated with fluid withdrawal (principally water, and to a lesser extent, oil and gas). A causative link between subsidence and fluid extraction has been convincingly reported in the published literature. However, the degree to which fluid withdrawal affects fault movement in the Texas Gulf Coast, and the mechanism(s) by which this occurs are as yet unclear.

Faults that offset the ground surface are not confined to the large (>6000-km^2) subsidence "bowl" centered on Houston, but rather are common and characteristic features of Gulf Coast geology. Current observations and conclusions concerning surface faults mapped in a 35,000-km^2 area between Victoria and Beaumont, Texas (which area includes the Houston subsidence bowl) may be summarized as follows:

(1) Hundreds of faults cutting the Pleistocene and Holocene sediments exposed in the coastal plain have been mapped. Many faults lie well outside the Houston—Galveston region; of these, more than 10% are active, as shown by such features as displaced, fractured, and patched road surfaces, structural failure of buildings astride faults, and deformed railroad tracks.

(2) Complex patterns of surface faults are common above salt domes. Both radial patterns (for example, in High Island, Blue Ridge, Clam Lake, and Clinton domes) and crestal grabens (for example, in the South Houston and Friendswood—Webster domes) have been recognized. Elongate grabens connecting several known and suspected salt domes, such as the fault zone connecting Mykawa, Friendswood—Webster, and Clear Lake domes, suggest fault development above rising salt ridges.

(3) Surface faults associated with salt domes tend to be short (<5 km in length), numerous, curved in map view, and of diverse trend. Intersecting faults are common. In contrast, surface faults in areas unaffected by salt diapirism are frequently mappable for appreciable distances (>10 km), occur

singly or in simple grabens, have gently sinuous traces, and tend to lie roughly parallel to the ENE—NE "coastwise" trend common to regional growth faults identified in subsurface Tertiary sediments.

(4) Evidence to support the thesis that surface scarps are the shallow expression of faults extending downward into the Tertiary section is mostly indirect, but nonetheless reasonably convincing. Certainly the patterns of crestal grabens and radiating faults mapped on the surface above salt domes are more than happenstance; analogous fault patterns have been documented around these structures at depth. Similarly, some of the long surface faults not associated with salt domes seem to have subsurface counterparts among known regional growth faults documented through well logs and seismic data. Correlations between surface scarps and faults offsetting subsurface data are not conclusive because of the large vertical distances (1900—3800 m) involved in making the most of the inferred connections. Nevertheless, the large number of successful correlations — in trend, movement sense, and position — suggests that many surface scarps represent merely the most recent displacements on faults formed during the Tertiary.

(5) Upstream-facing fault scarps in this region of low relief can be significant impediments to streams. Locally, both abandoned, mud-filled Pleistocene distributary channels and, more commonly, Holocene drainage lines still occupied by perennial streams reflect the influence of faulting on their development. Some bend sharply near faults and have tended to flow along or pond against the base of scarps; others meander within topographically expressed grabens. Such evidence for Quaternary displacement of the ground surface is widespread in the Texas Gulf coast. In the general, however, streams in areas now offset by faulting show no disruption of their courses where they cross fault scarps. Such scarps are probably very young, and where they can be demonstrated to partly or wholly predate fluid withdrawal, very recent natural fault activity is indicated.

(6) Early aerial photographs (1930) of the entire region and topographic maps (1915—16 surveys) of Harris County (Houston and vicinity) show that many faults had already displaced the land surface at a time when appreciable pressure declines in subjacent strata were localized to relatively few areas of large-scale pumping. Prehistoric faulting of the land surface, as noted above, appears to have affected much of the Texas Gulf Coast.

(7) A relation between groundwater extraction and current motion on active faults is suspected because of the increased incidence of ground failure in the Houston—Galveston subsidence bowl. This argument is weakened somewhat by recognition of numerous surface faults, some of them active today, far beyond the periphery of the strongly subsiding area. Moreover, tiltbeam records from two monitored faults in northwest Houston and accounts of fault damage from local residents demonstrate a complex, episodic nature of fault creep which can only partially be correlated with groundwater production. Nevertheless, although specific mechanisms are in doubt, the extraction of groundwater from shallow (<800-m) sands is pro-

bably a major factor in contributing to current displacement of the ground surface in the Houston—Galveston region. Within this large area, the number of faults recognizable from aerial photographs has increased at least tenfold between 1930 and 1970. Elsewhere in the Texas Gulf Coast only a moderate increase has been noted, some of which is possibly attributable to oil and gas production. Surface fault density in the Houston—Galveston region is far greater than in any other area of the Texas Gulf Coast investigated to date. A plausible explanation for these differences is that large overdrafts of groundwater over an extended period of time in the Houston—Galveston region have stimulated fault activity there. Throughout the Texas Gulf Coast, however, a natural contribution to fault motion remains a distinct possibility.

HOLOCENE DISPLACEMENT ALONG BRANCH AND SECONDARY FAULTS IN THE SAN ANDREAS FAULT ZONE, SOUTHERN CALIFORNIA *

JAMES E. KAHLE, ALLAN G. BARROWS and DAVID J. BEEBY

California Division of Mines and Geology, Los Angeles, Calif. 90012 (U.S.A.)

(Accepted for publication May 22, 1978)

ABSTRACT

Detailed geologic mapping of the San Andreas fault zone in Los Angeles County since 1972 has revealed evidence for diverse histories of displacement on branch and secondary faults near Palmdale. The main trace of the San Andreas fault is well defined by a variety of physiographic features. The geologic record supports the concept of many kilometers of lateral displacement on the main trace and on some secondary faults, especially when dealing with pre-Quaternary rocks. However, the distribution of upper Pleistocene rocks along branch and secondary faults suggests a strong vertical component of displacement and, in many locations, Holocene displacement appears to be primarily vertical. The most recent movement on many secondary and some branch faults has been either high-angle (reverse and normal) or thrust. This is in contrast to the abundant evidence for lateral movement seen along the main San Andreas fault. We suggest that this change in the sense of displacement is more common than has been previously recognized.

The branch and secondary faults described here have geomorphic features along them that are as fresh as similar features visible along the most recent trace of the San Andreas fault. From this we infer that surface rupture occurred on these faults in 1857, as it did on the main San Andreas fault. Branch faults commonly form "Riedel" and "thrust" shear configurations adjacent to the main San Andreas fault and affect a zone less than a few hundred meters wide. Holocene and upper Pleistocene deposits have been repeatedly offset along faults that also separate contrasting older rocks. Secondary faults are located up to 1500 m on either side of the San Andreas fault and trend subparallel to it. Moreover, our mapping indicates that

---

* Based on work sponsored in part by U.S. Geological Survey Grants Nos. 14-08-0001-G-74, 133, 263 or 344, and in cooperation with the County of Los Angeles, Department of County Engineer. Mapping done by Drew Smith was very helpful in the early phases of this work.

some portions of these secondary faults appear to have been "inactive" throughout much of Quaternary time, even though Holocene and upper Pleistocene deposits have been repeatedly offset along other parts of these same faults. For example, near 37th Street E. and Barrel Springs Road, a limited stretch of the Nadeau fault has a very fresh normal scarp, in one place as much as 3 m high, which breaks upper Pleistocene or Holocene deposits. This scarp has two bevelled surfaces, the upper surface sloping significantly less than the lower, suggesting at least two periods of recent movement. Other exposures along this fault show undisturbed Quaternary deposits overlying the fault. The Cemetery and Little Rock faults also exhibit selected reactivation of isolated segments separated by "inactive" stretches.

Activity on branch and secondary faults, as outlined above, is presumed to be the result of sympathetic movement on limited segments of older faults in response to major movement on the San Andreas fault. The recognition that Holocene activity is possible on faults where much of the evidence suggests prolonged inactivity emphasizes the need for regional, as well as detailed site studies to evaluate adequately the hazard of any fault trace in a major fault zone. Similar problems may be encountered when geodetic or other studies, which depend on stable sites, are conducted in the vicinity of major faults.

Tectonophysics, 52 (1979) 378—379
© Elsevier Scientific Publishing Company, Amsterdam — Printed in The Netherlands

QUATERNARY CRUSTAL DEFORMATION ALONG A MAJOR BRANCH OF THE SAN ANDREAS FAULT IN CENTRAL CALIFORNIA

G.E. WEBER [1], K.R. LAJOIE [2] and J.F. WEHMILLER [3]

[1] *U.S. Geological Survey, Menlo Park, Calif. 94025 (U.S.A.) and University of California, Santa Cruz, Calif. 95064 (U.S.A.)*
[2] *U.S. Geological Survey, Menlo Park, Calif. 94025 (U.S.A.)*
[3] *University of Delaware, Newark, Del. 19711 (U.S.A.)*

(Accepted for publication May 22, 1978)

ABSTRACT

Deformed marine terraces and alluvial deposits record Quaternary crustal deformation along segments of a major, seismically active branch of the San Andreas fault which extends 190 km SSE roughly parallel to the California coastline from Bolinas Lagoon to the Point Sur area. Most of this complex fault zone lies offshore (mapped by others using acoustical techniques), but a 4-km segment (Seal Cove fault) near Half Moon Bay and a 26-km segment (San Gregorio fault) between San Gregorio and Point Año Nuevo lie onshore.

At Half Moon Bay, right-lateral slip and N—S horizontal compression are expressed by a broad, synclinal warp in the first (lowest: 125 ka?) and second marine terraces on the NE side of the Seal Cove fault. This structure plunges to the west at an oblique angle into the fault plane. Linear, joint-controlled stream courses draining the coastal uplands are deflected toward the topographic depression along the synclinal axis where they emerge from the hills to cross the lowest terrace. Streams crossing the downwarped part of this terrace adjacent to Half Moon Bay are depositing alluvial fans, whereas streams crossing the uplifted southern limb of the syncline southwest of the bay are deeply incised. Minimum crustal shortening across this syncline parallel to the fault is 0.7% over the past 125 ka, based on deformation of the shoreline angle of the first terrace.

Between San Gregorio and Point Año Nuevo the entire fault zone is 2.5—3.0 km wide and has three primary traces or zones of faulting consisting of numerous en-echelon and anastomozing secondary fault traces. Lateral discontinuities and variable deformation of well-preserved marine terrace sequences help define major structural blocks and document differential motions in this area and south to Santa Cruz. Vertical displacement occurs on all of the fault traces, but is small compared to horizontal displacement. Some blocks within the fault zone are intensely faulted and steeply tilted.

One major block 0.8 km wide east of Point Año Nuevo is downdropped as much as 20 m between two primary traces to form a graben presently filling with Holocene deposits. Where exposed in the sea cliff, these deposits are folded into a vertical attitude adjacent to the fault plane forming the southwest margin of the graben. Near Point Año Nuevo sedimentary deposits and fault rubble beneath a secondary high-angle reverse fault record three and possibly six distinct offset events in the past 125 ka.

The three primary fault traces offset in a right-lateral sense the shoreline angles of the two lowest terraces east of Point Año Nuevo. The rates of displacement on the three traces are similar. The average rate of horizontal offset across the entire zone is between 0.63 and 1.30 cm/yr, based on an amino-acid age estimate of 125 ka for the first terrace, and a reasonable guess of 200—400 ka for the second terrace. Rates of this magnitude make up a significant part of the deficit between long-term relative plate motions (estimated by others to be about 6 cm/yr) and present displacement rates along other parts of the San Andreas fault system (about 3.2 cm/yr).

Northwestward tilt and convergence of six marine terraces northeast of Año Nuevo (southwest side of the fault zone) indicate continuous gentle warping associated with right-lateral displacement since early or middle Pleistocene time. Minimum local crustal shortening of this block parallel to the fault is 0.2% based on tilt of the highest terrace. Five major, evenly spaced terraces southeast of Año Nuevo on the southwest flank of Mt. Ben Lomond (northeast side of the fault zone) rise to an elevation of 240 m, indicating relatively constant uplift (about 0.19 m/ka and southwestward tilt since Early or Middle Pleistocene time (Bradley and Griggs, 1976).

Tectonophysics, 52 (1979) 380

ANOMALOUSLY HIGH UPLIFT RATES ALONG THE VENTURA—SANTA BARBARA COAST, CALIFORNIA — TECTONIC IMPLICATIONS

J.F. WEHMILLER [1], A. SARNA-WOJCICKI [2], R.F. YERKES [2] and K.R. LAJOIE [2]

[1] *University of Delaware, Newark, Del. 19711 (U.S.A.)*
[2] *U.S. Geological Survey, Menlo Park, Calif. 94025, (U.S.A.)*

(Accepted for publication May 22, 1978)

ABSTRACT

The NW—SE trending segments of the California coastline from Point Arena to Point Conception (500 km) and from Los Angeles to San Diego (200 km) generally parallel major right-lateral strike-slip fault systems. Minor vertical crustal movements associated with the dominant horizontal displacements along these fault systems are recorded in local sedimentary basins and slightly deformed marine terraces. Typical maximum uplift rates during Late Quaternary time are about 0.3 m/ka, based on U-series ages of corals and amino-acid age estimates of fossil mollusks from the lowest emergent terraces.

In contrast, the E—W-trending segments of the California coastline between Point Conception and Los Angeles (200 km) parallel predominantly northward-dipping thrust and high-angle reverse faults of the western Transverse Ranges. Along this coast, marine terraces display significantly greater vertical deformation. Amino-acid age estimates of mollusks from elevated marine terraces along the Ventura—Santa Barbara coast imply anomalously high uplift rates of between 1 and 6 m/ka over the past 40 to 100 ka. The deduced rate of terrace uplift decreases from Ventura to Los Angeles, conforming with a similar trend observed by others in contemporary geodetic data.

The more rapid rates of terrace uplift in the western Transverse Ranges reflect N—S crustal shortening that is probably a local accommodation of the dominant right-lateral shear strain along coastal California.

Tectonophysics, 52 (1979) 381—387

REMAINS OF PREHISTORIC HUMAN IN STRATA DEFORMED BY THE SAN ANDREAS FAULT NEAR STONE CANYON, SAN BENITO COUNTY, CALIFORNIA

DONALD J. STIERMAN [1]*, EZRA ZUBROW [2]** and LACY ATKINSON [1]

[1] *Department of Geophysics, Stanford University, Stanford, Calif. 94305 (U.S.A.)*
[2] *Department of Anthropology, Stanford University, Stanford, Calif. 94305 (U.S.A.)*

(Accepted for publication May 22, 1978)

ABSTRACT

Stierman, D.J., Zubrow, E. and Atkinson, L., 1979. Remains of prehistoric human in strata deformed by the San Andreas fault. In: C.A. Whitten, R. Green and B.K. Meade (Editors), Recent Crustal Movements, 1977. Tectonophysics, 52: 381—387.

Human bones discovered in a river terrace about 20 m from an actively creeping trace of the San Andreas fault place an upper limit on the age of a sand deposit which has been deformed by fault-related movements. Amino-acid racemization infers these bones are about 5000 years old. No evidence was found associating this burial with cultural activity. The relationship between these bones and the surrounding soils, as well as the disarticulation and mixing of the bones, suggests that these human remains were buried by stream action. Sorted materials in this terrace dip 20° to the southwest, and other features of this exposure show that fault-related vertical displacements occur along this segment of the San Andreas fault. This style of deformation is consistent with dip-slip movement postulated by published results of dynamic modeling of creep-related strain events at the Stone Canyon Geophysical Observatory, only a few hundred meters from the burial, and with published models of tilt events recorded at Melondy Ranch, 6 km to the southeast. However, extrapolation of deformation rates based on these models and with the short-term creep rate cannot be reconciled with deformation observed in the terrace. The bone fragments lie in the zone of complex deformation within the San Andreas fault zone. Holocene movements in this zone have occurred at least 20 m beyond the fault trace currently monitored for fault creep.

INTRODUCTION

While examining deformed rocks exposed in a quarry near the San Andreas fault, R.W. Simpson and D.J. Stierman of Stanford University found

---

* Present Address: Department of Earth Sciences, University of California, Riverside, Calif. 92521 (U.S.A.)
** Present Address: Department of Anthropology, State University of New York, Buffalo, N.Y. (U.S.A.)

fragments of human bone in unconsolidated sand about 20 m from the actively creeping trace of the fault. This find is interesting because it presents an opportunity to study deformation near the San Andreas fault at a scale in time lying somewhere between the historic record and the long-term geological record. The purpose of our study was to determine how these remains were emplaced, whether or not the bones limit the age of the deposit in which they rested, and how this deposit is related structurally to the adjacent deformed rocks.

The site is located on the 101 Ranch, about 30 km south of Hollister, California (Fig. 1), near the Stone Canyon Geophysical Observatory (Fig. 2). The deposit containing the bones is a poorly sorted, angular, coarse arkosic sand containing many well-rounded pebbles up to 4 cm in diameter, as well

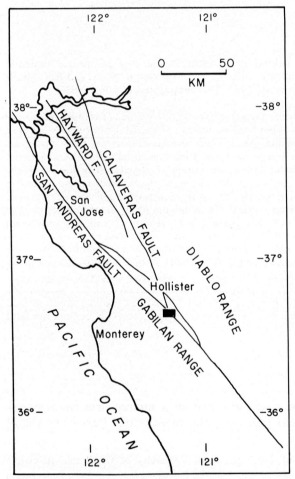

Fig. 1. Dark square toward center of regional map shows location of Stone Canyon with respect to central California landmarks.

383

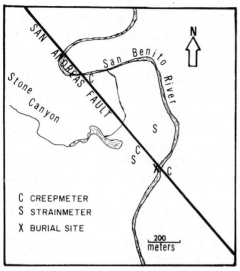

Fig. 2. Map showing location of burial with respect to instruments of the Stone Canyon Geophysical Observatory (Bufe et al., 1973).

as occasional rounded granitic boulders from 15 to 40 cm in diameter. Structures and rock types present in the sand clearly identify this deposit as a terrace of the San Benito River. Mixing of boulders with river sand suggests deposition occurred during a flood.

ARCHAEOLOGY

The bone fragments lay about 90 cm below the ground surface, nearly 10 m above the streambed of the San Benito River. Prior to excavation, the entire quarry face was inspected for other signs of cultural materials or human remains. No further items of archaeological value were found. A test face was cleared and analyzed, and the stratigraphy recorded. Because it was impossible to determine beforehand if the bones extended horizontally or vertically into the hillside, a modified isolated block method of excavation was used to collect the bone fragments. This excavation consisted of two units, the smaller (20-cm square) unit, immediately surrounding the exposed bones, nested within the large (1-m square) unit. When excavation of the larger unit yielded no cultural remains, the inner unit, consisting of the bone fragments intact in the matrix, was removed and transported to the laboratory for study.

The bones are human and include: 5 skull fragments, 1 occipital fragment with foramen magnum, 1 incomplete mandible, 1 femur fragment, 1 pelvis fragment and 1 tibia fragment. None of the bones is complete and some bone chips remain unidentified. An osteological examination revealed that the incisors are cusped and the third molars had not developed. Cusped

384

incisors is a typological indicator used by some archaeologists to identify remains of native Americans. Undeveloped third molars places the age at the time of death at between 13 and 20 years. Finally, there was no evidence that more than one individual need be involved.

These bones were probably emplaced by moving water. There is no evidence of grave goods, intrusive pits or backfill. The excavated unit was completely devoid of, aside from the bone fragments, anything other than natural deposits. Disarticulation of the bones could have resulted from separation during transport to this location, and the massive fragmentation of the bones indicates considerable trauma. This is feasible, considering the current which moved the nearby boulders operating on a decomposed body or previously buried skeleton.

Preliminary age determination of the bones by means of amino-acid racemization suggests the bones are about 5000 years old. Taken at face value, this measurement places a maximum on the time since emplacement at this site. The disarticulation yet close proximity of the bones suggests reburial by the river of a skeleton transported a short distance by the current. Stream action sufficient to break apart a body and place the femur near the mandible would most likely scatter the bones rather than deposit them together. Likewise, transportation of a skeleton for a long distance would scatter the bones. Therefore, this deposit is probably younger, rather than the same age as, the bones found there, although the bones probably originated only a short distance away.

The terrace containing the bones lies in fault contact with an arkosic conglomerate (Fig. 3) of Upper Miocene age, described by Wilson (1943) and recently mapped by Dibblee (1974). Thin (1—6-cm), discontinuous layers

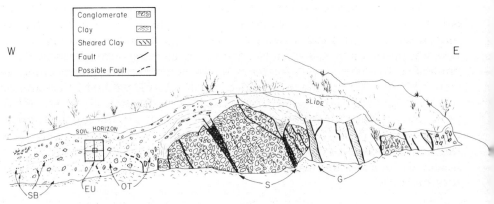

Fig. 3. Diagramatic sketch of quarry face. *EU* denotes excavated unit. *SB* sorted beds, *OT* older terrace, *S* shear zones. *G* denotes gouge zones marking currently creeping trace of the San Andreas fault. Square marking excavated unit is 1 m high. Sketch made from photo mural which did not preserve linear scale because of parallax and loss of depth perspective. Debris from quarry forms foreground.

of well-sorted imbricated pebbles and clean sand in the terrace dip about 20°S and strike about N45°W. Although sand lenses occur stratigraphically above the bones, no clearly defined beds were found within a meter of the bones. Rapid lateral variation makes tracing a bed for more than 3 m difficult. However, it appears that the main volume of the sand surrounding the bones was deposited as a single unit. No cut-and-fill structures, old soils or other signs of multiple episodes of deposition are evident in this sand unit. An A-horizon soil has developed to a depth of up to 70 cm in the terrace. This soil formation indicates that the terrace is ancient and, together with the human bones in the terrace, places age constraints on the terrace independent of the age suggested by the amino-acid racemization. The terrace overlies older terrace deposits containing highly weathered granitic boulders and sand showing evidence of previous soil formation. The old terrace is cut by faults which do not appear to extend into the younger terrace.

The conglomerate is highly fractured but remains largely unweathered except for gouge formed along shear zones (Fig. 3). Some faults extend into the terrace and possibly into the soil. Vertically trending slickensides on clay faulted against the conglomerate imply recent vertical movement. The whole mass of conglomerate has a domed-shaped look suggestive of vertical movement. Such vertical movement explains the consistent strike and dip (N40—50°W, 20°S) measured for the five segments of sorted layers in the terrace. Although streams can deposit materials in nonhorizontal layers, the beds in this deposit dip in the upstream direction and there is no apparent crossbedding. Thus, it appears that the sand containing the bone fragments was deposited in a horizontal attitude and subsequently tilted. Both the vertically trending slickensides and the dipping beds show that motion on the San Andreas fault here includes a dip-slip component.

DISCUSSION

Major displacements observed along the San Andreas fault are strike-slip in nature, and earthquake focal mechanisms near Stone Canyon are consistent with right-lateral strike-slip motion along a nearly vertical fault striking N30—50°W, parallel with the surface trace of the fault (Ellsworth, 1975). However, dynamic modeling of strain events recorded by strainmeters at Stone Canyon (Fig. 2) requires that the dislocations associated with these strains events have a dip-slip as well as a strike-slip component (Stewart et al., 1973). McHugh and Johnson (1976) found that, in dynamic modeling of tilt events recorded at Melondy Ranch, 6 km southeast of Stone Canyon, dip-slip dislocations were needed to explain the details of tilt events recorded there. In both the strain and tilt models, the southwest block moves up and to the right with respect to the northeast block, and the ratio of strike-slip to dip-slip is about 3 : 1. These strain and tilt events are thought to be associated with aseismic creep on the San Andreas fault. Records from nearby creepmeters (Fig. 2) indicate that this segment of the fault moves from 7 to 16 mm/yr, averaging about 12 mm/yr (Nason et al., 1974). If all creep includes such a dip-slip component, vertical movements of about 4 mm/yr may explain the vertical displacements observed in the quarry.

Uncertainties exist concerning the use of amino-acid racemization as a precision method of determining age. However, this was the only method available to us because standard ^{14}C dating requires a greater mass of bone than was collected. But, if the terrace containing the bones is 5000 years old, and if the creep rates and strike-slip : dip-slip ratio detected in the short term can be extrapolated back 5000 years, total inferred vertical displacement during that time span is 20 m. Assuming the dipping layers were once horizontal, an average tilt rate of 60 μrad/yr is required to yield the observed dip. This extrapolated rate exceeds the rate measured by tiltmeters within a few kilometers of this site by at least a factor of 6 (M.J.S. Johnson, personal communication, 1976). Because the bone rested only 10 m above the bed of the San Benito River, 10 m is probably the maximum uplift since deposition. Owing to the possible redeposition of the bones, this terrace could be considerably younger than the dated remains buried there. Decreasing the age of the deposit would reconcile the maximum probable uplift with the extrapolated elevation change. However, if the terrace is younger than 5000 years old, the average tilt rate would be even greater than the already high value. Therefore the deformation observed in this exposure is probably not directly applicable to the large-scale tectonic picture of the San Andreas fault, but, rather, is a local feature confined to the fault zone itself and has thus been subjected to accelerated deformation as part of the San Andreas fault zone.

Movement along multiple fault traces has been associated with severe earthquakes (Clarke et al., 1972) and with episodic prehistoric events on major faults (Taylor and Cluff, 1973). However, alignment array studies have detected aseismic displacements distributed across fault zones up to 80 m wide (Raleigh and Burford, 1969). The San Andreas fault near Stone Canyon is currently characterized by moderate earthquakes and aseismic creep. Aside from the deformation observed in this quarry, nearby topographic scarps and outcrops of fault gouge suggest that the zone of recent deformation here might easily extend several tens of meters either side of the line connecting the Stone Canyon creepmeters. If the wide zone of deformation observed at this site is entirely due to aseismic creep, the creepmeters installed here measure only part of the slip occurring along the fault. If, on the other hand, all current creep-related deformation is occurring along a narrow zone of well-developed fault gouge spanned by the creepmeters, the complex fault pattern evident here may serve as a warning that severe earthquakes can occur along this segment of the San Andreas fault.

CONCLUSIONS

Human remains found in this river terrace were buried by natural forces and thus limit the time elapsed since deposition of the terrace. This terrace was subsequently cut and tilted by activity related to movements of the San Andreas fault. Dip-slip as well as strike-slip motion occurs along this segment of the San Andreas fault. However, long-term extrapolations of movements based on short-term creep data and dynamic models of creep-related tilt and strain events are difficult to reconcile with deformation observed here. We conclude that the terrace lies within the zone of accelerated deformation encompassed by active strands of the San Andreas fault. Thus, the fault zone in Stone Canyon, the zone of Holocene deformation, extends at least 20 m beyond the trace of the San Andreas fault currently monitored for creep.

ACKNOWLEDGEMENTS

Keith Kvenvolden and David Blunt of the U.S. Geological Survey provided the amino-acid racemization data. Discussions with M.J.S. Johnson and Stuart McHugh of the U.S. Geological Survey and with Kerry Sieh of Stanford University were especially helpful in this study.

REFERENCES

Bufe, C.G., Bakum, W.H. and Tocher, D., 1973. Geophysical studies in the San Andreas fault zone at the Stone Canyon Observatory, Stanford Univ. Publ., Geol. Sci., XIII: 86—93.
Clarke, M.M., Grantz, A. and Rubin, M., 1972. Holocene activity of the Coyote Creek fault as recorded in sediments of Lake Cahuilla. The Borrego Mountain Earthquake of April 9, 1968. U.S. Geol. Surv., Prof. Pap., 787: 112—130.
Ellsworth, W.L., 1975. Bear valley, California earthquake sequence of February—March 1972. Bull. Seismol. Soc. Am., 65: 483—506.
Dibblee, T.W., 1974. Geologic map of the San Benito Quadrangle, California. U.S. Geol. Surv., Open-file Map.
McHugh, S. and Johnson, M.J.S., 1976. Short-period nonseismic tilt perturbations and their relation to episodic slip on the San Andreas fault in central California. J. Geophys. Res., 81: 6341—6346.
Nason, R.D., Philippsborn, F.R. and Yamashita, P.A., 1974. Catalog of creepmeter measurements in central California from 1968 to 1972. U.S. Geol. Surv., Open-file Rep. 74-31.
Raleigh, C.B. and Burford, R.O., 1969. Tectonics of the San Andreas fault system strain studies. Trans. Am. Geophys. Union, 50: 380—381.
Stewart, R.M., Bufe, C.G. and Pfluke, J.H., 1973. Creep-caused strain events at Stone Canyon, California. Stanford Univ. Publ., Geol. Sci., XIII: 286—293.
Taylor, C.L. and Cluff, L.S., 1973. Fault activity and its significance assessed by exploratory excavation. Stanford Univ. Publ., Geol. Sci., XIII: 239—247.
Wilson, I.F., 1943. Geology of the San Benito quadrangle, California. Calif. Div. Mines Rep., 39: 183—270.

Tectonophysics, 52 (1979) 389—405
© Elsevier Scientific Publishing Company, Amsterdam — Printed in The Netherlands

STRUCTURE AND NEOTECTONICS OF THE WESTERN SANTA YNEZ FAULT SYSTEM IN SOUTHERN CALIFORNIA

ARTHUR G. SYLVESTER [1] and ARTHUR C. DARROW [2]

[1] *Department of Geological Sciences, University of California, Santa Barbara, Calif. 93106 (U.S.A.)*
[2] *Dames and Moore, 1100 Glendon Avenue, Los Angeles, Calif. 90024 (U.S.A.)*

(Accepted for publication May 22, 1978)

ABSTRACT

Sylvester, A.G. and Darrow, A.C., 1979. Structure and neotectonics of the western Santa Ynez fault system in southern California. In: C.A. Whitten, R. Green and B.K. Meade (Editors), Recent Crustal Movements, 1977. Tectonophysics, 52: 389—405.

Geologic, geomorphic and seismologic data indicate that west of Lake Cachuma the Santa Ynez fault branches into several major W- and NW-trending splay faults. Two of the faults bracket the wedge-shaped Santa Maria basin. The most compelling evidence for the existence of these two faults is the fact that the Santa Maria basin is floored by Franciscan basement overlain only by Miocene and younger sedimentary rocks, whereas across the inferred traces of each of these faults, the adjacent terrains consist of Franciscan basement overlain by thick sequences of Early Tertiary strata, as well as by Miocene and younger rocks. The third splay fault strikes northwestward through the central Santa Maria basin. Narrow zones of tightly appressed, left-stepping en-echelon folds are locally adjacent to the faults along the south edge, and through the center of the basin. The geometrical arrangement of these folds is indicative of formation over buried sinistral wrench faults. Evidence for Holocene surface rupturing is lacking or nebulous at best, but epicenters of damaging historical earthquakes are spatially, and by inference, genetically related to the central Santa Maria basin faults, indicating that they comprise the presently active strands among the several splay faults.

INTRODUCTION

Geography and structural geometry

The Santa Ynez fault is a major, E—W-trending structure on the north edge of the western Transverse Ranges province of southern California (Fig. 1). As presently mapped (e.g., Jennings, 1975), the fault has a gently sinuous trace, 135 km long, from its very complicated intersection with the San Gabriel and related faults at its eastern end (Gordon, 1978) to the Pacific coastline at its western end. East of Gaviota Pass the fault lies at the base of the steep, north-facing escarpment of the Sante Ynez Mts; at Gaviota Pass it bifurcates into the South branch which cuts southwestward across the Santa Ynez Mts., and the North and Pacifico branches which strike westward through the axis of the range.

The fault is generally characterized in structure sections as a nearly vertical or steep, south-dipping reverse fault. The steep dip is interpreted from the generally straight course of the fault across variable topography and from isolated exposures, and from tunnel and well intersections (Dibblee, 1950, 1966; Page et al., 1951; Trefzger, 1966). The fault trace itself is commonly obscured by landslide debris, colluvial deposits, and thick vegetation. Locally it is mapped as a zone up to 150 m wide comprised of short, en-echelon segments and anastomosing strands bounding tectonic slivers of high fractured and gouged rocks.

Separation

The Santa Ynez fault juxtaposes rocks of different age and lithology, ranging from the Franciscan Formation of Mesozoic age to thick, stratified sequences of Late Mesozoic and Cenozoic sedimentary rocks. Maximum dip separation of Eocene sedimentary rocks is 2900 m in the central segment and 500 m in the western segment (Pacifico and North branches; Dibblee, 1950, 1966). However, palinspastic reconstructions of dip separations juxtapose rocks of different age and type. Most investigators have resolved this problem by invoking from 1 to 45-km sinistral separation on that part of the fault east of Lake Cachuma (Table I). But west of Lake Cachuma estimated horizontal separations are considerably less, perhaps by an order of magnitude (Table I).

It is unlikely that the strike separation simply decreases abruptly westward on the presently known fault traces without some compensatory struc-

TABLE I

Estimated horizontal separation, Santa Ynez fault

Amount and sense of separation	Ages of offset rock units	Reference
East of Lake Cachuma		
1—2 miles (1.6—3 km) left lateral	Eocene (Matilija Fm.)	Link (1971)
7—9 miles (11—14 km) left lateral	Eocene (Sierra Blanca Limestone)	Page et al. (1951)
22 miles (35 km) left lateral	Oligocene (Sespe Fm.)	McCracken (1969)
37 miles (60 km) left lateral	Early Miocene (Vaqueros Fm.)	Edwards (1971)
30 miles (48 km) right lateral	Eocene (Matilija Fm).	Schmitka (1973)
West of Lake Cachuma		
2 miles (3 km) left lateral	?	Roubanis (1963)
800 feet (244 m) left lateral	Oligocene and Early Miocene (Sespe and Vaqueros Fms.)	Dibblee (1950) O'Brien (1973)

tures. Structural adjustments, such as thrust faults, folds and splay faults are found at the east end of the Santa Ynez fault where it intersects the San Gabriel fault (J.C. Crowell, personal communication, 1977; Gordon, 1978), but similar structures are not found west of Lake Cachuma where strata are deformed only into a broad homocline.

The apparent variable magnitude of horizontal separation along the western Santa Ynez fault can be explained by postulating that other splay faults branch off the main fault west of Lake Cachuma (Fig. 1), analogous to the way splay faults branch off the ends of major wrench faults elsewhere. In this hypothesis, from 15 to 20 km of sinistral separation could be accommodated distributively on these postulated splays.

Redwine (1963, 1975) and Hall (1977) suggested the existence of major faults in the Santa Maria basin. Geologic, geomorphic and seismologic data presented below are consistent with Hall's conclusions and show that three major faults branch from one another in the vicinity of Lake Cachuma and fan westward to form important structural discontinuities in the Santa Maria basin (Fig. 1).

As shown in Fig. 1, the present course of the Santa Ynez River is approximately coincident with the trace of one of the faults, herein named the Santa Ynez River fault. We consider it to be the main extension of the Santa Ynez fault. In the central part of the Santa Maria basin, a variety of data provides strong evidence for the existence of a major, throughgoing, NW-trending fault which links up the discontinuous traces of the Baseline, Los Alamos, Casmalia and Pezzoni faults. The northeastern boundary of the Santa Maria basin is bounded by the Little Pine and Foxen Canyon faults, and their extension to the north, the Santa Maria River fault (C.A. Hall, personal communication, 1977). We have not studied this latter-mentioned set of faults in detail and can add little to Hall's information.

SANTA YNEZ RIVER FAULT

Several lines of evidence point to the existence of a major, west-trending fault which branches off the presently mapped trace of the Santa Ynez fault at Lake Cachuma. Foremost is the distribution of Tertiary rocks in the western Santa Ynez Mts. and the Santa Maria basin (Fig. 2). Outcrop and oil-well data show that a thick succession of Eocene, Oligocene and Miocene sedimentary rocks overlies the Franciscan Formation in the Santa Ynez Mts. south of the Santa Ynez River. However, immediately north of the Santa Ynez River, the Franciscan Formation is overlain only by Miocene and younger rocks. Similarly, the Tranquillon Volcanics of Early Miocene age crop out in the western Santa Ynez Mts. (Fig. 1), but not in the Santa Maria basin, judging from records of oil wells that have reached the basement in the Santa Maria basin. This striking difference in stratigraphic sequence indicates that a major structural discontinuity flanks the southern Santa Maria basin, a discontinuity which we interpret to be the principal extension of the

(For Figs. 1 and 2, see pp. 397—400)

Santa Ynez fault west of Lake Cachuma. In contrast, previously mapped western extensions of the Santa Ynez fault, including the South, North and Pacifico branches, lie wholly within the Santa Ynez Mountain block. These faults show comparatively small offsets of stratified rocks and no compelling evidence of recent activity; thus, we regard them as minor strands of the fault splay system.

Additional evidence for the existence of the Santa Ynez River fault is present at several locations along the Santa Ynez River (Fig. 2):

Loc. A. Trefzger (1966) reports a "zone of fault and fracturing" 300 m wide about 1000 m south of the intake portal of the Tecolote Tunnel. This zone of faulting separates the Rincon Formation of Early Miocene age on the south from the Late Miocene Monterey Formation on the north, indicating an apparent south-side-up dip separation. This zone of faulting is 2000 m north of the presently mapped trace of the Santa Ynez fault.

Loc. B. In the Santa Ynez riverbed, south of Solvang and Buellton (Fig. 2), Woodring et al. (1945) and Upson and Thomasson (1951) mapped a fault with apparent south-side-up dip separation, consistent with that of the fault segment discussed above.

Loc. C. South of Solvang, along the south bank of the Santa Ynez River, we have mapped major zones of fault gouge in rocks of the Monterey Formation.

Loc. D. Woordring et al. (1945) mapped an E—W-trending fault at Locality *D* (Fig. 2) which separates Tranquillon Volcanics on the south from Pliocene Careaga sandstone on the north.

Loc. E. Evenson and Miller (1963) postulated the existence of faults in the terrace west of Lompoc and north of the Honda fault. In this area Dibblee (1950) shows a south-dipping outcrop of the Monterey Formation just south of the fault, but 1 km north of the fault, a well encountered the top of the Monterey Formation at a depth of 716 m. Whereas this relationship can be explained by the presence of an anticline, it is equally explicable by offset on a west-trending fault whose south block is upthrown.

Loc. F. Dibblee (1950) mapped a fault along the Santa Ynez River at the south edge of the Santa Rita Hills. The fault is a west-trending, south-side-up structure.

Loc. G. Further evidence for the existence of a major fault is the presence of numerous, tightly appressed, WNW-trending, left-stepping, en-echelon folds along the south side of the Santa Ynez River between Lake Cachuma and the Santa Rita Hills, and especially south of the Santa Rita faults, north of the Honda fault, and between the Honda and Santa Rite faults (Dibblee, 1950). Dibblee (1950, p. 56) postulated that these en-echelon folds are related in space and origin to left-slip wrench faults, analogous to the relation of similar structures in clay model laboratory studies of wrench faulting (Wilcox et al., 1973), as well as to similar, but oppositely arranged folds observed along right-slip wrench faults elsewhere (e.g., Harding, 1973, 1976; Barrows, 1974; Sylvester and Smith, 1976).

The evidence presented above and illustrated in Figs. 1 and 2 point strongly toward the existence of a major, throughgoing fault. The magnitude and direction of separation have not been determined; however, the apparent differences in lateral separation on the Santa Ynez fault east and west of Lake Cachuma may be resolved, at least in part, by invoking sinistral separation on the Santa Ynez River fault. As mentioned above, sinistral separation is structurally indicated by the presence of left-stepping en-echelon folds adjacent to the postulated trace of the fault.

CENTRAL SANTA MARIA BASIN FAULTS

The Central Santa Maria basin faults strike discontinuously as WNW-trending segments across the basin from the vicinity of Lake Cachuma to the Pacific Ocean in the vicinity of Pt. Sal (Fig. 1). The strongest evidence that they form a continuous fault zone through the central basin lies along a zone which includes, from southeast to northwest, the Baseline, Los Alamos, Casmalia and Pezzoni faults. However, the presence of NW-trending, left-stepping en-echelon folds in the Purisima Hills suggests a possible extension of the Baseline fault through the Lompoc oil field west-northwestward to the Lion's Head fault.

Baseline—Los Alamos—Casmalia—Pezzoni fault

North and northwest of Lake Cachuma, the Baseline fault is marked by a prominent, north-facing scarp. The scarp, which is deeply dissected and cut by major throughgoing stream courses, is as high as 10 m, based on the estimated offset of a Quaternary alluvium surface west of Santa Cruz Creek (Loc. *H*, Fig. 2). The scarp decreases in height northwestward until it is lost in the agricultural area east of Ballard (Loc. *I*, Fig. 2). Although the fault itself is not exposed, the fault trace is locally marked by aligned springs, and it evidently constitutes an important groundwater barrier, judging from the locations of water wells north of the western half of the scarp. In addition, terrace deposits of Middle to Late-Pleistocene age dip steeply close to the scarp, but elsewhere they are nearly flat-lying.

The salient facts and interpretations for the existence of the other faults along this trend are as follows:

(1) East of Los Alamos (Loc. *J*, Fig. 2), in the upper part of Los Alamos Valley, Woodring and Bramlette (1950) mapped a NW-trending fault with SW-side-up dip separation.

We infer that this fault segment is connected at depth with the Baseline fault, because they are aligned with one another, they have similar orientation and sense of displacement and exhibit youthful geomorphic features.

(2) Approximately 13 km northwest of the fault in Los Alamos Valley, structure sections across the Orcutt oil field (Loc. *K*, Fig. 2) show a major southwest-dipping reverse fault with 2400 m of dip separation of the Mon-

terey Formation (Woodring and Bramlette, 1950; Krammes and Curran, 1959). Several other faults with significantly less separation have also been identified in the oil field.

We infer that the major fault in the Orcutt oil field is connected with the fault in the Los Alamos area on the basis of similarities in sense of separation, the youthful age of units cut by the fault segments, and the distribution of historic seismicity as discussed below.

(3) Approximately 5.6 km northwest of the Orcutt oil field along the same trend, the Pezzoni fault exhibits 2400 m of dip separation of the Miocene Monterey Formation; the southwest block is uplifted relative to the northeast block (Woodring and Bramlette, 1950). Strike separation is not known. We infer that the Pezzoni fault segment and the major fault in the Orcutt oil field are connected at depth on the basis of similarities in the amount and sense of separation. The Pezzoni fault is also on trend with the Casmalia fault in the Casmalia oil field (Fig. 1). As shown by Woodring and Bramlette (1950), the fault in the Casmalia oil field cuts the Careaga Formation and exhibits a similar sense of separation.

We infer that the Pezzoni fault segment bends toward the west and intersects the coastline at a point about 4 km north of Pt. Sal (Fig. 1) where Woodring and Bramlette (1950) mapped a fault with the same sense of dip separation. Hoskins and Griffiths (1971) infer that this fault continues offshore.

Baseline—Purisima Hills—Lompoc oil field—Lion's Head fault

The strongest evidence for a throughgoing fault along this trend is the presence of a zone of left-stepping, en-echelon folds that extends through the Purisima Hills (Loc. *L*, Fig. 2) to the Lompoc oil field (Loc. *M*, Fig. 2). As discussed previously, the geometry of these folds suggests proximity of a buried left-slip wrench fault. In the Lompoc oil field Dibblee (1943, 1950) identified subsurface faults showing relative north-side-up dip separation, and the Paso Robles Formation of Plio—Pleistocene age is cut by some of the faults (Dibblee, 1943; Krammes and Curran, 1959).

The Lion's Head fault (Figs. 1 and 2) extends approximately 8 km southeast from the coast near Pt. Sal. Near Pt. Sal, Woodring and Bramlette (1950) found that dip separation of the Franciscan Formation exceeds 1300 m with the northeast side uplifted with respect to the southwest side. This fault juxtaposes Pliocene sedimentary rocks against the Franciscan Formation. Strike separation is not known.

In the vicinity of Santa Maria, Canfield (1939) and Worts (1951) identified the Santa Maria fault (Loc. *N*, Fig. 2) and Bradley Canyon fault (Loc. *O*, Fig. 2) which may also be related to the Central Santa Maria basin faults. These faults trend slightly west of north and offset the Paso Robles Formation. According to Canfield (1939) the Santa Maria fault is a high-angle reverse fault that extends into "alluvium and stream gravels" overlying the Paso Robles Formation.

On trend with the Bradley Canyon fault is a fault in the subsurface of the West Cat Canyon oil field (Loc. *P*, Fig. 2), and both show the same west-side-up dip separation (Huey, 1954). The fault in Cat Canyon cuts the base of the Pliocene Sisquoc Formation.

SANTA MARIA RIVER—FOXEN CANYON—LITTLE PINE FAULT

This group of faults trends northwestward from a structurally complex intersection with the Santa Ynez fault east of Lake Cachuma to the Santa Maria River northeast of Santa Maria (Fig. 1).

The Little Pine fault segment north and east of Lake Cachuma is a NE-dipping reverse fault which thrusts the Franciscan Formation on the northeast over the Paso Robles Formation on the southwest. The Loma Alta fault is probably related to the Little Pine fault, because it thrusts the Monterey Formation on the northeast over the Paso Robles Formation on the southwest. Dibblee (1950, 1966) shows Holocene alluvial deposits undeformed by the Little Pine and Loma Alta faults.

The Foxen Canyon and Santa Maria River fault segments have been mapped by Hall (1977) and are believed by him to be parts of the same fault system. The Santa Maria River fault segments truncate the West Huasna fault (Fig. 1) which, in turn, displaces Orcutt Sand, suggesting possible Holocene age of faulting on the Santa Maria River fault (C.A. Hall, personal communication, 1977).

RECENCY OF FAULTING

Santa Ynez River and related faults

Even though our data document the existence of the Santa Ynez River fault, recency of faulting has not been established. West of Lompoc, the base of the Orcutt Sand of Late Pleistocene age is impressively warped (J.I. Ziony, personal communication, 1977), but it has not been shown to be offset.

The South branch fault truncates Quaternary terrace deposits, but the overlying soil is undisturbed (Bortugno, 1977). In 1970 a trench was excavated across the trace of the South branch fault about 300 m north of the coastline. The relationships demonstrated a Late Pleistocene age of faulting, but not Holocene activity (J.I. Ziony and D.W. Weaver, personal communications, 1977; observations by the senior author, 1970).

The youngest rocks cut by either the Pacifico or North branch faults are Late Tertiary in age (Ziony et al., 1974), and we find no evidence to suggest displacement of Quaternary deposits. Roubanis (1963) reported sag ponds along the Pacifico fault, but our studies of these features revealed no compelling evidence to conclude that they are tectonic in origin. Ziony et al. (1974) assign the activity of this fault to the Late Pleistocene but not Holocene.

Approximately 25 km east of Lake Cachuma, "strands of the Santa Ynez fault in Blue Canyon northeast of Santa Barbara (not shown in Figs. 1 or 2), appear to displace elevated river terrace gravels of probable late Pleistocene age" (Ziony, 1971, p. A168). If the shallow furrows or fault scarps and deflected streams observed there (Page et al., 1951, p. 1768) are fault controlled, our mapping indicates left-lateral separation of about 65 m and dip separation of about 6 m. We found no fault planes in terrace deposits, and Holocene deposits are not disturbed by faulting.

Central Santa Maria Basin faults

The available data for the recency of movement of these faults are summarized in Table II. With the possible exception of the fault in the Orcutt oil field that cuts the Orcutt Sand of Late Pleistocene age, most of the fault activity is older than Holocene. However, Ziony et al. (1974) inferred a Holocene age for the most recent displacement on the fault segment southeast of Los Alamos, based on ephemeral geomorphic features found along this fault trace.

TABLE II

Recency of faulting, central Santa Maria basin faults

Fault	Youngest offset rock unit	Reference
Baseline	Paso Robles Formation (Plio—Pleistocene)	this paper
Unnamed fault east of Los Alamos	Paso Robles Formation	Woodring and Bramlette (1950)
Fault in Orcutt oil field	Orcutt Sand (Late Pleistocene)	Woodring and Bramlette (1950), Krammes and Curran (1959)
Casmalia	Careaga Formation (Late Pliocene)	Woodring and Bramlette (1951)
Pezzoni	Sisquoc Formation (Pliocene)	Woodring and Bramlette (1950)
Lompoc oil field	Paso Robles Formation	Dibblee (1943), Krammes and Curran (1959)
Lion's Head	Terrace (Quaternary)	Miller (1977)
Santa Maria	Terrace (Quaternary)	Canfield (1939)
Bradley Canyon	Orcutt (Sand(?))	Canfield (1939), Worts (1951)
West Cat Canyon oil field	Sisquoc Formation	Huey (1954)

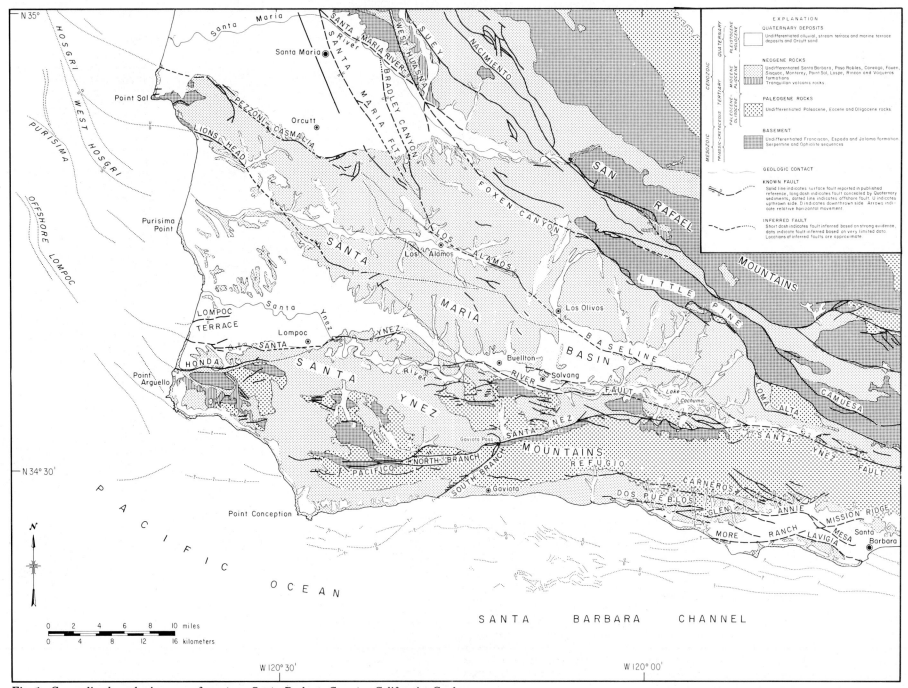

Fig. 1. Generalized geologic map of western Santa Barbara County, California. Geology adapted from Jennings (1975) and Jennings and Strand (1969). Faults modified from many sources, chiefly Ziony et al. (1974), and J.M. Buchanan-Banks (personal communication, 1977). Offshore faults adapted from Earth Science Associates (1975), U.S. Geological Survey (1974), Hoskins and Griffiths (1971) and Dames and Moore (1977).

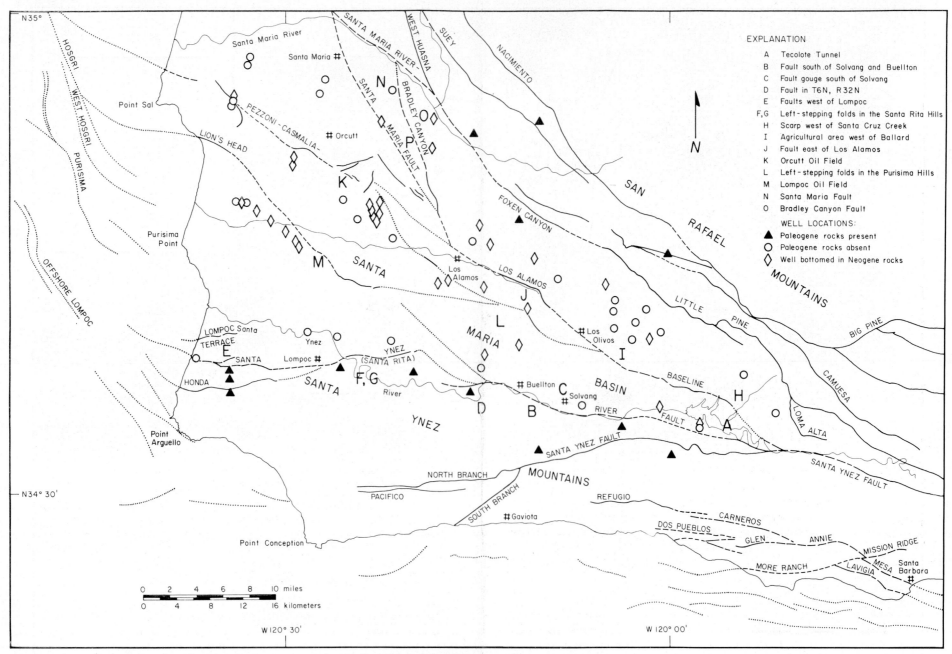

Fig. 2. Map showing geometry of major faults, contrasts in stratigraphic sequence as revealed by oil-well data, and locations of features referred to in the text. Subsurface data on file at California Division of Oil and Gas, Santa Maria.

Woodring and Bramlette (1950) reported that Pleistocene terrace deposits overlying the Lion's Head fault are not displaced, but recent field investigations have found that they are displaced (D.G. Miller, personal communication, 1977).

We excavated three trenches across the inferred trace of the central part of the Baseline scarp to evaluate the recency of faulting. There alluvium of Holocene age unconformably overlies fanglomerate and fluvial deposits of the Plio-Pleistocene Paso Robles Formation. The fault was not exposed within the depths of the trenches. A carbon sample obtained from a depth of 2 m in one trench yielded a radiometric date (^{14}C) of 1800 ± 130 years B.P., indicating a Late Holocene age for the alluvium. The youthfulness of the unfaulted alluvium precludes a definitive statement about the recency of faulting, but the data suggest that the fault has not been active for at least a few thousand years, perhaps much longer. Thus, the most youthful-appearing fault feature observed during our study of this area is at least 1800 years old.

HISTORIC SEISMICITY

Locations of felt and instrumentally located earthquakes are shown in Fig. 3. Azimuthal distribution of seismographs has not been optimal for locating earthquakes in the Santa Maria basin in the period 1934—69; nevertheless, a clustering of events is evident in the center of the basin.

Foremost of the Santa Maria basin earthquakes are those that occurred in the vicinity of Los Alamos in 1902 and 1915 (Townley and Allen, 1939; Beal, 1915). Although early newspaper accounts of the 1902 earthquake (summarized by Gawthrop, 1975) report "a strip of country 15 miles (25 km) long by 4 miles (6 km) wide rent with gaping fissures and dotted with hills and knolls that sprung up during the night . . .", subsequent accounts reported: "Brick and adobe houses and chimneys fared badly, and a good deal of damage was done in the way of destroying goods and household equipment; but the numerous great rents and fissures in the earth and the toppling of hills are largely a myth . . ." (*Ventura Daily Democrat*, Sunday, August 3, 1902). We concur with Gawthrop (1975) that the fissuring was probably a soil failure phenomenon rather than tectonic surface rupturing, and that the 1902 and 1915 earthquakes suggest "a seismically active fault exists under or near Los Alamos Valley".

Minor felt earthquakes have been reported persistently in the central and northern Santa Maria basin (Townley and Allen, 1939), but their meaning is uncertain. In addition, small localized earthquakes have been reported to us by residents of the Lompoc area since 1968, but the tectonic significance of the earthquakes is questionable (C.F. Richter, personal communication, 1977).

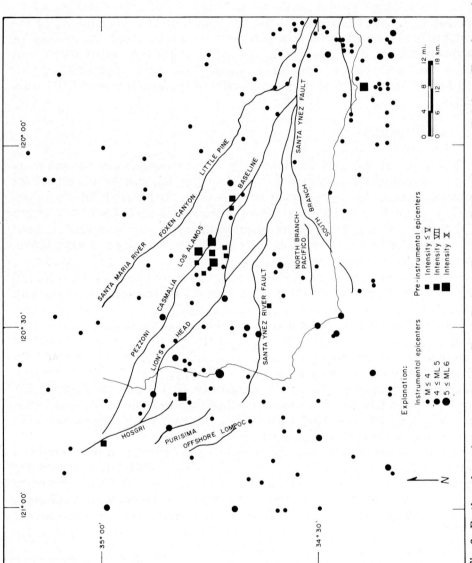

Fig. 3. Earthquake epicenters, western Santa Barbara County, California, compiled from Townley and Allen (1939), Coffman and Hake (1973), California Institute of Technology (1977).

CONCLUSIONS

The regional distribution of major rock sequences and the geometry of structural and geomorphic features demonstrate the presence of a major, heretofore unrecognized system of faults in the Santa Maria basin which branch as splay faults off the Santa Ynez fault. Their geometry is a more reasonable interpretation of the structural transition from the Transverse Ranges to the California Coast Ranges than is shown on available published maps.

En-echelon folds are related spatially to the splay faults and indicate that the faults have had left-slip displacement at depth during at least part of their history. Although this does not mean that they are left-slip faults today, it does mean that these splay faults may have absorbed much of the 10—15 km of left separation documented on the eastern segment of the fault.

Geomorphic features show that some of the faults may have been active in Holocene time, but the evidence for surface rupturing in Holocene time is lacking or nebulous on most of the faults.

The short record of historic seismicity merely suggests that the central Santa Maria basin faults are presently active; field studies and review of historic records reveal no compelling evidence of surface rupturing.

The significance of the results of this study is that important and potentially active seismogenic structures are identified in the Santa Maria basin, and they alter considerably the most recent assessments of seismic hazards (Moore and Taber, 1974) of an area undergoing moderately rapid urbanization.

ACKNOWLEDGMENTS

This study was initiated by the senior author in 1969 with support through 1972 from the National Sea Grant Program of the National Science Foundation and the U.S. Department of Commerce under Grants GH-43 and GH-95 and U.S.D.C. 2-35208-7. Most of the recent work reported here was the result of a study done by Dames and Moore for Western L.N.G. Terminal Company. Michael Hoover, Gordon Haxel, Jeff Keaton and Eric Simison helped gather and collate data. The authors acknowledge with pleasure the support and help of these agencies and people.

REFERENCES

Barrows, A.G., 1974. A review of the geology and earthquake history of the Newport—Inglewood structural zone, southern California. California Div. Mines Geol., Spec. Rep. 114, 115 pp.
Beal, C.H., 1915. Earthquake at Los Alamos, California, January 11, 1915. Seismol. Soc. Am. Bull., 5: 14—25.

404

Bortugno, E.J., 1977. Santa Ynez fault (South branch): California Div. Mines Geol., Fault
Evaluation Rep. FER-12, 9 pp.
California Institute of Technology, 1977. Computer Plot of Epicenters, unpubl.
Canfield, C.R., 1939. Subsurface stratigraphy of Santa Maria Valley oil field and adjacent
parts of Santa Maria Valley, California. Am. Assoc. Pet. Geol. Bull., 23: 45—81.
Coffman, J.L. and Hake, C.A. von, 1973. Earthquake History of the United States. U.S.
Dept. Commer., 208 pp.
Dames and Moore, 1977. Final Report, Offshore Geological and Geophysical Study, Pro-
posed LNG terminal, Point Conception, California, for Western LNG Terminal Co.
Dames and Moore, Job No. 0011-195-02, pp. 427—429.
Dibblee, Jr., T.W., 1943. Lompoc oil field. Geol. Formations and Econ. Development of
Oil and Gas Fields in California, Calif. Div. Mines Geol., Bull., 118: 427—429.
Dibblee, Jr., T.W., 1950. Geology of southwestern Santa Barbara County: Point Arguello,
Lompoc, Point Conception, Los Olives, and Gaviota quadrangles. Calif. Div. Mines
Geol. Bull., 150: 95 pp.
Dibblee Jr., T.W., 1966. Geology of the central Santa Ynez Mountains, Santa Barbara
County, California. Calif. Div. Mines Geol. Bull., 186: 99 pp.
Earth Science Associates, Inc., 1975. Additional geologic and seismologic studies — 1975.
In: Final Safety Analysis Report, units 1 and 2, Diablo Canyon site. Prepared for
Pacific Gas and Electric Company and submitted to U.S. Nuclear Regulatory Comm.
Docket Nos. 50-275 and 50-323, app. 2.5E.
Edwards, L.N., 1971. Geology of the Vaqueros and Rincon Formations, Santa Barbara
Embayment, California. Ph.D. thesis, Univ. California, Santa Barbara, Calif.
Evenson, R.E. and Miller, G.A., 1963. Geology and groundwater features of Point Ar-
guello Naval Missile Facility, Santa Barbara County, California. U.S. Geol. Surv. Water-
Supply Pap., 1619-F: 35 pp.
Gawthrop Jr., W., 1975. Seismicity of the central California coastal region. U.S. Geol.
Surv. Open-file Rep. 75-134, 90 pp.
Gordon, S.A., 1978. Relations among the Santa Ynez, Pine Mountain, Agua Blanca, and
Cobblestone Mountain faults, Transverse Ranges, California. M.A. Thesis, Univ. Cali-
fornia, Santa Barbara, Calif.
Hall, C.A., 1977. Origin and development of the Lompoc—Santa Maria pull-apart basin
and its relation to the San Simeon—Hosgri fault, California (abstr.). Geol. Soc. Am.
9(4): 428.
Harding, T.P., 1973. Newport—Inglewood trend, California — an example of wrenching
style of deformation. Am. Assoc. Pet. Geol. Bull., 57: 97—116.
Harding, T.P., 1976. Tectonic significance and hydrocarbon trapping consequences of
sequential folding synchronous with San Andreas faulting, San Joaquin Valley, Califor-
nia. Am. Assoc. Pet. Geol. Bull., 60: 356—378.
Hoskins, E.G. and Griffiths, J.R., 1971. Hydrocarbon potential of northern and central
California offshore. In: I.H. Cram (Editor), Future Petroleum Provinces of the United
States — their geology and potential. Am. Assoc. Pet. Geol. Mem., 15: 121—228.
Huey, W.F., 1954. West Cat Canyon area of Cat Canyon oil field. Summ. Oper., Calif.
Div. Oil and Gas, 40: 15—22.
Jennings, C.W., 1975. Fault map of California. Calif. Div. Mines Geol., Geol. Data Map
Ser., Map. No. 1.
Jennings, C.W. and Strand, R.G., 1969. Geologic map of California, Los Angeles sheet.
Calif. Div. Mines Geol., scale 1 : 250,000.
Krammes, K.F. and Curran, J.F., 1959. Correlation section across Santa Maria Basin. Am.
Assoc. Pet. Geol. (Pacific Sec.), 1 sheet.
Link, M.H., 1971. Sedimentology, petrography, and environmental analysis of the Ma-
tilija sandstone north of the Santa Ynez fault, Santa Barbara Co., California. M.A.
thesis, Univ. California, Santa Barbara.
McCracken, W.A., 1969. Environmental reconstruction-Sespe Formation, Ventura basin,

California (abstr.). Geol. Soc. Am. Annu. Meet., pt. 7, 145—146.

Moore and Taber, 1974. Seismic safety element. Santa Barbara County Comprehensive Plan, 93 pp.

O'Brien, J., 1973. Narizian—Refugian (Eocene—Oligocene) Sedimentation, Western Santa Ynez Mountains, Santa Barbara County, California. M.A. thesis, Univ. California, Santa Barbara.

Page, B.M., Marks, J.G. and Walker, G.W., 1951. Stratigraphy and structure of the mountains northwest of Santa Barbara, California. Am. Assoc. Pet. Geol. Bull., 35: 1727—1780.

Redwine, L.E., 1963. Morphology, sediments, and geological history of basins of Santa Maria area, California (abstr.). Am. Assoc. Pet. Geol. Bull., 47: 1775.

Redwine, L.E., Ryall, P., Suchsland, R. and Whaley, H., 1975. Roadlog. In: D. Oltz and R. Suchsland (Editors), Geologic Field Guide of the Eastern Santa Maria Area. Soc. Econ. Paleontologists and Mineralogists, Pacific Section, pp. 19—20.

Roubanis, A.S., 1963. Geology of the Santa Ynez fault, Gaviota Pass-Point Conception Area, Santa Barbara County, California. M.A. thesis, Univ. California, Los Angeles.

Schmitka, R.O., 1973. Evidence for major right-lateral separation of Eocene rocks along the Santa Ynez fault, Santa Barbara and Ventura Counties, California (abstr.). Geol. Soc. Am., 104.

Schroeter, C., 1972. Stratigraphy and Structure of the Juncal Camp — Santa Ynez Fault Sliver, Southeastern Santa Barbara County, California. M.A. thesis, Univ. California, Santa Barbara.

Sylvester, A.G. and Smith, R.R., 1976. Tectonic transpression and basement-controlled deformation in San Andreas fault zone, Salton Trough, California. Am. Assoc. Pet. Geol. Bull., 60: 2081—2102.

Townley, W.D. and Allen, M.W., 1939. Descriptive catalog of earthquakes of the Pacific Coast of the United States, 1796—1928. Seismol. Soc. Am. Bull., 29: 1—297.

Trefzger, R.E., 1966. Tecolote Tunnel. In: R. Lung and R.J. Proctor (Editors), Engineering Geology in Southern California. Assoc. Eng. Geol. Spec. Publ. (Los Angeles Sect.), pp. 109—113.

Upson, J.E. and Thomasson Jr., H.G., 1951. Geology and water resources of the Santa Ynez River basin, Santa Barbara County, California. U.S. Geol. Surv. Water-Supply Pap. 1107, 194 pp.

U.S. Geological Survey, 1974. Final environmental statement, proposed plan of development, Santa Ynez unit, Santa Barbara Channel, off California. U.S. Geol. Surv., FES 74-20, 3 v.

Wilcox, R.E., Harding, T.P. and Seely, D.R., 1973. Basic wrench tectonics. Am. Assoc. Pet. Geol. Bull., 57: 74—96.

Woodring, W.P. and Bramlette, M.N., 1950. Geology and paleontology of the Santa Maria district, California. U.S. Geol. Surv., Prof. Pap., 222: 185 pp.

Woodring, W.P., Loofbourow Jr., J.W. and Bramlette, M.N., 1945. Geology of the Santa Rosa Hills district, Santa Barbara County, California. U.S. Geol. Surv. Oil and Gas Invest., Prelim. Map, 26, 1 : 48,000.

Worts Jr., G.F., 1951. Geology and groundwater resources of the Santa Maria Valley area, California. U.S. Geol. Surv. Water-Supply Pap. 1000: 169 pp.

Ziony, J.I., 1971. Quaternary faulting in coastal southern California. In: Geological Survey Research for 1971. U.S. Geological Surv., Prof. Pap., 750-A, Ch. A, pp. 167—168.

Ziony, J.I., Wentworth, C.M., Buchanan-Banks, J.M. and Wagner, H.C., 1974. Preliminary map showing recency of faulting in coastal southern California. U.S. Geol. Surv. Misc. Field Studies Map MF-585.

Tectonophysics, 52 (1979) 407—408
© Elsevier Scientific Publishing Company, Amsterdam — Printed in The Netherlands

MARINE TERRACE DEFORMATION, SAN DIEGO COUNTY, CALIFORNIA

P.A. McCRORY and K.R. LAJOIE

U.S. Geological Survey, Menlo Park, Calif. 94025 (U.S.A.)

(Accepted for publication May 22, 1978)

ABSTRACT

The NW—SE trending southern California coastline between the Palos Verdes Peninsula and San Diego roughly parallels the southern part and off-shore extension of the dominantly right-lateral, strike-slip, Newport—Inglewood fault zone. Emergent marine terraces between Newport Bay and San Diego record general uplift and gentle warping on the northeast side of the fault zone throughout Pleistocene time. Marine terraces on Soledad Mt. and Point Loma record local differential uplift (maximum 0.17 m/ka) during middle to late Pleistocene time on the southwest side of the fault (Rose Canyon fault) near San Diego.

The broad Linda Vista Mesa (elev. 70—120 m) in the central part of coastal San Diego County, previously thought to be a single, relatively unde-formed marine terrace of Plio—Pleistocene age, is a series of marine terraces and associated beach ridges most likely formed during sea-level highstands throughout Pleistocene time. The elevations of the terraces in this sequence gradually increase northwestward to the vicinity of San Onofre, indicating minor differential uplift along the central and northern San Diego coast during Pleistocene time. The highest, oldest terraces in the sequence are obliterated by erosional dissection to the northwest where uplift is greatest.

Broad, closely spaced (vertically) terraces with extensive beach ridges were the dominant Pleistocene coastal landforms in central San Diego County where the coastal slope is less than 1% and uplift is lowest. The beach ridges die out to the northwest as the broad low terraces grade laterally into nar-rower, higher, and more widely spaced (vertically) terraces on the high bluffs above San Onofre where the coastal slope is 20—30% and uplift is greatest. At San Onofre the terraces slope progressively more steeply toward the ocean with increasing elevation, indicating continuous southwest tilt accompanying uplift from middle to late Pleistocene time. This southwest tilt is also recorded in the asymmetrical valleys of major local streams where strath terraces occur only on the northeast side of NW—SE-trending valley segments.

The deformational pattern (progressively greater uplift to the northwest with slight southwest tilt) recorded in the marine and strath terraces of central and northern coastal San Diego County conforms well with the historic pattern derived by others from geodetic data. It is not known how much of the Santa Ana structural block (between the Newport—Inglewood and the Elsinore fault zones) is affected by this deformational pattern.

Tectonophysics, 52 (1979) 409—410

TERTIARY AND HOLOCENE DEVELOPMENT OF THE SOUTHERN SIERRA NEVADA AND COSO RANGE, CALIFORNIA

P. St.-AMAND and G.R. ROQUEMORE

Naval Weapons Center, China Lake, Calif. 93555 (U.S.A.)
University of Nevada, Reno, Nev. 89507 (U.S.A.)

(Accepted for publication May 22, 1978)

ABSTRACT

Eroded remnants of lacustrian sediments, in part pyroclastic units, and containing Middle Pleistocene diatoms, are scattered on an erosion surface on parts of the Coso Range. The surface is, in part, covered by basaltic lavas (K/Ar age 2—3 m.y.) that overlie some sediments (K/Ar age 2.5—3 m.y.) and underlie others. The whole is faulted with some scarps in modern alluvium. The Plio-Pleistocene Coso Formation surrounds the range from Haiwee on the west to Darwin Wash on the north and northeast. Similar deposits are exposed in Panamint Valley, on Argus Range and near Searles Pass. Near Haiwee, the Coso Formation has been displaced vertically more than 3000 and possibly as much as 5000 ft. The White Hills Formation of Blancan age is found well above the floor of Indian Wells Valley. Lacustrian sediments underlie the valley. On the west flank of the Inyo Mts. the Plio—Pleistocene Waucobi beds appear to have been deposited over the present site of the Inyos. From the above, and from field, areal and photographic work, it can be inferred that:

(1) An erosion surface, probably the Chagoopa, extended from the vicinity of Mt. Whitney, as far east as the Panamints and east of the Inyos, at the beginning of the Pleistocene.

(2) A lake, containing some islands, extended from at least Bishop on the north to Cantil on the south, and from at least the present front of the Sierra to the Panamints during parts of the Pliocene and Early to Middle Pleistocene. It may be the offset equivalent of the Lava Mt. Lake.

(3) Uplift of the southern Sierra and the development of Basin and Range topography in this area took place mainly since the eruption of the basalts, and probably since the Middle Pleistocene.

(4) Long, linear, N—S-trending, probably right-slip faults pre-existed the uplift and continued south into the Mojave block.

(5) These faults were cut off by movement of the Garlock fault.

(6) Land north of the Garlock was stretched in an E—W direction from

the present Sierra front to beyond Death Valley. To the west of the Sierra it was shoved westward, bending the San Andreas fault.

(7) Separation along some of the N—S and E—W faults during the extension led to outpouring of the lava. Basalt filled the voids left by the stretching, transferring heat to the country rock and remelting it in part, giving rise to the andesites and rhyolites found in the area.

(8) Normal faulting developed, as extension continued, and differential uplift began.

(9) Uplift along the Sierran front first occurred by vertical displacement along the N—S faults and a set of N—W-trending left-slip faults. Segments of the frontal faults are found on the upper block, showing no recent movement there, but upon transition to the foot of the range, they become the active frontal faults with movement continuing to this day. Large landslides developed in the crushed rock at the intersections of the fault systems.

(10) The E—W spreading probably began in the Miocene in Washington and Oregon and moved southward, displacing the mass of Sierra westward by the Middle Pleistocene. The activity is still continuing.

Tectonophysics, 52 (1979) 411—415
411
© Elsevier Scientific Publishing Company, Amsterdam — Printed in The Netherlands

HOLOCENE OFFSET AND SEISMICITY ALONG THE PANAMINT VALLEY FAULT ZONE, WESTERN BASIN-AND-RANGE PROVINCE, CALIFORNIA

ROGER S.U. SMITH

Geology Department, University of Houston, Houston, Tex. 77004 (U.S.A.)

(Accepted for publication May 22, 1978)

ABSTRACT

Smith, R.S.U., 1979. Holocene offset and seismicity along the Panamint Valley fault zone, western Basin-and-Range province, California. In: C.A. Whitten, R.Green and B.K. Meade (Editors), Recent Crustal Movements, 1977. Tectonophysics, 52: 411—415.

Holocene right-slip along the central segment of the Panamint Valley totals 20 m and dip-slip is somewhat less. The most recent offset, about 2 m right-slip, probably occurred at least several hundred years ago. If a comparable amount of slip occurred during earlier earthquakes, mean seismic recurrence intervals would have been about 700—2500 years during the Holocene.

INTRODUCTION

The Panamint Valley fault zone, a major structural feature in the western Basin-and-Range structural province of California, extends 100 km northwards from the left-lateral Garlock fault along the steep western escarpment of the Panamint Range. Noble (1926) described and named the fault zone, which he felt belonged to a class of faults distinguished from those of the San Andreas type by the great amount of topographic relief across them. This suggested that most of the fault offset had been dip-slip. Subsequent workers (e.g. Hopper, 1947; Maxson, 1950) recognized a major right-slip component in the fault zone's Quaternary displacement. This paper describes Holocene displacement and seismicity along a 20-km segment where recent right-slip is several times greater than the dip-slip.

Evidence for Holocene faulting is seen best where faults transect Holocene fan segments. Along the segment of the Panamint Valley fault zone between Ballarat and Goler Wash (Fig. 1), Holocene alluvial fans envelop remnants of older surfaces into which were cut shorelines of the last shallow pluvial lake to occupy Panamint Valley. The age of these shorelines is probably latest Pleistocene (10,000—20,000 years B.P.), as judged by the degree of shoreline

412

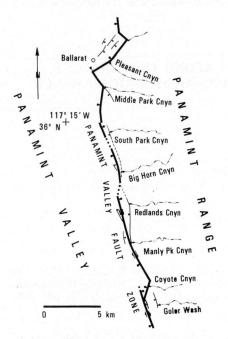

Fig. 1. Index map of the central segment of the Panamint Valley fault zone.

degradation, dissection and deformation relative to older, better-dated shore-lines of much deeper lake stands (Smith, 1975). Deposits of the shallow lake stand have yielded no material suitable for radiocarbon dating.

FAULT DISPLACEMENT

This segment of the fault zone, 1—3 km wide, is marked by discontinuous ground traces and by NW-facing range-front embayments (Fig. 1). Where parallel fault traces occur, right-slip seems largely confined to the western-most trace and dip-slip characterizes all others, most of whose scarps face west. Some scarps along the westernmost trace face west and others face east.

Right-slip

Scarps attributed to right-slip are commonly lower and less distinct than those attributed to dip-slip. In some places, scarps are absent and the fault trace follows a shallow furrow or an indistinct line marked only by channel offsets and contrast in the textural character of materials on either side of the fault trace. Even the freshest fault trace cannot be followed continuously across the rough, bouldery surface of some young alluvial fan deposits.

Most observed right-slip falls into the range 2.0 ± 0.6 m, and the minimum seen is 0.9 and 1.4 m on opposite banks of the same channel south of Goler Wash. Although smaller offsets may have escaped notice, this 2-m figure probably characterizes the latest event of ground breakage along this segment. Larger offsets seem to have a common denominator of about 2 m, although the 1.2 m uncertainty in this figure precludes extrapolation of the common denominator to offsets of more than 6 m. The maximum observed right-slip is 20.0 m on mudflow levees at the mouth of Manly Peak Canyon.

Dip slip

Abundant fresh scarps attributed to dip-slip are 0.6—1.8 m tall and have maximum slopes of 27—31°, although tiny patches of vertical slope are preserved locally along some scarps, mostly under boulders (Fig. 2). In general, taller scarps are steeper than lower ones, probably because more debris must be moved to degrade a tall scarp than to degrade a low scarp by the same

Fig. 2. Two-meter fault scarp at the mouth of Manly Peak Canyon. Note the relict groundline band of desert varnish which marks the position of an earlier, gentler scarp profile. Note also the nearly vertical surface beneath the granitic boulder; the rock hammer resting against this surface gives scale.

amount. Many of these scarps do not lie along the main trace of the fault zone and probably represent single events of ground breakage subsidiary to, or incidental to, movement along the main trace.

Scarps between 2 and 6 m tall probably formed by repeated offset along the same line. Some granitic boulders on these scarps are marked by an inclined band which appears to represent the groundline band between the above-surface (back) and below-surface (orange) facies of desert varnish (Fig. 2). This suggests that the boulders lingered on scarps gentler than modern ones long enough for desert varnish to form, and that subsequent offset has heightened and steepened the scarps, whose retreat by erosion has exposed the older groundline bands. These 2—6 m scarps commonly retain larger patches of vertical slope than do lower ones. At the mouth of Coyote Canyon, some granitic boulders protruding from the upper part of a 3-m scarp display a nearly vertical relict groundline band which has been exposed by parallel scarp retreat. This relation indicates that some nearly vertical surfaces can persist long enough in the same position for desert varnish to form. Precipitous scarps taller than several meters may persist from one episode of offset to the next, to be renewed and heightened by each succeeding episode of offset.

SEISMICITY

The age of the youngest event of offset is not known. No offset is recognized in fan deposits adjacent to modern channels. At the mouth of Goler Wash, a well-defined trail of unknown age shows probable right-slip of 1.8 m, suggesting possible historic offset. However, three nearby trails show no convincing evidence for offset where they cross the fault trace, nor does an adjacent road shown on the Searles Lake one degree topographic sheet surveyed during 1911 to 1913. These relations and the absence of historical reports of any damage to buildings at Ballarat, some made of adobe, suggest that the earthquake of November 4, 1908 in the Death Valley region did not occur along this segment of the Panamint Valley fault zone. Richter (1958, p. 469) assigned a queried location in the southern Panamint Range and a queried magnitude of $6\frac{1}{2}$ to this earthquake.

Scarp morphology suggests that dip-slip has not occurred on most scarps for hundreds of years or more. This estimate is based on the paucity of retained surfaces steeper than the angle of repose. Wallace (1977, pp. 1272—1273 inferred that some patches steeper than the angle of repose ("free faces") would linger for at least 300 years, perhaps for 1000—2000 years. He based this on a study of the degradation of scarps, 3—6 m tall, produced during the 1915 and 1954 earthquakes in Nevada. These findings can be applied only loosely to the lower scarps of Panamint Valley where the climate is more arid.

The mean recurrence interval between earthquakes along this segment of the fault zone can be estimated roughly by comparing the magnitude of the youngest offset with the magnitude of the greatest observed Holocene offset. If the latest and all earlier events each produced right-slip of 1.4—2.6 m, the 20-m total offset represents 8—14 events during the interval since the oldest deposits buried the 10,000—20,000-year shoreline of pluvial Lake Panamint. These relations suggest that the mean recurrence interval between earthquakes along this segment is on the order of 700—2500 years. Recurrence intervals along the entire 100-km length of the fault zone may be shorter if the ground did not break along the zone's entire length during each earthquake.

Evidence from scarp morphology and from relict groundline bands of desert varnish is imprecise, but suggests recurrence intervals of the same order of magnitude. Once some scarps formed, no subsequent offset occurred along them until they were degraded and desert varnish formed on their surficial boulders. Desert varnish may take as little as 25 years to develop under ideal conditions, but probably takes hundreds to thousands of years in most cases (Blackwelder, 1948; Engel and Sharp, 1958; Hunt, 1961). This interval, added to the hundreds to thousands of years probably required to degrade scarps, suggests that dip-slip along these scarps has recurred only every 1000 years or more. Recurrence intervals along tall, steep scarps may be somewhat shorter.

ACKNOWLEDGEMENTS

This research was supported by Grant No. 14-08-0001-G-368 from the Office of Earthquake Studies of the U.S. Geological Survey. Field work was facilitated by large-scale air photos lent by D.B. Slemmons.

REFERENCES

Blackwelder, Eliot, 1948. Historical significance of desert lacquer (abs). Geol. Soc. Am. Bull., 59: 1367.
Engel, C.G. and Sharp, R.P., 1958. Chemical data on desert varnish. Geol. Soc. Am. Bull., 69: 487—518.
Hopper, R.H., 1947. Geologic section from the Sierra Nevada to Death Valley, California. Geol. Soc. Am. Bull., 58: 393—423.
Hunt, C.B., 1961. Stratigraphy of desert varnish. U.S. Geol. Surv., Prof. Pap. 424-B, 194—195.
Maxson, J.H., 1950. Physiographic features of the Panamint Range, California. Geol. Soc. Am. Bull., 61: 99—114.
Noble, L.F., 1926. The San Andreas Rift and some other active faults in the desert region of southern California. Carnegie Inst. Wash. Yearb., 25: 415—428.
Richter, C.F., 1958. Elementary Seismology. W.H. Freeman, San Francisco, Calif., 768 pp.
Smith, R.S.U., 1975. Late-Quaternary Pluvial and Tectonic History of Panamint Valley, Inyo and San Bernardino Counties, California. Calif. Inst. Technol., unpubl. Ph.D. diss., 295 pp.
Wallace, R.E., 1977. Profiles and ages of young fault scarps, north-central Nevada. Geol. Soc. Am. Bull., 88: 1267—1281.

Tectonophysics, 52 (1979) 417—430
© Elsevier Scientific Publishing Company, Amsterdam — Printed in The Netherlands

TWO AREAS OF PROBABLE HOLOCENE DEFORMATION IN SOUTHWESTERN UTAH

R.E. ANDERSON and R.C. BUCKNAM

U.S. Geological Survey, MS 966, Box 25046, Denver Federal Center, Denver, Colo. 80225 (U.S.A.)

(Accepted for publication May 22, 1978)

ABSTRACT

Anderson, R.E. and Bucknam, R.C., 1979. Two areas of probable Holocene deformation in southwestern Utah. In: C.A. Whitten, R. Green and B.K. Meade (Editors), Recent Crustal Movements, 1977. Tectonophysics, 52: 417—430.

Recent geologic studies in southwestern Utah indicate two areas of probable Holocene ground deformation.

(1) A narrow arm of Lake Bonneville is known to have extended southward into Escalante Valley as far as Lund, Utah. Remnants of weakly developed shoreline features, which we have recently found, suggest that Lake Bonnevile covered an area of about 800 km² beyond its previously recognized limits near Lund. Shoreline elevations show a gradual increase from 1553 m near Lund to 1584 m at a point 50 km further southwest, representing a reversal of the pattern that would result from isostatic rebound. The conspicuously flat floor of Escalante Valley covers an additional 100 km² southward toward Enterprise, where its elevation is greater than 1610 m, but no shoreline features are recognizable; therefore, the former presence of the lake is only suspected. The measured 31-m rise over 50 km and the suspected 57-m rise in elevation over 70 km apparently occurred after Lake Bonnevile abandoned this area. The abandonment could have occurred as recently as 13,000 years ago, in which case the uplift is mainly of Holocene age. It probably has a deep-seated tectonic origin because it is situated above an inferred 9-km upwarp of the mantle that has been reported beneath the southern part of Escalante Valley on the basis of teleseismic *P*-wave residuals.

(2) Numerous closed topographic basins, ranging from a few hundred square meters to 1 km² in area, are found at various elevations along the west margin of the Colorado Plateau northeast of Cedar City. Geologic mapping in that area indicates that the basins are located over complex structural depressions in which the rocks are faulted and folded. Several of the depressions are perched along the walls of the West Fork of Braffits Creek, one of a few north-draining creeks that have incised deeply into the plateau margin. Extremely active modern erosion by the creek has produced a 6-km-long gorge along which excellent exposures provide good evidence that the topographic depressions, as well as the entire valley, are located over a north-trending structural graben in which rocks of Cretaceous, Tertiary, and Quaternary age are complexly deformed. The trough appears to be actively subsiding, as evidenced by inward-dipping youthful scarps and V-shaped trenches found along both walls of the valley. The scarp on the east side is continuous for 1.5 km, and that on the west is discontinuous for the same distance. Charcoal-bearing alluvium from a sequence of faulted sedimentary debris in the inner gorge has yielded discordant

dates by the ^{14}C technique, but the dates suggest that at least 6 m of fault displacement occurred during the Late Holocene.

INTRODUCTION

Topical studies conducted as part of a project designed to map faults of Late Quaternary age in southwestern Utah have led to the discovery of two areas of probable Holocene deformation, one a broad area of uplift in the southwestern Escalante Valley and the other a small area of complex fault-ing northeast of Cedar City (Fig. 1). In an earlier abstract (Anderson and Bucknam, 1977), we reported on *three* areas of Holocene deformation in southwestern Utah. Subsequent study indicates that the evidence cited sug-gesting Holocene uplift in the bed of Lake Bonneville near Delta, Utah (*3* in Fig. 1) is not valid. In particular, the reported deflection of topographic con-tours is probably related to sedimentation at the west margin of the Sevier River delta in ancient Lake Bonneville and not to uplift as was assumed; and secondly, the dips recognized in postfault lake beds probably resulted from prograde sedimentation at the delta margin and not to upwarping as was assumed. For these reasons, the title and text of this report are different from those of Anderson and Bucknam (1977).

Studies of the Late Cenozoic structural history of southwestern Utah are

Fig. 1. Index map showing the location of the southern Escalante Valley area *1*, the Braf-fits Creek area *2*, and the Delta area *3*, relative to some of the major tectonic boundaries of the intermountain west (heavy lines), the generalized areas where igneous rocks of Cenozoic age are found (stipple), and the intermountain seismic belt (*ISB*) and southern Nevada seismic belt (*SNB*) (shown by cross-ruled patterns). Illustration adapted from Smith and Sbar (1974, Figs. 2 and 3).

continuing, and it is hoped that they will be augmented by geodetic and geo-physical studies designed to evaluate the nature and extent of any current or historic deformation or seismicity in areas where Holocene deformation is indicated.

HOLOCENE UPLIFT INDICATED IN ESCALANTE VALLEY

The southernmost arm of ancient Lake Bonneville projected southward and southwestward into Escalante Valley. Shoreline features were not well developed along much of this arm, possibly because of generally shallow water depths and prevailing wind directions parallel to the arm. Also, prevailing winds from the southwest have caused vigorous eolian erosion and deposition that tend to obliterate features of historic age and must have been a major factor in obliterating or subduing prehistoric features. As a result, since the time of Gilbert's (1890) classical study, there has been more uncertainty about the exact location of the southernmost lake boundary in Escalante Valley than about the location of the boundary elsewhere in the Bonneville basin.

Shoreline features thought to have formed during a high stand of Lake Bonneville were identified in the southern part of Escalante Valley as far south as Enterprise (Fig. 2) by G.K. Gilbert's coworkers (Gilbert, 1890). Though Gilbert (1890, p. 366) reported the identification and compiled his map of the extent of Lake Bonneville accordingly, he clearly had some doubt as to whether shorelines existed in that area and as to whether the elevations measured on them were valid, as has been pointed out by Dennis (1944) and Crittenden (1963). The features reported by Gilbert, though difficult to locate precisely, were thought by Dennis (1944) to be either alluvial terraces or pediments.

Dennis (1944) reported that after a diligent search of the southern Escalante Valley for various types of shoreline features within the elevation range where they had been reported by Gilbert (1890), he found none. He was able to follow shorelines southward to a point several kilometers south of Nada (Fig. 2), whereas Gilbert had reported them 70 km further to the south-southwest in the Beryl—Enterprise area of the valley. Fix, et al. (1950) endorsed the more limited extent as determined by Dennis. Crittenden (1963), though agreeing with Dennis that the features for which elevations were reported by Gilbert are not shorelines, showed the southernmost extent of Lake Bonneville to be about 20 km southwest of Nada near Lund (Fig. 2), where he assigned an elevation of 1552 m to the highest stand of the ancient lake. Our studies reaffirm Dennis' conclusion that the features on which Gilbert's assistants measured elevations were not formed by Lake Bonneville. Some are fault scarps and some are stream terraces, and all are well above the highest lake stand as we identify it. In the following paragraphs, we present evidence suggesting that Lake Bonneville extended far into the Beryl—Enterprise area of Escalante Valley, but probably not as far as Gilbert (1890) sug-

420

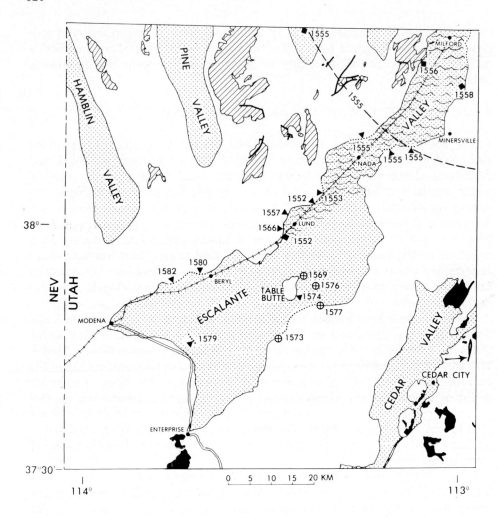

Fig. 2. Map of Escalante Valley and environs showing the location of the major valleys (stippled) relative to outcrops of Paleozoic rocks (rules) and Quaternary basalt (black); the unpatterned areas are underlain mostly by rocks of Mesozoic and Tertiary age. Heavy solid lines are faults; heavy dashed line is a portion of the southernmost isobase showing elevation (1555 m) of maximum high stand of ancient Lake Bonneville (from Crittenden, 1963); dotted lines mark locations where continuous or quasi-continuous shoreline features can be seen on aerial photographs; hachured line is Union Pacific Railroad. Solid squares mark localities at which Crittenden (1963) measured maximum elevations of Bonneville shorelines; solid triangles mark localities where we determined shoreline elevations, using an altimeter; open circles with cross where we determined the elevations from $7\frac{1}{2}$-min topographic maps; all elevations are in meters. Wave pattern indicates extent of ancient Lake Bonneville as reported by Crittenden (1963). Heavy arrow points to faults along the West Fork of Braffits Creek where we recognize evidence of Holocene displacement.

gested. The position of the southernmost edge of the lake is still not known.

Study of black-and-white aerial photographs (approx. scale 1 : 60,000) reveals a systematic southwesterly decrease in the degree of development of shoreline features in Escalante Valley. Near Milford and to the north of Milford, the highest shoreline is marked by wave-cut embankments that are formed on unsorted fanglomerate and by beach ridges and deltas of various shapes that are formed from alluvium. Though the features are conspicuous, they have been locally subdued or obliterated by erosion. At the approximate latitude of Nada (Fig. 2), the highest shoreline is represented by sparse remnants of low beach ridges. Further to the southwest, where the extent of Lake Bonneville is controversial, the most conspicuous indicators of the highest shoreline show up in the aerial photographs as contrasts in gray tone across discontinuous curved lines that approximately follow topographic contours. In general, the zones higher than the discontinuous curved lines exhibit an orderly drainage pattern typical of gentle fan surfaces in the semi-arid part of the southwestern United States, whereas drainages in the zones that are lower than the curved lines tend to be disorderly and highly modified by eolian features. Subdued sinuous to sublinear ridges are seen on the low side of the curved lines in some areas, especially in the area near Table Butte (Fig. 2). At Table Butte, they are composed of a capping about 1 m thick of calcareous sand with sparse pebbles, underlain by pebble-free calcareous sand interstratified with layers of laminated clay and cleanly washed, well-sorted, medium-grained sand. We interpret these deposits to be of lacustrine origin and the ridges to be looped and intersecting offshore bars. The deposits are found at an elevation of 1564 m on the west side of Table Butte and at 1572 m on the east side. Their stratigraphic position and age are not known, but they are inferred to correlate with the Alpine formation of Hunt (1953).

The constructional features between Milford and Table Butte that mark the highest shoreline (bars, ridges, deltas) are composed of reworked alluvial sand and gravel that contain little, if any, record of wave-formed rounding produced during the reworking. This absence of significant wave-formed rounding is consistent with the observation that shoreline features were not well developed in Escalante Valley.

Elevations were measured with an altimeter at selected sites on the crests of known and suspected bars and at the base of known and suspected wave-cut embankments in the area between Minersville and Enterprise, and additional elevations for similar features identified on aerial photographs of that area were obtained from recent 1 : 24,000 topographic maps (Fig. 2). The altimeter measurements are subject to errors of about ±1 m, and those taken from topographic maps to errors of about ±3 m.

Included on Fig. 2 is the southernmost isobase of Bonneville shoreline taken from Crittenden (1963). The position of the isobase is in excellent agreement with our measurements. The elevations that we measured with an altimeter in the vicinity of Lund are not in good agreement with the one that

Crittenden measured from a different feature using a topographic map. The difference of about 14 m probably arises from the fact that our measurements were taken on higher shoreline features west of that measured by Crittenden. In any case, our altimeter elevations suggest a southerly and southwesterly increase in elevation of the highest recognizable shoreline features. The rise amounts to about 30 m from the lowest measured point located about 13 km northeast of Lund to the highest measured point located 15 km northeast of Modena. This increase is opposite in sense to that produced by isostatic rebound related to the disappearance of the waters of ancient Lake Bonneville. We interpret these data as indicating an uplift of as much as 30 m in a distance of about 40 km. The indicated uplift may have affected an area of at least 800 km^2.

The poorly developed highest recognizable shoreline features in the Beryl—Enterprise area merge northeastward with better developed features undisputedly produced during the highest lake stand in the Milford area; both the extent of development and the elevations of the features appear gradational between the two areas. The age of the highest lake stand has not been firmly established, but it is probably about 13,000 years (Morrison and Frye, 1965). The lake probably did not stand at that level for a long time, which is consistent with the weak development of shoreline features and the absence of significant rounding in wave-reworked material in Escalante Valley. In any case, we infer that the indicated uplift of 30 m took place in the last 13,000 years and is, therefore, at least in part, Holocene in age.

In the area near Beryl and Enterprise that we believe was inundated by Lake Bonneville, the floor of Escalante Valley is conspicuously flat, suggestive of an ancient lake bed. There is only 20 m of relief in the southernmost 400 km^2 of the area. This conspicuously flat valley floor extends almost to Enterprise, and if it records the former presence of the lake that far south, the indicated total uplift is about 57 m in about 70 km. No shoreline features are recognizable in the vicinity of Enterprise so the former presence of the lake there is only suspected. The boundary of Escalante Valley between Modena and Enterprise in Fig. 2 does not correspond to the shore of Lake Bonneville, whereas elsewhere in the figure there is an approximate correspondence.

Whether or not an area of active uplift exists in the Beryl—Enterprise portion of Escalante Valley could be evaluated if a first-order level line that was established along the Union Pacific Railroad in 1908 (Fig. 2) were resurveyed. The line crosses the uplift that we infer from elevations of ancient shoreline features, and it has not been resurveyed since it was established. A systematic search for the 1908 monuments revealed that they remain at only eight localities along the 75 km of railroad northeast of Modena.

Figure 3 shows contours of teleseismic P-wave residuals indicating an anomalous area of high velocity that Smith et al. (1975) have interpreted as a 9-km upwarp in the mantle. The seismic anomaly is centered on the Beryl—Enterprise area where we infer Holocene uplift. This spatial coincidence of

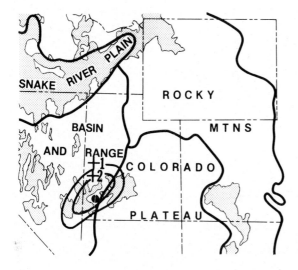

Fig. 3. Map showing contours in seconds of teleseismic *P*-wave residuals indicating an anomalous area of seismic velocity that Smith et al. (1975) have interpreted as a feature produced by a 9-km upwarp in the mantle. The dot inside the +2-sec contour marks the center of the geophysical anomaly as well as the location of the area of probable Holocene uplift in Escalante Valley.

the geophysical and geodetic anomalies suggests that the two may be genetically related.

HOLOCENE FAULTING INDICATED NORTHEAST OF CEDAR CITY

The west margin of the Colorado Plateaus northeast of Cedar City (Fig. 1) is sharply defined by an abrupt NW-facing mountain front that overlooks Cedar and Parowan Valleys of the adjacent Basin-and-Range Province. The mountain front is the western margin of the Markagunt Plateau. Threet (1952, 1963) has described this mountain front as a NE-trending monoclinal flexure that has been fragmented by N- and NE-trending normal faults that form a series of horst and graben structures as depicted in Fig. 4. We have found evidence that indicates Holocene displacement on faults of the horst-and-graben system along the West Fork of Braffits Creek, one of several structurally controlled north-flowing creeks that drain the western Markagunt Plateau. Detailed geologic mapping of the Braffits Creek area is not yet complete so only a preliminary account of the geology can be given.

The West Fork of Braffits Creek flows in a narrow graben produced by faults that are exposed on the valley walls to the east and west of the creek (see Fig. 9B). For convenience, they are referred to here as the east and west faults of the Braffits Creek graben. These faults are shown on the geologic map of Utah (Hintze, 1963) and on Fig. 2 of this report. The graben terminates to the north against a cross-fault that is exposed in Braffits Creek. For

424

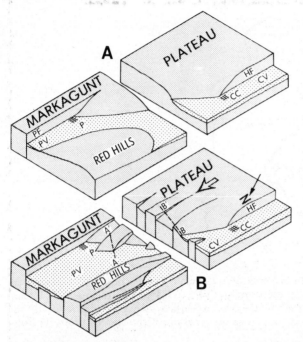

Fig. 4. Highly simplified and diagrammatic block sketches of the edge of the Markagunt Plateau between Cedar City (*CC*) and Parowan (*P*) as viewed in a southeasterly direction. Here, *PV* is the Parowan Valley; *CV* is the Cedar Valley (from Threet, 1952). A. Plateau margin depicted as a NE-trending monoclinal flexure that serves as a structural bridge between the Paragonah fault (*PF*) and Hurricane fault (*HF*). B. Monocline is fragmented by northerly-trending horst and graben structures. The two diagrams are not intended to represent the sequential development of the plateau margin; flexing and faulting probably developed simultaneously. The large arrow indicates the direction of view shown in Fig. 8, and the dashed lines mark the approximate trace of the diagrammatic cross-sections shown in Fig. 9.

distances of more than 1.5 km along the east and west faults, youthful displacement is evidenced by: (1) raw unvegetated scarps a few centimeters to several meters high with slope angles at or greater than the angle of repose (Fig. 5); (2) V-shaped collapse trenches as much as 4 m deep, some with little or no sediment fill, located at the base of and parallel to the scarps (Fig. 6); and (3) toppled dead trees and toppled or disrupted live trees along the scarps (Fig. 7). Numerous closed basins ranging in size from a few hundred square meters to 1 km² are found in the Braffits Creek area and provide additional evidence of youthful displacements. Some of these basins are perched on the flanks of Braffits Creek (Fig. 8).

A recently eroded inner gorge of Braffits Creek provides excellent exposures, several of which show that some of the closed basins and other basins that have been breached by the creek are controlled by faults with steep to gentle dips. In the inner gorge near one of these closed basins two very dif-

Fig. 5. View northward along western fault of Braffits Creek graben showing free face of vertical scarp formed on very weakly cemented monolithologic debris composed of angular clasts of silicic ash-flow tuff. Highest part of scarp is approximately 4 m. The playa of Little Salt Lake in Parowan Valley is visible in upper right.

Fig. 6. View northward along eastern fault of Braffits Creek graben showing trench at base of scarp. Steepness of trench is masked by vegetation, but is suggested by upper part of 4-m surveying rod that is being held by a man standing out of sight in bottom of trench. Unvegetated scarp is visible in upper right and the playa of Little Salt Lake in Parowan Valley is visible in middle distance.

426

Fig. 7. View northwestward across trench at base of scarp along western fault at Braffits Creek graben showing a toppled dead tree (most distant), a toppled life tree (center), and a live tree that is tilted and has had its trunk split by displacement of its root system which is apparently attached to both fault blocks. Young, undisturbed trees as high as 2 m grow along the fault trace nearby.

Fig. 8. View to the northeast across the West Fork of Braffits Creek. Bare area in the lower left is one of several closed basins that are found along this drainage. The light-colored rocky cliffs across the creek are composed of Cretaceous sedimentary rocks that dip steeply as part of the NW-facing monocline that forms the margin of the Markagunt Plateau. The dark rocks in the middle ground are basalt lavas and cinder beds dated by the K—Ar method at about 1 m.y. They rest paraconformably on Cretaceous rocks as shown in Fig. 9.

TABLE I

Age data from the Braffits Creek area

Sample no.	Lab. no.	Analyst	Technique	Location	Description	Age (yrs B.P.)
C346-67 [a]	W-3784	MR [e]	[14]C	37°44'32"N, 112°58'05"W	organic sediment	1240 ± 200
C346-67B [a]	W-3781	MR	[14]C	37°44'32"N, 112°58'05"W	organic sediment	810 ± 200
C346-33 [b]	DKA3479	HM [f]	K–Ar	37°45'35"N, 112°58'22"W	basalt lava	1.08 ± 0.1 m.y.
C1730-9 [c]	DKA3353	HM	K–Ar	37°46'24"N, 112°57'35"W	basalt dike	0.98 ± 0.16 m.y.
C1730-8A [d]	DKA3352	HM	K–Ar	37°46'12"N, 112°57'33"W	basalt lava	0.93 ± 0.14 m.y.

[a] Samples are of charcoal-bearing alluvium; sample 67 is about 2 m higher stratigraphically than 67B despite its older apparent age.

[b] Olivine basalt taken from eastern extreme of capping basalt depicted schematically in Fig. 9B.

[c] Olivine basalt taken from dike that cuts cinder beds at Cinder Hill as depicted schematically in Fig. 9A.

[d] Olivine basalt taken from lava beneath cinders of Cinder Hill as depicted schematically in Fig. 9A.

[e] MR, Meyer Rubin.

[f] HM, Harald Mehnert.

428

ferent sequences of strata are juxtaposed along a high-angle fault. Strata east of the fault are steeply tilted and sheared coarse monolithologic sediments, whereas strata west of the fault are undeformed flat-lying heterolithic alluvium, some beds of which contain concentrations of charcoal and humus. Charcoal samples taken from two intervals in the heterolithic alluvium yielded discordant ^{14}C dates (Table I) which suggest a Late Holocene age for the alluvium. These isotopic data tend to confirm that widespread faulting has occurred in the Braffits Creek area quite recently, as is suggested by geomorphic evidence. The magnitude of displacement is uncertain. If it is pure dip-slip the apparent offset is about 6 m.

Many of the physiographic features in the Braffits Creek area suggest that the terrain has been shaped by landsliding and slumping. However, our geologic mapping suggests that although landsliding and slumping have been common processes in the area in the past, the majority of modern landforms actually result from movements on faults and folds. In general, exposures in

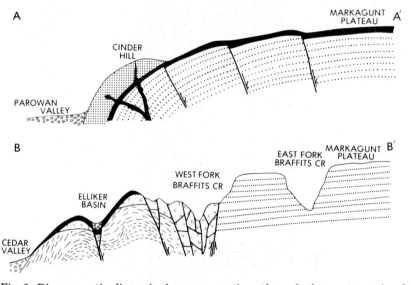

Fig. 9. Diagrammatic dimensionless cross-sections through the west margin of the Markagunt Plateau along lines approximated in Fig. 4. In A—A', the margin north of Braffits Creek is depicted as a slightly faulted monocline that affects basalt lava and overlying cinder beds (dot pattern), both of which we have dated at about 1 m.y. In B—B', the graben structure along the West Fork of Braffits creek is depicted as a mosaic of jumbled heterolithic blocks with relatively undeformed Cretaceous rocks to the east (short-dashed lines) and folded and faulted Tertiary and Quaternary rocks to the west (ruled patterns). The Quaternary rocks are basalts equivalent in age to those in A—A'. The Tertiary rocks include sedimentary strata of the upper part of the Claron formation and volcanic rocks of the Great Basin volcanic province as defined by Anderson et al. (1975). The Cretaceous rocks in both sections belong to the Tropic Shale and to the Straight Cliffs and Wahweap formations as defined by Averitt (1962). Quaternary alluvium in both sections shown by open circles.

the inner gorge display a complex assemblage of coarse to extremely coarse sedimentary breccia, landslide breccia, intact landslide blocks consisting of gently to steeply dipping strata as much as 100 m thick, and sparse clastic sedimentary strata, all of which are faulted and folded into a confusing to completely chaotic array. This highly complex zone is bounded sharply by the eastern fault, east of which are flat-lying to gently dipping upthrown Cretaceous sedimentary strata. The western fault forms the western boundary which is not so sharply defined because highly folded and faulted strata that include large volumes of coarse breccia and landslide material locally continue westward to the base of the mountain front. These relationships are depicted schematically on Fig. 9. Though we recognize the widespread occurrence of the products of landsliding, those products are displaced by the faults of the Braffits Creek graben and are therefore older than the graben. The age of the tectonic event that produced the accumulation of thick deposits of debris and landslide blocks is not known, but it is probably Late Tertiary.

Quaternary basalt dated by the K—Ar method at about 1 m.y. (Table I) is common in the vicinity of Braffits Creek. The basalt is monoclinally flexed, faulted, and folded along the mountain front (Figs. 8 and 9), indicating that at least 250 m of uplift is younger than 1 m.y. We interpret the graben structure as a feature related to tensional collapse along the rising and spreading mountain front. It is probably younger than 1 m.y., as evidenced by the presence of fault-bounded basalt blocks along the cross-fault against which the graben terminates to the north. The graben has apparently been active during the Late Holocene.

Present plans call for the installation of a series of bridged quadrilaterals for high-accuracy trilateration measurements along an E—W transect across the graben. Periodic measurements should provide data adequate to determine whether or not the graben is active and, if so, how that activity is related to uplift at the mountain front.

ACKNOWLEDGEMENTS

We are grateful to Meyer Rubin and Harald Mehnert for providing isotopic age data and to David Russ and William Scott for helpful technical reviews.

REFERENCES

Anderson, R.E. and Bucknam, R.L., 1977. Three areas of Holocene deformation in southwestern Utah, U.S.A. Int. Symp. Recent Crustal Movements, Palo Alto, Calif., July 25—30, 1977, Abstr., p. 3.
Anderson, J.J., Rowley, P.D., Fleck, R.J. and Nairn, A.E.M., 1975. Cenozoic geology of southwestern high plateaus of Utah: Geol. Soc. Am. Spec. Pap. 160: 88 pp.
Averitt, Paul, 1962. Geology and coal resources of the Cedar Mountain quadrangle, Iron County, Utah. U.S. Geol. Surv. Prof. Pap., 389: 72 pp.
Crittenden Jr., M.D., 1963. New data on the isostatic deformation of Lake Bonneville: U.S. Geol. Surv. Prof. Pap. 454-E: E1-31.

Dennis, P.E., 1944. Shorelines of the Escalante Bay of Lake Bonneville (abstr.): Utah Acad. Sci. Proc., 1941—43, 19—20: 121—124.

Fix, P.F., Nelson, W.B., Lofgren, B.E. and Butler, R.G., 1950. Ground water in the Escalante Valley, Beaver, Iron and Washington Counties, Utah. Utah State Eng. Tech. Publ., 6: 109—210.

Gilbert, G.K., 1890. Lake Bonneville. U.S. Geol. Surv. Mon., 1: 438 pp.

Hintze, L.F., compiler, 1963. Geologic map of southwestern Utah. In: Geology of southwestern Utah, 1963. Utah Geol. Mineral. Surv., Intermt. Assoc. Pet. Geol Guideb., 12th Annu. Field Conf., 232 pp.

Hunt, C.B., 1953. General geology, In: C.B. Hunt, H.D. Varnes and H.E. Thomas (Editors), Lake Bonneville — geology of northern Utah Valley, Utah. U.S. Geol. Surv. Prof. Pap., 257-A: 11—45.

Morrison, R.B. and Frye, J.C., 1965. Correlation of the middle and late Quaternary successions of Lake Lahontan, Lake Bonneville, Rocky Mountain (Wasatch Range), Southern Great Plains, and Eastern Midwest areas. Nevada Bur. Mines Rep. 9: 45 pp.

Smith, R.B. and Sbar, M.L., 1974. Contemporary tectonics and seismicity of the western United States with emphasis on the Intermountain seismic belt. Geol. Soc. Am. Bull., 85: 1205—1218.

Smith, R.B., Braile, L.W. and Keller, G.R., 1975. Upper crustal low-velocity layers: Possible effect of high temperatures over a mantle upwarp at the Basin Range—Colorado Plateau transition. Earth Planet. Sci. Lett., 28: 197—204.

Threet, R.L., 1952. Geology of the Red Hills area, Iron County, Utah. Ph. D. thesis, Univ. Wash., Seattle, 107 pp.

Threet, R.L., 1963. Structure of the Colorado Plateau margin near Cedar City, Utah, In: Geology of Southwestern Utah, 1963. Utah Geol. Mineral. Surv., Intermt. Assoc. Pet. Geol. Guideb., 12th Annu. Field Conf., pp. 104—117.

Tectonophysics, 52 (1979) 431—445

© Elsevier Scientific Publishing Company, Amsterdam — Printed in The Netherlands

QUATERNARY FAULTING ALONG THE CARIBBEAN—NORTH AMERICAN PLATE BOUNDARY IN CENTRAL AMERICA

DAVID P. SCHWARTZ [1], LLOYD S. CLUFF [1] and THOMAS W. DONNELLY [2]

[1] *Woodward-Clyde Consultants, Three Embarcadero Center, San Francisco, Calif. (U.S.A.)*

[2] *Department of Geological Sciences, State University of New York, Binghamton, N.Y. (U.S.A.)*

(Revised version accepted May 22, 1978)

ABSTRACT

Schwartz, D.P., Cluff, L.S. and Donnelly, T.W., 1979. Quaternary faulting along the Caribbean—North American plate boundary in Central America. In: C.A. Whitten, R. Green and B.K. Meade (Editors), Recent Crustal Movements, 1977. Tectonophysics, 52: 431—445.

Recent detailed mapping along the Motagua fault zone and reconnaissance along the Chixoy—Polochic and Jocotán—Chamelecón fault zones provide new information regarding the nature of Quaternary deformation along the Caribbean—North American plate boundary in Central America.

The southern boundary of the Motagua fault zone is defined by a major active left-slip fault that ruptured during the February 4, 1976 Guatemala earthquake. The recurrent nature of slip along the fault is dramatically demonstrated where stream terraces of the Río El Tambor show progressive left-slip and vertical (up-to-the-north) slip. Left-slip increases from 23.7 m (youngest mappable terrace) to 58.3 m (oldest mappable terrace) and vertical slip increases from 0.6 m to 2.5 m. The oldest mappable terrace crossed by the fault appears to be younger than 40,000 years and older than 10,000 years.

Reconnaissance along the Chixoy—Polochic fault zone between Chiantla and Lago de Izabal has located the traces of a previously unmapped major active left-slip fault. Geomorphic features along this fault are similar to those observed along the active trace of the Motagua fault zone. Consistent and significant features suggestive of left-slip have so far not been observed along the Guatemala section of the Jocotán—Chamelecón fault zone.

In Central America, the active Caribbean—North American plate boundary is comprised of the Motagua, Chixoy—Polochic, and probably the Jocotán—Chamelecón fault zones, with each accommodating part of the slip produced at the mid-Cayman spreading center. Similarities in geomorphic expression, apparent amount of left-slip, and frequency and magnitude of historical and instrumentally recorded earthquakes between the active traces of the Motagua and Chixoy—Polochic fault zones suggest a comparable degree of activity during Quaternary time; the sense and amount of Quaternary slip on the Jocotán—Chamelecón fault zone remain uncertain, although it appears to be an active earthquake source. Uplift of major mountain ranges on the north side of each fault zone reflects the small but consistent up-to-the-north vertical component (up to 5% of the lateral component) of slip along the plate boundary. Preliminary findings, based on offset stream terraces, indicate a late Quaternary slip rate along the Caribbean—North American plate boundary of between 0.45 and 1.8 cm/yr. Age dating of offset Quaternary terraces in Guatemala will allow refinement of this rate.

INTRODUCTION

The transcurrent fault zones that cross Guatemala are generally regarded as the landward extensions of the Cayman Trough (Hess and Maxwell, 1953; Donnelly et al., 1968) (see inset, Fig. 1). Infrequent, moderate, shallow focus earthquakes that range in magnitude from approximately 4.0 to 6.0 occur along the Cayman Trough. These earthquakes have focal plane solutions that indicate left-slip motion (Molnar and Sykes, 1969). Holcombe et al. (1973) show that earthquake epicenters located between the mid-Cayman spreading center and the Central American mainland are restricted to the south wall of the Cayman Trough. A bathymetric survey along the western end of the Cayman Trough (Banks and Richards, 1969) indicates that both the Motagua and Chixoy—Polochic fault zones represent the landward continuation of the southern escarpment (wall) of the trough (Fig. 1). The relationship between the Jocotán—Chamelecón fault zone and the Cayman Trough is less clear.

Detailed geologic mapping along the Motagua fault zone prior to (Schwartz, 1976a) and after the February 4, 1976 Guatemala earthquake (Richter magnitude 7.5), plus aerial and surface reconnaissance and photo-interpretation along the Chixoy—Polochic and Jocotán—Chamelecón fault zones provide new data on Quaternary faulting along these transcurrent fault zones. This paper presents the preliminary results of an ongoing investigation of these faults and discusses their implications regarding the distribution of strain, slip rates, and earthquake recurrence intervals along this portion of the Caribbean—North American plate boundary.

MOTAGUA FAULT ZONE

Previous work

The Motagua fault zone extends at least 300 km from Chichicastenango to the Caribbean coast (Fig. 1). This fault zone is a suture between small crustal plates (Schwartz and Newcomb, 1973) that formed as a result of plate convergence and closure of a small ocean basin in the Late Cretaceous (Lawrence, 1976; Donnelly, 1977). The time of the onset of subsequent strike-slip faulting along the Motagua fault zone is not well constrained but appears to be between Eocene and Miocene time (Schwartz, 1976a). Faulting on the north side of the fault zone is characterized by south-dipping thrust and high-angle reverse faults, and possible strike-slip faults, which separate generally continuous and linear belts of cataclastic gneiss, serpen-

433

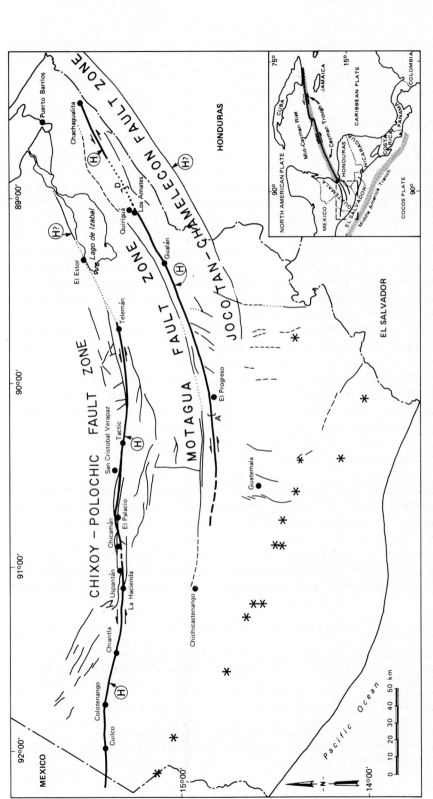

Fig. 1. Faults along plate boundary and localities referred to in text. Faults modified from Bonis et al. (1970). Heavy lines with H designate faults having known Holocene displacement; H is queried where Holocene displacement is inferred. Asterisks indicate volcanoes. A—D bracket area shown on Fig. 2.

434

tinite, and Tertiary continental clastic rocks (Schwartz, 1976b). Quaternary displacements have been looked for along the faults on the north side of the zone: however, none have been observed.

Quaternary faulting

The southern boundary of the Motagua fault zone is defined by an active left-slip fault, the Motagua fault *, which ruptured and produced the magnitude 7.5 earthquake of February 4, 1976. The Motagua fault extends at least 240 km along south of the Río Motagua. With the exception of a 30-km segment where it crosses water-saturated sand and silt of the active floodplain of the Río Motagua east of Los Amates, the fault exhibits continuous youthful geomorphic expression characterized by sag ponds, shutter ridges, scarps, springs, offset streams and river terraces, and a well-defined fault rift. Fig. 2 shows the locations of significant geomorphic features and Quaternary displacements along a 110-km segment of the Motagua fault between El Progreso and Los Amates.

Quaternary deformation along the Motagua fault occurs across a zone as wide as 30 m. Within this zone, recurrent slip has produced a distinct zone of intense shearing and gouge 3—5 m wide in a variety of rock types that include Paleozoic marble and schist, Tertiary sandstone, and Quaternary pumice. Secondary deformation within the 30-m-wide zone occurs north of the gouge zone and is characterized by minor warping and south-dipping normal faults having displacements measured in centimeters.

Lateral displacements

Pre-earthquake left-slip of Quaternary age has been measured at 20 localities along the fault between El Progreso and Los Amates. Measurements were made at 17 offset streams, 2 shutter ridges, and a sequence of offset stream terraces at the confluence of the Río El Tambor with the Río Motagua (Fig. 2). Left-lateral stream offsets range from 25 to 140 m, with offsets of 30—50 m the most commonly observed. These streams are developed on metamorphic and sedimentary rocks, as well as on Quaternary

* The fault that ruptured on February 4, 1976 has been referred to by two names. It was named the "Cabañas fault" by Eric Bosc on a preliminary geologic map of the San Augustín Acasaquastlán quadrangle (scale 1 : 50,000) submitted to the Guatemala Instituto Geográfico Nacional in 1965 and incorporated in the first geologic map of the republic (Bonis et al., 1970). The name "Cabañas fault" was subsequently used in the dissertations of Bosc (1971) and Schwartz (1976a). The first published reference using the name "Motagua fault" for the specific trace that ruptured in 1976 was a geologic map of the Chiquimula quadrangle (scale 1 : 250,000) published in 1969 by the Instituto Geográfico Nacional. Since the 1976 earthquake, several papers and popular usage have equated "Motagua fault" with the active trace in the Motagua fault zone. This usage is followed in this paper.

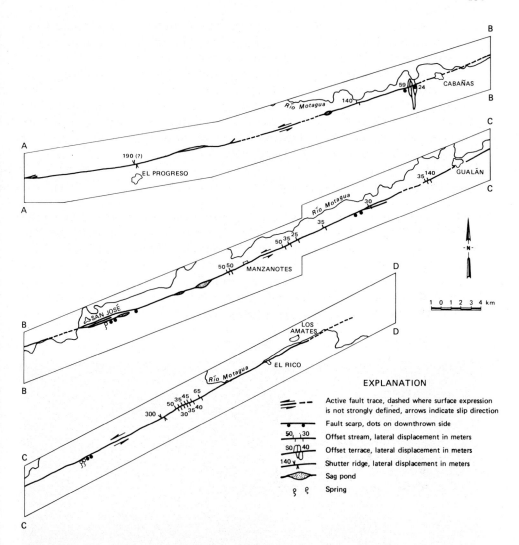

Fig. 2. Quaternary displacements and geomorphic characteristics along Motagua fault. Location of this segment of fault is between brackets A and D on Fig. 1.

alluvial and colluvial deposits. Offsets of 190 and 300 m are observed on shutter ridges just north of El Progreso and west of El Rico, respectively. Stream terraces along the Río El Tambor show progressive left-slip of 23.7—58.3 m.

Vertical displacements

Along most of its length the Motagua fault has little or no topographic relief; however, fault scarps can be observed at several localities (Fig. 2).

These scarps range from less than one to several meters in height and from several hundred meters to 2 km in length. Scarp morphology is variable, and both steep, well-defined scarps and broad breaks in slope are observed.

The scarps are invariably south-facing and indicate a minor up-to-the-north vertical component to the net slip. Gravity traverses across segments of the fault that have no surface relief show a down-to-the-south gravity step (Monges et al., 1976) that is consistent with an up-to-the-north vertical component of slip. A north-facing scarp on the eastern segment of the fault west of Chachagualita has been reported (Plafker, 1976).

Progressively offset stream terraces

The most significant locality for interpreting the history of slip along the Motagua fault is at the confluence of the Río El Tambor with the Río Motagua, where a sequence of stream terraces on the western bank of the Río El Tambor is displaced along the Motagua fault (Figs. 3 and 4). The terrace deposits are composed of poorly stratified, medium to coarse-grained gravel and minor interbedded sand derived from the metamorphic and volcanic terrane to the south. The terraces are unpaired and appear to have

Fig. 3. Aerial view of Río El Tambor terraces. Number indicates terrace level; dashed line is Motagua fault. View is south.

437

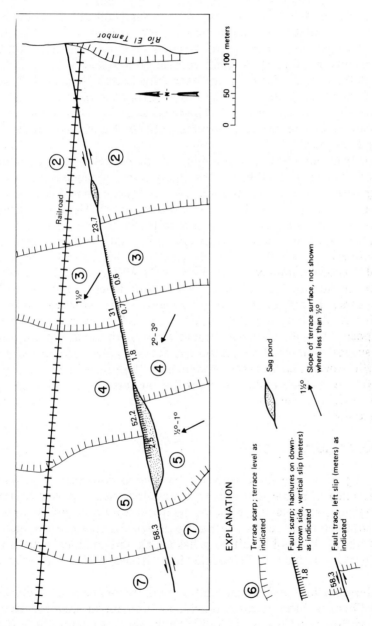

Fig. 4. Progressive displacement of Río El Tambor terraces along Motagua fault.

EXPLANATION

⑥ Terrace scarp; terrace level as indicated

Fault scarp; hachures on down-thrown side, vertical slip (meters) as indicated

58.3 Fault trace, left slip (meters) as indicated

Sag pond

1½ Slope of terrace surface, not shown where less than ½°

been formed by progressive downcutting of a large alluvial fan by the Río El Tambor.

There are at least seven distinct terrace levels along the Río El Tambor in the immediate area of the Motagua fault, and higher terraces occur south of the fault. The cumulative left-slip on the scarp of terrace 3 is 23.7 m, and the cumulative left-slip on the scarps of terraces 4, 5, and 7 is 31.0, 52.2, and scarp to the toe of the corresponding terrace scarp on the opposite side of the fault. The cumulative left-slip on the scarp of terrace 3 is 23.7 m and the cumulative left-slip on the scarps of terraces 4, 5, and 7 is 31.0, 52.2 and 58.3 m, respectively (Fig. 4). A small vertical component of slip is indicated by a well defined south-facing scarp extending across terraces 3, 4, and 5; scarp height increases gradually from 0.6 m across the fault on terrace 3 to 2.5 m across the fault on terrace 5. The terraces are also deformed by warping, and small streams on the terraces drain into the large sag pond developed on terraces 4 and 5.

There has been no systematic mapping of the complex alluvial, colluvial, and volcanic deposits that comprise the Quaternary sequence in the Motagua Valley, and little is known about their ages. However, some constraints can be placed on the age of the Río El Tambor terraces. Deposits of layered white pumice crop out just east of and south along the Río El Tambor. Reconnaissance mapping indicates this pumice is an eastward continuation of the widespread H ash-flow tuff mapped by Koch and McLean (1975). Bonis et al. (1966) obtained a ^{14}C date of 31,000 years B.P. on charcoal from the H ash-flow tuff. Additional ^{14}C dates by Koch and McLean (1975) of 40,000 years B.P. suggest that this is probably a minimum age for the unit. At the Río El Tambor, the coarse alluvial gravel unconformably overlies pumice deposits. If the ^{14}C dates accurately represent the age of the pumice, then they suggest a limiting maximum age for the gravel of approx. 40,000 years B.P.; however, the terraces eroded into the gravel may be considerably younger. A well-developed soil profile on terraces 5 and 7 and other geomorphic evidence suggest at least these upper terraces are older than 10,000 years B.P.

CHIXOY—POLOCHIC FAULT ZONE

The Chixoy—Polochic fault zone can be traced for more than 345 km across Guatemala as a sharply defined feature that truncates and juxtaposes folded Cretaceous sedimentary rock on the north against complex crystalline and sedimentary terrane to the south (Figs. 1 and 5). Like the Motagua fault zone, the Chixoy—Polochic fault zone has a long and complex structural history, possibly dating back to Paleozoic time (Kesler, 1971; Anderson et al., 1973).

Photointerpretation and aerial and ground reconnaissance between El Estor and Chiantla have located previously unmapped major active left-slip faults within the Chixoy—Polochic fault zone. At least three distinct

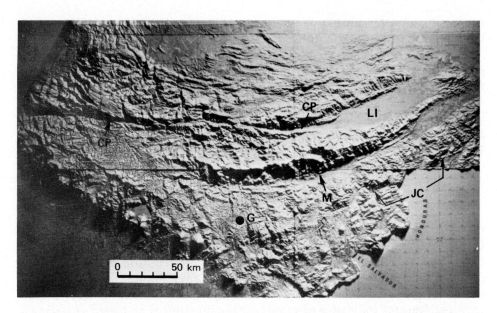

Fig. 5. Faults along plate boundary as seen on relief map of Guatemala; low-angle illumination from north. CP = Chixoy—Polochic fault zone; M = Motagua fault zone; JC = Jocotán—Chamelecón fault zone; LI = Lago de Izabal; G = Guatemala City. Note distinct up-to-the-north displacement on north side of each fault zone.

segments, each containing an active fault trace, define the zone between Lago de Izabal and the Mexican border (Fig. 1). Within the eastern segment, the active trace extends a minimum of 110 km from just east of Telemán to El Palacio. It strikes N80°E to N80°W, forming a gentle arc convex to the south. The fault is defined by an almost continuous narrow linear trough, sidehill ridges, left-laterally offset streams, springs, slumps, and small landslides. Quaternary stream offsets, which are especially well developed east of Tactic, range consistently from 35 to 100 m. Local south-facing scarps indicate a small up-to-the-north vertical component to the net slip. The active trace is poorly defined where the Polochic River Valley widens east of Telemán; however, apparent left-lateral stream offsets west of El Estor suggest that an active fault trace may continue along the north side of Lago de Izabal.

The active fault trace in the central segment extends approximately 110 km from El Palacio to Colotenango. It is also convex south, arcing from N75°E in the eastern portion to N80°W on the west. Youthful geomorphic expression of recurrent Quaternary slip along this trace of the fault is characterized by sag ponds, a narrow linear trough, and left-lateral stream offsets. These features are particularly well developed between La Hacienda and Uspantán. The fault can be observed in reworked Quaternary pumice west of Chicamán, and in Late Pleistocene or Holocene river deposits west of Uspantán. Kupfer and Godoy (1967) reported beheaded valleys, shutter ridges, chains of scarplets, and streams left-laterally displaced 65—130 m along this segment of the fault.

In the western segment, the active trace extends a minimum distance of 45 km from Colotenango into Mexico. In-progress photointerpretation and field mapping along this section of the fault (T.H. Anderson and R.J. Erdlac Jr., personal communication, 1977) show that a single active trace trends west from Colotenango, splays into at least three en-echelon traces oriented N80°E in the vicinity of Cuilco, and then continues as a single trace into Mexico.

JOCOTAN—CHAMELECON FAULT ZONE

The Jocotán—Chamelecón fault zone juxtaposes Paleozoic—Mesozoic metamorphic terrane on the north against Tertiary volcanic rock and Cretaceous limestone on the south. Mapping along various sections of the fault zone (Crane, 1965; Clemons, 1966) suggests up-to-the-north normal faulting. Quaternary faulting has not been reported. Photointerpretation and aerial reconnaissance have not revealed any consistently well-defined geomorphic features indicative of recent strike-slip faulting along the Guatemala section of the fault zone. However, the topographic expression of the fault zone is much stronger in Honduras. In addition, the Jocotán—Chamelecón fault zone appears to be a source of historical seismicity along the section in Honduras (Schwartz, 1976a). It is likely that future work along the Jocotán—Chamelecón fault zone will show evidence of active left-slip faulting.

HISTORICAL SEISMICITY

An earthquake catalog for Central America (Carr and Stoiber, 1977) lists 118 damaging earthquakes in Central America since 1528. Of these, Carr and Stoiber (1977) associated only four events with the transcurrent fault zones in Guatemala and Honduras. These four earthquakes occurred in Guatemala at Chinique in 1881, Puerto Barrios in 1929, and Quiriguá in 1945, and in Honduras at Omoa in 1856. The instrumental seismic record prior to the February 4, 1976 Guatemala earthquake shows at least 26 earthquakes of magnitude 5 or less associated with the Motagua, Chixoy—Polochic, and Jocotán-Chamelecón fault zones, indicating that each fault zone is active (Schwartz, 1976a). Eight of these earthquakes are clustered along the Motagua fault zone, 13 are clustered along the Chixoy—Polochic fault zone, 3 appear to have occurred along the Jocotán—Chamelecón fault zone, and 2 are associated with either the Motagua or Chixoy—Polochic fault zones. Harlow (1976) also conluded that each of these fault zones is seismically active and noted concentrations of earthquakes during the past 30 years at Chiantla on the Chixoy—Polochic fault zone and in the general area of Quiriguá along the Motagua fault zone.

TECTONICS OF THE PLATE BOUNDARY

Slip rates

The offset stream terraces at the Río El Tambor provide data for estimating upper and lower bounds of the rate of slip along the Motagua fault. If the upper terrace (terrace 7) is close in age to the maximum age of the alluvial fan material (approximately 40,000 years B.P.), a cumulative lateral offset of 58.3 m in that time would give a slip rate of approximately 0.15 cm/yr for this segment of the Motagua fault. Because the upper terrace is probably younger than 40,000 years, this is a minimum rate. Using 10,000 years B.P. as the minimum age of the upper terrace gives a slip rate of approximately 0.6 cm/yr; because terrace 7 may be older, 0.6 cm/yr is considered to be a reasonable maximum slip rate.

The evidence for Quaternary faulting and the record of historical seismicity indicate that the active plate boundary in Central America is defined by the Motagua, Chixoy—Polochic, and probably the Jocotán—Chamelecón fault zones (Schwartz, 1976a; 1977). Each of these fault zones accommodates a portion of the total strain produced by spreading in the Cayman Trough. The rate of slip along the Chixoy—Polochic and Jocotán—Chamelecón fault zones is not known and could differ from the rate of slip along the Motagua fault. However, the similarity in the amount of Quaternary stream offset observed along the Motagua and Chixoy—Polochic faults and relative similarity in the degree of development of youthful fault-related geomorphic features, suggest that their slip rates may be comparable. Assuming the rate is similar for all three fault zones, and using the values obtained for the Motagua fault, the late Quaternary slip rate along the segment of the plate boundary in Central America is estimated to range from 0.45 to 1.8 cm/yr. The actual rate is probably closer to the higher value.

Previous estimates of the relative rate of slip between the Caribbean and North American plates have ranged from 0.4 to 2.2 cm/yr. Pinet (1972) inferred a rate of 1.5—2.0 cm/yr, and Malfait and Dinkelman (1972) suggested a rate of 0.5 cm/yr. Both estimates were based on averaging apparent displacement of geologic structures of uncertain correlation over a long and poorly constrained period of geologic time. Molnar and Sykes (1969) obtained a maximum rate of 2.2 cm/yr by calculating the closure of the relative velocity triangle about the Caribbean—Cocos—North American triple junction. Holcombe et al. (1973) used bathymetric data to compute a spreading rate of 2.2 cm/yr ± 50% across the mid-Cayman rise, and Jordan (1975) calculated a spreading rate of 2.1 cm/yr across the mid-Cayman rise using data on poles of rotation and angular velocities. Perfit (1977) suggests that the presence of Miocene—Pliocene limestone east of the axial valley indicates that E—W spreading at the mid-Cayman rise has been relatively slow for at least the past 5 m.y. and has averaged 0.4 cm/yr. The preliminary estimate of 0.45—1.8 cm/yr, using data from on-land active faults in Central

America, is in close agreement with other spreading rates calculated for the mid-Cayman rise. The ongoing investigation of the displaced terraces along the Río El Tambor and of other active faults in Guatemala should provide refinement of this rate.

Earthquake recurrence intervals

The historical seismicity data suggest that, although each fault zone is seismically active and produces small and moderate earthquakes, surface fault rupture and earthquakes of the magnitude of the February 4, 1976 earthquake (magnitude 7.5) are infrequent events. Earthquakes as large as that which occurred in 1976 may not have occurred along any of the transcurrent fault zones in more than 450 years. Kelleher and Savino (1975) have suggested that large but infrequent earthquakes occur away from spreading centers of transform faults in regions where lithospheric thickness increases. This appears to be the relationship in Guatemala where the continental crust of Central America is appreciably thicker than the oceanic crust of the Cayman Trough.

An average of 1.1 m of the left-slip and a very localized maximum of 3.4 m were measured along the Motagua fault after the February 4, 1976 earthquake (Bucknam et al., 1976; Plafker, 1976). If 1.1 m is characteristic of the average slip for this magnitude earthquake along this fault, 53 earthquakes of this size would be necessary to produce the 58.3 m of slip on the upper terrace (terrace 7). Using the minimum (10,000 years) and maximum (40,000 years) ages for this terrace gives a recurrence interval of from 190 to 755 years for large earthquakes with surface fault rupture along the Motagua fault. A recurrence interval of from 180 to 730 years is obtained using the maximum (0.6 cm/yr) and minimum (0.15 cm/yr) slip rates for the Motagua fault and an average displacement of 1.1 m. Some intermediate recurrence interval is reasonable and is in general agreement with the infrequent occurrence of large earthquakes along these faults suggested by the historical seismic record. These recurrence intervals may also apply to the Chixoy—Polochic and Jocotán—Chamelecón fault zones.

Cumulative slip

The Caribbean—North American plate boundary in Central America is a former collision boundary along which younger strike-slip faulting has been superimposed, and there are few geologic structures that can be confidently correlated to measure lateral slip across the transcurrent fault zones. Cumulative slip may be estimated by extrapolating slip rates, but uncertainties in this method include the time of initiation of strike-slip faulting and the constancy of the slip rate. Geologic relationships in the Motagua fault zone indicate that strike-slip faulting began between Eocene and Miocene time in probable response to opening of the Cayman Trough (Schwartz, 1976a), but they do not allow for the time to be more precisely defined. Ladd (1976), using the rotation of Africa with respect to North America and of South America with respect to Africa, shows that eastward slip of the Caribbean plate relative to North America began about 38 m.y. ago (late Eocene). Holcombe et al. (1973) suggest that the present E—W spreading regime at the mid-Cayman rise has probably been in existence for at least 15 m.y. (middle Miocene). If all motion across Central America has been left lateral strike-slip since the Late Eocene, and the slip rate has been constant, preliminary minimum and maximum Quaternary slip rates derived in this study give cumulative displacement values of 170—685 km; spreading rates of Holcombe et al. (1973) and Jordan (1975) give cumulative displacements of 800 km ± 50% and 835 km, respectively; the spreading rate of Perfit (1977) gives a displacement of 150 km. However, if periods of oblique slip or N—S extension across the Cayman Trough occurred between the late Eocene and the onset of E—W spreading, the cumulative left-slip may be less than the values given above. It is also not certain how far back a Quaternary slip rate — or any rate — can be reliably extrapolated, and small variations would affect the total displacement. The present data do not allow a definitive assessment of the amount of cumulative left-slip along the plate boundary; however, there is some suggestion that cumulative slip is less than the upper values given above and may not exceed 500 km. This slip would be distributed along the Motagua, Chixoy—Polochic, and Jocotán—Chamelecón fault zones.

A consistent up-to-the-north vertical component of slip also occurs along this plate boundary. This can be observed as well-defined up-to-the-north topographic steps across the Motagua, Chixoy—Polochic, and Jocotán—Chamelecón fault zones (Fig. 5). These steps, which define the south-facing fronts of the major mountain ranges of Guatemala, are essentially large fault scarps formed by a small but consistent vertical slip component over time. This vertical component could represent up to 5% of the lateral component of slip along the plate boundary.

CONCLUSIONS

Evidence of Quaternary faulting along the Caribbean—North American plate boundary in Central America reveals that slip along the Cayman Trough is being accommodated along distinct active left-slip faults within the Motagua and Chixoy—Polochic fault zones in Guatemala and probably along the Honduras portion of the Jocotán-Chamelecón fault zone. Preliminary analysis of progressively offset terraces along the Motagua fault indicates a slip rate of from 0.45 to 1.8 cm/yr across this segment of the Caribbean—North American plate boundary in late Quaternary time. Evidence of faulted Quaternary terraces, observations from the February 4, 1976 Guatemala earthquake, and historical seismicity data suggest a recurrence interval of from 180 to 755 years for large earthquakes associated with surface fault rupture along the Motagua fault.

444

ACKNOWLEDGEMENTS

The authors wish to thank Oscar Salazar, Chief of the Geology Division of the Instituto Geográfico Nacional, for logistical support in Guatemala. Research was supported by N.S.F. Grant G458 and an applied research grant from Woodward-Clyde Consultants.

REFERENCES

Anderson, T.H., Burkart, B., Clemons, R.E., Bohnenberger, O.H. and Blount, D.N., 1973. Geology of the western Altos Cuchumatanes, northwestern Guatemala. Geol. Soc. Am. Bull., 84: 805—826.
Banks, N.G. and Richards, M.L., 1969. Structure and bathymetry of western end of the Bartlett Trough, Caribbean Sea. In: A.R. McBirney (Editor), Tectonic Relations of Northern Central American and the Western Caribbean: The Bonacca Expedition. Am. Assoc. Pet. Geol. Mem., 11: 221—228.
Bonis, S.B., Bohnenberger, O. and Dengo, G., 1970. Mapa geologico de la Republica de Guatemala (1st ed.). Guatemala Instituto Geográfico Nacional.
Bonis, S.B., Bohnenberger, O., Stoiber, R.E. and Decker, R.W., 1966. Age of pumice deposits in Guatemala. Geol. Soc. Am. Bull., 77: 211—212.
Bosc, E., 1971. Geology of the San Augustín Acasaquastlán Quadrangle, East Central Guatemala. Ph.D dissertation, Rice Univ., Houston, 131 pp.
Bucknam, R.C., Plafker, G., Sharp, R.V. and Bonis, S.B., 1976. Surface displacement in the Motagua fault zone during the 4 February 1976 Guatemala earthquake (abstr.). EOS, 57: 946.
Carr, M.J. and Stoiber, R.L., 1977. Geologic setting of some destructive earthquakes in Central America. Geol. Soc. Am. Bull., 88: 151—156.
Clemons, R.E., 1966. Geology of the Chiquimula Quadrangle, Guatemala, Central America. Ph.D dissertation, Univ. of Texas, Austin, 123 pp.
Crane, D.C., 1965. Geology of the Jocotán and Timushán Quadrangles, Southeastern Guatemala. Ph.D. dissertation, Rice Univ., Houston, 85 pp.
Donnelly, T.W., 1977. Metamorphic rocks and structural history of the Motagua suture zone, eastern Guatemala. Abstr. 8th Caribbean Geol. Conf., Curaçao, pp. 41—42.
Donnelly, T.W., Crane, D. and Burkart, B., 1968. Geologic history of the landward extension of the Barttlett Trough — some preliminary notes. Proc. 4th Caribbean Geol. Conf., Trinidad.
Harlow, D.H., 1976. Instrumentally recorded seismic activity prior to the main event. In: A.F. Espinosa (Editor), The Guatemalan Earthquake of February 4, 1976, Preliminary Report. U.S. Geol. Surv. Prof. Pap. 1002: 12—16.
Hess, H.H. and Maxwell, J.C., 1953. Caribbean research project. Geol. Soc. Am. Bull., 64: 1—6.
Holcombe, T.L., Vogt, P.R., Matthews, J.E. and Murchison, R.R., 1973. Evidence for sea-floor spreading in the Cayman Trough. Earth Planet. Sci. Lett., 20: 357—371.
Jordan, T.H., 1975. The present-day motions of the Caribbean plate. J. Geophys. Res., 80: 4433—4439.
Kelleher, J. and Savino, J., 1975. Distribution of seismicity before large strike-slip and thrust-type earthquakes. J. Geophys. Res., 80: 260—271.
Kesler, S.E., 1971. Nature of ancestral orogenic zone in Nuclear Central America. Am. Assoc. Pet. Geol. Bull., 55: 2116—2129.
Koch, A.J. and McLean, H., 1975. Pleistocene tephra and ash-flow deposits in the volcanic highlands of Guatemala. Geol. Soc. Am. Bull., 86: 529—541.

445

Kupfer, D.H. and Godoy, J., 1967. Strike-slip faulting in Guatemala (abstr.). Am. Geophys. Union Trans., 48: 215 pp.

Ladd, J.W., 1976. Relative motion of South America with respect to North America and Caribbean tectonics. Geol. Soc. Am. Bull., 87: 969–976.

Lawrence, D.P., 1976. Tectonic implications of the geochemistry and petrology of the El Tambor formation: probable oceanic crust in central Guatemala. Geol. Soc. Am. Abstr. with Programs, 8: 973–974.

Malfait, B.T. and Dinkelman, M.G., 1972. Circum-Caribbean tectonic and igneous activity and the evolution of the Caribbean plate. Geol. Soc. Am. Bull., 83: 251–272.

Molnar, P. and Sykes, L.R., 1969. Tectonics of the Caribbean and middle America regions from focal mechanisms and seismicity. Geol. Soc. Am. Bull., 80: 1639–1684.

Monges, J., Jachens, R.C., Andrade, J. and Case, J.E., 1976. Comparison of gravity measurements before and after the 1976 Guatemala earthquake (abstr.). EOS, 57: 949.

Perfit, M.R., 1977. Petrology and geochemistry of mafic rocks from the Cayman trench: evidence for spreading. Geology, 5: 105–110.

Pinet, P.R., 1972. Diapir-like features offshore Honduras: implications regarding tectonic evolution of Cayman Trough and Central America. Geol. Soc. Am. Bull., 83: 1911–1922.

Plafker, G., 1976. Tectonic aspects of the 4 February 1976 Guatemala earthquake – a preliminary evaluation. Science, 193: 1201–1208.

Schwartz, D.P., 1976a. Geology of the Zacapa Quadrangle and vicinity. Ph.D dissertation, State Univ. of New York, N.Y., Binghamton, 191 pp.

Schwartz, D.P., 1976b. The Motagua fault zone, Guatemala: Tertiary and Quaternary tectonics (abstr.). Geol. Soc. Am. Abstr. with Programs, 8: 1092–1093.

Schwartz, D.P., 1977. Active faulting along the Caribbean–North American plate boundary in Guatemala (abstr.). Abstr. 8th Caribbean Geol. Conf., Curaçao: 180–181.

Schwartz, D.P. and Newcomb, W.E., 1973. Motagua fault zone: a crustal suture (abstr.). EOS, 54: 477.

Tectonophysics, 52 (1979) 447—455
© Elsevier Scientific Publishing Company, Amsterdam — Printed in The Netherlands

EL PILAR FAULT ZONE, NORTHEASTERN VENEZUELA: BRIEF REVIEW

C. SCHUBERT

Instituto Venezolano de Investigaciones Científicas, Apartado 1827, Caracas 101 (Venezuela)

(Accepted for publication June 8, 1978)

ABSTRACT

Schubert, C., 1979. El Pilar fault zone, northeastern Venezuela: brief review. In: C.A. Whitten, R. Green and B.K. Meade (Editors), Recent Crustal Movements, 1977. Tectonophysics, 52: 447—455.

The El Pilar fault zone extends for about 700 km in an approximately E—W direction, from the Cariaco Trench to a point about 200 km northeast of Trinidad. It marks the southern boundary of the Araya—Paria peninsulas (eastern Caribbean Mts.) and of the Northern Range of Trinidad. It is characterized along its length by straight valleys, fault wedges, fumaroles, thermal springs, and sulfur deposits. The displacement along the El Pilar fault zone has been the subject of much controversy. Various authors recognize the following types of displacement: (1) southward thrust; (2) normal or graben faulting; (3) right-lateral strike-slip. The El Pilar fault zone probably represents a transform fault between the Caribbean and South America plates, and a hinge fault at its contact with the subduction zone east and south of the Lesser Antilles. It is planned to investigate the present-day movement along the El Pilar fault zone by high-precision geodetic methods, at the following localities: across the Gulf of Cariaco and at Casanay (central-southern Araya—Paria peninsulas).

INTRODUCTION

The Caribbean Sea region has been interpreted as a crustal block which is moving eastward with respect to North and South America (Hess, 1938; Hess and Maxwell, 1953). These authors, and Bucher (1952), Rod (1956a), and Alberding (1957), suggested that this relative eastward movement of the Caribbean block took place mainly along great strike-slip fault zones, located along the northern boundary of South America (Oca, Boconó, and El Pilar faults) and along the southern boundary of the Greater Antilles. With the rise of the "new global tectonics" (Isacks et al., 1968), the Caribbean region was interpreted as a zone of interaction between two lithosphere plates (Fig. 1), the Caribbean and Americas plates (Molnar and Sykes, 1969). According to this hypothesis, a crustal subduction zone is located east of the Lesser

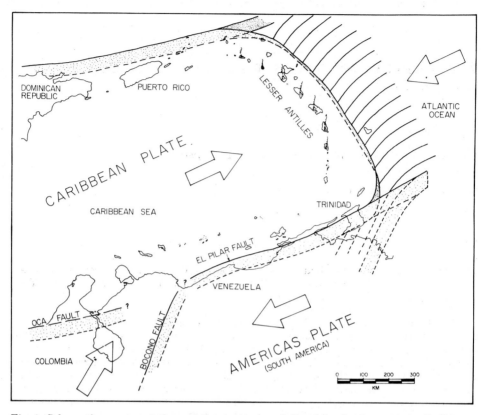

Fig. 1. Schematic representation of the tectonic relationships in the eastern Caribbean region (modified after Molnar and Sykes, 1969; Schubert, 1970). The large arrows indicate relative movement of the plates. East of the Lesser Antilles there is a subduction zone; northeast of Trinidad there is a zone of hinge faulting, which extends westward (along the northern boundary of South America) as a transform fault.

Antilles volcanic arc, where the Atlantic crust sinks beneath the Caribbean crust, producing the seismicity and volcanism of the Lesser Antilles. Along the northern coast of South America there is a transform fault, between Trinidad and the Nazca Plate in the eastern Pacific Ocean. Consequently, it was postulated that the Caribbean Plate represents a wedge of Pacific crust which penetrates the Americas Plate, as the latter moves westward from the Mid-Atlantic Ridge, due to sea-floor spreading (these interpretations were reviewed by Schubert, 1970 and 1976). In detail, these movements are much more complicated, as was demonstrated by Ladd (1976), who proposed various periods of extension and compression between the Caribbean and Americas plates, since the Jurassic to the present. Jordan (1975) has presented evidence that the Caribbean Plate is stationary and that it is the Americas Plate which is moving westward. In addition, Jordan calculated that the present velocity of displacement between the Caribbean and South

America is 1.5 ± 0.6 cm/yr in a N75° ± 15°W direction. This implies that the displacement is not purely in an E—W direction (as postulated earlier along the strike-slip faults), but that there is compression in a southeast direction (as shown, for example, by the southeast tectonic transport direction derived from macro- and mesoscopic structures in the metamorphic rocks of the Araya Peninsula; Schubert, 1971) and strike-slip movement in a northwest direction along the contact between the Caribbean and Americas (South America) plates.

The seismicity of the Caribbean Sea region was studied by Sykes and Ewing (1965) and an extensive catalog of the historic seismicity of the region was compiled by Grases (1971). The seismicity of Venezuela was compiled and analyzed by Centeno-Graü (1969) and Fiedler (1961, 1972). Three seismically active regions are clearly defined in the maps of the last two authors: Cumaná, Caracas, and the Andean cordillera; a clear relationship between the location of epicenters and the great fault zones mentioned above is shown in Fiedler's maps.

The present report is a review of what is known about the El Pilar fault zone, particularly concerning its importance as a possible part of a plate boundary. This review was written as a preliminary phase in a project concerning the measurement of the present-day rate of movement along active Venezuelan faults. It is based mainly on previous work and reports widely scattered throughout the literature. However, I am solely responsible for the way in which the data are presented and interpreted.

EL PILAR FAULT ZONE

Probably the first mention of a major geological and structural accident in northeastern Venezuela was made by Alexander von Humboldt and Amadëe Bonpland, at the beginning of the 19th century, in the first volume of their "Voyage to the Equinoctial Regions of the New Continent", published in 1816 (Humboldt, 1956, Vol. 1, p. 320). In this citation, Humboldt wrote: "the Gulf of Cariaco owes its existence to a tearing apart of the land accompanied by inflow of the ocean" (author's translation from the Spanish). This author attributed the frequent earthquakes in the Cumaná region to this "tearing apart". Over a century later, Liddle (1928, p. 493) considered that the discordant contact between the metamorphic and sedimentary rocks of the Caribbean Mts., between the Araya—Paria peninsulas and the Northern Range of Trinidad, probably represented a fault. Later, the same author (Liddle, 1946, pp. 569—570) formally, and for the first time, described a fault zone along this contact, between Cumaná and Yaguaraparo (Fig. 2), with a western continuation into northwestern Venezuela (Barquisimeto region), and an eastern continuation along the southern boundary of the Northern Range of Trinidad. Lidz et al. (1968), demonstrated the western continuation of the El Pilar fault zone in the Cariaco Trench.

The name El Pilar fault was first published by Rod (1956a), in a descrip-

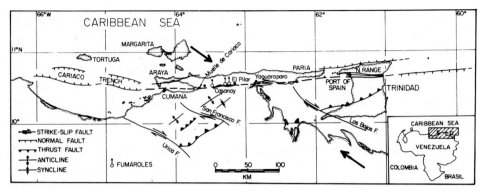

Fig. 2. Tectonic sketch map of the El Pilar fault zone (modified and simplified after Metz, 1968; Ball et al., 1971; Bassinger et al., 1971; Murany, 1972; Lau and Rajpaulsingh, 1976). The heavy arrows indicate the direction of the primary compressive system. The faulting and folding, associated with this system, located south of the fault zone, is not complete and represents only the general tectonic pattern.

tion of the main fault zones of northern Venezuela (Oca, Boconó, and El Pilar faults), and which this author considered as responsible for Venezuelan seismicity (Rod, 1956b), due to strike-slip movement along them. This type of displacement had already been proposed by Bucher (1952, p. 76). Up to the present, the most detailed study of the central part of the El Pilar fault zone (across the isthmus between the eastern Venezuelan mainland and the Araya—Paria peninsulas) was made by Metz (1964, 1968). This author mapped the geology of this region and concluded that the displacement along the fault zone was mainly southward-thrusting, with a minor right-lateral strike-slip component. More recently, Vierbuchen (1977) reinterpreted Metz's (1968) evidence for displacement along some of the branches of the El Pilar fault zone, and concluded that, according to detailed structural analysis, there is evidence for significant right-lateral strike-slip movement. Vignali (1977), on the contrary, did not find evidence of strike-slip movement, and concluded that the present fault activity could be very young (Pleistocene—Holocene) and may not represent a long history of activity.

The submarine extensions of the El Pilar fault zone, to the west and east, were determined by geophysical methods by Ball et al. (1971) and Bassinger et al. (1971). These authors clearly show vertical displacement of graben type within the fault zone. However, no strike-slip movement is shown, most probably because of the orientation of the geophysical survey lines which were essentially perpendicular to the trace of the fault zone. At present, the El Pilar fault zone is recognized between the western end of the Cariaco Trench and a point located approximately 200 km east of the northeastern end of Trinidad, a length of about 700 km.

The El Pilar fault zone crops out above sea level in only two regions (Fig. 2): between Muelle de Cariaco and the region south of Yaguaraparo

(State of Sucre, across the isthmus south of the Araya—Paria peninsulas), and between Port of Spain and Matura (Trinidad). In the first region, it is characterized by straight river valleys (Río Casanay and Río Chaguaramas), fault wedges containing Jurassic—Cretaceous metamorphic rocks (Tunapuy and Güinimita formations), Cretaceous sedimentary rocks (Barranquín and San Antonio formations), and serpentinite (Christensen, 1961; Metz, 1968; Seijas, 1972). In addition, the active fault trace is marked by fumaroles, hot-water springs, and sulfur deposits. According to Metz (1968), the fault zone consists of several fracture zones which bifurcate, forming the wedges; these fractures frequently contain broad zones of fault gouge and crushed rocks (cataclasite and mylonite).

In Trinidad, the El Pilar fault zone crops out in a long, straight depression (Caroní syncline or Northern Basin), oriented in an E—W direction, between the Northern and Central ranges (Kugler, 1959). According to this author, the fault zone contains several faults, of which the most important is the Arima fault, the northern boundary of the graben structure which represents the southern limit of the Northern Range.

DISPLACEMENT ALONG THE EL PILAR FAULT ZONE

The displacement along the El Pilar fault zone has been analyzed by numerous authors, based on displacement of similar lithologies, geomorphic features, and earthquake slip-vector calculations. Table I shows a compilation of the postulated types of displacement, estimates of displacement

TABLE I

Summary of data on the El Pilar fault zone

Author	Type of fault [*]	Rate	Total displacement (km)	Age [*]
Rod (1956a)	RLSS	—	>100	—
Alberding (1957)	RLSS	—	~475	TQ
Lidz et al. (1968)	N	—	3.5—5	PH
Metz (1968)	T-RLSS	—	<15	TQ
Potter (1968)	N	—	~1.8	—
Salvador and Stainforth (1968)	T	—	Not significant	M (?)
Molnar and Sykes (1969)	RLSS	1 cm/yr	—	—
Bassinger et al. (1971)	N	—	2.3	—
Saunders (1974)	N	—	>3	PH
Lau and Rajpaulsingh (1976)	N	—	—	MR
Vierbuchen (1977)	RLSS	—	>25	—
Vignali (1977)	—	—	0	Q

[*] RLSS = right-lateral strike-slip; N = normal; T = thrust; TQ = Tertiary—Quaternary; PH = Paleocene—Holocene; M = Miocene; MR = Miocene—Recent; Q = Quaternary.

velocities, estimates of total displacement, and the age of the fault. From this table it is obvious that there is no agreement between different authors on these problems. The El Pilar fault zone has been considered as:

(1) A southward thrust, which pushed Jurassic—Cretaceous metamorphic rocks of the Caribbean Mts. and the Northern Range of Trinidad, over Cretaceous and Tertiary unmetamorphosed sediments (Metz, 1968; Salvador and Stainforth, 1968).

(2) A zone of normal faulting or a graben structure (Lidz et al., 1968; Potter, 1968; Bassinger et al., 1971; Saunders, 1974; Lau and Rajpaulsingh, 1976).

(3) A zone of right-lateral strike-slip faulting (Rod, 1956a; Alberding, 1957; Metz, 1968; Molnar and Sykes, 1969; Vierbuchen, 1977).

Only Metz (1968) proposed two types of displacement, a southward-thrusting main one, and a minor one of right-lateral strike-slip faulting with a total displacement of less than 15 km.

Tomblin (1972) analyzed the seismicity of the eastern Caribbean and concluded that it was compatible with crustal subduction east of the Lesser Antilles (sea Introduction) and with strike-slip movement along the northern and southern boundaries of the Caribbean Plate. Consequently, east or northeast of Trinidad begins a zone of transform faulting which extends along the northern boundary of South America. This transform fault is connected to the subduction zone by a hinge fault (Molnar and Sykes, 1969) (Fig. 1). Lau and Rajpaulsingh (1976) analyzed the tectonic characteristics of Trinidad and adjacent regions in terms of plate tectonics, and postulated that the subduction zone continues north and northwest of Trinidad, giving rise to a zone of tension in the Northern Range and a zone of compression south of the El Pilar fault zone. This model corresponds to the analysis by Murany (1972) of the Serranía del Interior faults (south of the Araya—Paria peninsulas, Venezuela), in which there is a system of first-order compression in a N55°W direction; this system is responsible for the E—W right-lateral strike-slip movement along the El Pilar fault zone. In this region there are also other secondary compression systems which give rise to other faults in eastern Venezuela and Trinidad, such as the San Francisco, Urica, and Los Bajos strike-slip faults, and the Pirital thrust. In this way, the El Pilar fault zone seems to be, at the same time, the southern transformational boundary between the Caribbean and Americas plates and, in its eastern end, a hinge fault in contact with a crustal subduction zone.

Recently, a project was begun to measure present-day movement along active Venezuelan faults (Schubert and Henneberg, 1975), by means of high-precision geodetic measurements across them. So far, two such geodetic systems have been constructed across the Boconó fault (Venezuelan Andes), and the initial measurements have been carried out. Future geodetic measurements will determine the existence and rate of movement along this fault. As an extension of this work, it is proposed to install similar high-precision geodetic systems across the El Pilar fault zone. Two sites have been

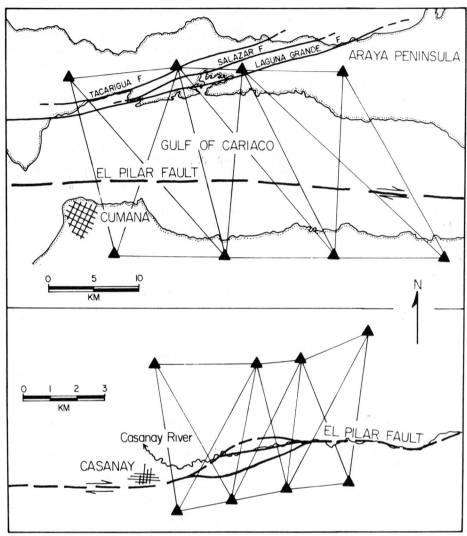

Fig. 3. High-precision geodetic systems to measure present-day strike-slip displacement along the El Pilar fault zone, proposed for the Cumaná (above) and Casanay (below) regions (Fig. 2). The triangles indicate reinforced concrete geodetic stations (Schubert and Henneberg, 1975) and the lines connecting them are the simple visuals. The fault traces were derived from Rod (1956a) and Schubert (1971).

selected (Fig. 3): one across the Gulf of Cariaco, between Cumaná and the Araya Peninsula, and another east of Casanay.

ACKNOWLEDGEMENTS

A project on measurement of present-day movement along active Venezuelan faults is being supported by CONICIT (Grant No. 31.26.S1-

0374). The first phase of this project was carried out on the Boconó fault in close cooperation with the School of Geodesy of the Universidad del Zulia (Maracaibo), which carried out the geodetic measurements. The present report represents the initiation of a second phase of this project, concerning the present-day movement along the El Pilar fault zone.

REFERENCES

Alberding, H., 1957. Application of principles of wrench-fault of Moody and Hill to northern South America. Geol. Soc. Am. Bull., 68: 785—790.

Ball, M.M., Harrison, C.G.A., Supko, P.R., Bock, W. and Maloney, N.J., 1971. Marine geophysical measurements on the southern boundary of the Caribbean Sea. Geol. Soc. Am. Mem., 130: 1—33.

Bassinger, B.G., Harbison, R.N. and Weeks, L.A., 1971. Marine geophysical study northeast of Trinidad-Tobago. Am. Assoc. Pet. Geol. Bull., 55: 1730—1740.

Bucher, W.H., 1952. Geologic structure and orogenic history of Venezuela. Geol. Soc. Am. Mem., 49: 113 pp.

Centeno-Graü, M., 1969. Estudios sismológicos. Acad. Cienc. Fis. Mat. Nat. (Venez.), 8: 365 pp.

Christensen, R.M., 1961. Geology of the Paria-Araya Peninsula, Northeastern Venezuela. Doctoral Thesis, Univ. Nebraska, Lincoln, 112 pp.

Fiedler, G., 1961. Areas afectadas por terremotos en Venezuela. Bol. Geol., Publ. Espec. No. 3, 4: 1791—1810.

Fiedler, G., 1972. La liberación de energía sísmica en Venezuela, volumenes sísmicos y mapa de isosistas. Bol. Geol., Publ. Espec. No. 5, 4: 2441—2462.

Grases, J., 1971. La sismicidad histórica del Caribe. 1. Documentos de trabajo. Inst. Mater. Modelos Estructurales, Univ. Cent. Venez., Caracas.

Hess, H.H., 1938. Gravity anomalies and island arc structure with particular reference to the West Indies. Am. Philos. Soc. Proc., 79: 71096.

Hess, H.H. and Maxwell, J.C., 1953. Caribbean research project. Geol. Soc. of Am. Bull., 64: 1—6.

Humboldt, A. von, 1956. Viaje a las regiones equinocciales del nuevo continente. Ed. Minist. Educ., Caracas, Vol. 1, 338 pp.

Isacks, B., Oliver, J. and Sykes, L.R., 1968. Seismology and the new global tectonics. J. Geophys. Res., 73: 5855—5900.

Jordan, T.H., 1975. The present-day motion of the Caribbean Plate. J. Geophys. Res., 80: 4433—4439.

Kugler, H.G., 1959. Geological Map of Trinidad and Geological Sections through Trinidad. Pet. Assoc. Trinidad, Füssli, Zürich.

Ladd, J.W., 1976. Relative motion of South America with respect to North America and Caribbean tectonics. Geol. Soc. Am. Bull., 87: 969—976.

Lau, W. and Rajpaulsingh, W., 1976. A structural review of Trinidad, West Indies in the light of current plate-tectonics and wrenchfault theory. In: R. Causse (Editor), Trans. 7th Caribbean Geol. Conf., Guadeloupe, pp. 473—483.

Liddle, R.A., 1928. The Geology of Venezuela and Trinidad. MacGowan, Fort Worth, 552 pp.

Liddle, R.A., 1946. The Geology of Venezuela and Trinidad (2nd ed.). Paleontological Research Institution, Ithaca, 890 pp.

Lidz, L., Ball, M.M. and Charm, W., 1968. Geophysical measurements bearing on the problem of the El Pilar Fault in the northern Venezuelan offshore. Bull. Mar. Sci., 18: 545—560.

Metz, H.L., 1964. Geology of the El Pilar Fault Zone, State of Sucre, Venezuela. Doctoral thesis, Princeton Univ., Princeton, 102 pp.

Metz, H.L., 1968. Geology of the El Pilar Fault Zone, State of Sucre, Venezuela. In: J.B. Saunders (Editor), Trans. 4th Caribbean Geol. Conf., Trinidad, pp. 193—198.

Molnar, P. and Sykes, L.R., 1969. Tectonics of the Caribbean and Middle America regions from focal mechanisms and seismicity. Geol. Soc. Am. Bull., 80: 1639—1684.

Murany, E.E., 1972. Structural analysis of the Caribbean coasteastern Interior Range of Venezuela. In: C. Petzall (Editor), Trans. 6th Caribbean Geol. Conf., Porlamar, pp. 295—298.

Potter, H.C., 1968. A preliminary account of the stratigraphy and structure of the eastern part of the Northern Range, Trinidad. In: J.B. Saunders (Editor), Trans. 4th Caribbean Geol. Conf., Trinidad, pp. 15—20.

Rod, E., 1956a. Strike-slip faults of northern Venezuela. Am. Assoc. Pet. Geol. Bull., 40: 457—476.

Rod, E., 1956b. Earthquakes of Venezuela related to strike-slip faults? Am. Assoc. Pet. Geol. Bull., 40: 2509—2512.

Salvador, A. and Stainforth, R.M., 1968. Clues in Venezuela to the geology of Trinidad, and vice versa. In: J.B. Saunders (Editor), Trans. 4th Caribbean Geol. Conf., Trinidad, pp. 31—40.

Saunders, J.B., 1974. Trinidad. In: A.M. Spencer (Editor), Mesozoic—Cenozoic orogenic Belts: Data for Orogenic Studies. Geol. Soc. Lod., Spec. Publ. No. 4, pp. 671—682.

Schubert, C., 1970. Venezuela y la "nueva tectónica global". Acta Cient. Venez., 21: 13—16.

Schubert, C., 1971. Metamorphic rocks of the Araya Peninsula, eastern Venezuela. Geol. Rundsch., 60: 1571—1600.

Schubert, C., 1976. Investigaciones neotectónicas en Venezuela: objetivos y resultados. Interciencia, 1: 159—169.

Schubert, C. and Henneberg, H.G., 1975. Geological and geodetic investigations on the movement along the Boconó Fault, Venezuelan Andes. Tectonophysics, 29: 199—207.

Seijas, F.J., 1972. Geología de la región de Carúpano. Bol. Geol., Publ. Espec. No. 5, 3: 1887—1923.

Sykes, L.R. and Ewing, M., 1965. The seismicity of the Caribbean region. J. Geophys. Res., 70: 5065—5074.

Tomblin, J.F., 1972. Seismicity and plate tectonics of the eastern Caribbean. In: C. Petzall (Editor), Trans. 6th Caribbean Geol. Conf., Porlamar, pp. 277—282.

Vierbuchen, R., 1977. New data relevant to the tectonic history of the El Pilar Fault. Abstr. 8th Caribbean Geol. Conf., Curaçao, pp. 213—214.

Vignali, M., 1977. Geology between Casanay and El Pilar (El Pilar Fault Zone), Edo. Sucre, Venezuela. Abstr. 8th Caribbean Geol. Conf., Curaçao, pp. 215—216.

Tectonophysics, 52 (1979) 457—467
© Elsevier Scientific Publishing Company, Amsterdam — Printed in The Netherlands

Observed Horizontal Crustal Deformation

STATISTICAL ANALYSIS OF GEODETIC MEASUREMENTS FOR THE INVESTIGATION OF CRUSTAL MOVEMENTS

J. VAN MIERLO

Department of Geodesy, Delft University of Technology, Thijsseweg 11, Delft (The Netherlands)

(Accepted for publication June 8, 1978)

ABSTRACT

Van Mierlo, J. 1979. Statistical analysis of geodetic measurements for the investigation of crustal movements. In: C.A. Whitten, R. Green and B.K. Meade (Editors), Recent Crustal Movements, 1977. Tectonophysics, 52: 457—467.

Geodetic networks are designed to obtain data that can be used to monitor crustal movements. The relative position on the earth's surface is determined from these networks by means of coordinates. The coordinates of stations and its variance—covariance matrix are based on the computational model. In spatial networks at least three points, the base points, should be chosen to define the coordinate system "fixed" to the earth. In monitoring crustal movements these base points are considered to be stationary over the time span of the motion involved. A procedure for testing the stability of the base points, together with other stable points, is described.

The coordinate differences between two time epochs, t_0 and t_1, are considered to investigate crustal movements. A statistical test is introduced to determine whether crustal movements have actually occurred.

The reliability, i.e., the influence, of nondetected errors in the observations or computations, should be considered. Two types of decisions can be made which may lead to incorrect conclusions. These conclusions are as follows:

(1) That no movement has taken place, although a nondetected error leads to the opposite conclusion.

(2) That a movement has occurred, although a nondetected error in the observations leads to the opposite conclusion.

The chance of arriving at these conclusions can be computed. Boundary values for assumed crustal motion in specified latitudinal and longitudinal directions give a better insight into the desired specifications for geodetic networks.

The testing procedure and the above-mentioned method of computing boundary values can be used for all types of networks, e.g., those obtained by conventional triangulation or by a satellite-borne ranging system.

INTRODUCTION

The problems associated with crustal movement are extremely complex and involve many disciplines. The geodesist is able to contribute to the solution of some of these problems through his ability of accurately measuring the deformation which takes place. A technique for monitoring crustal movement is to repeat precise geodetic surveys at regular intervals. Geodesists generally present their results in a coordinate reference frame "fixed" to the earth. There is now the problem of anchoring these networks to stable points. The uncertainties due to the selection of stable points may lead to erroneous interpretations. A basic advantage of the coordinate comparison method is that it gives the integrated or total movement pattern over the entire survey.

Because geodetic observations are stochastic, differences between successive surveys, they do not always represent crustal movements. The "noise" of the measurement is often approximately equivalent to the "signal" or amount of movement: the signal-to-noise ratio is often unfavorable. One task for geodesists is to develop and apply statistical tests of significance for apparent motion and to assign valid specifications of precision and reliability to such motions. If a model or hypothesis is assumed for the crustal movement, a testing procedure has to be designed for accepting or rejecting this model, using geodetic observations. The statistic to be used in the investigations whether or not any movement has occurred, should be chosen independent of the coordinate reference frame. Geodetic data obtained from periodical repeated surveys of the network would supplement the data which are obtained for crustal-movement studies. These data would provide valuable information for a long-range earthquake warning system.

COORDINATE SYSTEMS

The variation-of-coordinates method is accepted as being suitable for the adjustment and analysis of geodetic networks. When dealing with geodetic observations which do not depend on a coordinate system like angles, ranges, length ratios, range rates (Doppler) or range difference (integrated Doppler), an adjustment in terms of coordinates cannot be carried out unless a certain coordinate system is adopted. The number of elements to be specified in order to define a Cartesian coordinate system is seven: three parameters to define the origin, three to define its orientation and one parameter to define the scale. This could be effectively done by selecting seven coordinates (distributed over three points) in the network to be held fixed in the adjustment. These seven parameters are called *a minimal set of constraints*. Adjusted values of the coordinates of all the points involved, as well as their variance—covariance matrix, would then refer to this minimal set of constraints.

There are many ways in which a coordinate system could be chosen when an arbitrary choice of the parameters is necessary or desirable. All the nominal approximate coordinates are defined to be the same constants for any such choice. This means that the adjusted values of the coordinates corresponding to different choices of a coordinate system will only differ by differential quantities. Consequently, the respective coordinate systems can also only differ by differential quantities or all such coordinate systems can be considered differentially similar.

The angles and length ratios, as well as their variance—covariance matrix, are not affected by a (differential) similarity transformation. Functions of length ratios and angles are the only estimable quantities when solutions are obtained with inner constraints for origin, orientation and scale.

$S_{rs;t}$ system

The coordinate system was defined by selecting seven coordinates distributed over three network points P_r, P_s, P_t, e.g.:

$$\left.\begin{array}{ccc} x_r^0 & x_s^0 & \\ y_r^0 & y_s^0 & \\ z_r^0 & z_s^0 & z_t^0 \end{array}\right\} \text{ minimal set of constraints}$$

The points P_r, P_s and P_t are called *base points*. Devised by Baarda (1975) P_r and P_s can be called base points of the first type and P_t a base point of the second type. It is always possible to transform one coordinate system, defined by a minimal set of constraints, into another system by use of the so-called S-transformation devised by Baarda (1973, 1975). An astronomically oriented coordinate system may be derived by rotation.

A theoretically better approach of the definition of "coordinates" has been given by Baarda (1975). His theory leads to an invariant under a differential similarity transformation. This invariant is a vector whose components can be considered as Cartesian coordinates in a so-called $S_{rs;t}$ system. This $S_{rs;t}$ system leads to the same coordinate system as treated thus far, if the adopted coordinates of the base points P_r, P_s and P_t are identical.

As previously stated, the coordinates of all points involved, as well as their variance—covariance matrix, would refer to the adopted coordinate system or $S_{rs;t}$ system. Since the constraints uniquely define a $S_{rs;t}$ system of which the observations are invariant, theoretically the same adjusted observations and therefore the same set of residuals should be obtained from an adjustment, no matter which particular minimal set of constraints is used. The unique residuals obtained from such adjustment may then be used for analyzing the data (Baarda, 1968).

The variance—covariance matrix of the coordinates refers to the $S_{rs;t}$ system. Some geodesists consider the trace of the variance—covariance matrix as a reasonable measure of "goodness" in the determination of coordinates with respect to the coordinate system chosen. This trace may be excessively large in some cases (depending on the position of the base points). To obviate this situation, the "best" solution is arrived at in a coordinate system defined through the use of a set of inner adjustment constraints where the trace of the variance—covariance matrix for the coordinates would be minimum compared to any other solution. The resulting adjustment is termed "free". This approach defines an optimal coordinate system where the adjusted coordinates preserve the mean positions and orientation of the initially adopted approximate coordinates with no station or stations being preferred over any other station. This coordinate concept has been introduced by Meissl (1962). It depends on the number of stations of the network. It can, however, be obtained from a $S_{rs;t}$ system by means of a so-called *generalized S-transformation*. The angles and length ratios, as well as their variance—covariance matrix, are not affected by this generalized S-transformation. For that reason angles and length ratios, the estimable quantities, are most suited to the investigation of crustal movements. The "coordinates" should be used with care.

A further problem to be considered is the unit of length to be employed in the coordinate system. In a $S_{rs;t}$ system the scale is defined by the adopted (approximate) coordinates of the base points P_r and P_s. The shape of the network can be determined in a $S_{rs;t}$ system. The derived length of the chords between the stations are dependent on this choice:

$$\{(x_r^0 - x_s^0)^2 + (y_r^0 - y_s^0)^2 + (z_r^0 - z_s^0)^2\}^{1/2}$$

To avoid this dependence the unit of length of the coordinate system will, in practice, often be taken equal to that of the adjusted range measurements of the network. Such a coordinate system can be arrived at through a simplified S-transformation of the $S_{rs;t}$ system.

A problem is that the "instrument unit of length" varies from instrument to instrument, and within one instrument from period to period. It is possible to measure instantaneous length ratios with higher precision and reliability. Measuring of length ratios also has the advantage that the choice of a unit of length in a particular coordinate system becomes arbitrary. A conventional unit of length may be chosen, if possible in line with the instrument unit of length (Baarda, 1975).

Changes in the relative positions of the stations due to the crustal motion can be studied in a $S_{rs;t}$ system. As this system is closely connected to the base points, the stability of these points has to be checked.

ERROR CONTROL

It is well known that (systematic) observational errors are unavoidable. Some of these can be traced and rectified by statistical tests. Others, however, cannot be detected in this way. They will result in errors in the coordinates, which may lead to wrong conclusions regarding crustal movements. For this reason it is of paramount importance that the observations are carefully checked for possible errors. The probability of detecting errors by a specific statistical testing procedure is called the *power of the test*, denoted by β. The probability of error detection increases with the magnitude of the error. It is therefore necessary to compute the power of the test for different magnitudes of errors for all observations. These computations should be carried out prior to the measurements. They determine the *internal reliability* of a geodetic network (Baarda, 1968).

The influence of observational errors on the coordinates of points in the network can also be determined during the planning of the survey. This is the so-called *external reliability* of a network. The external reliability is very important to the purpose of the survey. It depends, however, on the definition of the coordinate system, the level of significance of the test and the adopted power. The external reliability should therefore be used with care.

It is possible to develop an invariant with respect to the choice of the base points. Further research on this invariant is necessary for analysing geodetic networks designed for the detection of crustal movements (Baarda, 1977).

Rejecting correct observations on the basis of statistical tests also introduces errors. This type of error is called a *type I error*. The probability of making type I errors is a-priori determined: significance level α. Type I errors are less harmful than *type II errors* which involve the acceptance of wrong observations or incorrect hypothesis. The probability of making a type II error is $1 - \beta = \gamma$ (β is called the power of the test). The power depends among other things on the design of the network, the variance–covariance matrix of the observations, the significance level and the magnitude of the errors in the observations.

The error control is independent of the definition of the coordinate system. For that reason there is no need to design a network which minimizes only the trace of the variance–covariance matrix of the coordinates. The choice of a survey plan should also be directed to a minimization of $1 - \beta$. It can be proved that a high precision of the geodetic network (i.e., the variance–covariance matrix of the coordinates), does not guarantee a good error control and vice versa. A network with poor error control is not suitable for monitoring crustal movements because the risk of making wrong decisions here becomes too great.

Reliability of stability of base points

After the observations are tested the coordinates of the network points are computed in a $S_{rs;t}$ system. The stability of the base points P_r, P_s and P_t should therefore be checked. The choice of the base points is not fully arbitrary; they must lie on one plate or block. Their mutual positions are presumed not to be disturbed by local or regional crustal movements. If on that same plate more points can be found eligible as base points, the stability of all these points can be tested (Mierlo, 1977). In this paper these points are denoted as reference points. Repeated geodetic surveys at regular intervals are used to test the stability of the reference points.

The developed testing procedure examines the movement of one reference point at a time in specified directions. During the planning of the network, the probability of detecting displacements of given magnitude in the specified directions can be computed (power of the tests). These computations give information about the reliability of the stability of the reference points.

462

If three widely spaced stable reference points can be found they are chosen as base points.

As in all statistical tests type I and type II errors can be made. A type II error involves the acceptance of a nonstable reference point which is more harmful than a type I error, i.e., rejection of a stable reference point.

To investigate crustal movements the coordinates of the remaining stations, defined in a $S_{rs;t}$ system, are then compared with each other. Another statistical test is introduced to determine whether crustal movements could actually have occurred. This test is also independent of the choice of the base points. Now type I and type II errors can also be made. A type I error would favor crustal movements which in reality did not exist.

Nondetected errors in the observations may also lead to wrong conclusions. The same holds for the instability of the base points which was not detected in the testing procedure. It is therefore very important that the geodetic observations, as well as the stability of the base points, be adequately checked. The computations of the power of the described statistical test can be used to design a survey plan in such a way that the chance of making a wrong decision is minimized. For practical reasons boundary values corresponding to a fixed power are computed.

If a model with respect to crustal motions has been utilized, a statistical test is again used to determine whether the differences between two or more successive surveys indicate a motion pattern according to this model. If this is the case the magnitude of the motion may be determined. Naturally, the test used to verify the crustal-motion model must have optimum power. In planning the network this should be taken into consideration.

A simplified crustal-motion simulation is shown in Fig. 1. The stations in area I east of the Calaveras fault were considered to be moving in a southeasterly direction at a constant unknown rate as indicated in Fig. 1. The sta-

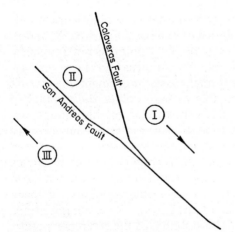

Fig. 1. Simplified crustal-motion pattern.

tions in area II between the Calaveras and San Andreas faults were assumed to be stationary. The reference points are therefore situated in area II. The stations in area III, west of the San Andreas fault, were considered to be moving in a northwesterly direction at the same unknown rate. The testing procedure developed at Delft (Baarda, 1968) can be used for testing this crustal-motion model. In this particular case a one-sided test can be used. It can be proved that the statistical test corresponding with this motion pattern has optimum power. The test can also be used to determine if the difference between two or more surveys indicates a significant motion.

Independent of the measurement itself the probability of the displacement of the stations in areas I and III of a certain given magnitude can be assessed, i.e., the power of the test can be determined. The power not only depends on the magnitude of the movement and the adopted level of significance but also on the way in which the mutual positions of the stations in the network were determined. This makes it possible to choose an optimal survey plan from various possible networks and observational techniques.

Consequences of nondetected errors

So far it has been implicitly assumed that no observational errors have been made. If observational errors are made that are not detected in the testing procedure two types of decisions can be made which may lead to wrong conclusions about crustal movements. These conclusions are as follows:

(1) That no movement has taken place although a nondetected error leads to a reverse conclusion. The probability of such a wrong conclusion can be computed prior to the measurements.

(2) That a movement has occurred, although a nondetected error in the observations leads to the reverse conclusion. The probability of arriving at this conclusion can also be computed.

In computing the probability of making errors of this type, no account is taken of the probability of making observational errors because it is unknown (Mierlo, 1975). These computations can be used to design a survey plan in such a way that the chance of making a wrong decision is minimized.

EXAMPLE OF A SURVEY OF THE STRAITS OF MESSINA

The network across the Straits of Messina (Fig. 2) is a quadrilateral with two points on the mainland of Calabria and two in Sicily (Caputo et al., 1974). The points in the Calabria and Sicily are thought to lie on opposite sides of a fault. The aim is to detect relative displacements between Sicily and Calabria. The hypothesis that Sicily belongs to the African continental plate, and its displacement in a northerly direction along the contact line with the Eurasian plate, can be tested.

In the period between September 1970 and September 1973 six sets of measurement were taken. In the first survey all angles and distances of the

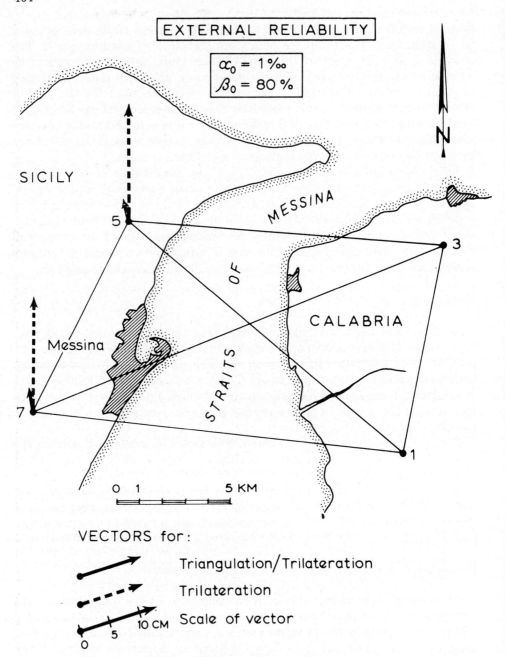

Fig. 2. The external reliability of the network for the detection of crustal movements across the Straits of Messina, indicated for a 1‰ significance level α and a power β of 80%. These are given for two different methods of surveying. The influence of a nondetected error in the distance between the points *1* and *5*, on the location of the points *5* and *7* is shown by means of vectors. This error is derived from α and β.

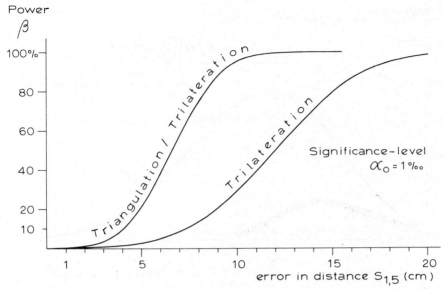

Fig. 3. The probability (power β) of detecting errors of a certain magnitude in the distance $S_{1,5}$ in Fig. 2, established by means of Baarda's (1968) testing method. These are indicated for the first survey, based on triangulation/trilateration, and for the second survey based exclusively on trilateration.

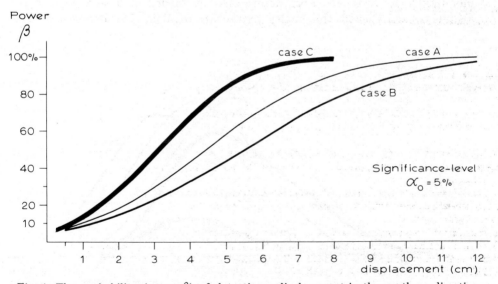

Fig. 4. The probability (power β) of detecting a displacement in the northern direction between Sicily and Calabria (see Fig. 2) by means of Baarda's (1968) testing method, indicated for successive surveys of the following types. Case A — triangulation/trilateration—trilateration; case B — trilateration in both surveys; case C — triangulation/trilateration in both surveys.

466

Probability of
taking wrong conclusions

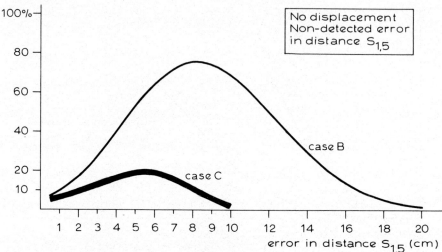

Fig. 5. The probability of incorrectly concluding that movements occurred on the basis of undetected errors in the distance $S_{1,5}$. The probability is indicated for the types of successive surveys known as case B and case C (Fig. 4).

quadrilateral were measured, whereas in later surveys only distances were measured. The distances were determined by means of a laser geodimeter (Model 8, AGA) and the angles by means of Wild T3 theodolites. The

Probability of
taking wrong conclusions

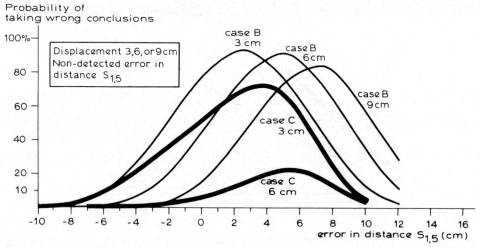

Fig. 6. The probability of concluding that no movement has occurred on the basis of undetected errors in the distance $S_{1,5}$, in successive surveys of types known as case B and case C (Fig. 4). The probability is established for different amounts of displacement, i.e., 3, 6 and 9 cm.

a-priori standard deviations of the distances S and the angles are respectively 10^{-6} S and 1^{cc}.

The results of the computations (for planning purposes) are given in Figs. 2—6. Three cases were considered:

(A) Combination triangulation/trilateration—trilateration.

(B) Combination trilateration/trilateration.

(C) Combination triangulation/trilateration—triangulation/trilateration.

The stations in Calabria were held fixed in the adjustment. They are the base points of the two-dimensional network. It is impossible to control the stability of the base points.

CONCLUSION

Computations of the probability of taking a wrong decision can and should be applied in planning networks, so that before the measurements are made an idea is obtained about the magnitude of the displacement which can be detected by statistical testing. It is very important that the geodetic observations, as well as the stability of the base points, are adequately checked.

REFERENCES

Baarda, W., 1968. A testing procedure for use in geodetic networks. Neth. Geod. Comm., Publ. Geodes., New Ser., 2: 1—97.

Baarda, W., 1973. S-transformations and criterion matrices. Neth. Geod. Comm. Publ. Geodes., New Ser., 5: 1—168.

Baarda, W., 1975. Difficulties in establishing a model for testing crustal movements. In: G.J. Borradaile, A.R. Ritsema, H.E. Rondeel and O.J. Simon (Editors), Progress in Geodynamics. North-Holland, Amsterdam/New York, pp. 45—51.

Baarda, W., 1977. Measures for the accuracy of geodetic networks. I.A.G. Symp. Sopron, Hungary, pp 1—26.

Caputo, M., Folloni, G., Pieri, L. and Unguendoli, M., 1974. Geodimetric control across the Straits of Messina. Geophys. J.R. Astron. Soc., 38: 1—8.

Meissl, P., 1962. Die innere Genauigkeit eines Punkthaufens. Oesterr. Z. Vermess. 50: 159—165, 186—194.

Van Mierlo, J., 1975. Statistical analysis of geodetic networks designed for the detection of crustal movements. In: G.J. Borradaile, A.R. Ritsema, H.E. Rondeel and O.J. Simon (Editors), Progress in Geodynamics. North-Holland, Amsterdam/New York, pp. 52—61.

Van Mierlo, J., 1977. Systematic Investigations on the stability of control points. Arch. 15th Int. Congr. Surv., Stockholm, Pap. 606.2: 169—182.

Tectonophysics, 52 (1979) 469—478

469

© Elsevier Scientific Publishing Company, Amsterdam — Printed in The Netherlands

HORIZONTAL DEFORMATION OF THE CRUST IN WESTERN JAPAN REVEALED FROM FIRST-ORDER TRIANGULATION CARRIED OUT THREE TIMES

TAKEHISA HARADA and MICHIYOSHI SHIMURA

Geographical Survey Institute, Tokyo (Japan)

(Accepted for publication June 8, 1978)

ABSTRACT

Harada, T. and Shimura, M., 1979. Horizontal deformation of the crust in western Japan revealed from first-order triangulation carried out three times. In: C.A. Whitten, R. Green and B.K. Meade (Editors), Recent Crustal Movements, 1977. Tectonophysics, 52: 469—478.

The first-order triangulation, in which the length of the average side is 45 km, was carried out twice over all of Japan, but the third repetition covered only the western part of Japan. At that time, a new geodetic survey system took the place of first-order triangulation in order to make the results more easily available for the prediction of earthquakes. Accordingly, in western Japan, the first-order triangulation was carried out three times: the first in 1886—1894, the second in 1948—1955 and the third in 1968—1972. During the past 60 years, deformation of the crust has been clarified by means of crustal strains and displacement vectors, which were found through precisely analyzing the first and second triangulations. Recently, the third triangulation was compared with the previous surveys by several methods. Some peculiar crustal movements found as the results of these studies are as follows:

(1) A vortex movement in the western Japanese crust seems to exist similar to that occurring in typhoons and hurricanes in the atmosphere.

(2) In local areas along the Japan Sea, deformed violently by earthquakes in 1927 and 1943, postseismic readjusting crustal movements still seem to be continuing.

(3) Crustal movements in coastal areas along the Pacific Ocean, due to the two severe earthquakes of 1944 and 1946, can be explained by the plate tectonics.

(4) Migration of crustal deformation can be seen in two areas.

INTRODUCTION

Figure 1 is the network of first-order triangulation in Japan which consists of about 350 stations. The average distance between adjacent stations is 45 km and the network has been observed two and a half times. The first set of observations was completely carried out all over Japan, including the southern chain of islands, from 1883 to 1913. The second set was done over most of Japan, excluding the southern islands, from 1948 to 1967. Imme-

470

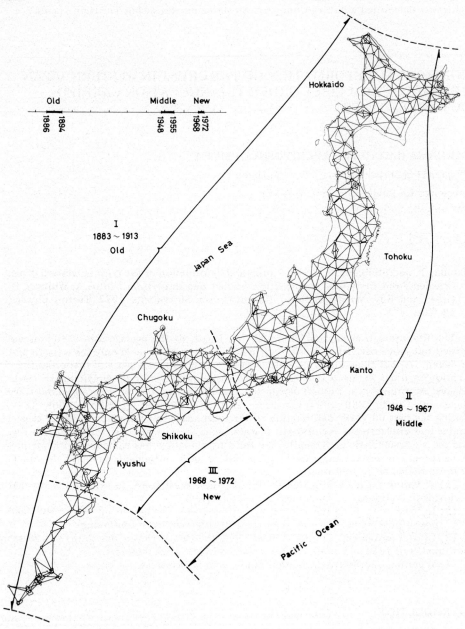

Fig. 1. Old, middle and new observations in the network of first-order triangulation. Heavy line shows observed distances in the middle survey.

diately after the second set of observations was finished, a repeat survey was started again in 1968 from south toward north; however, it was stopped in 1972 when it had covered the western half of Japan. In this paper these three sets of observations are called old, middle and new in order of time.

Following the 1972 survey in western Japan a precise geodetic survey was started instead of the first-order triangulation. In the precise geodetic survey, plans are to measure frequently, and as fast as possible, the first- and second-order network consisting of about 6000 stations with an average distance of 8 km, mainly by trilateration, using precise light-wave distance-measuring instruments. Originally, the first-order triangulation was repeated in an attempt to find the crustal deformation that might be valuable for earthquake prediction. A study has been made of horizontal crustal deformation

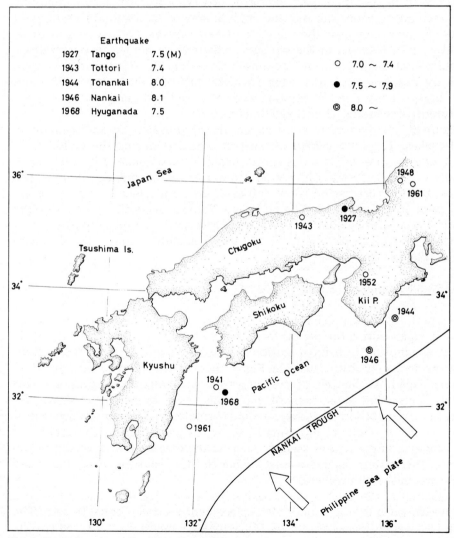

Fig. 2. Recent main earthquakes in western Japan.

from a comparison between the old and middle surveys. The important conclusion was that the network of first-order triangulation has a mesh too coarse to be able to record the precursors of an earthquake of magnitude 7. This is why first-order triangulation was stopped. Accordingly, in western Japan, the network of first-order triangulation was observed three times. Several illustrations on horizontal crustal deformation found from these analyses in western Japan are shown.

BACKGROUND IN SURVEY AND SEISMOLOGY

As shown in the upper part of Fig. 1, observing western Japan alone, the time interval between the old and middle surveys is about 60 years, with about 20 years between the middle and new surveys. Peculiarities in those surveys are as follows: in the old survey there were numerous observations of horizontal angles, but no astronomical azimuths. Four distances were given by base-line networks using an invar staff of 4 m. In the middle and new surveys many Laplace stations were observed and many distances were measured, using Models 2 or 8 geodimeters.

Figure 2 is a map of western Japan alone, and also shows recent main earthquakes. The only earthquake which occurred during the period from the middle to new surveys was the Hyuganada earthquake. All other earthquakes occurred during the period from the old to middle surveys. The Tango and Tottori earthquakes deformed the crust locally but violently. Two great earthquakes, at Tonankai and Nankai, made the crust along the Pacific Ocean move violently over an extensive area. Tectonically, the Philippine Sea Plate approaching from the south is sinking under the Japanese Islands' Plate at the Nankai Trough.

ANALYSIS BY DISPLACEMENT VECTORS

Figure 3, by Harada (1967), shows horizontal crustal displacement almost all over Japan during the period from the old to middle surveys. It was made on the assumption of a fixed station in central Japan, far from the western part. As indicated by vectors in Fig. 3 there is a clockwise movement in western Japan. At the time of this analysis it was thought this rotation might be false movement as a result of accumulated errors in western Japan, far from the fixed station. However, a similar rotation appeared again in Fig. 4 which shows the crustal movement during the period from the middle to new surveys. These figures seem to suggest the existence of a vortex movement in the western Japanese crust similar to that occurring in typhoons and hurricanes in the atmosphere.

Incidentally, the crustal deformation shown in Fig. 4 is a solution of a free network where the sum of all displacement vectors is equal to zero. The same free network was solved by Mittermayer's method (1972), where the sum of the squares of displacement vectors is a minimum. Differences

Fig. 3. Horizontal displacements at first-order triangulation stations over almost all of Japan during the period from the old to middle surveys. A station in the central part is assumed to be fixed.

between the two methods of solution are small — to an extent of only a few degrees in azimuth and a few centimeters in length for a vector. Accordingly, there are no clear distinctions between the figures of displacement vectors by the two methods. Various techniques of net adjustment, advanced in the past decade, have been programmed for the computer in the Geographical Survey Institute.

In order to avoid the effect of accumulated errors, Harada and Isawa (1969) calculated the displacement vectors during the period from the old to middle surveys by means of multi-fixed stations. The western part is shown in Fig. 5 where three stations were assumed as fixed. No significant changes occurred in the observed horizontal angles around these stations from the old to middle surveys. The movements toward the Pacific Ocean in the coastal areas from Shikoku to the Kii Peninsula were believed to be the result of strain release by the great earthquakes of Tonankai and Nankai, of magnitude >8. The vectors in western Shikoku encircled by the dotted

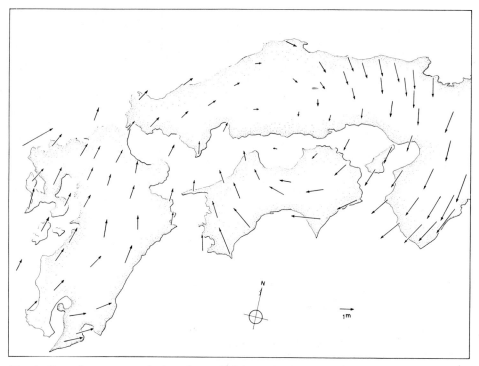

Fig. 4. Crustal movement during the period from the middle to new surveys, obtained by solving the free network where the sum of all displacement vectors is equal to zero.

line point to north. From this fact it was claimed that strain release did not reach to the extreme west of Shikoku. Fig. 6 shows displacement vectors during the period from the middle to new surveys, solved by the same means of multi-fixed stations. The coastal areas in Shikoku and the Kii Peninsula are already under pressure from the Philippine Sea Plate. In Fig. 6 strain in the western part of Shikoku has been released. Tada, of the Geographical Survey Institute, reports that the local area, including the channel between Shikoku and Kyushu was moved by a slow deep dislocation along the Nankai Trough after the Nankai earthquake. He also states that the Hyuganada earthquake of magnitude 7.5 was too small to cause the movement of this area. Flow and concentration of vectors in the Chugoku area are difficult to understand. Here, the selection of fixed stations could produce some uncertainty.

ANALYSIS BY STRAIN

Strain is free from the accumulation of errors as compared with displacement vectors. In the analysis of the observational data, strain components are computed for each triangle. Figs. 7 and 8 show the dilatations during the period from the old to middle surveys and from the middle to new surveys, respectively. White circles denote expansion and black circles contraction. In

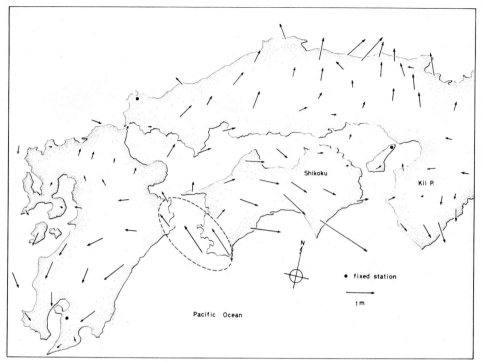

Fig. 5. Displacement vectors of first-order triangulation stations during the period from the old to middle surveys by means of multi-fixed stations.

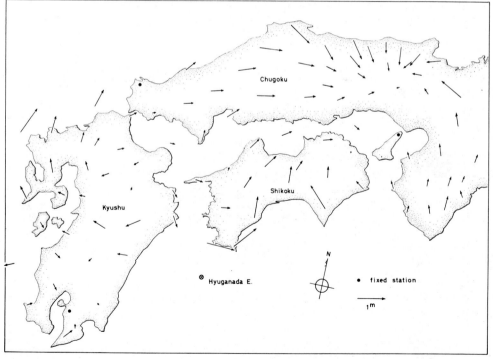

Fig. 6. Displacement vectors during the period from the middle to new surveys, solved by the same means of multi-fixed stations as in Fig. 5.

476

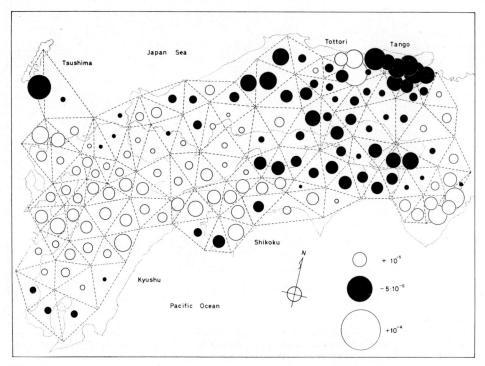

Fig. 7. Dilatations during the period from the old to middle surveys.

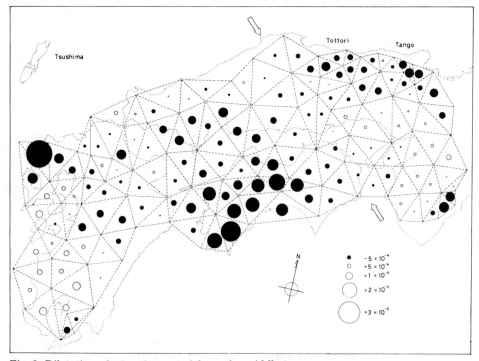

Fig. 8. Dilatations during the period from the middle to new surveys.

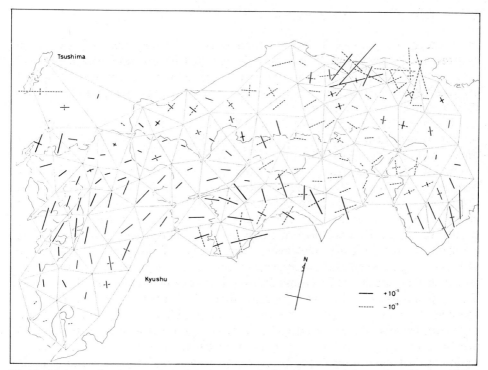

Fig. 9. Major and minor principal strain axes during the period from the old to middle surveys.

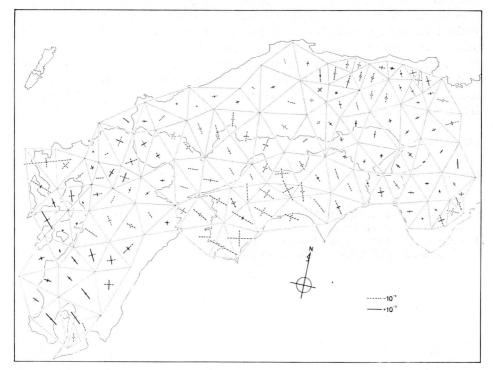

Fig. 10. Major and minor principal strain axes during the period from the middle to new surveys.

478

both areas, Tango and Tottori, which were deformed violently by earthquakes in 1927 and 1943, postseismic readjusting crustal movements still
seem to be continuing. We are surprised to see such a long duration of about
40 years in the readjusting crustal movement after the earthquake of magnitude 7.5 in the Tango district.

All dilatations are small in the neighborhood of the zone from southeast
to northwest shown by the two arrows in Fig. 8, but the same zone was
once a rather large dilatation zone in Fig. 7.

Previously, it was strange to see only one triangle with a big contraction
between Kyushu and Tsushima Island in Fig. 7, and yet to be unable to find
the corresponding earthquake. At that time it seemed to be due to observed
errors. Observations on Tsushima Island were not made in the new survey.
It is unfortunate that we do not know how much this large contraction has
changed since that time. However, surprisingly a large similar contraction
suddenly appeared in the northwestern part of Kyushu, as shown in Fig. 8,
which is just south of Tsushima Island and was once the expanded area in
Fig. 7. It appears that contraction has shifted from north to south. However,
the occurrence of a single large strain in the margin of a network is difficult
to prove, because an observed value with a large error at such a station cannot be corrected, even by a strict net adjustment unless there are numerous
redundant observations. Nevertheless, the successive large contractions do
not seem to be due to observed errors. When we review the principal axes of
strain as shown in Figs. 9 and 10 for the periods from the old to middle and
from the middle to new surveys, the shift mentioned above can be explained.
In these figures the solid shows extension and the dotted line shows contraction. In the area in question, it can be seen there is good coincidence in the
directions of the principal strain axes in both figures. There is one point of
interest as indicated in Figs. 9 and 10, i.e., that azimuths of principal strain
axes are almost fixed for a long period, even though contraction happens
to alternate with expansion, and vice versa.

REFERENCES

Harada, T., 1967. Precise readjustment of old and new first-order triangulations, and the
    result in relation with destructive earthquakes in Japan. Bull. Geogr. Surv. Inst. Jap.,
    12: 5—64.
Harada, T. and Isawa, N., 1969. Horizontal deformation of the crust in Japan — result
    obtained by multiple fixed stations. J. Geod. Soc. Jap., 14: 101—105.
Mittermayer, E., 1972. A generalisation of the least-squares method for the adjustment of
    free networks. Bull. Geodes., 104: 139—157.

Tectonophysics, 52 (1979) 479

STRAIN MEASUREMENTS AND TECTONICS OF NEW ZEALAND

R.I. WALCOTT

Geophysics Division, DSIR, Box 1320, Wellington (New Zealand)

(Accepted for publication June 8, 1978)

ABSTRACT

Measurements of shear strain from triangulation data have been made at 30 locations in New Zealand. The standard error of measurement in terms of strain rate is about $\pm 1 \cdot 10^{-7} \, y^{-1}$ and values of up to $7 \cdot 10^{-7} \, y^{-1}$ are observed. Together with 22 fault-plane solutions for crustal earthquakes the measurements indicate broad-scale patterns of deformation. Between the Hikurangi and Flordland active margins is a 100-km-wide belt, the axial tectonic belt, with shear strain rate averaging $5 \pm 1 \cdot 10^{-7} \, y^{-1}$ and an azimuth of the principal axis of compression of $114 \pm 8°$. The rate of movement (45 mm y^{-1}) and direction (085°) between the Pacific and Indian plates from the Minster et al. pole can be accounted for by the measured strain in the axial tectonic belt through simple shear parallel to, and compression normal to, the belt. The similarity in the rates determined from triangulation data averaged over 20—100 years and from plate movement averaged over 5 m.y. indicates plate movement to be uniform in time. West of the axial tectonic belt in Nelson and Fiordland are two zones in which movement is highly oblique to plate movement, and can be explained by slip line deformation analogous to the deformation of Asia. The azimuth of the principal axis of compression in the Taupo rift and East Cape region is NE—SW, perpendicular to its direction in the axial tectonic belt, suggesting extension in the rift and East Cape region normal to the subduction zone.

Tectonophysics, 52 (1979) 481—496   481
© Elsevier Scientific Publishing Company, Amsterdam — Printed in The Netherlands

A COMPARISON OF LONG-BASELINE STRAIN DATA AND FAULT CREEP RECORDS OBTAINED NEAR HOLLISTER, CALIFORNIA

L.E. SLATER and R.O. BURFORD

Applied Physics Laboratory, University of Washington, Seattle, Wash. 98195 (U.S.A.)
Office of Earthquake Studies, U.S. Geological Survey, Menlo Park, Calif. 94025 (U.S.A.)

(Accepted for publication June 8, 1978)

ABSTRACT

Slater, L.E. and Burford, R.O., 1979. A comparison of long-baseline strain data and fault
    creep records obtained near Hollister, California. In: C.A. Whitten, R. Green and B.K.
    Meade (Editors), Recent Crustal Movements, 1977. Tectonophysics, 52: 481—496.

    A comparison of creepmeter records from nine sites along a 12-km segment of the
Calaveras fault near Hollister, California and long-baseline strain changes for nine lines in
the Hollister multiwavelength distance-measuring (MWDM) array has established that
episodes of large-scale deformation both preceded and accompanied periods of creep
activity monitored along the fault trace during 1976. A concept of episodic, deep-seated
aseismic slip that contributes to loading and subsequent aseismic failure of shallow parts
of the fault plane seems attractive, implying that the character of aseismic slip sensed
along the surface trace may be restricted to a relatively shallow (~1-km) region on the
fault plane. Preliminary results from simple dislocation models designed to test the concept demonstrate that extending the time-histories and amplitudes of creep events sensed
along the fault trace to depths of up to 10 km on the fault plane cannot simulate adequately the character and amplitudes of large-scale episodic movements observed at
points more than 1 km from the fault. Properties of a 2—3-km-thick layer of unconsolidated sediments present in Hollister Valley, combined with an essentially rigid-block
behavior in buried basement blocks, might be employed in the formulation of more
appropriate models that could predict patterns of shallow fault creep and large-scale
displacements much more like those actually observed.

INTRODUCTION

Surface displacements across the Calaveras fault zone in Hollister Valley
have been monitored simultaneously at distinctly different length scales
since September 1975, when a multiwavelength distance-measuring (MWDM)
instrument system was installed near the Calaveras fault trace in the city of
Hollister, California, to augment several other previously existing systems for
monitoring of tectonic activity (Slater, 1975; Slater and Huggett, 1976; Huggett et al., 1977). Broad-scale movements are tracked by the MWDM instrument, while slip activity along the trace of the Calaveras fault is measured by

a network of creepmeters (invar rod or wire extensometers), each spanning a zone of several meters width across the fault trace (Nason et al., 1974). The configurations of the two networks and the positions of major faults in the study area are shown in Fig. 1.

Most of the sites in the creepmeter network were established during 1971, and nearly continuous records from at least eight sites covering a 12-km segment of the fault trace are available for study. Creepmeter XCD, put into operation August 2, 1976, is the most recent installation within the network shown in Fig. 1. Events of a few millimeters' displacement and of several hours' to a few days' duration generally occur once or twice a year within the network, and the events are separated by several months of little or no

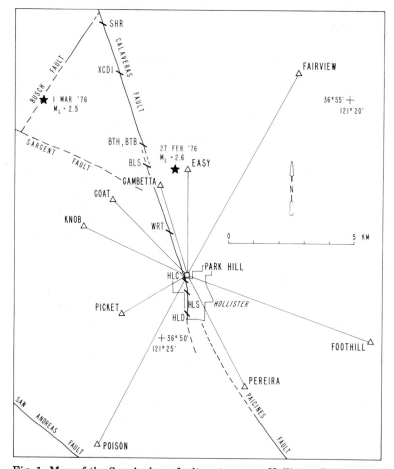

Fig. 1. Map of the San Andreas fault system near Hollister, California showing creepmeter sites along the Calaveras fault, the MWDM array, and the epicenters of two earthquakes that were possibly associated with a strain episode and a period of creep activity during February—March, 1976.

483

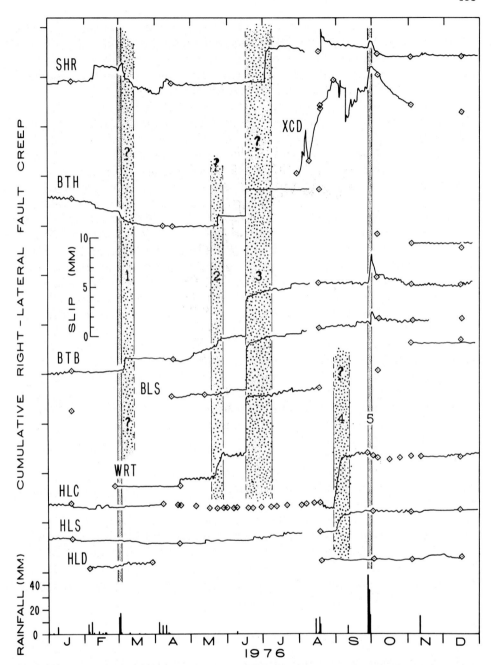

Fig. 2. Plot of 1976 fault-creep records and Hollister Valley rainfall. Continuous lines represent creep data sampled daily from on-site strip-chart records (blank for missing data). Diamond symbols enclose points obtained by manual readings. Shaded bars 1—5 identify periods of creep activity, and darker shading indicates that the period of activity was associated with heavy rainfall.

slip activity. However, a cluster of minor events and an increase in slip rate during the period April—May, 1976 apparently culminated in a major 6-mm creep event on June 16 (Fig. 2). Although such cluster activity is unusual for the Calaveras fault in Hollister Valley, similar sequences have been detected nearby along the San Andreas fault (Burford, 1977).

The new MWDM instrument site is located on Park Hill in Hollister at a point 150 m east of the main trace of the Calaveras fault and 1.5 km north of the HLD site, which is at the south end of the creepmeter network. The Hollister MWDM array consists of nine primary baselines 3—9 km long arranged in a radial pattern from the instrument site (Fig. 1). The system is capable of better than 10^{-7} length-measurement precision (ratio of the standard deviation of a set of length determinations to their mean value) at the maximum range monitored (Slater and Huggett, 1976). The baselines terminate at permanent retro-reflector sites that range from 0.1 to 7 km in distance from the Calaveras fault trace. Three of the reflector stations on the northeast side of the fault in the area north of Hollister lie along a path that is nearly straight and approximately normal to the strike of the Calaveras fault. Included in this group are the pair of stations closest to and furthest from the fault (Gambetta and Fairview), and an intermediate site (Easy) 1.5 km from the fault (Fig. 1). The focus of the discussion will be to explore the relations between length changes for these three baselines and aseismic slip activity monitored by the creepmeter network.

Creep events measured along fault traces within the San Andreas fault system represent episodes of propagating aseismic failure that are probably restricted to the uppermost kilometer or so of the fault planes, with varying amplitude and propagation velocity from site to site, as suggested by comparison of creep data and near-field strain or surface tilt measurements at a few sites on the San Andreas fault (Johnston et al., 1976; McHugh and Johnston, 1976; Mortensen et al., 1977; Goulty and Gilman, 1978). A model proposed for a 1971 creep event on the Calaveras fault in the present study area predicted a maximum depth penetration of 2.8 km beneath the center of a surface break 5.6 km long near the WRT creepmeter site (King et al., 1973). Answers to questions thus raised regarding the relation of fault-creep activity detected at the surface to slip activity deeper on the fault plane are of fundamental importance in the continuing effort to evaluate the role of fault creep in the general process of strain release along the plate boundary formed by the San Andreas fault system.

FAULT CREEP DATA

The fault creep records presented in Fig. 2 represent 1976 surface slip activity along a segment of the Calaveras fault trace approximately 12 km in length, from the Shore Road site (SHR) on the north to the D Street site (HLD) in Hollister on the south. This segment of the Calaveras fault is relatively straight and uncomplicated except for two features: a sharp inflection between the BTB and BLS creepmeters that may be indicative of an en-eche-lon overlap and related structural complications (Radbruch-Hall, 1974), and a 20° change in strike in the vicinity of the HLS creepmeter site in Hollister to a nearly N—S trend at the south end of the creepmeter network (Rogers and Nason, 1971). At least four periods of creep-event activity can be identi-fied in the creepmeter records from this segment (Table I and Fig. 2).

The period of greatest creep activity during 1976 started on June 16 with the sharp onset of a 6.0-mm maximum recorded slip at the WRT site that was followed within about 2 hr by creep events with decreasing amplitudes to the northwest through the BTH site (2.7 mm). For purposes of compari-son with the MWDM results, a later onset of creep at the SHR site (3.5 mm) on July 2 is included in the period of activity (Fig. 2), although it is not clear that slip was continuous throughout the gap in fault coverage between BTH and SHR. The fault segment involved in aseismic slip during the extended period starting June 16 thus has a minimum length of about 7 km. At about the time of the June 16 creep activity, the HLC creepmeter recorded a subtle change in trend from gradual contraction to gradual exten-sion (extension is equivalent to right-lateral displacement).

Another important episode of creep-event activity started on August 29 with the sharp onset of a 5.3-mm maximum recorded slip at the HLC site, adjacent to the MWDM instrument station. Loss of continuous coverage during this period at WRT prevents a definite identification of the event north of Hollister, but the change in the mechanical readings at WRT before

TABLE I

Slip (mm) during periods of creep-event activity on the Calaveras fault in Hollister Valley 1976

Creep periods (month/day)	Site names and relative positions along the fault southeastward from station SHR (km)								
	SHR 0.0	XCD 2.0	BTH 4.9	BTB 5.0	BLS 5.8	WRT 8.6	HLC 10.4	HLS 10.9	HLD 11.9
2/29— 3/13 (r)	−1.8	NR	−1.2	1.2	NR	0.0	0.2 *	0.0	0.5 *
05/19—05/27	0.1 *	NR	1.0	1.0	0.2	2.4	0.0	0.0	NR
06/16—07/07	3.5	NR	2.7	4.7	5.4	6.0	0.0	0.5 *	NR
08/29—09/11	0.0	−3.7 *	NR	0.0	0.0	NR	5.3	1.9	−0.2 *
9/28—10/1 (r)	0.7	2.3	NR	2.4	1.1	NR	−0.4	0.0	0.0

Number entries indicate amplitudes of right-lateral slip at each site during the designated period of creep activity. An (r) entry following the last date indicates the displacements occurred during a period of heavy rainfall. Negative entries designate retrograde displace-ment (apparent left-lateral slip). An asterisk following an entry indicates that slip at the site during the period may not have been spatially continuous with slip at sites showing the main event response. An entry of 'NR' indicates that no on-site record was obtained during the period of creep activity.

and after the event indicate a possible right-lateral offset of up to 1.7 mm during the active period. The displacement amplitude decreased sharply to the southeast to a value of only 1.9 mm at HLS, and no slip was recorded at the HLD site. A possible fault length of about 4 ± 2 km for the August 29 event is thus suggested by the network records. The maximum recorded slip at HLC was very nearly equal to that of the June 16 event at WRT, which provides a hint of delayed continuation of the June 16 event on the next section of the fault plane to the southeast, possibly with a common overlap zone in the vicinity of WRT. The HLD site was not involved in any of the 1976 creep events, which may mean a temporary dying-out of creep activity between HLS and HLD, near the change in fault strike. However, field tests to determine the reliability of the HLD response have shown that the creepmeter covers only a fraction of the total slip-zone width (R.D. Nason, personal communication, 1978), and past records show that the HLD instrument characteristically responds with considerably reduced slip amplitudes for events that propagate into that section of the fault (Schulz et al., 1976).

Two earlier periods of fault-creep activity during 1976 started on February 29 and May 19, and were of limited slip amplitude and extent along the fault. The first period of activity may have been related to heavy rainfall or, in a complex manner, to a pair of local earthquakes that occurred on February 27 ($M_{\mathrm{L}} = 2.6$) and on March 1 ($M_{\mathrm{L}} = 2.5$). The first shock had an epicenter between Easy and Gambetta within 1.6 km of creepmeter sites BLS, BTB, and BTH, and the epicenter for the second shock was located about 5.5 km north of Goat (Fig. 1). None of the creepmeter records show any significant coseismic slip.

The February—March fault-creep activity appeared as an apparent left-lateral offset at SHR and BTH, while BTB, only 80 m southeast of BTH, recorded a normal right-lateral movement of similar amplitude. The possibility of reversed polarity in the recording systems at SHR and BTH has been eliminated by comparison of the on-site records with the occasional mechanical readings. A gap in the record for the BLS creepmeter precludes positive identification of the event there, but the right-lateral offset indicated by mechanical readings before and after the data gap is 1.5 mm, similar to the event amplitude at BTB. The WRT station shows no response, hence the minimum fault length possibly involved in the February—March activity is about 7 ± 2 km.

The 1976 rainfall record for Hollister Valley is shown at the bottom of Fig. 2. The creep records at several sites were probably affected by rainfall starting on March 1, but the strongest such influence during the year is seen for the September 28—30 period of heavy rainfall. However, several of the periods of minor as well as major rainfall influenced the results from the SHR site.

Of the two early periods of minor creep activity, the second one in particular might be a forerunner of the June period of major creep events: so-called fore-creep activity (King et al., 1973). The suspected fore-creep of

1.0-mm amplitude occurred on a limited section of about 5 ± 3 km length, a part of the same section that produced the June 16 offset (Table I).

LONG-BASELINE STRAIN DATA

The general character of episodic line-length changes discussed in an earlier publication (Huggett et al., 1977) continues to be evident in the 1976 data shown in Fig. 3. The previous discussion was based on data obtained before April 18, 1976, and details concerning those results will not be repeated here except where they are obviously related to fault creep activity.

Considering 1976 MWDM data alone, there are only a few major episodes of strain activity defined by almost simultaneous strain-rate changes on several lines, and these are most clearly shown by length records for lines to Gambetta, Easy, and Fairview. Approximate dates defining these periods, and the corresponding total length changes for each line are listed in Table II.

Dramatic extension measured on the lines to Easy and Goat during the period March 1 to March 10 was accompanied by smaller length changes on lines to Knob and Fairview, as well as possible minor changes on other lines (Fig. 3). The extension reached a value of about 1 μstrain (10^{-6}) on the 4.1-km line to Easy (+4.4 mm) by March 10, and a value of 0.6 μstrain was measured on the line to Goat. Length changes measured during this period are apparently either reflections of strain changes accompanying the pair of small earthquakes (February 27 and March 1) or responses to the period of heavy rainfall, both mentioned above with regard to the first period of creep-event activity.

The second and third episodes of strain changes within the MWDM network are best defined by a period of contraction from April 30 to about July 21 on the lines to Easy, Fairview, and Gambetta, followed by a period of extension on the same lines that ended about September 4 (Figs. 3—5). The record for the line to Poison shows nearly the opposite strain-change pattern during the same two periods (Fig. 3). The selection of an April 30 start is also supported by the onset of gradual contraction on the line to Goat on that day and by an increase in the extension rate on the line to Knob during the first few days of May.

In contrast to the relatively smooth trends for longer lines terminating at reflector stations more than 1 km from the fault, contraction on the Gambetta line started later with sudden, step-like decreases on May 17 and 20 (−0.6 and −1.2 mm, respectively), followed by a period of gradual contraction that ended with a final step decrease (−2.2 mm) on June 16 (Figs. 3 and 4). At about the same time as the final step, the contraction rate decreased on the lines to Easy and Fairview, and the line to Goat began to show a slight extensional trend.

From mid-June until about September 4 the records for lines to Gambetta and Goat are almost identical, and until August 9 the record for the line to

488

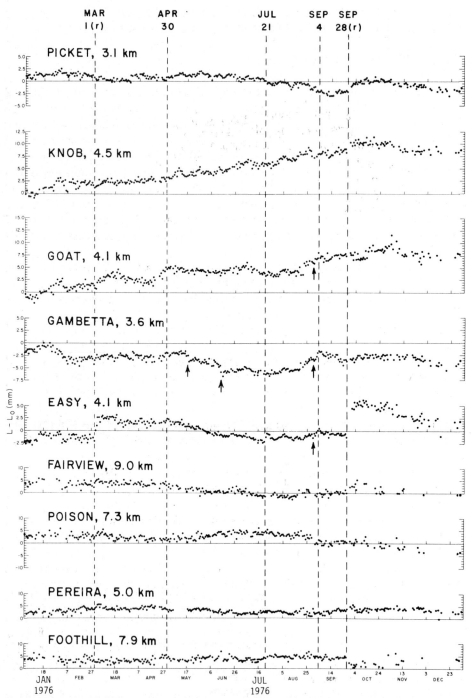

Fig. 3. MWDM long-baseline length changes, 1976. Time limits for periods of strain activity (Table II) and onset times for heavy rainfall (r) are indicated by dashed vertical lines. Arrows under the Gambetta record point to May 20, June 16, and August 29 line-length changes associated with creep events sensed at WRT or HLC.

TABLE II

Total length changes (mm) on MWDM lines during 1976 strain episodes

Strain periods (month/day)	Retro-reflector site names								
	Picket	Knob	Goat	Gamb.	Easy	Fair.	Foot.	Pere.	Pois.
03/01—03/10	−0.5	1.5	2.4	0.6	4.4	1.2	0.0	2.0	1.2
04/30—07/21	−0.7	2.4	−1.5	−4.3	−4.5	−5.1	−0.6	−1.0	3.1
07/21—09/04	−2.1	2.0	3.5	4.5	2.7	1.2	0.0	0.0	−3.4
1976 totals	−2.7	8.3	8.5	−4.0	4.5	−4.0	0.5	0.5	−6.5

Number entries indicate baseline length changes for each MWDM line during the selected period indicated in the left column. Positive and negative entries designate line extension and contraction, respectively.

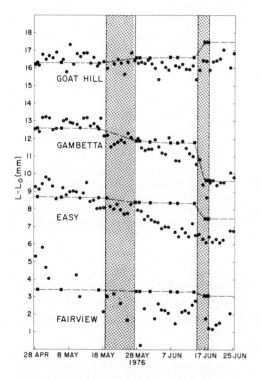

Fig. 4. Length-changes predicted for selected MWDM lines by a constrained 10 × 10 km Volterra dislocation loop with a southern limit halfway between WRT and HLC creep-meter sites. Time-history for a 6-mm right-lateral slip in the model was patterned after the creep record from WRT, and shading shows the durations of the two creep events. Details of measured length changes are shown by dots, for comparison with predicted changes shown as squares connected by dashed lines.

490

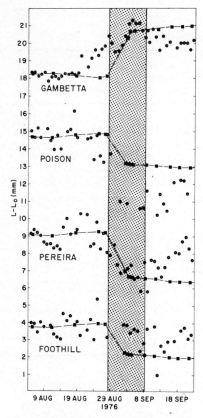

Fig. 5. Length-changes predicted for selected MWDM lines by a constrained 5 × 5 km Volterra dislocation loop with a northern limit coincident with the southern limit of the larger loop of Fig. 4. Time-history for a 5-mm right-lateral slip in the model was patterned after the creep record from HLC, and shading shows the duration of the event. Predicted and measured length changes are shown for comparison in the same manner as in Fig. 4.

Knob also shows a very similar pattern superimposed on a continuing extensional trend (Fig. 3). All three lines show extension that started (or increased) about June 18 and peaked on about July 4, followed by a contractional trend that continued until about July 21.

Each of the records for lines to Fairview, Easy, Gambetta, Goat, and Knob shows an onset of extension during the period July 21 to August 9, and by August 20 all, except for the line to Fairview, were undergoing rapid extension (Fig. 3). The record for the line to Goat shows a sudden length increase of approximately 2 mm on about August 22, and a smaller step increase of 1 mm on about August 30. August 29—30 also marks the onset time for a short period of rapid extension on the lines to Gambetta and Easy, accompanied by a contractional step on the line to Poison (Figs. 3 and

5). The pattern of changes during August, especially August 29—30, indicates an independent southward movement of the Park Hill instrument site.

COMPARISON OF LONG-BASELINE STRAIN CHANGES AND FAULT CREEP ACTIVITY

Even a causal comparison of the 1976 data-sets produced by the creepmeters and the MWDM array clearly demonstrates that periods of fault-creep activity and episodes of long-baseline length changes are closely related. The most obvious relation is shown by results for the period during the late spring and summer, when contractional trends on the lines to Gambetta, Easy, and Fairview were accompanied by major creep events along the Calaveras fault from WRT to SHR, and extensional trends on the same lines were accompanied by a large creep event centered near Park Hill at the HLC creepmeter site. Another aspect of the close relation between the two data sets is evident in comparisons of the apparent responses to periods of heavy rainfall (Figs. 2 and 3).

Several of the creepmeters showed significant responses to periods of rainfall during 1976 (Fig. 2), especially the SHR instrument. For example, most of the creepmeter records obtained during the period March 1—13 are probably contaminated by the effects of rainfall, which casts considerable doubt on any suggestion concerning a possible tectonic relation between the February 29 to March 13 creep activity, the associated MWDM line changes, and local earthquakes. A widespread response of the creepmeter network to the second period of heavy rainfall that started on September 28 is similar at several of the sites to the rather high-amplitude response of the XCD instrument. Only the HLD and HLS instruments failed to respond to the September rainfall, perhaps because these instruments were installed almost entirely under asphalt pavements in the city of Hollister (R.D. Nason, personal communication, 1978). The remaining sites that produced records during the storm clearly show responses indicating sharp onsets of apparent slip early in the rainfall period, possibly due either to expansion of soil or fault-gouge material (XCD), or perhaps to surficial slumping of soil or to relaxation of residual stress in shallow material near the fault (HLC). The length of the subsequent period of decay seems to be related to the amplitude of the initial response at each site, and in most cases a return to pre-rain signal levels takes place within a few days (Fig. 2).

Significant changes in apparent lengths (or in relative positions of reflector stations) within the MWDM array were associated with at least the two periods of heaviest rainfall beginning on March 1 and September 28. Although the possible tectonic effects of local earthquakes could be involved in the line changes measured during March 1—10, the similarity of the two responses measured on the line to Easy for both rainfall periods indicates the presence of a strong rainfall influence in the March results. Apparent length changes of significant amplitudes associated with the March rain were also measured on lines to Knob, Goat, Fairview, and Poison; the lines to Picket, Knob, Fairview, Pereira, and Foothill also show responses to the late September rain. In a manner similar to the creepmeter results associated with the second period, the line-length to Easy shows a tendency for a return to pre-rain values that is established soon after the end of the rain. However, decay of the apparent rainfall effects on the line to Easy requires up to several weeks rather than the few days of readjustment typically shown by the creepmeters. The effects of rainfall on the results from both monitoring systems illustrate the need for additional studies designed to investigate more thoroughly the mechanics and the areal or volume scale of the rainfall response. The relative stabilities of various creepmeter pier and bench-mark emplacement designs in a wide range of slope and soil-type environments also should be tested.

The complexities of instrumental and/or soil and fault-gouge response to rainfall obviously are not well understood, and a complete discussion of these phenomena is beyond the scope of the present study. However, as emphasized by Wood (1977), the response of various geophysical field instruments to rainfall and other surface influences is of great potential importance. It is usually impossible to separate clearly the effects of rainfall from signals generated in response to tectonic activity, even where a clear association with rainfall exists (Wood, 1977), but it is essential to identify the association if a more complete understanding of the full range of possible rainfall responses is ever to be gained.

Contraction steps on the line to Gambetta during May 20 and June 16, 1976 closely correspond in time to the onsets of creep events recorded on nearby creepmeters (May 19 and June 16, WRT), and decreases in length during the steps are nearly equal to the half-amplitudes of the maximum creep offsets recorded at WRT (Figs. 3 and 4). The lack of corresponding step responses on the longer lines to reflector stations Easy, Goat, and Fairview, which are all 1.5 km or more from the fault, indicates that slip was probably restricted to depths of about 1 km on the fault plane, at least during the periods when creep events were sensed along the trace. However, the onset of contraction on the longer lines preceded the onset of the first creep event by about 18 days — a possible indication of a deep-seated fault-creep episode culminated by emergent, observed surface offsets. The total length decreases on each of the lines to Gambetta, Easy, and Fairview during the period April 30 to July 21 were very nearly equal (−4.3 to −5.1 mm), and closely matched the 4.2-mm half-amplitude of the maximum, total creep offset at WRT during the same period. There is good evidence from other MWDM lines and from the creep records produced at HLC and HLS that the instrument station on Park Hill was not subjected to the amount of displacement necessary to account for the contraction on the lines to Gambetta, Easy, and Fairview. However, the extension on the line to Poison and the slight contraction on the line to Goat during this period perhaps could have been produced in part by a small amount of northward translation (~1 mm) of the terrain containing the Park Hill—Foothill—Pereira triangle.

As implied during the discussion of fault-creep data, creep activity on the Calaveras fault north of Hollister during May and June must have contributed to the load on the section of the fault adjacent to Park Hill and may have hastened its subsequent aseismic failure. A mid-June increase in load across the fault near Park Hill is evident from the change in sense of strain (or perhaps the onset of gradual right-lateral slip) seen on the HLC record. Supporting evidence is found in the similar extensional patterns developed in records for lines to Knob, Goat, and Gambetta, as well as minor contractions on lines to Foothill and Pereira during the period mid-June to July 4, indicating a common source probably attributable to gradual southward movement of the Park Hill instrument station during that period.

Results of an attempt to model the surface displacements for the periods of major surface creep activity are presented in Figs. 4 and 5. Volterra dislocations (Chinnery, 1961) operating over square areas on the fault plane were chosen in a manner to match the creepmeter results while also providing maximum horizontal displacement at the MWDM reflector stations. Choices for size and position of the simple dislocation loops were based on combined results from the entire creepmeter network. However, amplitudes of slip for the two loops were based on an approximate value interpolated to a point opposite Gambetta between total offset at WRT and BLS for the May—June activity, and the maximum value recorded at HLC near Park Hill for the August—September activity.

The model for the May and June events consists of a 6-mm displacement distributed uniformly over a vertical 10×10 km dislocation loop in the position of the Calaveras fault with a southern end located midway between WRT and HLC. The time-history for slip on the model fault was patterned after the recorded slip history at WRT for the period April 28—June 25. Relative displacements predicted for critical MWDM reflector stations from the constrained dislocation model are shown in Fig. 4, along with the actual measured displacements. The model indicates that baselines to Gambetta, Easy, and Fairview should all have shown some contraction, but only the predicted change for the line to Gambetta agrees closely in amplitude with the measured displacement.

The results shown in Fig. 4 illustrate several important points. First, the close agreement between measured and predicted displacements at Gambetta indicate that the shallow fault behavior is represented reasonably well by the model, as expected. The two creep events from the model match very closely the two discontinuous length changes measured on the Gambetta line, except for slightly earlier observed onsets and a somewhat smoother shift in the measured length associated with the larger event. Second, the early onsets of length-change and smooth trends for the lines to Easy and Fairview, with no sudden steps or abrupt changes in rate near the creep event times, suggest that the signals recorded on the longer lines were generated by deeper aseismic slip which preceded the initial surface creep activity by about 18 days and the large June 16 event by as much as 47 days. Third, all three reflector stations (Gambetta, Easy, Fairview) exhibited very similar total relative displacements that match very closely the predicted displacement for Gambetta and the half-amplitude of creep offset between BLS and WRT. The total measured displacement at Fairview was about an order of magnitude greater than that predicted by the simple dislocation model (-4.2 mm observed compared with -0.3 mm predicted), and the ratio for observed to predicted relative displacement at Easy is 3 : 1.

The model for the August—September creep activity consists of a 5-mm displacement distributed uniformly over a vertical 5×5 km dislocation loop on the Calaveras fault, positioned with a northern limit adjoining the southern end of the larger loop midway between WRT and HLC. The time distribution for slip in the model was matched to the August 4—September 23 creep record from HLC. Fig. 5 shows the predicted line-length changes resulting from the model, superimposed on the details of observed length changes measured on the lines to Gambetta, Poison, Pereira, and Foothill. Predicted and observed relative displacements at reflector stations Gambetta and Pereira, both located very near the fault trend on the northeast side, are reasonably well matched, except that observed extension on the line to Gambetta began several days before the onset of creep at HLC. The observed relative displacements at these two reflector stations during the period August 19—September 6 demonstrate that most, if not all of the displacement was occurring at the Park Hill instrument site during that time, the total amount matching almost perfectly the half-amplitude of the total creep-event offset (2.7 mm). However, predictions from the model fail to match the observed relative displacements of Foothill and Poison, 5.5 and 6.0 km distant from the Calaveras fault trend. The observed and predicted relative movements at Foothill and Poison match in sense (contraction), but the amplitude of the observed displacement at Poison during the period August 19—September 6 is larger than the predicted value by a factor of nearly 2.5 and more nearly matches the total slip amplitude at HLC, while observed relative displacement at Foothill during the same period (about -0.5 mm) reaches less than a third of the predicted amplitude. If the shallow dislocation model were to provide an adequate representation of the total deformation during the period of creep activity, then measured relative displacements at the more distant points should also reflect isolated movement at the Park Hill instrument station. Obviously, the model fails adequately to represent the behavior at greater distances from the fault. Rigid-block displacements could explain this lack of agreement, as well as the general aspects of measured relative displacements for the two terrains separated by the Calaveras fault zone.

CONCLUSIONS

The comparison of 1976 MWDM strain data and fault creep records obtained near Hollister, California, has provided a fresh opportunity to consider the possible relations between crustal deformation, displacement, and shallow fault creep on a major discontinuity in the earth's crust. Examination of both data sets has clearly established that creep measured across the Calaveras fault trace near Hollister is preceded by episodes of gradual length change on lines that terminate at points 1.5—7 km distant from the fault. Although the associations between periods of surface fault-creep activity and episodes of long-line changes are fairly obvious, the correct choice between various possible interpretations is much less clear. One possibility is that the pre-creep line-length changes reflect episodes of gradual aseismic fault slip, of about the same final amplitude as that measured by the creepmeters, starting deep on the fault plane ($\sim$10 km) and eventually emerging along the fault trace. The problem with this explanation is that simple dislocation models based on the concept do not predict the large magnitudes of relative displacement observed during the episodes for points far from the fault, and the antisymmetric patterns of line-length changes predicted from the models also do not match the observed patterns.

Comparisons of the predicted and observed changes indicate that a more appropriate model might consist of a very thick, stiff elastic layer overlain by a relatively thin, soft layer. Both layers would be cut by a vertical fault plane, and two separate dislocation loops matched to the thicknesses of the contrasting layers might be assigned individual histories and amplitudes of slip. The relative displacements measured by the MWDM system at points far from the fault could perhaps be matched by introducing an appropriate amplitude and history of slip across the lower dislocation loop (essentially rigid-block displacement), and the relative displacements of points near the fault (including those indicated by the creepmeter results) by a different history and amplitude of slip across the shallow loop. Geologic evidence for a 2—3-km-thick layer of unconsolidated sedimentary material deposited in the vicinity of Hollister more or less continuously since Early Pliocene time is well documented (Allen, 1946; Rogers, 1978). The lower half to two-thirds of the soft sedimentary layer might act as a transition zone between the more extreme properties and behavior of the deep and near-surface layers.

If a general model of the type described here is appropriate, the following preliminary conclusions are suggested:

(1) Episodic fault-creep behavior corresponding to large-scale, essentially rigid-block movements extends to considerable depth ($\sim$10 km) on the Calaveras fault near Hollister.

(2) Deeper fault creep is characterized by episodes of accelerated aseismic slip of several weeks' duration, in contrast to the shorter periods of up to several days' duration for the typical near-surface creep response.

(3) Onsets of deep fault-creep episodes can precede the shallow creep response by several weeks.

(4) Clusters of otherwise seemingly isolated creep events sensed across the fault trace apparently may be related to a single long-term episode of deep-seated aseismic movement.

ACKNOWLEDGEMENTS

We wish to thank the city of Hollister and the many people who have kindly allowed continued access to their property. Sandra S. Schulz helped with the preparation of creepmeter records. The original manuscrupt was improved with the help of critical reviews contributed by Phillip Harsh and Wayne Thatcher. Operation of the MWDM system was supported by the U.S. Department of the Interior, Geological Survey, under contracts 14-08-0001-15263 and 14-08-0001-15877.

REFERENCES

Allen, J.E., 1946. Geology of the San Juan Bautista quadrangle, California. Calif. Div. Mines Bull., 133: 9—75.

Burford, R.O., 1977. Bimodal distribution of creep event amplitudes on the San Andreas fault at Melendy Ranch, California. Nature, 268: 424—426.

Chinnery, M.A., 1961. The deformation of the ground around surface faults. Bull. Seismol. Soc. Am., 51: 355—372.

Goulty, N.R. and Gilman, R., 1978. Repeated creep events on the San Andreas fault near Parkfield, California recorded by a strainmeter array. J. Geophys. Res., 83 (in press).

Huggett, G.R., Slater, L.E. and Langbein, J., 1977. Fault slip episodes near Hollister, California: Initial results using a multiwavelength distance-measuring instrument. J. Geophys. Res., 82: 3261—3368.

Johnston, M.J.S., McHugh, S. and Burford, R.O., 1976. On simultaneous tilt and creep observations on the San Andreas fault. Nature, 260: 691—693.

King, C.-Y., Nason, R.D. and Tocher, D., 1973. Kinematics of fault creep. Philos. Trans. R. Soc. Lond., Ser. A, 274: 355—360.

McHugh, S. and Johnston, M.J.S., 1976. Short-period nonseismic tilt perturbations and their relation to episodic slip on the San Andreas fault in central California. J. Geophys. Res., 81: 6341—6346.

Mortensen, C.E., Lee, R.C. and Burford, R.O., 1977. Observations of creep-related tilt, strain, and water-level changes on the central San Andreas fault. Bull. Seismol. Soc. Am., 67: 641—649.

Nason, R.D., Phillippsborn, F.R. and Yamashita, P.A., 1974. Catalog of creepmeter measurements in central California from 1968 to 1972. U.S. Geol. Surv., Open-file Rep., 74—31.

Radbruch-Hall, D.H., 1974. Map showing recently active breaks along the Hawyard fault zone and the southern part of the Calaveras fault zone, California. U.S. Geol. Surv., Misc. Invest. Map I-813.

Rogers, T.H., 1978. Geology and seismicity at the convergence of the San Andreas and Calaveras fault zones near Hollister, San Benito County, California. Calif. Div. Mines and Geol. Spec. Rep. (in press).

Rogers, T.H. and Nason, R.D., 1971. Active displacement on the Calaveras fault zone at Hollister, California. Bull. Seismol. Soc. Am., 61: 399—416.

Schulz, S.S., Burford, R.O. and Nason, R.D., 1976. Catalog of creepmeter measurements in central California from 1973 through 1975. U.S. Geol. Surv., Open-file Rep. 77—31.

Slater, L.E., 1975. A multiwavelength distance-measuring instrument for geophysical experiments. Ph.D. thesis, Univ. of Washington, Seattle, Washington.

Slater, L.E. and Huggett, G.R., 1976. A multiwavelength distance-measuring instrument for geophysical experiments. J. Geophys. Res., 81: 6299—6306.

Wood, M.D., 1977. Relation between earthquakes, weather, and soil tilt. Science, 197: 154—156.

Tectonophysics, 52 (1979) 497—503

497

MEKOMETER MEASUREMENTS IN THE IMPERIAL VALLEY, CALIFORNIA

R.G. MASON, J.L. BRANDER and M.G. BILL

Geology Department, Imperial College, London SW7 2BP (England)

(Accepted for publication June 8, 1978)

ABSTRACT

Mason, R.G., Brander, J.L. and Bill, M.G., 1979. Mekometer measurements in the Imperial Valley, California. In: C.A. Whitten, R. Green and B.K. Meade (Editors), Recent Crustal Movements, 1977. Tectonophysics, 52: 497—503.

A network of about 130 bench marks, mostly about 800 m apart at road intersections and forming a continuous chain of quadrilaterals crossing the Imperial Valley in the vicinity of El Centro, has been measured three times, in 1971, 1973 and 1975. The instrument used was a Mekometer, a superior short-range electronic-distance-measuring (EDM) instrument having a sensitivity of 0.1 mm and a potential accuracy of better than 1 ppm. In 1975 the network was extended, particularly where it crosses the Imperial fault, to about 200 bench marks defining about 500 lines.

For various reasons, it has not been possible to achieve standard errors of much less than 3 ppm in the distance measurements. Even so, about one line in three shows significant change at the 95% level in one or more of the three intervals 1971—73, 1973—75 and 1971—75. Adopting stringent consistency criteria, the number of places where significant changes occurred can be reduced to seven. One of these is almost certainly associated with the Heber geothermal anomaly, one with the Imperial fault, and the others with one or other of the faults identified on the ground further to the north. The largest and most consistent changes occur on the Imperial fault, and imply an aseismic slip on it averaging about 8 mm/yr.

The data also lend themselves to regional strain analysis. The results are consistent with a linear strain rate of about 0.3 ppm/yr, which is comparable with the rate deduced from large-scale geodetic surveys of the area.

INTRODUCTION

The Imperial Valley network (Fig. 1) was established in 1970, using funds made available to the University of California at Riverside by the U.S. Bureau of Reclamation, for geothermal studies. Initially, it comprised about 130 stations, mostly about 800 m apart at road intersections, forming a continuous chain of quadrilaterals stretching from the abandoned airfield east of Holtville, on the sand dunes on the east side of the valley, to a point south of Seeley on the west side, detouring El Centro en route and spanning a 45-km (28-mile) width of the valley.

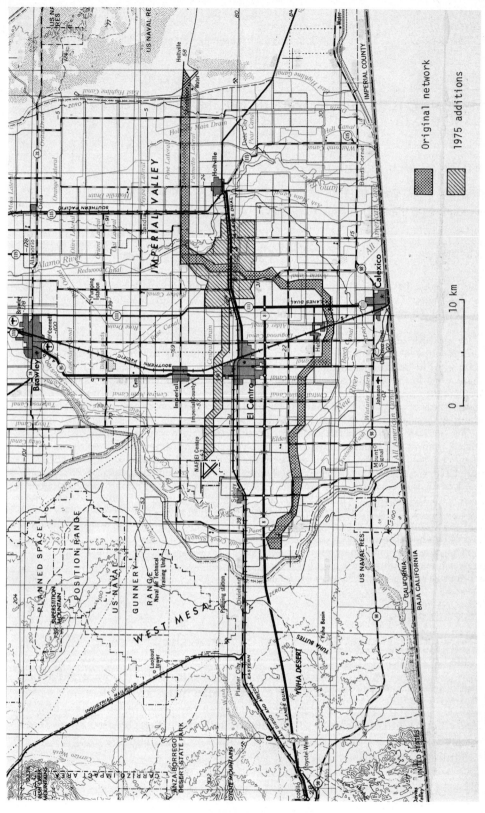

Fig. 1. The Imperial Valley Mekometer network.

Considered as a single geodetic unit, the network is clearly very weak. However, its purpose was to look for movements that might be associated with the Imperial fault, and with the several conjectured extensions of known faults thought to run subparallel to it through this part of the valley, also with the centre of anomalous heat flow near Heber, which it crosses. It was intended to be measured by trilateration, and the results studied in terms of changes in line lengths. There was no expectation of being able to determine relative displacements over the network as a whole, though it was expected that this would be possible over distances of a few kilometres.

The network was measured completely for the first time in 1971 by a team from Imperial College, using in-house funds, and again in 1973, using funds provided by the U.S.G.S. In both cases the measurements were made with a prototype Mekometer (Froome, 1971), loaned by the U.K. National Physical Laboratory. This is a superior short-range modulated light pulse instrument in which the modulation frequency is derived from a cavity resonator, and where phase-matching of outgoing and returning pulses is achieved by means of an internal variable light path. It has a sensitivity of 0.1 mm and an accuracy of better than one part per million (with appropriate corrections applied), and its availability was one of the factors that led to the setting up of the project.

These first two sets of measurements provided evidence for movement at a number of places. At the Imperial fault in the vicinity of Highway 80 they showed that, while points about 2½ km apart on opposite sides had moved right laterally relative to one another by about 10 mm during the two years, across the fault itself the total displacement exceeded 20 mm (see Fig. 4), implying not only a slip of that amount, possibly induced by the Superstition Hills earthquake of September 30, 1971, but also a release of previously accumulated strain. They showed that with proper planning a considerable amount of detail about the strain processes taking place across active faults can be obtained from measurements of this kind.

THE 1975 FIELD PROGRAM

For the third measurement, made in 1975, the network was extended where it crosses the Imperial fault in the neighbourhood of Highway 80, to form a solid block of stations about 8 km × 8 km (Fig. 2), also westward from the northwest corner of the above block to form a new E—W line of quadrilaterals passing to the north of El Centro. The purpose of the latter was to look for movements that might be matched with movements on the western part of the original network.

The new stations were constructed from 1-m lengths of 12.5-mm diameter stainless steel rod driven into the ground by hand and grouted in with cement. This appears to be a perfectly satisfactory alternative to the more expensive, capped, 3-m lengths of steel barrel used in 1970, which were cemented into holes drilled with a truck-mounted rig. Although carefully

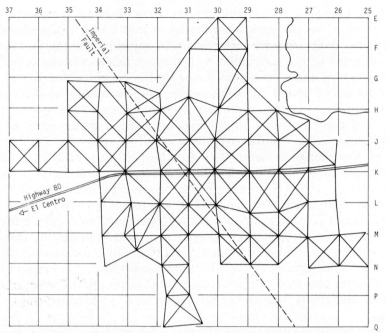

Fig. 2. Lines in the vicinity of the Imperial fault measured in 1975. The grid spacing is 800 m ($\frac{1}{2}$ mile). Stations are designated by the nearest grid intersection.

sited, stations of both types have been lost (and replaced) at a rate of about 5% p.a., mainly as a result of earth-moving operations connected with realinement of roads and irrigation channels. Unfortunately, one such station lost between 1973 and 1975 was L31, the station nearest the Imperial fault. With the latest additions, the network now contains 205 stations. The approximate location of the extended network is shown in Fig. 1.

The 1975 measurements were made with a commercial Kern ME3000 Mekometer. This has a more powerful light source than the prototype instrument, and a more effective reflector system, which enabled the measurements to be made in the daytime instead of at night, as was necessary previously. Apart from the obvious advantage of working in daylight, it became possible to estimate the mean air temperature along the optical path (which is required for making the atmospheric correction) to the necessary accuracy of rather better than 1°C simply by monitoring the temperature at the Mekometer end of the line during the 15—20 min taken to make a set of observations.

The 1975 measurements were sponsored by the U.S.G.S. under Contract No. 14-08-0001-14800, and the results, together with changes in lengths in the various intervals, are contained in the final report on that contract (Mason, 1976), which also includes station descriptions. Altogether, 478 lines were measured. In addition, 175 repeat measurements were made, including 45 of the baseline (K26—K27) and repeats of lines measured before and during the January 1975 Brawley earthquake swarm (Johnson and Hadley, 1976; Sharp, 1976), which occurred shortly after the survey started. Unfortunately, although some lines in the eastern part of the network had been measured before January 23, they were not suitably disposed for significant conclusions to be reached about movements that might be associated with the swarm.

ACCURACY OF MEASUREMENTS

The main sources of error are (1) incorrect estimation of the mean air temperature along the optical path, (2) short-term fluctuations in air and instrument temperatures, (3) fluctuations of the cavity-derived standard frequency, and (4) incorrect positioning of instrument and reflector over their respective stations. Except for the last, errors are proportional to the length being measured.

Two kinds of information are available on which the accuracy can be assessed — the standard deviations of the individual readings making up each set of length measurements, and the standard deviation of the baseline measurements. The former averaged about 0.6 mm, for all lines, and 46 measurements of the baseline (K26—K27, length 786.5246 m), followed a normal distribution and gave a standard deviation of 2.1 mm. The difference between these two figures arises, in probable order of importance, from fluctuations in the standard frequency, mal-positioning of instrument and reflector, and errors in estimating the air temperature. It may be noted that a frequency standard accurate to considerably better than one part in 10^7 is readily available in the field, in the form of the carrier wave of a radio station such as WWV. A method of monitoring the modulation frequency in the field using an off-air frequency standard has been described by Bradsell (1977).

From the repeated baseline measurements it may be concluded that the standard error of an individual measurement is a little less than 3 ppm, i.e., there is a 68% probability of the error being less than this, and a 95% probability of it being less than 5 ppm. In terms of significance of apparent changes in length, these figures imply a 68% probability of a change of >4 ppm, and a 95% probability of a change of >7 ppm, being significant.

BRIEF DISCUSSION OF RESULTS

Statistics of length changes

On average, lengths increased by 0.78 and 0.55 ppm respectively in the intervals 1971—73 and 1973—75, and by 1.07 ppm between 1971 and 1975. The last differs from the sum of the first two because the three intervals involve different combinations of lines. Taking into account the number of lines involved (150—200), these results are just significant at the 95% level, and indicate a linear strain rate of about 0.3 ppm/yr, which is within a factor of two of the rate deducible from the results of Miller et al. (1970). An attempt to obtain a value for regional shear strain was unsuccessful, possibly because of the biased nature of the network.

Movements since 1971

210 of the lines have been measured in at least two of the three years, and about one in three show significant change at the 95% level (i.e. >7 ppm) in one or more of the three intervals. However, some of these changes may be the result of "bad" measurements. Considering only the interval 1971—75, and rejecting all cases where changes are significant in the four years but of opposite sign in the separate intervals, the number of places where significant changes have occurred reduces to six, labelled *A—F* in Fig. 3, which shows also the directions of maximum extension and maximum positive shear at the six places. Of these, either *A* or *B* might be associated with the Calipatria fault, and *C* with the Brawley; *D* is the Imperial fault, and *E* might mark the Superstition Mountain fault; *F* coincides with the Heber geothermal area.

Of the six places, the most consistent changes occur at the Imperial fault. Fig. 4 shows displacement vectors for the five stations nearest to the fault

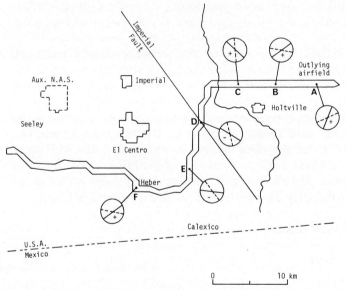

Fig. 3. The six places (*A—F*) crossed by the network at which the most consistently significant changes have taken place between 1971 and 1975. Firm lines in insets show directions of maximum extension, broken lines show directions of maximum positive shear. The positive and negative signs indicate positive or negative dilatation.

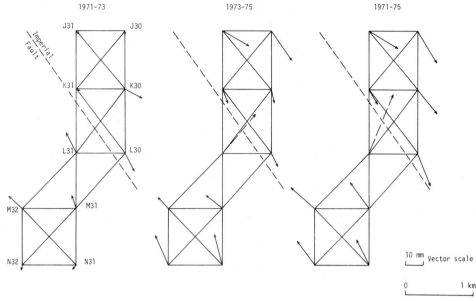

Fig. 4. Displacements across the Imperial fault between 1971 and 1975, "balanced" arbitrarily about the fault. Note that Station L31 had been disturbed some time between 1973 and 1975. The 1973—1975 and 1971—1975 displacement vectors include this accidental movement and have no tectonic significance.

on each side, for the three intervals. These were derived by first assuming that Station J31 had not moved and line J31—K30 had not rotated. A common vector was then added to each such as to make their vector sum zero, i.e., to make the mean displacement of the 10 points zero. The procedure is arbitrary, but valid, and it has the effect of visually balancing the vectors about the fault. The most interesting fact to emerge is the large rate of shear across the fault relative to points further away. Disregarding Station K31, which is almost on the fault, this could indicate an aseismic slip on the fault itself averaging about 8 mm/yr over the four-year period.

REFERENCES

Bradsell, R.H., 1977. A simple calibrator for the Mekometer. Surv. Rev., 24: 1—4.
Froome, K.D., 1971. Mekometer III: E.D.M. with sub-millimetre resolution. Surv. Rev., 21: 98—112.
Johnson, C.E. and Hadley, D.M., 1976. Tectonic implications of the Brawley earthquake swarm, Imperial Valley, California, January 1975. Bull. Seismol. Soc. Am., 66: 1133—1144.
Mason, R.G., 1976. Mekometer measurements in the Imperial Valley, California. U.S.G.S. Contract 14-08-0001-14800, Final Rep.
Miller, R.W., Pope, A.J., Stettner, H.S. and David, J.L., 1970. Crustal movement investigations — triangulation: Imperial Valley. Oper. Data Rep. C. & G.S. Dr-10, Coast Geod. Surv., U.S. Dept. of Commerce, Rockville, Md.
Sharp, R.V., 1976. Surface faulting in the Imperial Valley during the earthquake swarm of January—February, 1975. Bull. Seismol. Soc. Am., 66: 1145—1154.

Tectonophysics, 52 (1979) 505—518
© Elsevier Scientific Publishing Company, Amsterdam — Printed in The Netherlands

INSTRUMENTATION AND FIELD MONITORING OF GROUND CRACKING

J.E. O'ROURKE, A.P. RIDLEY and J.R. McCONNELL

Woodward-Clyde Consultants, Three Embarcadero Center, Suite 700, San Francisco, Calif. 94111 (U.S.A.)

(Accepted for publication June 8, 1978)

ABSTRACT

O'Rourke, J.E., Ridley, A.P. and McConnell, J.R., 1979. Instrumentation and field monitoring of ground cracking. In: C.A. Whitten, R.Green and B.K. Meade (Editors), Recent Crustal Movements, 1977. Tectonophysics, 52: 505—518.

The results of a one-year-duration precision measurement program for monitoring horizontal and vertical movements of bedrock surfaces adjacent to a ground-cracking phenomenon are presented. The ground cracking extends in a relatively straight line about 1200 m across ridges and draws in a foothill area of northern California. The site is in a seismically active region, but the closest faults, about 0.8 km away, are considered inactive. Cracking generally followed the contact of a steeply dipping claystone bed with a massive sandstone bed. Precision monitoring lines, 30 m in length, were established at two distinctive topographical features along the cracking, a saddle area and a side hill area. The precision lines were supplemented by a 230-m control line and several isolated points to be monitored. Instrumentation included a portable micrometer beam extensometer and mechanical tiltmeter, a tape extensometer, and piezometers. Seasonal and artificially stimulated movements were observed during the monitoring period. Vertical ground movements of 75 mm and horizontal ground extensions of 55 mm in 3 m were observed during the study period. Data suggest moisture/volume change properties of the claystone unit as the principal mechanism.

INTRODUCTION

Geological field studies for a proposed large land development in northern California revealed the presence of a series of discontinuous ground cracks trending for over 1200 m along the west side of a synclinal valley. The bedrock strata consist of sandstones, siltstones and claystones. These sedimentary strata generally dip steeply toward the east, but in some areas have been mapped vertical or slightly overturned and dipping west. Faults have been mapped on either side of the site. One fault is located about 800 m west of the site and the other fault is located about 600 m east of the site. These fault features are roughly subparallel to the syncline axis and there is no available evidence that either fault is active. The nearest known active fault is located about 8 km northeast of the site.

Field mapping showed that the ground cracks follow the contact between a massive sandstone bedrock unit and an interbedded strata composed primarily of claystone similar to a hard-fissured clay. Exploration trenches indicated that the cracks occur about 1.5—3.0 m on the uphill or west side of the contact in the sandstone unit. Cracking was observed to extend into bedrock below the bottom of the trenches.

The presence of the consistent linear cracking feature across such a great distance, and to significant depth, pointed to the need to define the mechanism for the cracking so that the impact on proposed development in the area could be evaluated. The candidate mechanisms were believed to be either moisture/volume change properties of expansive clay in the claystone interbed, or possibly some form of deep-seated tectonic movement. Downhill creep was also believed to be a possible operative in either case, but not the principal mechanism.

In view of the importance of demonstrating and documenting data on which an assessment could be based in as short a study period as possible, a decision was made to install instrumented monitoring lines across the crack at two widely separated points, and to monitor the horizontal and vertical movements over a one-year period. The one-year period was estimated to be necessary to observe the effect of seasonal changes in surface-movement patterns.

FIELD INSTRUMENTATION

Two precision monitoring lines, L-1 and L-2, were established across the crack in April 1976 as shown in Fig. 1. The lines were approximately 30 m in length, with bedrock monuments installed every 3 m along the line. A "star" cluster of six monuments was centered at the approximately orthogonal intersection of the cracking line with the monitoring line, by adding an additional four monuments. This arrangement was used to monitor any relative lateral movement of the ground on either side of the crack, along the strike direction of the contact.

Precision monitor line instrumentation

Figure 2 shows the precision line instrumentation system. Note that the bedrock monuments are isolated from contact with the surface soil and a weak, strongly weathered upper layer of bedrock. Thus, changes in distance between monuments would represent movements in the less weathered bedrock material. The rigid beam extensometer used to measure distances between monuments is equipped with a 25-mm range micrometer, reading to the nearest 0.01 mm. A thermometer is mounted on the extensometer to obtain data necessary to correct extensometer readings for temperature changes.

507

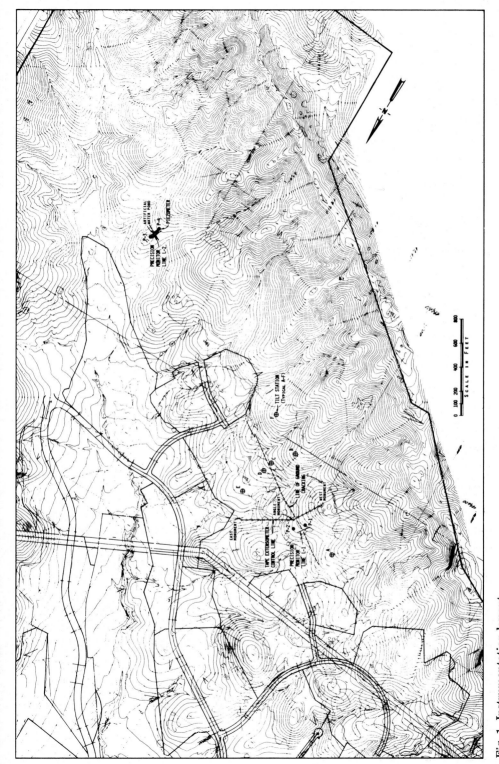

Fig. 1. Instrumentation layout.

508

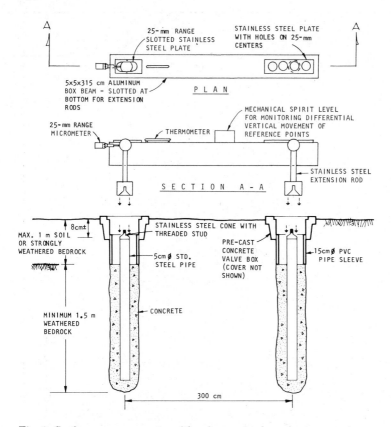

Fig. 2. Surface measurements with micrometer beam extensometer.

A mechanical spirit level, equipped with a micrometer adjustment and an optical prism system, is mounted on top of the extensometer to read changes in vertical angle of the extensometer beam to the nearest 1 sec, within a range of $\pm\frac{1}{2}°$. The change in vertical angle of the extensometer beam from one set of measurements to the next permits calculation of the differential vertical movement of the bedrock monuments. The matching, machined tapers on the bedrock monument cone and the removable extension rod seat ensure repeatable, consistent positioning of the extension rod.

At the end of 1976, piezometers, tilt stations and a tape extensometer control line were added in the region of line L-1 to supplement the existing precision monitoring line. Piezometers, and an artificial water pond were added near the precision monitor line L-2. The purpose of the pond was to induce wet-season data along the L-2 precision monitoring line.

Pore-water pressure instrumentation

The instrumentation used to measure pore-water pressure consisted of pneumatic piezometers. The piezometers were installed in 100-ft-deep drill

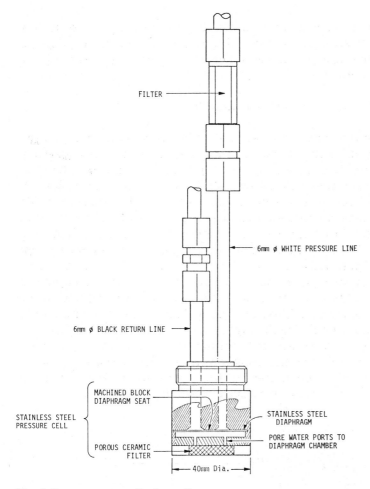

Fig. 3. Pore pressure cell schematic.

holes near each precision monitoring line. One piezometer was installed near the end bedrock monument in the claystone bed, and the other was installed near the end bedrock monument in the sandstone unit of each precision monitoring line.

The pneumatic piezometer used is shown in Fig. 3. The cell comprises a small stainless steel cylinder with an internal reservoir chamber, a porous stone groundwater inlet, and a pressure-sensitive flexible diaphragm. The cell is connected by two plastic tubes to the measuring station at ground surface. Groundwater enters the reservoir chamber through the porous stone and acts to seal the diaphragm against the gas ports on the opposite side of the diaphragm. To make a measurement of pore-water pressure, nitrogen is metered down to the cell under continuously rising pressure, through one of the plastic tubes. When the gas pressure against the diaphragm is equal to the pore-water pressure, the diaphragm lifts off the gas ports and a return flow to the measurement station takes place. At that moment, a pressure gauge on the readout set stops rising, and the indicated pressure, equal to the pore-water pressure, is recorded.

Tape extensometer control line instrumentation

A supplementary tape extensometer control line was established near the L-1 precision monitoring line as shown in Fig. 1. Bedrock monuments along this line are on approximate 15-m centers. Fig. 4 shows the tape extensometer system. The instrument has a 15-m steel tape, with a punched hole in the tape at every 50 mm. The tape extensometer has two ball sockets, which fit over the spherical-head extension rods seated on the bedrock monuments. There are three ball-bearing contact points within each socket which provide positive, repeatable seating on the extension rod heads when the tape extensometer is mechanically tensioned in place between two monuments. Tensioning force is adjusted by means of a knurled ring, which engages a threaded rod attached between the measuring tape and a proving ring on the extensometer. The operator adjusts the knurled ring until a reading on the proving ring dial indicator shows the tape is under 18 kg of tension. The distance between monuments is read as the distance shown at the punch mark in the tape plus the reading on the displacement dial gauge, shown in Fig. 4. Measured distances must be corrected for temperature. Manufacturer's specifications state the accuracy of the system as 0.08 mm.

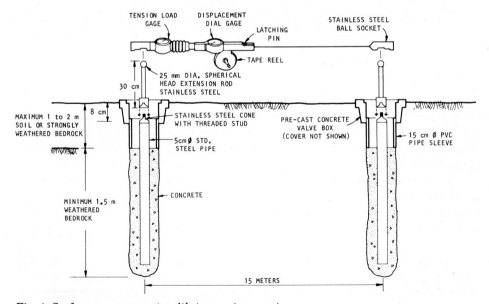

Fig. 4. Surface measurements with tape extensometer.

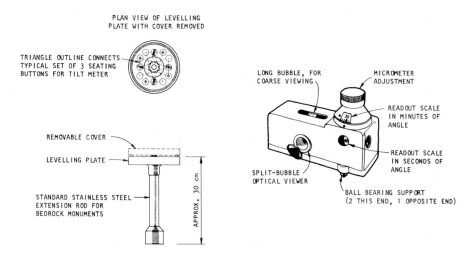

Fig. 5. Mechanical spirit level tiltmeter and levelling plate.

Tilt station instrumentation

Tilt stations were located at six points in the region of precision line L-1. The mechanical, spirit-level tiltmeter system is shown in Fig. 5. The tiltmeter is the same instrument as used in conjunction with the micrometer beam on the precision monitor line. The single point station use requires a leveling plate, fixed to one of the standard extension rods for the bedrock monuments. The leveling plate has four sets of steel seating buttons for the spirit level. Three seating buttons per set are arranged in triangular patterns along two orthogonal axes across the leveling plate.

Artificial ponding

The artificial water pond was excavated into the bedrock alongside the precision monitoring line L-2 and is shown in dashed outline in Fig. 1. The pond was filled with water at the beginning of February 1977, and kept filled to depths of about 0.3—1 m for approximately one month in order to observe the effect on movements of the nearby bedrock monuments.

DISCUSSIONS OF FIELD DATA

Precision monitoring line — horizontal movements

Horizontal movement data for precision monitor lines L-1 and L-2 are given in Figs. 6 and 7, respectively. The change in horizontal distance between adjacent monuments is plotted in millimeters on the vertical axis with the relevant monument pair indicated on the horizontal axis.

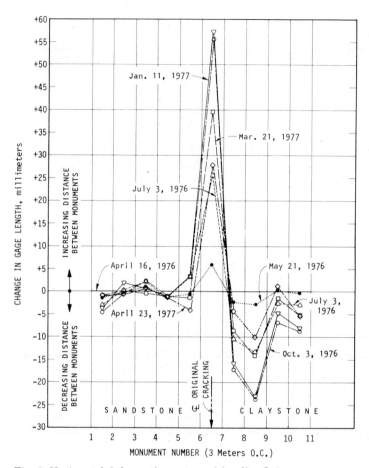

Fig. 6. Horizontal deformations at precision line L-1.

Readings were made every two weeks during the first three months, then at approximately three-month intervals over the remaining year. Horizontal deformations during the period April—October 1976 showed uniform consistent trends at both neighborhoods. Very slight movements of less than a couple of millimeters occur in the sandstone; abrupt increases of length occur in the interval spanning the crack; and shortening of length occurs for the several intervals in the claystone. In precision line L-1 the crack interval opened about 27 mm per three-month interval, to a total 54 mm in October 1976. At the same time, intervals in the claystone interbed shortened by as much as 23 mm. At precision line L-2 the pattern is similar, but magnitudes are about one-quarter those of precision line L-1 through the first six months.

The first heavy intensity rainfall occurred in early January, and the set of readings taken on January 11, 1977 suggest rainfall arrested the trend of

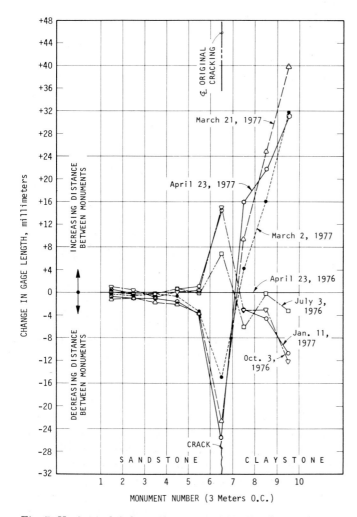

Fig. 7. Horizontal deformation at precision line L-2.

crack development and the shrinkage of surface lengths over the claystone bed. The October 1976 and January 1977 curves are significantly overlapped at both monitoring sites, demonstrating little change in data.

In early February the artificial pond was added alongside the precision monitoring line L-2. The effect on movements at precision line L-2 was dramatic. Not only did the "crack" interval close to the installed gauge length but the interval further shortened by 15 mm for the first reading after watering on March 2, 1977, and to 25 mm by April 23, 1977. The lengthening of distance between bedrock monuments in the claystone was especially significant, with 3-m gauge lengths increasing up to 50 mm over the original installed gauge length.

514

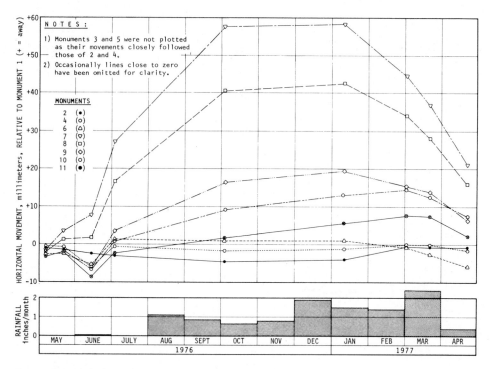

Fig. 8. Horizontal movement at precision line L-1.

A summary plot of the total horizontal movement with time for individual bedrock monuments at precision line L-1 with respect to bedrock monument 1 is shown in Fig. 8. The plot for precision line L-1 shows a gradual movement of all monuments east of the crack away from monument 1 through the January reading, and a gradual movement back toward monument 1 of all of these monuments through the April readings. The pattern is somewhat similar for precision line L-2 through the January reading, but the reaction of L-2 to the filling of the adjacent constructed water pond in January was very strong, as will be shown later. The "star" clusters of monuments centered on the crack at each precision line demonstrated relative lateral movement of adjacent sides of the crack up to 8 mm as a function of seasonal moisture conditions, and showed the strongest response rate as a result of the ponding experiment near precision line L-2.

Precision monitoring line — vertical movements

The vertical movements of bedrock monuments at precision monitoring lines L-1 and L-2 are shown in Figs. 9 and 10 respectively. Patterns of movement up to the time of ponding water next to precision monitoring line L-2 are similar. From April through October, monuments in the claystone

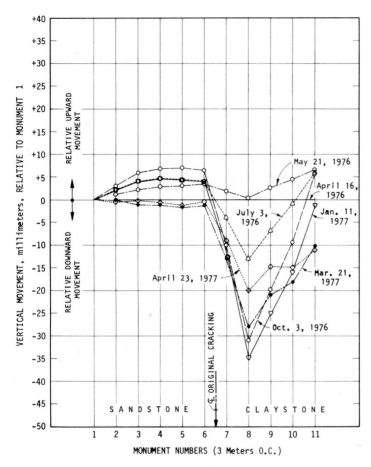

Fig. 9. Vertical movement at precision line L-1.

unit established a trend of downward movement at both monitoring lines.

In January—April 1977, data for precision line L-1 shows generally small rises in the vertical profile across the claystone. The trend of the October—January readings for precision line L-2 agree, but the change in the March readings, following the filling of the water pond in early February, are quite dramatic. The 20-mm-deep trough of the claystone unit heaved upward to a convex surface about 50 mm maximum above the installed profile.

Tape extensometer control line — horizontal movements

The January—April 1977 tape extensometer data generally confirm the closure trend for the crack interval as shown by the precision monitoring line on the far side of the local ridge. A decrease in distance of 7 mm was mea-

516

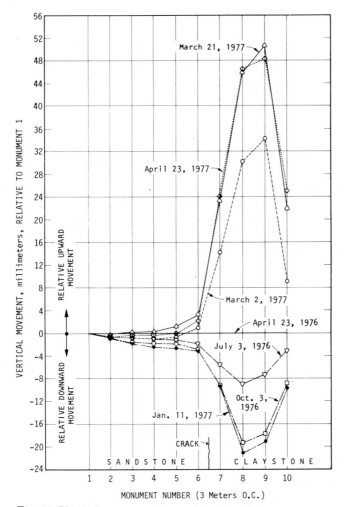

Fig. 10. Vertical movement at precision line L-2.

sured between monuments spanning the crack at 15-m spacing. Closure of the 3-m gauge interval spanning the crack on the precision line during the same period was about 26 mm on the precision line, but within the next 8 m of the precision line over into the claystone there was a lengthening of at least 20 mm. In general, the longer gauge lengths along the tape extensometer line confirm small activity in the crack area and no significant horizontal movement activity elsewhere along the 230-m line.

Tiltmeter stations

Five of the six tilt stations which were located in sandstone units near precision line L-1 exhibit a downhill tilt over the monitoring period.

The overall trend of small, increasing downhill tilt is possibly indicative of some downhill creep. Shallow trenches excavated for geological logging purposes indicate near-surface bending of the steep beds in a downhill direction.

Groundwater instrumentation

The piezometers were installed in late January, and over the entire period of monitoring they measured a constant piezometric head equivalent to a groundwater table 10—15 m below ground surface.

Survey monitoring data

The end bedrock monuments on each of the precision monitoring lines were referenced to a pump station slab-on-grade on the far side of the valley with an electronic distance measurement (EDM) instrument in August 1976. The distances were checked again in January and April of 1977. Movements of the end monuments on the precision monitoring line were a maximum of about 33 mm from an installed position, and were somewhat random.

CONCLUSIONS

The ground cracking was easily visible at the time the precision monitoring lines were installed in April 1976. During the first six months of field measurements, the weather was noticeably dry and the cracking became more apparent, crack opening being demonstrated by the horizontal movement data at both precision lines. A maximum crack opening of 58 mm occurred at precision line L-1 during the six-month period. The data presented demonstrate that the onset of the wetter period in 1977 generally arrested the strong uniform trend of crack opening and shortening of distances that had been occurring in the claystone bed, and started a reversal in movement trends. Artificial ponding of water beside precision line L-2 was initiated in February 1977 in order to more clearly demonstrate the role of moisture/volume change in the claystone bed relative to the ongoing ground deformations. The ponding results appear strongly to confirm the claystone bed moisture/volume change characteristics as the principal operative mechanism in the ground-cracking phenomena.

The vertical movement graphs show that the maximum dry period/wet period changes in elevation occur in the claystone within about 5—8 m from the crack (a maximum increase of almost 70 mm in elevation was measured at precision line L-2 following ponding). Fig. 11 shows the general nature of the relative moisture/volume change movements over the length of the precision monitoring line.

Data for the six tilt stations, show that ground tilt beyond the crack area at precision line L-1 does occur. Generally, it is shown that the near-surface

518

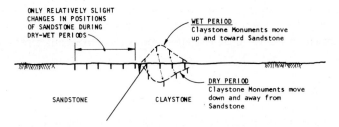

Fig. 11. Relative dry/wet movements of bedrock monuments.

bedrock at the sandstone tilt stations undergoes small increasing tilts in a downslope direction toward adjacent areas of lower relief.

It is believed that the data suggest moisture/volume change characteristics of clay material within the claystone interbed are sufficient to explain the presence of the cracking evident along large reaches of the contact of the interbed with the host sandstone unit, and that gravity movements, or downhill creep, might be a small component of residual crack opening.

Tectonophysics, 52 (1979) 519
© Elsevier Scientific Publishing Company, Amsterdam — Printed in The Netherlands

CHANGES IN RATE OF FAULT CREEP

P. HARSH

U.S. Geological Survey, Menlo Park, Calif. 94025 (U.S.A.)

(Accepted for publication June 8, 1978)

ABSTRACT

Aseismic slip or fault creep is occurring on many faults in California. Although the creep rates are generally less than 10 mm/yr in most regions, the maximum observed rate along the San Andreas fault between San Juan Bautista and Gold Hill in central California exceeds 30 mm/yr. Changes in slip rates along a 162 km segment of the San Andreas fault in this region have occurred at approximately the same time at up to nine alinement array sites. Rates of creep on the fault near the epicenters of moderate earthquakes (M_L 4—6) vary for periods of several years, decreasing before the main shocks and increasing thereafter, in agreement with prior observations based on creepmeter results. The change of surface slip rate is most pronounced within the epicentral region defined by aftershocks, but records from sites at distances up to 100 km show similar variations. Additionally, some variations in rate, also apparently consistent among many sites, have a less obvious relation with seismic activity and have usually taken place over shorter periods. Not all sites exhibit a significant variation in rate at the time of a regional change, and the amplitudes of the change at nearby sites are not consistently related. The time intervals between measurements at the nine array sites during a given period have not always been short with respect to the intervals between surveys at one site; hence, uneven sampling intervals may bias the results slightly. Anomalies in creep rates thus far observed, therefore, have not been demonstrably consistent precursors to moderate earthquakes; and in the cases when an earthquake has followed a long period change of rate, the anomaly has not specified time, place, or magnitude with a high degree of certainty. The consistency of rate changes may represent a large scale phenomenon that occurs along much of the San Andreas transform plate boundary.

Tectonophysics, 52 (1979) 520

DISLOCATION MODELING OF CREEP-RELATED TILT CHANGES

S. McHUGH * and M.J.S. JOHNSTON

U.S. Geological Survey, Menlo Park, Calif. 94025 (U.S.A.)

(Accepted for publication June 8, 1978)

ABSTRACT

Tilt changes associated with 1—5 mm of fault creep have been detected at several different locations on the San Andreas fault on tiltmeters within 500 m of the creep observation point. The creep-related tilts have amplitudes of $\lesssim 0.5$ μrad and durations comparable to the creep events. No changes $\gtrsim 10^{-2}$ μrad have been observed on tiltmeters at distances $\gtrsim 1$ km from the fault at the time of the creep events. Dislocation models capable of replicating the creep-related tilt events have been constructed to examine the relationship of the model parameters to details of the tilt waveforms. The tilt time histories and bounded assumptions of the source-station configurations, and the displacement time history, can be used to infer the type and amount of displacement, the propagation direction and depth of the slip zone. The shallow depth and finite size of the slip zone indicated by these models constrasts with the horizontal extent.

----

* Now at Stanford Research Institute, 333 Ravenswood Avenue, Menlo Park, Calif. 94025.

Tectonophysics, 521—532   521
© Elsevier Scientific Publishing Company, Amsterdam — Printed in The Netherlands

Seismology

THE MAY 1976 FRIULI EARTHQUAKE (NORTHEASTERN ITALY) AND INTERPRETATIONS OF PAST AND FUTURE SEISMICITY

THOMAS H. ROGERS and LLOYD S. CLUFF

Woodward-Clyde Consultants, Three Embarcadero Center, Suite 700, San Francisco, Calif. 94111 (U.S.A.)

(Accepted for publication June 8, 1978)

ABSTRACT

Rogers, T.H. and Cluff, L.S., 1979. The May 1976 Friuli earthquake (Northeastern Italy) and interpretations of past and future seismicity. In: C.A. Whitten, R. Green and B.K. Meade (Editors), Recent Crustal Movements, 1977. Tectonophysics, 52: 521—532.

The May 6, 1976 Friuli earthquake (surface wave magnitude 6.3; intensity VII—X, Mercalli—Sieberg scale) was one of the strongest earthquakes ever felt in south-central Europe. Seismologic and geologic data have indicated: (1) the fault rupture causing the earthquake most probably occurred within a fault zone striking E—NE, and dips gently N—NW under the Venetian Alps; and (2) the principal fault displacement was most probably thrust on the northward-dipping plane.

In recent years, many authors have proposed plate tectonic models for the Mediterranean region. A new variation on previously proposed models is postulated to form a basis for interpreting the Friuli earthquake and other northern Italy historical earthquakes. This model postulates an "Adriatic" microplate having a norhtern boundary that, in part, extends along the NE-trending front of the Venetian Alps. Relative motion of the Adriatic microplate northward and under the European megaplate is believed to be a reasonable explanation of the Friuli earthquake.

Based on known geologic and seismologic data, the northern boundary of the Adriatic microplate is interpreted to extend westward in an irregular manner through the central part of the Po Valley. Major earthquakes in the Po Valley have occurred along linear trends subparallel to the valley during the period between A.D. 1001 and 1781. These major earthquakes are possibly related to past motion of the Adriatic microplate, occurring along either microplate boundaries or intraplate zones related to movement along microplate boundaries.

Future seismicity is anticipated, based on these data and interpretations, and current and future siting studies for major critical facilities in northern Italy and south-central Europe should consider these factors. For instance, in the selection of design earthquakes and the evaluation of surface-faulting potential, available geologic and seismologic data should be reevaluated. Additional investigations should be conducted to gain a greater understanding of the tectonic processes involved and their evolution as imprinted in the Late Cenozoic geologic record.

INTRODUCTION

The Friuli earthquake of May 6, 1976 (surface wave magnitude 6.3) was on of the strongest earthquakes ever felt in south-central Europe. Maximum intensity ratings reported were VIII—X on the Mercalli—Sieberg scale (C.N.E.N./E.N.E.L., 1976). In severity, this earthquake is comparable to the Basel (Switzerland) earthquake of 1356 and the Villach (Austria) earthquakes of 1348 and 1690 (Drimmel et al., 1971, Pavoni, 1974a).

The purposes of this paper are to discuss various seismologic and geologic parameters of the earthquake in terms of a proposed relationship to local and regional conditions; to interpret the regional tectonic framework in terms of a possible plate tectonic model; and to utilize these data and this model to propose an interpretation of past and future seismicity. The implications of these data and interpretations for siting critical facilities are then discussed.

Interpretations presented in this paper are based on published data and on unpublished geologic and seismologic reports by Italian geoscientists. The unpublished reports were reviewed during recent siting studies by Woodward-Clyde Consultants for major critical facilities in northern and central Italy.

REGIONAL SETTING

The Friuli earthquake occurred within the Venetian Alps near the southern boundary of this NE-trending mountain mass (Fig. 1). The Venetian Alps rise abruptly above and north of the Venetian Plain, which slopes gradually southward into the Adriatic Sea.

A marked contrast in regional geologic conditions occurs at the southern boundary of the Venetian Alps. To the south, the Venetian Plain is underlain by a homoclinal, SE-dipping stratigraphic sequence of Cretaceous—Holocene age (Leonardi et al., 1971). To the north, within the Venetian Alps, a complexly deformed sequence of Cretaceous and pre-Cretaceous sedimentary rocks occurs. The tectonic boundary between these two terranes (Fig. 2) is a wide and complex zone of thrust faults, folds, and local normal faults located along the topographic front of the Venetian Alps (Leonardi et al., 1971). In a comprehensive study of the seismotectonics of northeastern Italy, Benvegnu et al. (1977a) demonstrate a clear seismologic relationship between earthquakes and faults in this region.

The Friuli earthquake occurred in a region that has experienced many historical earthquakes. Ambraseys (1977) reports that this region has (during the past 900 years) been "devastated by a series of large shocks"; the largest shocks occurred in 1348 and 1511. According to Schneider (1968), the Friuli region has experienced 17 strong earthquakes since A.D. 1000. Benvegnu et al. (1977b) report that the Friuli event was the highest intensity earthquake in northeastern Italy since 1500; and had somewhat lower intensity than earlier earthquakes of A.D. 1117 and 365.

Epicenter data from 1900 to 1970 (Consiglio Nazionale delle Ricerche, 1973) show a band or general alignment of epicenters located parallel to and partly coincident with the southern front of the Venetian Alps (Fig. 1).

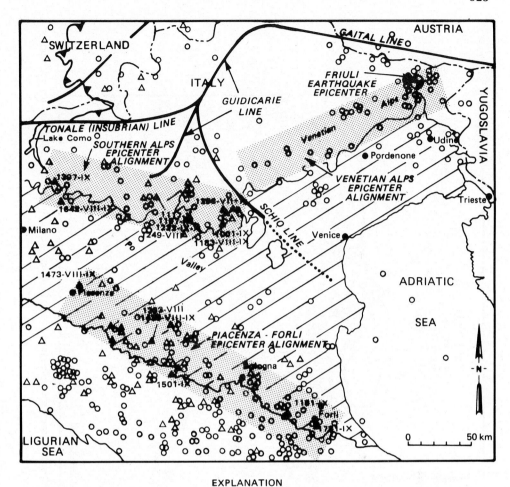

EXPLANATION

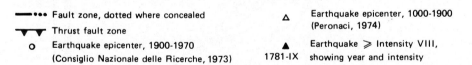

——••• Fault zone, dotted where concealed

▼▼▼ Thrust fault zone

o Earthquake epicenter, 1900-1970
   (Consiglio Nazionale delle Ricerche, 1973)

△ Earthquake epicenter, 1000-1900
   (Peronaci, 1974)

▲ Earthquake ⩾ Intensity VIII,
1781-IX showing year and intensity

Fig. 1. Location map showing Venetian Alps of northeastern Italy, seismicity, and epicenter alignments along Venetian Alps front and Po Valley margins.

This linear band of epicenters is called here the Venetian Alps epicenter alignment. Two other linear epicenter bands (Fig. 1) are located along the margins of the Po Valley and are referred to as the Piacenza—Forli epicenter alignment and the Southern Alps epicenter alignment. These two features are strongly defined by the epicenter data from Peronaci (1974), who compiled M—S intensity VIII and large events recorded from A.D. 1000 to 1900. The

524

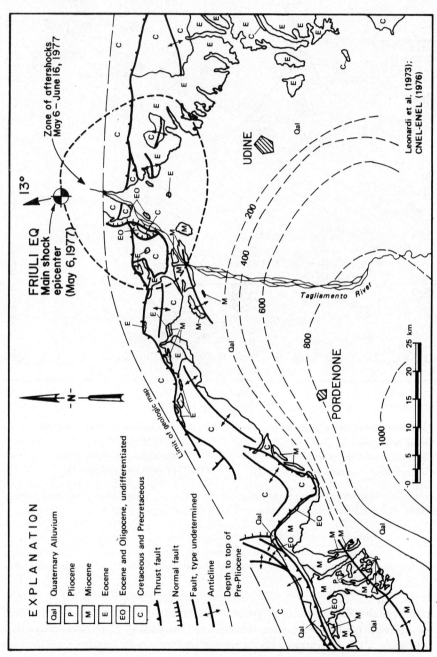

Fig. 2. Geologic map of vicinity of Friuli earthquake, showing epicenter of main shock, focal mechanism data, and aftershocks of May 6–June 16, 1977.

area of the Peronaci (1974) study did not extend eastward to the Venetian Alps epicenter alignment.

THE FRIULI EARTHQUAKE AND LOCAL GEOLOGY

The Friuli earthquake occurred at 21:00 (local time) on May 6, 1976, and was assigned a surface-wave magnitude of 6.3 (Ambraseys, 1977). It was preceded by one foreshock of magnitude 4.5. Mayer-Rosa et al. (1976) report the main shock as a double event having magnitudes of 4.5 and 6.5 Ambraseys also reports a possible precursory phenomenon — an observed "glow in the sky above the northern part of the epicentral region". This glow persisted from some time after the main shock.

The main shock epicenter was located at 46.24N, 13.28E (Ambraseys, 1977), which is approximately 20 km north of the Venetian Alps front. Drimmel estimated a focal depth of 10—11 km (reported in Van Bemmelen, 1977). Ambraseys (1977) reports a focal depth of less than 10 km. Mayer-Rosa et al. (1976) report a focal depth of 10 km for each shock of their postulated double event.

A focal mechanism study was made by Mueller (reported in Van Bemmelen, 1977), who reported a thrust-type mechanism and a preferred fault plane striking N78°E and dipping 13°N. Mayer-Rosa et al. (1976) report two possible focal mechanisms — one being strike-slip (nodal plane oriented NNE—SSW), and another being thrust (nodal plane oriented E—W to NE—SW). Both possibilities clearly indicate N—S shortening of the crust. Bosi et al. (1976) report one field example of strike slip (oriented E—W) on one fracture surface. Weber and Courtot (1977) believe that seismic and geologic data support left-slip for the main shock, and thrust motion for the after-shocks. Aftershocks recorded from May 6 to June 16 (C.N.E.N./E.N.E.L., 1976) were located in a crudely triangular zone extending southward from the main shock epicenter, as shown in Fig. 2.

The local geology along this part of the Venetian Alps front (Leonardi et al., 1972) is also shown in Fig. 2. The complex thrust fault system between the Venetian Plain and the Venetian Alps involves shallow north-dipping thrust faults that displace Cretaceous calcareous rocks southward over early Tertiary flysch and younger deposits. The spatial and geometric relationships between the Friuli seismic data and local geology (Fig. 2) strongly suggest the seismicity occurred within a shallow thrust fault system directly related to the fault system mapped along the front of the Venetian Alps. Data from geodetic resurveying performed after the earthquake (F.S. Muzzi, personal communication, February, 1977) support this hypothesis. It is reported that the resurvey data indicate both uplift of the Venetian Alps relative to the Venetian Plain, and N—S shortening across the Venetian Alps front.

Regarding surface fault rupture, no conclusive evidence of primary surface faulting has yet been found. Bosi et al. (1976) report that: (1) examination of pre-earthquake aerial photographs revealed geomorphic evidence suggesting active faults within the Venetian Alps frontal fault system; and (2) evidence obtained during surface-faulting investigations is most fully explained by a hypothesis relating surface fractures to seismogenic faulting at depth.

It seems reasonable, based on the above data, to hypothesize that the Friuli earthquake represents an increment of relative overthrusting of the Venetian Alps over the Venetian Plain, or underthrusting of the Venetian Plain block under the Venetian Alps. Mayer-Rosa et al. (1976) conclude similarly, that the 1976 Friuli sequence and an earlier, 1975 Friuli event having a similar focal mechanism are expressions "of the continuing process of north—south crustal shortening". Benvegnu et al. (1977b) report the 365 and 1117 (M—S intensity XI) events most likely occurred along this same thrust zone.

Furthermore, the geologic data and historical seismicity record suggest that earthquake-related underthrusting has occurred repeatedly in the past, and has perhaps continued for a significant part of late Cenozoic time.

PLATE TECTONIC INTERPRETATIONS

A postulated collision between the continental masses of Africa and Europe has long been invoked as a mechanism to produce N—S crustal shortening and the spectacular overthrust features of the Alps (e.g., Argand 1916; Dewey et al., 1973).

The concern of most authors has been focused on the late Mesozoic and Tertiary tectonic histories, when the major Alpine crustal shortening and overthrust processes occurred. As a result, the study of Pliocene—Quaternary tectonics ("neotectonics") has been neglected and today is a relatively new method of study in southern and central Europe. Because this paper addresses Pliocene—Quaternary tectonics, only models within this time period are discussed. The models proposed are tentative concepts, offered for discussion and debate. It is recognized that a relatively small amount of research has been directed toward neotectonic studies of the region, and limited data for model development exist. Models based on such limited data can be only preliminary in nature. New investigations in this area typically discover previously unrecognized evidence of Quaternary tectonic activity (Woodward-Clyde Consultants, 1975; Cremaschi and Papani, 1975; Angelier, 1976; Bousquet, 1976). A primary aim of this paper is to encourage comprehensive neotectonic studies.

POSTULATED PLATE TECTONIC MODEL

Several authors (for example, Dewey et al., 1973) have postulated convergence of the African and European megaplates, involving an intervening series of microplates that become progressively squeezed between the converging megaplates, and that have broken up and migrated in a complex manner in response to the N—S compression. Whether or not such microplates ever existed as separate masses in past geologic time, geologic and seismic data strongly indicate that, at present, a mosaic of distinct regions (or blocks) having more or less distinct boundaries occurs at least in some parts of southern Europe.

The model in this paper for the northeastern Italy/western Yugoslavia area is a variation on previously proposed models of Lort (1971), Finetti and Morelli (1972) and Dewey et al. (1973). This model postulates an Adriatic microplate (Fig. 3) that extends from the western Adriatic coastline of Yugoslavia to the Apennine Mts. of Italy, and north to the Po Valley and the Venetian Plain. This microplate consists of late Mesozoic—Holocene strata that are relatively undeformed except along the margins of the microplate (Consiglio Nazionale delle Ricerche, 1973; Leonardi et al., 1971; Biju-Duval et al., 1974). Seismicity is relatively concentrated along the margins of this region (Lort, 1971; Consiglio Nazionale delle Ricerche, 1973; Peronaci, 1974). Active surface faulting (normal, down to the west) has been recognized along the western margin in the Apennines, possibly associated with large 20th-century earthquakes (Woodward-Clyde Consultants, 1975). In the

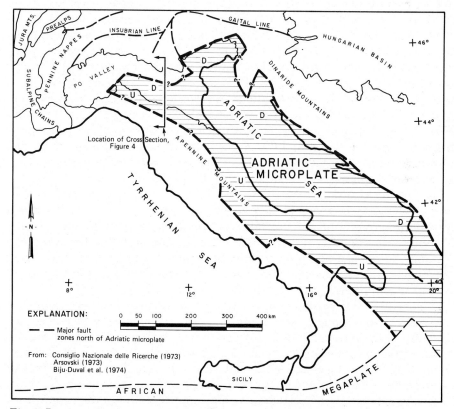

Fig. 3. Postulated Adriatic microplate, showing location of cross-section in Fig. 4.

Po Valley, major earthquakes since A.D. 1000 (Peronaci, 1974) have occurred in distinct alignments parallel to and possibly along part of the northern boundary of the Adriatic microplate. In western Yugoslavia, the eastern boundary of the Adriatic microplate is characterized by numerous earthquakes and geologically by a distinct zone of steeply east-dipping reverse faults having the east side up (Arsovski, 1973). The Friuli earthquake, the Venetian Alps epicenter alignment, and a coincident major north-dipping thrust fault zone (as previously described) are located along part of the northern margin of the Adriatic microplate.

Various data, including the Friuli earthquake (and interpreted northward underthrusting), suggest the general relative motion of the postulated Adriatic microplate is northward (underthrusting or impinging against the Alps) and having a tilt to the east and north.

INTERPRETATIONS OF PAST AND FUTURE SEISMICITY

In the interpretation of historical seismicity, an important question is whether the regional historical seismicity and associated tectonic forces represent a continuation of Cenozoic events and processes, or a departure from them. Studies by Pavoni (1974b) of focal mechanisms and late Cenozoic geologic structures in the Alps indicate a parallelism between stress fields deduced from historical earthquakes and stress fields indicated by post-Pliocene geologic structure. In the Central Alps, Pavoni (1974b) found a general N—S orientation of maximum horizontal stress that rotated to a NW—SE and then E—W orientation in the western Alps (generally perpendicular to the regional Alpine structural grain). Studies in Spain by Bousquet (1976) and in the Aegean area by Angelier (1976) show that N—S compression has occurred in Spain and the Aegean area during Pleistocene time. The results of these studies indicate that the Tertiary compressive regime is continuing in the Mediterranean region and southern Europe.

It is then reasonable to postulate that historical seismicity (and, in particular, concentrated zones or alignments of numerous large earthquakes) represent preferred zones of fault displacement that accommodate, in part, this continuing regional compression. Also, future large earthquakes can reasonably be postulated along such zone of preferred displacement — such as the Venetian Alps epicenter alignment, the Piacenza—Forli epicenter alignment and the Southern Alps epicenter alignment (Fig. 1).

The epicenter alignments are on or near the postulated northern boundary of the Adriatic microplate (Figs. 1 and 3). The Po Valley may be considered as a wide, asymmetric, faulted downwarp (or tectogene) that essentially forms the northern margin of the Adriatic microplate, as shown in cross-section on Fig. 4. The Piacenza—Forli epicenter alignment and the Southern Alps epicenter alignment occur on the margins of this tectogene; the large earthquakes along these alignments may represent periodic, significant accommodations to the regional (generally N—S) compressive stress. In a

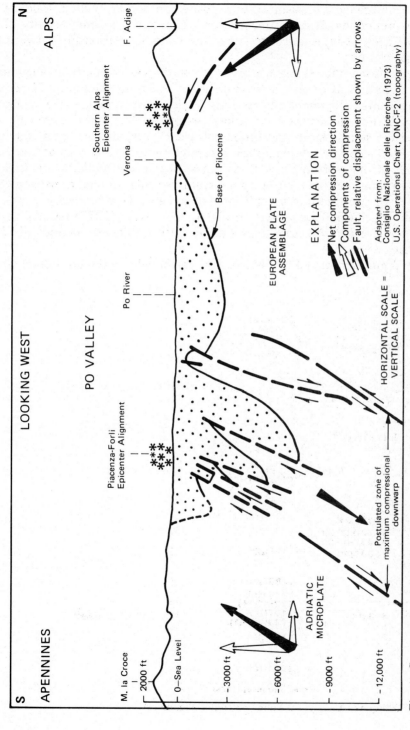

Fig. 4. Cross-section across Po Valley and northern boundary of postulated Adriatic microplate, along 11°E longitude.

recent comprehensive seismotectonic paper on the Po Valley, Caloi et al. (1970) present detailed documentation of preferred seismic activity along the Po Valley margins and suggest a seismotectonic relationship between the marginal features.

An additional interesting feature of these Po Valley epicenter alignments is the migration of seismic activity in space and time. Along the Piacenza— Forli epicenter alignment, successively younger earthquakes have migrated from near the southeastern end (oldest event) to the northwestern end, and back to the southeastern end (youngest event). Periods as long as 150 years of repeated strong earthquakes are separated by periods as long as 300 years with no strong earthquakes. Sometimes, strong earthquakes are isolated events in time, and the recurrence interval is irregular. A similar pattern characterizes the Southern Alps epicenter alignment. This similarity suggests a tectonic relationship, as shown on Fig. 4. Ambraseys (1977) reports a similar pattern for the Venetian Alps area, including quiescent periods of 150— 200 years.

The Friuli earthquake is the first large earthquake in the Po Valley—Vene-

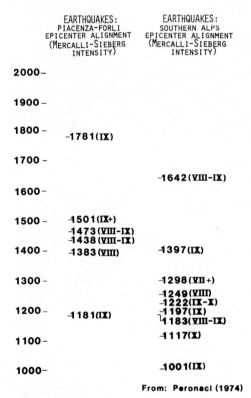

From: Peronaci (1974)

Fig. 5. Earthquake history, Po Valley epicenter alignments.

tian Plain area in this century. The clustered pattern of time histories (Fig. 5) raises the important question of whether the Friuli earthquake represents the beginning of a 100- or 150-year period of recurring destructive earthquakes, or is one of the isolated events in time. This suggests a related question of whether the Friuli earthquake (on the Venetian Alps epicenter alignment) suggests future earthquakes along the Po Valley alignments, given an indirect tectonic connection between these epicenter alignments along the northern margin of the Adriatic microplate.

IMPLICATIONS FOR SITING STUDIES FOR CRITICAL FACILITIES

The questions posed by the foregoing discussions are important to the siting of modern critical facilities such as nuclear power plants and dams. Such questions are addressed in evaluations of earthquake hazards, particularly in the selection of design earthquakes and in studies of the potential for surface fault rupture. Considering the limited neotectonic data available for southern Eruope, extensive investigations need to be made along the major tectonic features within significant distances from a given site. These investigations should reevaluate the available data and should include new coordinated geologic, seismologic, geophysical, and geodetic investigations. The aim of these investigations should be to gain an understanding of the complex tectonic processes and to interpret their evolution as imprinted in the Late Cenozoic geologic record.

REFERENCES

Ambraseys, N., 1977. Gemona, The reckoning. Nature, 265: 3—4.

Angelier, J., 1976. Sur l'existence d'une neotectonique en compression dans l'arc egeen meridional (Crete, Karpathos) et ses consequences. Bull. Soc. Géol. Fr. (7), XVIII (2): 373—381.

Argand, E., 1916. Sur l'arc des Alpes occidentala: Eclogae Geol. Helv., 14: 145—191.

Arsovski, M., 1973. General characteristics of the neotectonic structure of the territory of S.F.R. Yugoslavia. Inst. Earthquake Eng. and Eng. Seismol., Univ. Kiril and Metodij, Skopje, Publ. No. 34.

Benvegnu, F., Carrara, C., Iaccarino, E., Magri, G., Mittempergher, M. and Molin, D., 1977a. Seismotectonic investigations in northeastern Italy. Specialist Meeting on the 1976 Friuli Earthquake and the Antiseismic Design of Nuclear Installations, Rome, Italy, Oct. 11—13, 1977, 29 pp.

Benvegnu, F., Carrara, C., Iaccarino, E., Magir, G., Mittempergher, M., Molin, D., Sennis, C. and Zaffiro, C., 1977b. Considerations on the safety of possible nuclear installations in northeastern Italy in relation to local seismotectonics. Specialist Meeting on the 1976 Friuli Earthquake and the Antiseismic Design of Nuclear Installations, Rome, Italy, Oct. 11—13, 1977, 11 pp.

Biju-Duval, B., Letouzey, J., Montadert, L., Courrier, P., Mugniot, J. and Sancho, J., 1974. Geology of the Mediterranean Sea basins, In: C.A. Burk and C.L. Drake (Editors), The Geology of Continental Margins. Springer, New York, N.Y., pp. 695—721.

Bosi, C., Camponeschi, B. and Giglio, G., 1976. Indizi di possibili movimenti lungo faglie in occasione del terremoto del Friuli del 6 Maggio 1976. Paper presented at Rome, Italy, nella Seduta Scientifica del 2 Luglio, 1976, 28 pp.

Bousquet, J-C., 1976. Observations microtectoniques sur la compression nord-sud quaternaire des Cordilleres betiques orientales (Espagne meridionale-Arc de Gilbraltar). Bull. Soc. Géol. Fr. (7), XVIII (3): 711—724.

Caloi, P., Romualdi, G. and Spadea, M.C., 1970. Caratteristiche sismiche e geodinamiche della Val Padana quali risultano dall'attivita sismica ivi verificatasi dall'inizio dell'Era Volgare a tutto il 1969. Ann. Geofis., 23 (2—3).

C.N.E.N./E.N.E.L., 1976. Contributo allo studio del terremoto del Friuli del Maggi, 1976, Spec. Publ., Nov. 1976, 135 pp. Tectonophysics, 5 (6): 459—511.

Consiglio Nazionale delle Ricerche, 1973. Structural model of Italy, composite of lithostratigraphic—tectonic map, bathymetric map, total magnetic field intensity map, seismicity map; and composite of lithostratigraphic—tectonic map, bathymetric map, structural map of the base of the Pliocene, and gravity map, scale 1 : 1,000,000.

Cremaschi, M. and Papani, G., 1975. Contributo preliminare alla Neotettonica del margine Padano dell'Appennino: le forme terrazzate comprese tia Cavriage e Quattro Castella (Reggie E.). Ateneo Parmense, Acta Nat., 11: 335—371.

Dewey, J., Pitman III, W., Ryan, W. and Bonin, J., 1973. Plate tectonics and the evolution of the Alpine system. Geol. Soc. Am. Bull., 84 (10): 3137—3180.

Drimmel, J., Gangl, G. and Trapp, E., 1971. Kartenmässige Darstellung der Seismizität Österreichs. Österreichische Akad. Wiss., Mitt. der Erdbeben-Kommission, Neue Folge-Nr. 70, 8 pp.

Finetti, I. and Morelli, C., 1972. Wide-scale digital seismic exploration of the Mediterranean Sea. Boll. Geofis. Teor. Appl., XIV (56): 291—342.

Leonardi, P., Semenza, E. and Vuillerman, E., 1971. Geological problems of the Venice region. Boll. Geofis. Teor. Appl., XIII (49): 68—75.

Lort, J., 1971. The tectonics of the eastern Mediterranean: a geophysical review. Rev. Geophys. Space Phys., 9 (2): 189—216.

Mayer-Rosa, D., Pavoni, N., Graf, R. and Rast, B., 1976. Investigations of intensities, aftershock statistics, and the focal mechanisms of Friuli earthquakes in 1975 and 1976. Pageophysici, 114: 1095—1103.

Pavoni, N., 1974a. Maximale Erdbebenintensitäten im Gebiet der Schweiz. Jahresber. 1972 der Schweiz. Erdbebendienstes, Inst. Geophys. der ETH-2, pp. 164—166.

Pavoni, N., 1974b. Zur Seismotektonik des Westalpenbogens. Vermess., Photogramm. Kulturtech., Fachblatt III/IV: 185—187.

Peronaci, F., 1974. Seismic characteristics of some areas of the western pre-Alps. Report prepared for E.N.E.L., Rome, Italy.

Schneider, G., 1968. Erdbeben und Tektonik in südwest-Deutschland. Tectonophysics, 5 (6): 459—511.

Van Bemmelen, R.W., 1977. Note on the seismicity of northeastern Italy (Friuli Area). Tectonophysics, 39: T13—T19.

Weber, C. and Courtot, P., 1977. Le seisme du Frioul (Italie, 6 mai 1976) dans sou contexte sismotectonique. Specialist Meeting on the 1976 Friuli Earthquake and the Antiseismic Design of Nuclear Installations, Rome, Italy, Oct. 11—13, 1977, 13 pp.

Woodward-Clyde Consultants, 1975, Investigations and evaluations of geology, seismology, and ground response at T. Saccione nuclear power plant site, Molise Coast, Italy, Appendix B — Seismology and design earthquake evaluation. Report prepared for E.N.E.L. by Woodward-Clyde Consultants, Oakland, California.

Tectonophysics, 52 (1979) 533—547
© Elsevier Scientific Publishing Company, Amsterdam — Printed in The Netherlands

EVIDENCE FOR THE RECURRENCE OF LARGE-MAGNITUDE EARTHQUAKES ALONG THE MAKRAN COAST OF IRAN AND PAKISTAN

WILLIAM D. PAGE [1], JOHN N. ALT [1], LLOYD S. CLUFF [1], and GEORGE PLAFKER [2]

[1] *Woodward-Clyde Consultants, San Francisco, Calif. 94111 (U.S.A.)*
[2] *U.S. Geological Survey, Menlo Park, Calif. 94025 (U.S.A.)*

(Accepted for publication June 8, 1978)

ABSTRACT

Page, W.D., Alt, J.N., Cluff, L.S. and Plafker, G., 1979. Evidence for the recurrence of large-magnitude earthquakes along the Makran coast of Iran and Pakistan. In: C.A. Whitten, R. Green and B.K. Meade (Editors), Recent Crustal Movements, 1977. Tectonophysics, 52: 533—547.

The presence of raised beaches and marine terraces along the Makran coast indicates episodic uplift of the continental margin resulting from large-magnitude earthquakes. The uplift occurs as incremental steps similar in height to the 1—3 m of measured uplift resulting from the November 28, 1945 (*M* 8.3) earthquake at Pasni and Ormara, Pakistan. The data support an E—W-trending, active subduction zone off the Makran coast.

The raised beaches and wave-cut terraces along the Makran coast are extensive with some terraces 1—2 km wide, 10—15 m long and up to 500 m in elevation. The terraces are generally capped with shelly sandstones 0.5—5 m thick. Wave-cut cliffs, notches, and associated boulder breccia and swash troughs are locally preserved. Raised Holocene accretion beaches, lagoonal deposits, and tombolos are found up to 10 m in elevation. The number and elevation of raised wave-cut terraces along the Makran coast increase eastward from one at Jask, the entrance to the Persian Gulf, at a few meters elevation, to nine at Konarak, 250 km to the east. Multiple terraces are found on the prominent headlands as far east as Karachi. The wave-cut terraces are locally tilted and cut by faults with a few meters of displacement.

Long-term, average rates of uplift were calculated from present elevation, estimated elevation at time of deposition, and ^{14}C and U—Th dates obtained on shells. Uplift rates in centimeters per year at various locations from west to east are as follows: Jask, 0 (post-Sangamon); Konarak, 0.031—0.2 (Holocene), 0.01 (post-Sangamon); Ormara 0.2 (Holocene).

INTRODUCTION

The presence of Pleistocene beaches and marine terraces along the Makran coast of Iran and Pakistan (Fig. 1) was first mentioned by Blanford (1872). Recently, these features have been described in more detail because of their

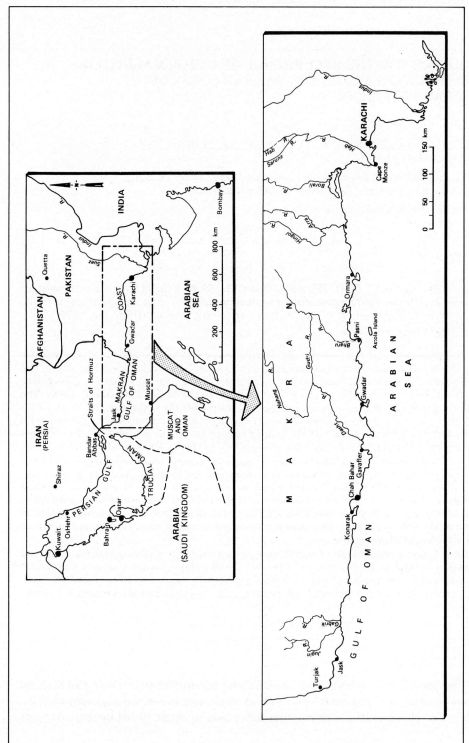

Fig. 1. Location maps of the Makran coast.

significance as indicators of an emerging coast by Snead (1969, 1970); Little (1972); and Vita-Finzi (1975). The present study of the raised beaches along the Makran coast was undertaken to determine the rate of uplift and evaluate whether or not the uplift occurred in discrete increments associated with earthquakes.

Six weeks from May to July 1975 were spent in the field. Air-photo analysis and aerial reconnaissance of the Iranian coastline in a fixed-wing aircraft preceded ground studies at Jask and Konarak. A special investigation and ground studies were made of the reported uplift associated with the November 28, 1945, earthquake off the coast of Pakistan at Ormara and Pasni.

THE MAKRAN COAST

The Makran coast extends east from the Strait of Hormuz in Iran to the mouth of the Indus River in Pakistan. The eastern two-thirds of the coastline is unprotected by other land masses, and during the monsoon the ocean and surfs are extremely rough. Low-lying alluvial plains make up the narrow coastal strip between the coast and the mountains. The plains, 5—20 km wide, occur where the streams have built deltas and eroded the siltstone and claystone bedrock of the area into low-lying pediments. Locally, they are covered by sand dunes. The Makran Ranges rise irregularly from the coastal plain in a series of ridges underlain by folded sedimentary rocks. The crest of the Makran Ranges is an E—W-trending divide up to 2000 m high and 120 km from the coast. North of the divide, the topography drops steeply into the Jaz Murian Depression.

The bedrock stratigraphy along the coast in Pakistan consists of 4000 m of Miocene and Pliocene siltstone and mudstone with minor interbeds of sandstone. The Late Tertiary units are unconformably overlain by 75 m of Plio—Pleistocene conglomerate, sandstone and shale (Ormara Formation). Up to 30 m of Pleistocene coquina limestone and sandstone (Jiwani formation) overlie the Ormara Formation (Photographic Survey Corporation, 1958). Similar rocks are found in Iran, but mapping has been limited and correlations are not confirmed.

The coast is marked by a series of prominent headlands separated by low areas. The headlands are generally flat-topped or gently tilted and locally stepped with terraces and are preserved from rapid erosion by cemented, resistant conglomerates or coquina cap rocks 0.5—5 m thick. These headlands are erosional remnants protected by the resistant cap of Pleistocene terrace deposits. Where these deposits or other resistant rocks are not present, the siltstone and claystone bedrock is rapidly weathered and eroded to low-lying pediments. Some terraces are 1—2 km wide and 10—15 km long. Wave-cut cliffs, notches, and related geomorphic features such as boulder breccia and swash troughs are locally found at the shore side of the wave-cut platform. The wave-cut terraces are locally tilted, as east of Chah Bahar. Small faults with a few meters' displacement cut some terraces. No evidence was observed to indicate that the headlands are uplifted fault blocks. Landward from and between the headlands are accretionary beaches, lagoonal deposits, and tombolos up to 10 m in elevation. Lateral correlation of particular terraces was not possible during this investigation because of the discontinuity of the terraces.

EARTHQUAKE OF NOVEMBER 28, 1945

An M 8.3 earthquake occurred on the Makran coast on November 28, 1945 (3:00 a.m., IST). The epicenter of the earthquake was located just seaward of the village of Pasni (Gates, et al., 1977). Four mud-volcanoes, rising 8—30 m above the Arabian Sea in water 4—7 fathoms deep, were formed as a result of the earthquake (Sondhi, 1947).

Ground failure was substantial and ground cracks that issued water were numerous at Pasni and Ormara. A submarine slide apparently caused the shoreward part of Pasni to subside beneath the water so that the shore today is about 100 m inland. Shaking caused many landslides and rock falls along the steep bluffs on the headland of Ormara. Local inhabitants report that three tsunamis hit the Makran coast $1\frac{1}{2}$—2 hours after the earthquake, reaching a height of 7—10 m.

Field investigation showed that there was no uplift of Pasni resulting from the earthquake, but the Ormara area rose about 2 m.

Ormara

The Ormara area is characterized by a steep, rocky headland connected to the mainland by a low tombolo (Fig. 2). The headland is 16 km long and 1—4 km wide, and rises to an elevation of 480 m near the northwest end. The tombolo is 7 km long and 2—4 km wide, and is marked by accretion beaches that reach an elevation of 8 m. The headland is underlain by siltstones, mudstones, sandstones, and conglomerates of the Ormara Formation (Plio—Pleistocene) and capped by coquina, limestones, and sandstones of the Jiwani Formation (Pleistocene) (Photographic Survey Corporation, 1958).

The 1945 earthquake caused the Ormara area to be uplifted approximately 2 m. Pre-1945 beaches are now 1—3 m above the highest tide and the tombolo broadened several hundred meters. A large sand flat was exposed to the east of the village of Ormara, and fishing boats that used to be beached at the edge of town are now kept on the flat, which is inundated only at high tide. The 1945 beach and beach berm can be traced on the inner edge of this sand flat from the headland northward to where it has been buried by advancing sand dunes. The formation of new sand spits and bars, particularly on the more sheltered east side of the tombolo, is still continuing in response to the uplift caused by the 1945 event.

Local fishermen report that the west side of the headland was impassable to foot traffic even at the lowest tides prior to the earthquake. After the

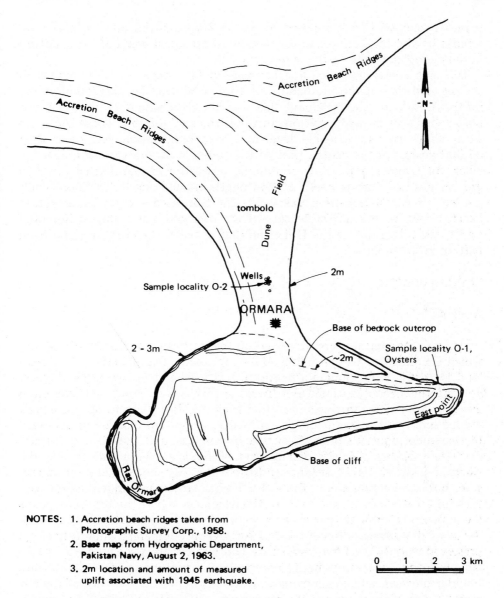

NOTES: 1. Accretion beach ridges taken from
Photographic Survey Corp., 1958.

2. Base map from Hydrographic Department,
Pakistan Navy, August 2, 1963.

3. 2m location and amount of measured
uplift associated with 1945 earthquake.

Fig. 2. Map of the Ormara headland and Tombolo.

earthquake this part of the headland was passable even at the highest tides. The maximum tidal range is 3.7 m between the highest high and the lowest low tide (National Ocean Survey, 1973), indicating that the amount of uplift was greater than 3.7 m. This part of the headland at present shows an elevated wave-cut platform 2—3 m above the present one, and the 1945 beach berm is about 2 m above the present beach berm. The difference in suggested sea-level changes (3.7 compared to about 2 m) may be due to inaccurate reports by the local fishermen or to downward adjustments of the headland in the period since the earthquake occurred.

Difficult access precluded a field check of Pleistocene marine terraces on the seaward side of the headland, but a raised alluvial fan was found on the northeast side of the headland that has boulders with slightly weathered oyster shells in growth position on the surface of the fan. These shells are 11 m above present low-tide level and have radiocarbon ages older than 37,000 years. On the central part of the tombolo at 8-m elevation near the village of Ormara, sand beds containing unweathered marine shells (not in growth position) are exposed at 2-m depth in hand-dug wells. These shells are 2710 ± 135 radiocarbon years old. The truncated and eroded accretion beach ridges on the northeast side of the tombolo may reflect differential uplift and tilting during the Holocene that changed the erosion—deposition pattern in that area.

IRANIAN COAST

Konarak

At least nine terraces are preserved on the south side of Konarak headland which is connected to the mainland by a tombolo (Figs. 3 and 4). The headland is about 18 km long and varies in width from 1 to 4 km. The highest area of the headland is the northwest corner, which reaches a maximum elevation of 105 m. In this area, tilted rocks of Mio—Pliocene age have been eroded and resistant sandstone is exposed as stripped structural surfaces protecting underlying siltstones. Elsewhere on the headland, marine coquina and sandstone terrace deposits of late Quaternary age overlie and protect the siltstone bedrock. The resistant sandstones that cap the headland retard erosion, but wave action erodes the softer Tertiary rocks and undercuts the cap, forming steep cliffs around most of the headland. Where the terrace deposits are exposed in cliffs, they overlie the bedrock with an angular unconformity. Some of the terraces are continuous for several kilometers while others are preserved as only scattered remnants. The terraces are associated with wave-cut platforms covered by marine coquina and sandstones, wave-cut cliffs and boulder breccias, and swash troughs. Little (1972) recognized two of the terraces, but he misinterpreted the raised wave-cut cliffs as faults. The terraces from lowest to highest show an increase in dissection and weathering, and shells in the deposits show increasing recrystallization.

The lowest two terraces (T1 and T2) occur as small remnants along the southeast side of the headland. Terrace T3 is the most extensive terrace and covers most of the southeast part of the headland; it is separated from terrace T4 by a scarp 10 m high that runs E—W parallel to the headland. The morphology of the scarp, the presence of swash troughs at regular intervals eroded into the rock in front of the scarp, and boulder breccia found at the

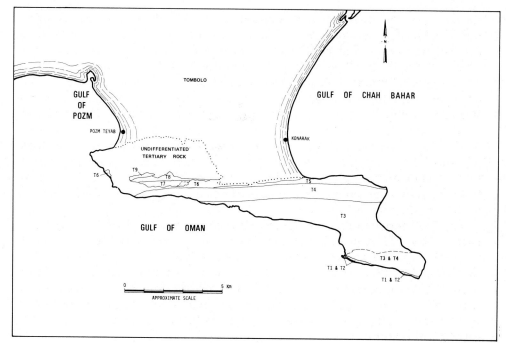

Fig. 3. Map of the Konarak headland and Tombolo.

base of the scarp indicate that it is a raised beach cliff. West of this scarp, but not continuous with it, is a fault scarp. Terrace deposits of T3 consist of about 4 m of limey sands and silts with numerous shells that unconformably overlie thick-bedded, bluish-gray claystone that dips 15° to the south. Radio-

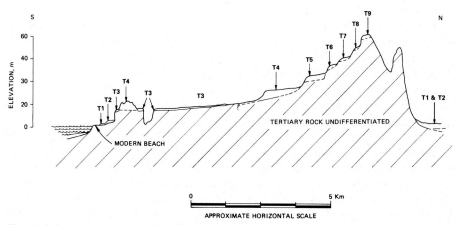

Fig. 4. Schematic section of the Konarak headland.

carbon dates on these samples and Little's (1972) date (Table I) are interpreted to mean the deposit is more than 31,500 years old. A portion of sample K-8 was also dated by the uranium series method. It dates 138,000 ±
12,000 years using ^{230}Th/^{234}U and 156,000 + 54,000—24,000, using ^{231}Pa/
^{235}U as an independent check, and is considered a reliable age (T.-L. Ku,
personal communication, 1975). This age agrees with the field evidence that
this is the first extensive terrace at Konarak older than the Holocene terraces.

The fourth terrace (T4) has a form similar to terrace T3, but all shells
examined from deposits on this terrace at the raised beach cliff were recrystallized, and not sampled. A distinct increase in slope along a linear E—W
trend represents an eroded and modified beach cliff separating T4 from the
next highest terrace, T5.

The fifth terrace (T5) extends across the eastern part of the headland in a
narrow strip. Though most of the shells in this deposit appeared recrystallized, three better-preserved clams and oysters gave a ^{230}Th/^{234}U series date
of 76,000 ± 12,000 years. No ^{231}Pa/^{235}U date was run. The date is considered
unreliable because it does not fit the field relationships which indicate that
this terrace is pre-Sangamon. Terraces T6—T9 are located on the southwest
part of the headland, do not have a large areal extent, and were not investigated for this study.

Two shell samples were collected at approximately 5-m elevation from a
borrow pit near the central part of the tombolo west of the town of Konarak. Radiocarbon dates for these samples are 5935 ± 165 years B.P. and
6255 ± 320 years B.P.

Jask

A single terrace is located on the south side of the village of Jask and is
5—6 m above present sea level. Extensive undercutting by wave action has
caused slumping of the terrace on its seaward side. The terrace deposits are
2—8 m thick and consist of horizontally bedded fossiliferous sandstone and
siltstone that overlie gray siltstone and sandstone bedrock of Tertiary age
with an angular unconformity. The upper 30 cm of the terrace deposit consists of resistant, cemented coquina with slight oxidation. The lower part is
softer sandy siltstone.

Six radiocarbon dates on shells indicate that the deposit is older than
34,000 years. Two uranium series dates are 133,000 ± 13,000 years (^{230}Th/
^{234}U) and 140,000 + 80,000—32,000 years (^{231}Pa/^{235}U), and 136,000 ±
14,000 (^{230}Th/^{234}U) and 114,000 + 80,000—20,000 (^{231}Pa/^{235}U). The first
date is considered reliable and the second date slightly less reliable because
of possible extraneous ^{230}Th and ^{231}Pa contamination (T.-L. Ku, personal
communication 1975). The consistency of the dates for the two samples
from the same deposit also suggest that the dates are reliable and that the
Jask terrace is Sangamon in age.

TABLE I

Dates of shell samples

Location (E—W)	Sample	^{14}C (years B.P.)	^{14}C Corrected for $^{13}C/^{12}C$	$^{230}Th/^{234}U$ (years B.P.)	$^{231}Pa/^{235}U$ (years B.P.)
Pakistan					
Ormara					
alluvial Fan	O-1 (oysters)	>37,000			
tombolo	O-2 (clams)	2710 ± 135			
Iran					
Konarak					
tombolo	K-4 (clams and gastropods)	5520 ± 165 [b]	5935 ± 165		
tombolo	K-6 (clams)	5880 ± 320 [b]	6255 ± 320		
terrace 2	K-10 (clams)	5190 ± 120 [b]	5595 ± 120		
terrace 2	K-11 (gastropods)	5330 ± 120 [b]	5795 ± 120		
lagoon	K-12 (clams)	5280 ± 160 [b]	5700 ± 160		
terrace 3	K-8 (barnacles)		25,675 ± 850 [a]	138,000 ± 12,000	$156,000\ ^{+54,000}_{-24,000}$
terrace 3	K-9 (clams and barnacles)		26,430 ± 910 [a]		
terrace 5	K-3 (clams and barnacles)			76,000 ± 12,000 [c]	
Jask					
terrace dep.	J-1 (clams and oysters)		34,310 ± 3000 [a]	133,000 ± 13,000	$140,000\ ^{+80,000}_{-32,000}$
terrace dep.	J-2 (shell frags)		28,010 ± 1660 [a]		
terrace dep.	J-3 (shell frags)		32,680 ± 2550 [a]		
terrace dep.	J-4 (shell frags)		26,025 ± 1050 [a]	136,000 ± 14,000 [d]	$114,000\ ^{+80,000}_{-20,000}$
terrace dep.	Vita-Finzi (1975)				
	HAR-1115	25,610 ± 640 [a]			
	HAR-1907	23,390 ± 400 [a]			

[a] ^{14}C minimum.
[b] Date by Isotopes; otherwise by Geochron.
[c] Insufficient material for a $^{231}Pa/^{235}U$ check; data not considered reliable.
[d] Possible extraneous ^{230}T and ^{231}Pa contamination; date may be too old.

RATES OF UPLIFT

Quaternary sea levels

Uplift amounts and rates at a particular place cannot be determined without estimates of the relative ("eustatic") position of sea level during the Late Quaternary at that place (Fig. 5). A working model of Late Pleistocene sea levels based on the record from Barbados (Steinen et al., 1973) and New Guinea (Bloom et al., 1974) using uranium series dating, is used in this study. Sea level was generally lower than −15 m between about 120,000 and about 6,000 years B.P. and was above present sea level about 125,000 years ago (called the Sangamon high stand). Published elevations of the Sangamon high stand from "non tectonic" areas fall mostly between 4 and 8 m above present sea level (Broecker and Van Donk, 1970). For the calculations in this study 6 m is used.

The Holocene rise of sea level, though studied in many places in the world and dated with hundreds of radiocarbon dates, is still debated (see e.g., Flint, 1971; Walcott, 1972). The controversy persists as to whether sea level rose rapidly to 1−3 m above its present position about 5000 years ago (the Holocene high stand) and then gradually dropped to its present position or rose to its present position without a high stand. Walcott (1972), taking an idea from Daly (1925), suggested that the earth responds to loading and unloading of continental glaciers and water in the ocean basins by the deformation

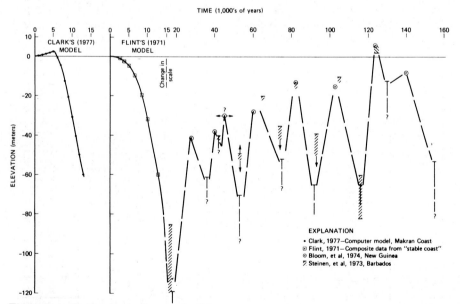

Fig. 5. Late Quaternary paleo-sea levels.

and lateral transfer of subcrustal materials. Any particular coast would then record a different relative sea level, depending on its proximity to the glacial ice caps and/or to the width of the continental shelf. Some coasts would subside and not record a Holocene high sea level while others would rise and record a high stand. Clark (1977), using a computer, divided the earth into zones of relative uplift and subsidence predicting reasonably accurately the Holocene sea levels measured in different parts of the world. Clark's (1977) relative sea-level curve for the Makran coast is used in this report, and for comparison Flint's (1971) composite curve is also used.

Age dating

Shells were collected for radiometric age determinations from terrace deposits and where possible shells in growth position were collected. The ^{14}C dates on 25 shell samples were determined by Geochron Laboratories and Teledyne Isotopes; uranium series dates on four samples were determined by T.-L. Ku at the University of Southern California Geochemical Laboratory (Table I). The shell samples were taken only from known stratigraphic relationships. If a freshly broken surface of a shell sparkled or appeared chalky or powdery it was presumed to be recrystallized, and the sample was rejected in the field. For the 11 samples checked carbon—isotope ratio indicated contamination was not a problem. Any shell date greater than 20,000 years was treated as a minimum date because small amounts of contaminants dramatically change the apparent age. Though mollusc shell dates using the uranium series methods have not always been considered reliable (Kaufman et al., 1971; Ku, 1976), no other materials were available for dating and four samples were run. Three $^{230}Th/^{234}U$ dates had similar $^{231}Pa/^{235}U$ dates and consistent field relationships and are considered reliable (T.-L. Ku, personal communication, 1975).

Long-term average uplift rates

The Holocene average rates of uplift using Flint's (1971) curve generally range between 0.2 and 0.3 cm/yr (Table II). The Holocene rates using Clark's (1977) curve are less, ranging between 0.01 and 0.2 cm/yr (Table II). At Konarak where the comparison can be made between the post-Sangamon rate and the Holocene rate, the Holocene rate is 20 times the post-Sangamon rate using Flint's (1971) curve, but only 1—6 times using Clark's (1977) curve. The best Holocene data, shells in situ from the raised lagoonal deposits, give a Holocene rate twice the post-Sangamon rate using Clark's (1977) curve, suggesting that the actual tectonic rate is 0.01—0.02 cm/yr. At Ormara the Holocene rate is higher, 0.2 cm/yr (based on Clark's curve) after the 1945 earthquake or 0.1 cm/yr before the earthquake. There is no post-Sangamon rate for comparison, but the headland and associated raised terraces are higher than at Konarak (Photographic Survey Corp.,

TABLE II

Uplift rates

Location (E–W)	Present elevation (m)	Estimated elevation at time of deposition (m)	Sea-level position at time of deposit [c] (m)		Amount of uplift (m)		Radiometric age [d] (years B.P.)	Uplift rate (cm/yr)	
			Flint (1971)	Clark (1977)	Flint (1971)	Clark (1977)		Flint (1971)	Clark (1977)
Pakistan									
Ormara									
tombolo (beach dep.)	6 [a]	0 [a]	−1	+1	7	5	2700	0.3	0.2
(prior to 1945 earthquake)	4 [a]	0 [a]	−1	+1	5	3		0.2	0.1
Iran									
Konarak									
terrace 2	2.5; 6	−1; +2	−6	+2.7	10	0.8; 2.3	5600–5800	0.2	0.01–0.04
lagoon	1.5	−1.5	−6	+2.3	9	0.8	5300	0.2	0.02
tombolo (beach dep)	5.5	0 to +2	−7.5	+1	11–13	2.5–3.5	5500–6000	0.2	0.05–0.06
terrace 3	14.5	<−1.5	+6 [b]		10 [b]		138,000	0.1 [b]	
Jask									
terrace	3.5–6	−1.5–3.5	+6 [b]		0 [b]		133–136,000	~0 [b]	

a Approximate.
b Bloom, et al. (1974).
c From Fig. 5.
d Average from Table I.

1958). This could indicate a faster long-term average rate of tectonic uplift for the Ormara area. At Jask the post-Sangamon uplift rate is zero.

Though the data are sparse, it appears that the long-term average uplift rate increases from Jask on the west to Ormara on the east. This reflects different amounts of uplift at different places on the coast, a fact supported by the tilting and faulting and discontinuous remnants of the pre-Holocene terraces seen all along the coast. In general the terraces increase in number and elevation from one at Jask to nine at Konarak up to 100 m elevation, and an unknown number at Ormara up to 500 m elevation.

TECTONIC IMPLICATIONS

If the 1945 Pasni—Ormara earthquake is typical, then the uplift occurs as discrete increments along different parts of the coast at different times. The lower two terraces at Konarak may have formed in this way. Using the measured uplift rate at Ormara of 0.1—0.2 cm/yr and an uplift of 2 m for the Pasni—Ormara earthquake, then the recurrence of such an earthquake along this particular part of the coast is 1000—2000 years. At Konarak, where the uplift rate is slower (0.01—0.06 cm/yr), the recurrence of a Pasni—Ormara-type earthquake would be 3000—20,000 years. Assuming that the Pasni—Oramara earthquake effected a strain release along the coast from Pasni to east of Ormara, a distance of 125 km, the length of the coast is 850 km, and the average uplift rate increases toward the east, the recurrence of a similar magnitude 8+ earthquake somewhere along the Makran coast is approximately 125—250 years, the east end of the coast appearing more likely to have a large-magnitude earthquake than the west end.

Although this coast has been civilized for the past 5000 years, population has been sparse along most of the coast and only recent earthquake records are known to have been kept. Large-magnitude earthquakes have been reported from the populated east end of the Makran (Pendse, 1948; Snead, 1969).

Stoneley (1974) first proposed a subduction zone off the Makran coast that formed the boundary between the Afro-Arabian Plate and the Eurasian Plate. Shearman (1977), and Farhoudi and Karig (1977) present data to support this hypothesis. We believe that the raised beaches along the coast confirm this tectonic model. The subduction zone connects to the Zagros thrust along the West Makran fault (the Zendan fault of Shearman (1977)), a thrust fault with a strike-slip component. The east end of the subduction zone probably connects with the Chaman fault in Pakistan (Stoneley, 1974).

The rate of uplift along this subduction zone compares to other uplifted coasts: Kikai Island, Ryukyu Islands, Japan, 0.1—0.2 cm/yr (Konishi et al., 1970); Middleton Island, Alaska, 1 cm/yr (Plafker 1969); Yakatat to Icy Bay and Lituya Bay, 0.9 cm/yr (Hudson et al., 1973); Huon Peninsula, New Guinea, 0.1—0.25 cm/yr (Bloom et al., 1974); Barbados, 0.04 cm/yr (Steinen et al., 1973).

ACKNOWLEDGEMENTS

This work was done as part of a siting study for the Atomic Energy Organization of Iran. The authors wish to thank the other geologists that worked on parts of the study for their assistance and critical discussion of the problem, particularly Shakir Zuberi of Woodward-Clyde Consultants, T.-L. Ku of the University of Southern California, and Jim Clark of Cornell University.

REFERENCES

Blanford, W.T., 1872. Note on the geological formations seen along the coasts of Baluchistan and Persia from Karachi to the head of the Persian Gulf, and on some of the Gulf Islands. Rec. Geol. Surv. of India, May, Par 2, pp. 41—45.

Bloom, A.L., Broecker, W.S., Chappell, J.M.A. Matthews, R.K. and Mesolella, K.J., 1974. Quaternary sea level fluctuations on a tectonic coast: new ^{230}Th/^{234}U dates from the Huon Peninsula, New Guinea. Quat. Res. (Tokyo), 4: 185—205.

Broecker, W.S. and van Donk, J., 1970. Insolation changes, ice volumes and the O^{18} record in deep-sea cores. Rev. Geophys. Space Phys., 8: 169—198.

Clark, J.A., 1977. Global Sea Level Changes since the Last Glacial Maximum and Sea Level Constraints on the Ice Sheet Disintegration History. Ph.D. Thesis, Univ. of Colorado, 150 pp.

Daly, R.A., 1925. Pleistocene changes of sea level. Am. J. Sci., 10: 281—313.

Farhoudi, G., and Karig, D.E., 1977. The Makran of Iran and Pakistan as an active arc system. Abstr., EOS Trans., Am. Geophys. Union, 58: 446.

Flint, R.F., 1971. Glacial and Quaternary Geology. Wiley, New York, N.Y., pp. 326—327.

Gates, G.O., Page, W.D., Savage, W.U. and Zuberi, S., 1977. The Makran earthquake November 28, 1945 (mag. 8.3) — a 1975 field appraisal of geological and cultural effects. Proc. 6th World Conf. Earthquake Eng., New Delhi, 2: 529.

Hudson, T., Plafker, G. and Rubin, M., 1973, Uplift rates of marine terrace sequences in the Gulf of Alaska. U.S. Geol. Surv. Circ. 733, pp. 11—13.

Kaufman, A., Broecker, W.S., Ku, T.-L. and Thurber, D.L., 1971. The status of U-series methods of mollusk dating. Geochim. Cosmochim. Acta, 35: 1155—1183.

Konishi, K., Schlanger, S.O. and Omura, A., 1970. Neotectonic rates in the central Ryukyu Islands derived from ^{230}Th coral ages. Mar. Geol., 9: 225—240.

Ku, T.-L., 1976. The uranium-series methods of age determination. Ann. Rev. Earth Planet. Sci., 4: 347—379.

Little, R.D., 1972. Terraces of the Makran Coast of Iran and parts of West Pakistan. M.A. Thesis, Univ. of Southern California, 151 pp.

Moore, W.S. and Somayajulu, B.L.K., 1974. Age determinations using fossil corals using ^{230}Th/^{234}Th and ^{230}Th/^{227}Th. J. Geophys. Res., 79: 5065—5068.

National Ocean Survey, 1973. Tide Tables, High and Low Water Predictions, 1974, Central and Western Pacific Ocean and Indian Ocean. U.S. Dep. Commer., National Oceanic and Atmospheric Administration, Washington, 344 pp.

Pendse, C.G., 1948. The Makran earthquake of the 28th November 1945. India Meteorol. Dep., Sci. Notes, 1 (3).

Photographic Survey Corporation, 1958. Reconnaissance Geology of Part of West Pakistan: a Columbo Plan Cooperative Project. Toronto, Canada, Geological Maps 3—5.

Plafker, G., 1969. Tectonics of the March 27, 1964 Alaska earthquake. U.S. Geol. Surv. Prof. Pap. 543-I, 74 pp.

Shearman, D.J., 1977. The geological evolution of southern Iran, the report of the Iranian Makran Expedition. Geogr. J., 142: 393—410.

Snead, R.E., 1969. Physical geography reconnaissance: West Pakistan coastal zone. University of New Mexico Publications in Geography, I, Univ. of New Mexico, Albuquerque, 55 pp.

Snead, R.E., 1970. Physical geography of the Makran coastal plain of Iran. Rep. Off. Nav. Res., Contract N00014 66 C D104, Task Order NR388 082, Univ. of New Mexico, Albuquerque, 715 pp.

Sondhi, U.P., 1947, The Makran earthquake, 28th November 1945 — the birth of new islands. Geol. Surv., Indian Miner., 1 (3).

Steinen, R.P., Harrison, R.S. and Matthews, R.K., 1973. Eustatic low stand of sea level between 125,000 and 105,000 B.P.: evidence from the subsurface of Barbados, West Indies. Geol. Soc. Am. Bull., 84: 63—70.

Stoneley, R., 1974. Evolution of the continental margins bounding a former Tethys. In: C.A. Burk and C.L. Drake (Editors), The Geology of Continental Margins. Springer, New York, N.Y., pp. 889—903.

Vita-Finzi, C., 1975. Quaternary deposits in the Iranian Makran. Geogr. J., 141: 415—420.

Walcott, R.I., 1972. Post sea levels, eustasy, and deformation of the earth. Quat. Res. (Tokyo), 2: 1—14.

Tectonophysics, 52 (1979) 549—559
© Elsevier Scientific Publishing Company, Amsterdam — Printed in The Netherlands

PREMONITORY CRUSTAL DEFORMATIONS, STRAINS AND SEISMOTECTONIC FEATURES (*b*-VALUES) PRECEDING KOYNA EARTHQUAKES

S.K.GUHA

Central Water and Power Research Station, Khadakwasla, Pune 411 024 (India)

(Accepted for publication June 8, 1978)

ABSTRACT

Guha, S.K., 1979. Premonitory crustal deformations, strains and seismotectonic features (*b*-values) preceding Koyna earthquakes. In: C.A. Whitten, R.Green and B.K. Meade (Editors), Recent Crustal Movements, 1977. Tectonophysics, 52: 549—559.

There have been instances of premonitory variations in tilts, displacements, strains, telluric current, seismomagnetic effects, seismic velocities (V_p, V_s) and their ratio (V_p/V_s), *b*-values, radon emission, etc. preceding large and moderate earthquakes, especially in areas near epicentres and along faults and other weak zones. Intensity and duration (T) of these premonitory quantities are very much dependent on magnitude (M) of the seismic event. Hence, these quantities may be utilised for prediction of an incoming seismic event well in advance of the actual earthquake. In the recent past, tilts, strain in deep underground rock and crustal displacements have been observed in the Koyna earthquake region over a decade covering pre- and postearthquake periods; and these observations confirm their reliability for qualitative as well as quantitative premonitory indices. Tilt began to change significantly one to two years before the Koyna earthquake of December 10, 1967, of magnitude 7.0. Sudden changes in ground tilt measured in a water-tube tiltmeter accompanied an earthquake of magnitude 5.2 on October 17, 1973 and in other smaller earthquakes in the Koyna region, though premonitory changes in tilt preceding smaller earthquakes were not so much in evidence. However, changes in strains in deep underground rock were observed in smaller earthquakes of magnitude 4.0 and above. Furthermore, as a very large number of earthquakes ($M = 1-7.0$) were recorded in the extensive seismic net in the Koyna earthquake region during 1963—1975, precise *b*-value variations as computed from the above data, could reveal indirectly the state of crustal (tectonic) strain variations in the earthquake focal region and consequently act as a powerful premonitory index, especially for the significant Koyna earthquakes of December 10, 1967 ($M = 7.0$) and October 17, 1973 ($M = 5.2$). The widespread geodetic and magnetic levelling observations covering the pre- and postearthquake periods indicate significant vertical and horizontal crustal displacements, possibly accompanied by large-scale migration of underground magma during the large seismic event of December 10, 1967 in the Koyna region ($M = 7.0$). Duration (T) of premonitory changes in tilt, strains, etc., is generally governed by the equation of the type $\log T = A + BM$ (A and B are statistically determined coefficients). Similar other instances of premonitory evidences are also observed in micro-earthquakes ($M = -1$ to 2) due to activation of a fault caused by nearby reservoir water-level fluctuations.

INTRODUCTION

For the last half a century or so there have been attempts to assess some premonitory changes which can be used as physical basis for the prediction of large seismic events. Thus, long-period changes in ground strain and tilt, geomagnetic and geoelectric fields, radon emission, micro-earthquake activity, even adjoining sea- and groundwater levels, etc., were observed in seismic zones, mainly in Japan, the U.S.A. and the U.S.S.R. for possible correlations of their premonitory changes with a succeeding significant seismic event or events. Until recently, the results of these investigations have confirmed only broad possibilities of prediction of significant seismic events due to physical changes in one or more of the above-mentioned quantities (Nersesov et al., 1969; Shamsi and Stacey, 1969; Suyehiro and Sekiya, 1972; Gupta, 1973; Aggarwal et al., 1975; Wyss, 1975; Osika et al., 1976; Smith, 1976). In view of the uncertainties in respect of both time and place of occurrence of large seismic events, these premonitory changes observed in a few cases could not be verified in a large number of cases to facilitate their early operational possibilities. These premonitory changes are principally due to the building up of stress in the potential focal regions, and the resulting premonitory period could be as long as several years for a large earthquake of magnitude 7.0 and as small as a few days for a low-magnitude earthquake ($M = 3.5$). These basic data, though scanty at present, have afforded some reliable basis for anticipating large seismic events, both in respect of size and premonitory period of impending seismic event. It is not as yet very clear if premonitory changes in all the above physical quantities could be equally effective in anticipating large seismic events with a sufficient premonitory period. It is quite possible that large-scale observations of the above physical quantities in potential seismic zones could, in course of time, decipher the relative efficiency of these physical quantities for their use as precursors. This will necessitate large-scale installations in potential seismic zones and systematic observations over long periods so as to obtain significant data to solve the problem of prediction effectively.

SEISMICITY, TILT AND STRAIN MEASUREMENTS IN THE KOYNA REGION

The closely spaced seismological net in the Koyna region (17°23′N, 73°45′E) afforded a unique opportunity for recording a large number of local earthquakes (Guha et al., 1974a, 1974b). Fig. 1 shows half-yearly and monthly b-values, lake levels, dam deflections, earthquake fracture volume, strain measurements at the dam foundation, and earthquakes of magnitudes 4.0 and above. The occurrence of earthquakes is generally found to be at low values of b. The b-values have also been observed to be inversely changing with the superimposed water load, which perhaps indicates that the level of regional tectonic stresses is such that the variations in water loads can bring about significant changes in the tectonic stresses. The significant earth-

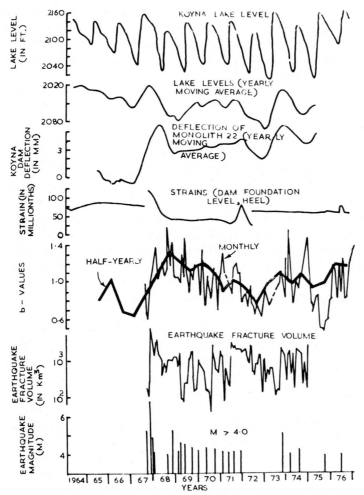

Fig. 1. Showing lake levels, deflections and strains of the Koyna dam, and corresponding earthquake data.

quakes of December 10, 1967 and October 17, 1973 occurred when the average levels increased sharply and the b-values were low.

The most frequently used statistical law in the study of earthquakes is log $N = a - bM$, developed by Gutenberg and Richter, relating to frequency of occurrence (N) of earthquakes with magnitude (M). The value of coefficient b obtained for a region from earthquake data covering an adequate time period has been generally regarded as the unchanging characteristic of the region representing its average seismic status. However, in a number of cases the values of b for foreshock and aftershock sequences are found to be different — that for the latter being higher following release of tectonic stress in the earthquakes (Gibowicz, 1973). Also, from the laboratory experiments,

Mogi (1963) and Scholz (1968) have established dependence of b on the amount of stress applied to rock samples. It can therefore be assumed that for a seismically active area the short-period regional stress changes which control earthquake activity can be reflected in periodic b values for the observed regional earthquakes.

From the arrival time data (t_i for $i = 1-5$) of five closely spaced observatories (x_i, y_i, z_i), for P_g/S_g phases, coordinates (x, y, z) of earthquake foci could be accurately determined using:

$$(x - x_i)^2 + (y - y_i)^2 + (z - z_i)^2 = V^2(t - t_i)^2$$

where V is the phase velocity and t the origin time. Most of the Koyna earthquakes which could be recorded on more than five stations are of magnitude <2.0. Hence, seismically active region determined from these foci could be considered as the earthquake volume active at any time period. Fig. 1 shows that, as may be expected, the earthquake volume is large when b-values are low. However, the b-values have better correlation with earthquake fracture volume if a time lag of two months is considered, suggesting that for the Koyna region there is a time lag of two months between the building up of the stress levels responsible for an earthquake and its actual occurrence.

Worldwide in-situ measurements of stresses in rocks to determine the tectonic stresses and to estimate the probability of earthquake occurrence have been made by a number of workers (Ranalli, 1975). These measurements have demonstrated the presence of large horizontal stresses in the earth's crust which are assumed to be the result of the gravitational and tectonic stresses, both current and remnant. Though the main drawback of in-situ measurements is their local nature, it has been ascertained that periodic stress measurements can illustrate the tectonic activity for any region. Hast (1973), from extensive studies, has pointed out that the knowledge of absolute stress in the upper earth's crust might be of value in seismological research not only for examining the cause and nature of earthquakes but ultimately as a technique for their prediction, especially in areas where they tend to recur at intervals. The earthquake studies in the Koyna region indicate that rock stress measurements, used in conjunction with the other premonitory phenomena, could prove to be a very useful tool for earthquake prediction and control.

Fig. 2 shows the variations of tilt (1967–70) as measured by a torsion pendulum fixed on Monolith No. 10 of Koyna Dam. Similar measurements of tilt on other monoliths were made on the Koyna Dam. The results obtained on different monoliths, though varying in details, depict a large preseismic tilt with a premonitory period of about a year or more. This preseismic tilt variation is in close conformity with the preseismic pendulum deflection shown in Fig. 1. Though there was large preseismic tilt before the main seismic event ($M = 7.0$), similar preseismic phenomena did not generally precede smaller earthquakes, say $M = 5.0$ and below (see Figs. 1

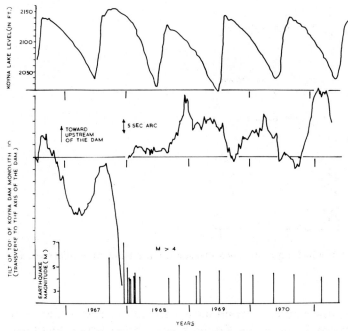

Fig. 2. Tilt of Koyna Dam top ("Monolith 10") as measured with torsion pendulum (T_0 = 4 sec), and occurrence of earthquakes.

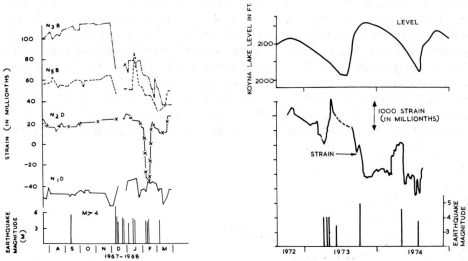

Fig. 3. Variation of strains in Carlson-type strain gauges embedded in the foundation level of the Koyna Dam.

Fig. 4. Variation of strain measured in rock at depths of about 1 km, and occurrence of earthquakes in the Koyna region.

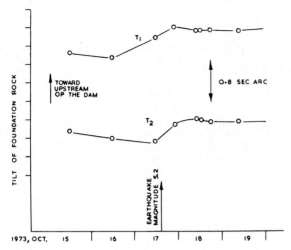

Fig. 5. Coseismic tilt variation in foundation rock measured with water-tube tiltmeters T_1 and T_2.

and 2). Figs. 3 and 4 show results of strain measurements at near surface and at depths of the order of 1 km in the Koyna region. These strain measurements show that strain variations are rather coseismic with a very small premonitory period of the order of a month or so (see Fig. 3). Strain drop in Fig. 3, especially in the N_3B strain gauge, corresponds to a stress drop of about 20 bars as expected in the main earthquake of $M = 7.0$. However, small earthquakes of magnitude below $M = 5.0$ are accompanied by almost coseismic strain variations measured at depths with at times a very small premonitory period, see Fig. 4. In the tectonic environment of the Koyna region, decrease in strain (i.e., tension) was immediately followed by earthquakes as shown in Fig. 4 — an observation which is very significant for prediction feasibility. However, no such strain observations at deeper horizons are available for the main Koyna seismic event of magnitude 7.0. Strain measurements at depths could perhaps also be used for prediction purposes

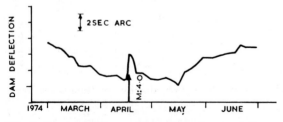

Fig. 6. Coseismic Koyna Dam deflection variation accompanying a local earthquake of magnitude (M) 4.0 on April 17, 1974.

for large seismic events. However, small coseismic tilt changes are evident in Figs. 5 and 6 for earthquakes M = 5.2 and M = 4.0. Such coseismic changes in smaller earthquakes should be limited to epicentral areas only.

GEOMAGNETIC, GEODETIC AND DEFLECTION STUDIES IN THE KOYNA REGION

From interpretations of changes in regional geomagnetic values both before and after the Koyna earthquake of December 10, 1967 Bhattacharji (1970) concluded that significant large-scale shift of magnetically susceptible rock or lava took place following internal readjustments due to the recent Koyna earthquake. This possible large-scale shift of magma associated with the Koyna earthquake is significant in view of widespread volcanism in the area in the recent geologic past. Similarly, plumb-line deflection observations and their interpretation by Bhattacharji (1970) (see Fig. 7) are most significant. Pre- and postearthquake changes in the deflections in Fig. 7 have been interpreted by him as due to internal relative changes in the subcrustal rock, including possibly magma. Furthermore, sharp changes in plumb-line deflections from the west coast to further east suggest major crustal discontinuity or faulting almost parallel to the coast. Incidentally, this assumed N—S direction of faulting or dislocation coincides with the hot-spring zone in the area.

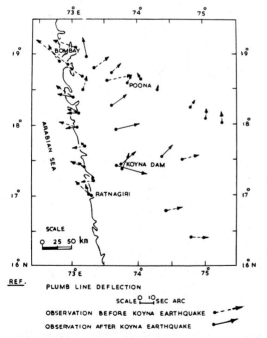

Fig. 7. Showing plumb-line deflection in the Koyna region (Bhattacharji, 1970).

The N—S strike of dislocation is also characterized by a significant broad gravity low. Some increases in the gravity values in the epicentral area of December 10, 1967 earthquake have been observed following the earthquake, and could be interpreted either as due to subsidence of the area or to the rise of heavier subcrustal magma during the seismic movement. Thus, magnetic, gravity and deflection measurements during pre- and postearthquake periods possibly indicate some large-scale relative changes in the subcrustal magma during the earthquake. The tectonophysical implications of such large-scale magmatic movements could be manifold — e.g., very slow expansion of earth, drift of the Indian Plate, intraplate tectonics, dormant volcanism, etc. It can thus be reasonably assumed that large-scale changes in plumb-line deflections in Fig. 7, and long-term tilting of the dam monoliths as observed in Figs. 1 and 2 prior to the earthquake, are related, and that

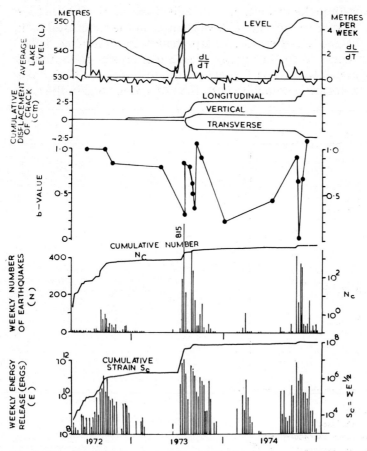

Fig. 8. Showing lake level variations, displacement at the ground crack, *b*-values and microearthquake activity at Mula Reservoir.

these measurements could be powerful premonitory phenomena, especially in major seismic events such as the Koyna earthquake of $M = 7.0$.

OTHER CASES OF RESERVOIR-INDUCED SEISMICITY

Similar observations have been extended to the microtectonic process associated with earthquakes of magnitude <1.0 and movements along the nearby fault at Mula Reservoir (19°22'N, 74°37'E), India, see Fig. 8 (Guha et al., 1974a). As the magnitudes of the earthquakes are small, the various premonitory quantities, such as b-values, relative displacement along nearby fault zones and rate of water-level fluctuations, are all naturally coseismic. Decrease in b-values is followed by enhanced seismic activity and fault displacements. The b-values are thus powerful premonitory indices, both for large and small seismic events and also ground tilts and displacements.

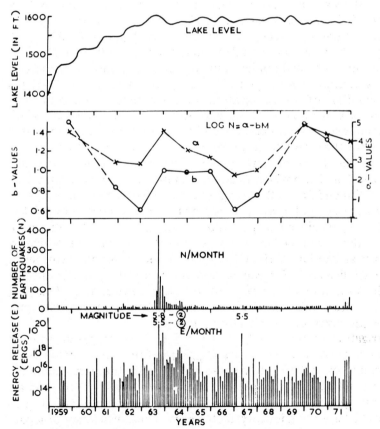

Fig. 9. Showing variation with time of lake levels, a- and b-values, frequency of occurrence and earthquake energy release (Kariba Lake region).

Analysis of seismicity of the Kariba Reservoir situated at the Zambia—Rhodesia border in Africa is cited in Fig. 9. Here again variation of b-values is a powerful premonitory index. The decrease in b-value is followed by enhanced seismicity. In this respect there is great similarity of tectonic behaviour on account of reservoir water load variations in these two reservoirs, namely Koyna and Kariba. Even microtectonic activity in Mula Reservoir (see Fig. 8) follows similar trends. It has thus to be seen whether the above characteristics are valid in reservoir-induced seismicity in general.

DISCUSSIONS AND CONCLUSIONS

Long-term observations of seismicity, tilt, deflections, magnetic and gravity changes in the Koyna earthquake region over about 10 years indicate that significant changes in the above parameters in the preseismic and postseismic periods could be used as powerful premonitory indices for predicting especially the large seismic events. Of all the above parameters, the most significant premonitory indices are b-values and tilt changes. Both these quantities began to change at least one to two years in advance of the main Koyna earthquake of $M = 7.0$ on December 10, 1967 — which is broadly in accordance with the generally accepted relationship $\log(T) = 0.68M - 1.31$ between the premonitory period (T in days) and M obtained from analysis of data of a large number of earthquakes and for various premonitory indices. In fact, b-value variations over a period of 10 years or so (1966 to 1976) confirm the reliability of its application in earthquake prediction. The decrease in b-values before the $M = 7.0$ earthquake on December 10, 1967 and $M = 5.2$ earthquake on October 17, 1967 are most significant. Coseismic levels of b-values for these two earthquakes are in accordance with release of tectonic stress in the form of earthquakes. Similarly, tilt and monolith deflections in Figs. 1 and 2 are significant premonitory indices for large seismic events of $M = 7.0$ and $M = 5.2$, though other premonitory indices such as changes in geomagnetic and gravity fields have broadly corroborated the results obtained from b-values and pendulum deflections. The b-values have also significantly decreased before enhanced seismic activity in reservoir-induced seismicity at Kariba (Africa) and Mula (India) reservoirs. This lends credence to the application of b-values as a premonitory index in diverse geological conditions and in a large range of seismic activities, with M varying from -2.0 to 7.0. The b-values and tilt changes are thus two powerful premonitory indices for earthquake-prediction purposes.

Though investigators in the U.S.A. and U.S.S.R. have found significant decrease in ratio of body-wave velocities (V_p/V_s) before some earthquakes, in the Koyna region, except for an isolated instance for the earthquake on June 27, 1969 ($M = 4.7$) this method has not shown much promise as an efficient premonitory index. Differences in tectonic environment in the Koyna region could be a contributory factor here. Thus, efficiency of a particular premonitory index may be dependent on geotectonic environment of the region, apart from other as yet unknown factors.

REFERENCES

Aggarwal, Y.P., Sykes, L.R., Simpson, D.W. and Richards, P.G., 1975. Spatial and temporal variations in t_s/t_p and in P-wave residuals at Blue Mountain Lake, New York: application to earthquake prediction. J. Geophys. Res., 80: 718—732.

Bhattacharji, J.C., 1970. A geodetic and geophysical study of the Koyna earthquake region. J. Indian Geophys. Union, 7: 17—27.

Gibowicz, S.J., 1973. Variation of the frequency—magnitude relation during earthquake sequences in New Zealand. Bull. Seismol. Soc. Am., 63: 517—528.

Guha, S.K., Gosavi, P.D., Nand, K., Agarwal, B.N.P., Padale, J.G. and Marwadi, S.C., 1974a. Artificially induced seismicity and associated ground motions. Proc. 2nd Int. Congr. Eng. Geol. Brazil, 1: II-PC 1.1—1.14.

Guha, S.K., Gcsavi, P.D., Nand, K., Padale, J.G. and Marwadi, S.C., 1974b. Koyna Earthquake (Oct. 1963 to Dec. 1973). Central Water and Power Research Station, Khadakwasla, Poona.

Gupta, I.N., 1973. Premonitory variations in S-wave velocity anisotropy before earthquakes in Nevada. Science, 182: 1129—1132.

Hast, N., 1973. Global measurements of absolute stress. Philos. Trans. R. Soc. Lond. Ser. A., 274: 409—419.

Mogi, K., 1963. The fracture of a semi-infinite body caused by an inner stress origin and its relation to earthquake phenomenon (2nd pap). Bull. Earthquake Res. Inst., 41: 595—614.

Nersesov, I.L., Semonova, A.N. and Simbireva, I.G., 1969. Physical Basis of Foreshocks, Nauka, Moscow.

Osika, D.G., Magomedov, A.M., Smirnova, M.N., Levkovich, R.A. and Megaev, A.B., 1976. Hydrodynamical and geochemical forerunners to the large earthquake in the northern Caucasus. Proc. Int. Symp. on Earthquake Forerunners Searching, 1974, Tashkent, pp. 65—68 (in Russian).

Ranalli, G., 1975. Geotectonic relevance of rock-stress determinations. Tectonophysics, 29: 49—58.

Scholz, C.H., 1968. The frequency—magnitude relation of microfracturing in rock and its relation to earthquakes. Bull. Seismol. Soc. Am., 58: 399—415.

Shamsi, S. and Stacey, F.D., 1969. Dislocation models and seismomagnetic calculations for California 1906 and Alaska 1964 earthquakes. Bull. Seismol. Soc. Am., 59: 1435—1448.

Smith, P.J., 1976. Radon to predict earthquakes. Nature, 261: 97—98.

Suyehiro, S. and Sekiya, H., 1972. Foreshocks and earthquake prediction. Tectonophysics, 14: 219—225.

Sykes, L.R., Fletcher, P.J., Armbruster, J. and Davis, J.F., 1972. Tectonic strain release and fluid injection at Dale, New York. EOS, Trans. Am. Geophys. Union, 53: 524.

Wyss, M., 1975. A search for precursors to the Sitka, 1972, earthquake: sea level, magnetic field, and P-residuals. Pure Appl. Geophys., 113: 297—309.

Tectonophysics, 52 (1979) 561—570
© Elsevier Scientific Publishing Company, Amsterdam — Printed in The Netherlands

RECENT CRUSTAL MOVEMENTS IN THE SIERRA NEVADA—WALKER
LANE REGION OF CALIFORNIA—NEVADA: PART I, RATE AND
STYLE OF DEFORMATION

DAVID B. SLEMMONS [1], DOUGLAS VAN WORMER [2], ELAINE J. BELL [1] and
MILES L. SILBERMAN [3]

[1] *Department of Geology, Mackay School of Mines, Univ. Nevada, Reno, Nev. 89557
(U.S.A.)*
[2] *Seismological Laboratory, Univ. Nevada, Reno, Nev. 89507 (U.S.A.)*
[3] *U.S. Geological Survey, Menlo Park, Calif. 94025 (U.S.A.)*

(Revised version accepted June 8, 1978)

ABSTRACT

Slemmons, D.B., Van Wormer, D., Bell., E.J. and Silberman, M.L., 1979. Recent crustal
movements in the Sierra Nevada—Walker Lane region of California—Nevada: Part I,
Rate and style of deformation. In: C.A. Whitten, R.Green and B.K. Meade (Editors),
Recent Crustal Movements, 1977. Tectonophysics, 52: 561—570.

This review of geological, seismological, geochronological and paleobotanical data is
made to compare historic and geologic rates and styles of deformation of the Sierra
Nevada and western Basin and Range Provinces. The main uplift of this region began
about 17 m.y. ago, with slow uplift of the central Sierra Nevada summit region at rates
estimated at about 0.012 mm/yr and of western Basin and Range Province at about 0.01
mm/yr. Many Mesozoic faults of the Foothills fault system were reactivated with normal
slip in mid-Tertiary time and have continued to be active with slow slip rates. Sparse data
indicate acceleration of rates of uplift and faulting during the Late Cenozoic. The Basin
and Range faulting appears to have extended westward during this period with a reduction in width of the Sierra Nevada.
The eastern boundary zone of the Sierra Nevada has an irregular en-echelon pattern of
normal and right-oblique faults. The area between the Sierra Nevada and the Walker Lane
is a complex zone of irregular patterns of horst and graben blocks and conjugate normal-
to right- and left-slip faults of NW and NE trend, respectively. The Walker Lane has at
least five main strands near Walker Lake, with total right-slip separation estimated at 48
km. The NE-trending left-slip faults are much shorter than the Walker Lane fault zone
and have maximum separations of no more than a few kilometers. Examples include the
1948 and 1966 fault zone northeast of Truckee, California, the Olinghouse fault (Part III)
and possibly the almost 200-km-long Carson Lineament.
Historic geologic evidence of faulting, seismologic evidence for focal mechanisms, geodetic measurements and strain measurements confirm continued regional uplift and tilting
of the Sierra Nevada, with minor internal local faulting and deformation, smaller uplift of
the western Basin and Range Province, conjugate focal mechanisms for faults of diverse
orientations and types, and a NS to NE—SW compression axis (σ_1) and an EW to NW—SE
extension axis (σ_3).

INTRODUCTION

Review of several recent studies, combined with earlier summary reports on the structural deformation of the Sierra Nevada and adjoining Basin and Range Provinces, permits an estimate to be made of rates of regional uplift and of fault-slip rates on major faults (Jennings, 1975). This report summarizes the type and rates of regional deformation and correlates historic evidence for activity with that provided by the Late Cenozoic record.

The regional relationships for uplift and tilting of the central Sierra Nevada area have been examined by Axelrod (1957, 1958 and 1962), Axelrod and Ting (1961), Christiansen (1966), Hudson (1955 and 1960), and Slemmons (1953 and 1966). More recently, many of the volcanic units that provide evidence for timing and style of deformation in the region have been studied in detail and carefully dated with many potassium—argon radiometric dates. This paper summarizes present data and estimates the rates and timing of regional uplift and faulting. Geologic mapping and review of historic earthquakes, geodetic changes, and strain measurements provide a basis for comparison.

The general regional style and rate of deformation is summarized in this part of our report; Part II documents the style and rate of faulting on the Pyramid Lake portion of the Walker Lane right-slip fault zone, and Part III reports on the 1869 surface faulting on the conjugate Olinghouse left-slip fault zone.

AREA STUDIED

The region discussed in this report is shown in Fig. 1, and includes the central section of the Sierra Nevada and the adjoining part of the Basin and Range Province to the west of Walker Lane. The two best-documented sectors include the Sonora Pass section of the range, with a diverse and well-dated series of mid-Tertiary to Quaternary volcanic deposits (Slemmons, 1953 and 1966; Halsey, 1953; and Noble et al., 1974), and the Donner Summit—Reno—Olinghouse—Pyramid Lake zone, with a series of unpublished studies by University of Nevada graduate students and work in progress by Ed Bingler and Harold Bonham of the Nevada Bureau of Mines and Geology.

RESULTS OF STUDY

Active faults and regional uplift

Central Sierra Nevada Province
The timing of uplift and initiation of faulting of the Sierra Nevada and Basin and Range Provinces has been the subject of many studies including Axelrod (1957 and 1962), Christiansen (1966), Dalrymple (1963 and 1964),

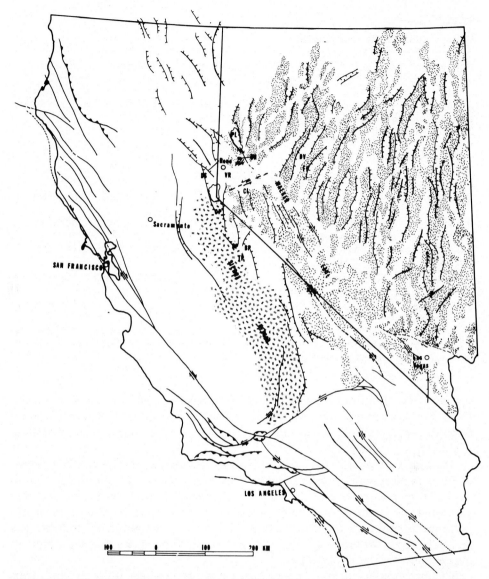

Fig. 1. Location of major structural, geological and geographic features of the Sierra Nevada—Basin and Range region. *BP* = Bridgeport; *CL* = Carson Lineament; *CP* = Carson Pass; *DS* = Donner Summit; *DV* = Dixie Valley; *FP* = Fairview Peak; *OH* = Olinghouse; *PL* = Pyramid Lake; *SP* = Sonora Pass; *TK* = Truckee—Hobart Mills area; *TP* = Tioga Pass; *VR* = Virginia Range; *W* = Winnemucca; *WL* = Walker Lake.

Noble (1972) and Slemmons (1953 and 1966). The Sierra Nevada block shows differential internal warping and tilting as summarized by Christiansen (1966). The Donner Summit area had maximum uplift of the gently sloping, Early Tertiary upland surfaces to present elevations of about 2400 m above

TABLE I

Rates of uplift in the areas studied

Reference	Unit	Est. age (m.y.) *	Uplift (m)	Uplift rate (mm/yr) (to present)
Uplift of the Donner Summit area				
Axelrod (1957) and this study	Mio—Pliocene	12—13	1616	0.13
Christiansen (1966)	Pre-andesite	12	1200—1500	0.10—0.12
Uplift of the Carson Pass—Sonora Pass area				
this study	Valley Springs formation	26	1524—2286	0.06—0.09
this study	Relief Peak formation	*c.* 19.6	1397—2096	0.07—0.11
Axelrod (1957) and this study	Mio—Pliocene andesites	12.2	1871	0.15
this study	Stanislaus Latite Group	9.3	889—1334	0.10—0.14
this study	erosional surface A	4—5 (est.)	635	0.11—0.16

* Estimates are based on K—Ar dates of volcanic rocks reported by Dalrymple (1963 and 1964), and Morton et al. (1978).

sea level. The Sonora Pass area was uplifted to about 3000 m above sea level. Using the most recent radiometric dates for the volcanic rocks of the Donner Summit area, estimates of amounts and rates of uplift were obtained (Table I).

Silberman et al. (1975) and Morton et al. (1978) document the occurrence of widespread andesitic volcanic activity between 17 and 8 m.y. ago in the western part of the Walker Lane and present Sierra Nevada. Stewart and Carlson (1977) show that volcanic rocks of this age are present in nearly continuous outcrops from near Reno to Bridgeport, California, and into the Sierra Nevada, where they are now about 2000 m above the same unit in nearby parts of the Great Basin.

The progressive increase in average rates for younger units or landforms of the Carson Pass—Sonora Pass area, indicated in Table I, and the review by Christiansen (1966) suggest progressive acceleration in rates of uplift from the Oligocene (about 0.01 mm/yr for 28 m.y. B.P. to 17 m.y. B.P.) to present (about 0.11—0.16 mm/yr for the last 4—5 m.y.).

The geodetic rate of uplift (4.5 mm/yr) at Donner Pass, suggested by the historic data of Bennett et al. (1977), is more than an order of magnitude greater, and the eastward tilt of the Sierra Nevada further south at Sonora Pass and Tioga Pass (West and Alt, 1977) indicate a short-term historic reversal of the long-term trend. A historic reversal in long-term trend of deformation is also shown by Priestley's (1974) instrumental observations of changes in strain axes in the western Basin and Range Province. This suggests that a longer record than the brief historic record is needed to provide valid rates of deformation in this region.

Sierra Nevada—Basin and Range boundary zone

The geologic record for the boundary zone of the Sierra Nevada provides sparse data on long-term rates of fault recurrence or timing of initial Basin and Range faulting. The thick and extensive volcanoclastic sedimentary rocks of about 5 m.y. age in the area between Truckee and Reno suggest that the basins were well developed between 5 and 12 m.y. ago. The high dips (commonly up to 30°) and variation in elevation of these deposits also verify large subsequent deformation. Perhaps one-half of the total deformation of the basement rocks is younger than 5 m.y. The flora of the Verdi and other formations of the region (Axelrod, 1957, 1958, 1962; Axelrod and Ting, 1960, 1961) suggest a surface of much lower relief than is now present and that the flora grew at an elevation of about 760 m above sea level (about 700 m lower than present). The long-term geologic data suggests slow uplift of the graben and major uplift of the horst blocks of the region.

The geodetic data of Bennett et al. (1977) suggest that most of the region between Donner Lake and Reno has been uplifted smaller amounts than the crest of the range for the period 1947 to present.

The Basin and Range Province appears to extend westward with a reduction in size of the Sierra Nevada. This encroachment is indicated by the recency of development of Owens Valley during the last 2 m.y. (Bachman, 1974; Roquemore, 1976), by the position of most 10—15 m.y. lacustrine deposits to the east of the present boundary zone, and the apparent inactivity of faults of the Sonora Pass region prior to 9.3 m.y. ago.

Western Basin and Range Province

The region between the Sierra Nevada eastern fault escarpment and the Walker Lane fault zone is referred to in this report as the western Basin and Range Province. The terrain has a general horst and graben pattern, but lacks the corrugated pattern of long NS to NNE—SSW topography that characterizes most of the Basin and Range Province to the east of the Walker Lane. Faults and fault blocks are generally short and although some northwest elongation parallel to the Walker Lane and the Sierra Nevada is apparent, it is not a dominant orientation. Within this zone, unlike the even more shattered pattern noted by Pease (1969) to the north, there is a conjugate set of Reidel shears (Tchalenko, 1970) with left-lateral separations along many NE to ENE—WSW faults. These Reidel shears are the two conjugate sets of en-echelon fractures that develop at low and high angles to a shear zone. The Carson Lineament, a zone of NE-trending faults, is the longest structure, about 200 km in length. The 1966 zone of faulting to the northeast of Truckee, California, had a left-oblique-slip fault plane solution (Tsai and Aki,

566

1966; Ryall et al., 1968), and left-slip surface faulting from the 1966 earthquake (Kachadoorian et al., 1967). The Olinghouse fault zone in Nevada (see Part III) has historic left-slip displacements and is part of a system of similarly oriented faults in the Virginia Range, which display left-lateral separations. The paleobotanical studies of Axelrod (1957, 1958, and 1962) support a slow regional rate of uplift of this region during the Late Cenozoic; geodetic data (Bennett et al., 1977) indicate a slow rate of historic uplift of at least part of this area.

Walker Lane fault zone

The Walker Lane fault zone is a major strike-slip fault zone that bounds the eastern edge of the study area. The zone has approximately 48 km of cumulative right-slip offset during the last 22 m.y. (Hardyman et al., 1975). The zone appears to become more distributed in northern California (Pease, 1969). The segments of this zone that have historic surface faulting include possible offset near Pyramid Lake during the mid-1800s (see Part II) and in the Cedar Mountain area in 1932 (Gianella and Callaghan, 1934). Geodetic and earthquake focal mechanisms are compatible with the geologic field data as will be discussed in a later section of this report.

The data of Hardyman et al. (1975) indicate a total strike-slip separation rate of over 2 mm/yr across the section of the Walker Lane opposite Walker Lake. At least five major strands of the Walker Lane fault zone are present in the Walker Lake area and the slip rates for individual branches are about 0.5 mm/yr, averaged over the last 22 m.y. Current rates have not been established, but may be higher.

Seismicity and earthquake fault-plane solutions

The regional seismicity has been summarized by Ryall et al. (1966), Thompson and Burke (1974), and Douglas and Ryall (1975). The historic earthquake record shows a zone of high seismicity to the east of the weakly seismic Sierra Nevada block. The zone consists of two trends, a zone along the Sierra Nevada—Basin and Range boundary, and a connected trend along part of the Walker Lane to near Winnemucca, Nevada. The focal mechanisms for earthquakes in this region are shown in Fig. 2, based on a new compilation by Smith and Lindh (in press). The mechanisms of Smith and Lindh are compatible with conjugate faulting on NS normal-slip, and NE—SW and NW—SE strike-slip faults of left-slip and right-slip, respectively.

Strain measurements

The only strain measurements within the study area are near the southeastern part of the area. The available data has been summarized by Priestley (1974) and Ryall and Priestley (1975). Their data suggest regional and temporal changes in rate and direction of strain, but with alternation in periods of NW—SE compression and extension. The orientation of the strain axes is in agreement with the earthquake focal mechanisms of the area.

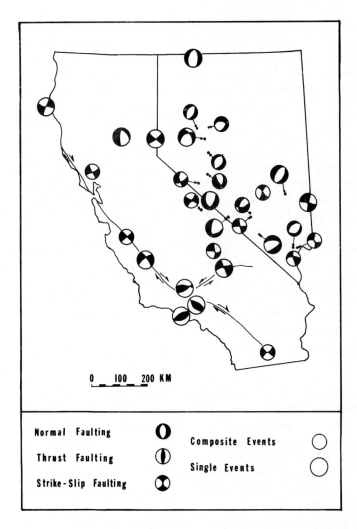

Fig. 2. Fault-plane solutions for the Sierra Nevada, western Basin and Range region, and adjoining parts of California and Nevada (modified from Smith and Lindh, in press).

Geodetic changes

The geodetic changes of the Sierra Nevada and adjoining Basin and Range region summarized by Bennett et al. (1977) indicate an irregular westward tilting of the Sierra Nevada block in the Donner Summit area. Maximum uplift of the summit of the range occurred at a rate of about 9.2 mm/yr for the period 1947—69. An average rate of uplift of about 4.5 mm/yr occurred in the western Basin and Range Province between Donner Summit and Reno, Nevada. Previous surveys in 1912 and 1947 suggest a lower maximum uplift rate of about 1.9 mm/yr for the Emigrant Gap area and about 0.5 mm/yr for the Donner Summit area. The post-1912 deformation of the western Basin and Range region between Truckee and Reno involved a slight subsidence, which indicates a historically nearly stable elevation for this area. Inflection points on the curves showing changes in elevation are commonly at or near major faults of the Foothills fault system. The bending at these points, rather than offseting of the curves indicates a change of warping rather than fault offset, with different uplift behavior for each block.

The geodetic changes associated with the 1954 earthquake series at the eastern edge of Walker Lane has been reviewed by Savage and Hastie (1969) with a reasonably close correlation between the surface faulting (Slemmons, 1957), which is partly of secondary origin, and the earthquake focal mechanisms. The Dixie Valley earthquake area of 1954 has been evaluated by Thompson and Burke (1974) for spreading rates and direction. Their data suggest an extension across the valley with an orientation of N55°W, and an extension rate of 1 mm/yr for 12,000 years and 0.4 mm/yr for 15 m.y. The slip direction shown by the geodetic data of the Fairview Peak area to the south is similar, about N65°W. The vertical-slip rates for the major boundary faults of the Dixie Valley graben have long-term rates of about 0.4—1 mm/yr.

RATES OF DEFORMATION

The historic geodetic data for the Sierra Nevada suggest irregular rates and directions of uplift with a maximum rate for the Donner Summit sector of the range that is between 4.5 and 9.2 mm/yr, a rate that is more than one order of magnitude greater than the geologic data indicated for this area (about 0.1—0.16 mm/yr). The geodetic data from the eastern edge of Walker Lane indicate similar mechanisms and directions of NW extension indicated by focal mechanism and geologic studies, with oblique-slip directions on moderate-angle normal faults.

The slip rates vary for major faults. About 48 km of slip across a zone of at least five major branches or splays of the Walker Lane during the last 22 m.y. indicates an average rate of 2.2 mm/yr. The deepest grabens of the region have geologic slip rates for the past 15 m.y. of about 0.5 mm/yr, with current Holocene rates of about 1 mm/yr. Rates of extension are about 1 mm/yr across the graben valleys, with extension directions that are regionally consistent.

STYLE OF DEFORMATION

The deformation of the Sierra Nevada Province is approximately that of a tilted block, but internal faulting and irregular geodetic deformation suggest a style of deformation that is complex in detail. The western Basin and Range is a complex zone bounded by en-echelon faults along the Sierra

Nevada on the west and the Walker Lane on the east. Conjugate faulting is indicated by the geologic and focal mechanism data and is suggestive of a complex mixture of the two types of deformation fields described by Wright (1976) for other parts of the province.

REFERENCES

Axelrod, D.I., 1957. Late Tertiary floras and the Sierra Nevada uplift. Geol. Soc. Am. Bull., 68: 19—46.

Axelrod, D.I., 1958. The Pliocene Verdi flora of western Nevada. Univ. Calif. Publ. Geol. Sci., 34: 91—160.

Axelrod, D.I., 1962. Post-Pliocene uplift of the Sierra Nevada, California. Geol. Soc. Am. Bull., 73: 183—198.

Axelrod, D.I. and Ting, W.S., 1960. Early Pleistocene floras from the Chagoopa surface, southern Sierra Nevada. Univ. Calif. Publ. Geol. Sci., 39: 1—18.

Axelrod, D.E. and Ting, W.S., 1961. Early Pleistocene floras from the Chagoopa surface, southern Sierra Nevada. Univ. Calif. Publ. Geol. Sci., 39: 119—194.

Bachman, S.B., 1974. Depositional and Structural History of the Waucobi Lake Bed Deposits, Owens Valley, California, M.S. Thesis, Univ. of California, Los Angeles, 129 pp.

Bennett, J.H., Taylor, G.C. and Toppozada, T.R., 1977. Crustal movement in the northern Sierra Nevada. Calif. Geol., 30: 51—57.

Christiansen, M.N., 1966. Late Cenozoic crustal movements in the Sierra Nevada of California. Geol. Soc. Am. Bull., 77: 163—182.

Dalrymple, G.B., 1963. Potassium—argon dates of some Cenozoic volcanic rocks of the Sierra Nevada, California. Geol. Soc. Am. Bull., 74: 379—390.

Dalrymple, G.B., 1964. Cenozoic chronology of the Sierra Nevada, California. Univ. Calif. Publ. Geol. Sci., 47: 1—41.

Douglas, B.M. and Ryall, A., 1975. Return periods for rock acceleration in western Nevada. Seismol. Soc. Am. Bull., 65: 1599—1611.

Gianella, V.P. and Callaghan, E., 1934. The earthquake of December 20, 1932, at Cedar Mountain, Nevada, and its bearing on the genesis of the Basin and Range structure. J. Geol., 42: 1—22.

Halsey, J.G., 1953. Geology of Parts of the Bridgeport, California, and Wellington, Nevada Quadrangles. Ph.D. thesis, California Univ. Berkeley.

Hardyman, R.F., Ekren, E.B. and Byers, F.M., 1975. Cenozoic strike-slip, normal and detachment faults in northern part of the Walker Lane, west-central Nevada. Geol. Soc. Am. Abstr., 7 (7): 1100.

Hudson, F.S., 1955. Measurement of the deformation of the Sierra Nevada, California. Geol. Soc. Am. Bull., 66: 835—870.

Hudson, F.S., 1960. Post-Pliocene uplift of the Sierra Nevada, California. Geol. Soc. Am. Bull., 71: 1547—1574.

Jennings, C.W. (compiler), 1975. Fault map of California. California Div. Mines Geol., scale: 1 : 750,000.

Kachadoorian, R., Yerkes, R.F. and Waananen, A.O., 1967. Effects of the Truckee, California earthquake of September 12, 1966. U.S. Geol. Surv. Circ., 537: 14 pp.

Morton, J.L., Silberman, M.L., Bonham, H.F. and Garside, L.J., 1978. Potassium—argon dates of volcanic rocks, plutonic rocks and ore deposits in Nevada — Determinations run under the U.S. Geological Survey—Nevada Bureau of Mines Cooperative Program: Isochron/West, to be published.

Noble, D.C., 1972. Some observations on the Cenozoic volcano-tectonic evolution of the Great Basin, western United States: Earth Planet. Sci. Lett., 17: 142—150.

Noble, D.C., Slemmons, D.B., Korringa, M.K., Dickinson, W.R., Al-Rawi, Y. and McKee, E.H., 1974. Eureka Valley tuff, east-central California and adjacent Nevada. Geology, 2: 139—142.

Pease, R.W., 1969. Normal faulting and lateral shear in northeastern California. Geol. Soc. Am. Bull., 80: 715—720.

Priestley, K.F., 1974. Crustal strain measurements in Nevada. Seismol. Soc. Am. Bull., 64: 1319—1328.

Roquemore, G.R., 1976. Cenozoic History of the Coso Mountains as Determined by Tuffaceous Lacustrine Deposits. M.A. Thesis, California State Univ., Fresno.

Rose, R.L., 1969. Geology of parts of the Wadsworth and Churchill Butte quadrangles, Nevada. Nev. Bur. Mines Geol., Bull., 71, 27 pp.

Ryall, A. and Priestley, K., 1975. Seismicity, secular strain, and maximum magnitude in the Excelsior Mountains area, western Nevada and eastern California. Geol. Soc. Am. Bull., 86: 1585—1592.

Ryall, A., Slemmons, D.B. and Gedney, L.D., 1966. Seismicity, tectonism and surface faulting in the western United States during historic time. Seismol. Soc. Am. Bull., 56: 1105—1135.

Ryall, A., Van Wormer, J-D. and Jones, A.E., 1968. Triggering of microearthquakes by earth tides, and other features of the Truckee, California, Earthquake sequence of September, 1966. Seismol. Soc. Am. Bull., 58: 215—248.

Savage, J.C. and Hastie, L.M., 1969. A dislocation model for the Fairview Peak, Nevada, earthquake. Seismol. Soc. Am. Bull., 59: 1937—1948.

Silberman, M.L., Noble, D.C. and Bonham, H.F., 1975. Ages and tectonic implications of the transition of calc-alkaline andesitic to basaltic volcanisms in the western Great Basin and Sierra Nevada. Geol. Soc. Am. Abstr., with Progr., Cordilleran Section, 7(3): 375.

Slemmons, D.B., 1953. Geology of the Sonora Pass Region, California. Ph.D. thesis, California Univ. Berkeley, 201 pp.

Slemmons, D.B., 1957. Geological effects of the Dixey Valley — Fairview Peak, Nevada, earthquakes of December 16, 1954. Seismol. Soc. Am. Bull., 47: 353—375.

Slemmons, D.B., 1966. Cenozoic volcanism of the central Sierra Nevada, California. Calif. Div. Mines Geol., Bull., 190: 199—208.

Smith, R.B. and Lindh, A., in press. A compilation of fault plane solutions of the western United States. Geol. Soc. Am. Spec. Pap.

Stewart, J.H. and Carlson, J.E., 1977. Geologic Map of Nevada. Nevada Bur. Mines and Geol., Map 57. Scale 1 : 1,000,000.

Tchalenko, J.S., 1970. Similarities between shear zones of different magnitudes. Geol. Soc. Am. Bull., 81: 1625—1640.

Thompson, G.A. and Burke, D.B., 1974. Regional geophysics of the Basin and Range province. Ann. Rev. Earth Planet. Sci., 2: 213—238.

Tsia, Yi-Ben and Aki, K., 1966. Source mechanism of the Truckee, California earthquake of Sept. 12, 1966. Seismol. Soc. Am. Bull., 60: 1199—1208.

West, D.O. and Alt, J.N., 1977. Analysis of recent crustal movement in the westcentral Sierra Nevada, California, using repeated geodetic leveling data. RCM 1977 Symposium, Tectonophysics, 52: 239—248.

Wright, L. 1976. Late Cenozoic fault patterns and stress fields in the Great Basin and westward displacement of the Sierra Nevada block. Geology, 4: 489—494.

Tectonophysics, 52 (1979) 571—583
© Elsevier Scientific Publishing Company, Amsterdam — Printed in The Netherlands

RECENT CRUSTAL MOVEMENTS IN THE CENTRAL SIERRA NEVADA—WALKER LANE REGION OF CALIFORNIA—NEVADA: PART II, THE PYRAMID LAKE RIGHT-SLIP FAULT ZONE SEGMENT OF THE WALKER LANE

ELAINE J. BELL and DAVID B. SLEMMONS

Department of Geological Sciences, Mackay School of Mines, University of Nevada-Reno, Reno, Nev. 89557 (U.S.A.)

(Revised version accepted June 8, 1978)

ABSTRACT

Bell, E.J. and Slemmons, D.B., 1979. Recent crustal movements in the Central Sierra Nevada—Walker Lane region of California—Nevada: Part II, The Pyramid Lake right-slip fault zone segment of the Walker Lane. In: C.A. Whitten, R.Green and B.K. Meade (Editors), Recent Crustal Movements, 1977. Tectonophysics, 52: 571—583.

The Pyramid Lake fault zone is within the Honey Lake—Walker Lake segment of the Walker Lane, a NW-trending zone of right-slip transcurrent faulting, which extends for more than 600 km from Las Vegas, Nevada, to beyond Honey Lake, California.

Multiscale, multiformat analysis of Landsat imagery and large-scale (1 : 12,000) low-sun angle aerial photography, delineated both regional and site-specific evidence for faults in Late Cenozoic sedimentary deposits southwest of Pyramid Lake. The fault zone is coincident with a portion of a distinct NW-trending topographic discontinuity on the Landsat mosaic of Nevada. The zone exhibits numerous geomorphic features characteristic of strike-slip fault zones, including: recent scarps, offset stream channels, linear gullies, elongate troughs and depressions, sag ponds, vegetation alignments, transcurrent buckles, and rhombohedral and wedge-shaped enclosed depressions. These features are conspicuously developed in Late Pleistocene and Holocene sedimentary deposits and landforms.

The Pyramid Lake shear zone has a maximum observable width of 5 km, defined by Riedel and conjugate Riedel shears with maximum observable lenghts of 10 and 3 km, respectively. P-shears have formed symmetrical to the Riedel shears and the principal displacement shears, or continuous horizontal shears, isolate elongate lenses of essentially passive material; most of the shears are inclined at an angle of approximately $4°$ to the principal direction of displacement. This suggests that the shear zone is in an early "Pre-Residual Structure" stage of evolution, with the principal deformation mechanism of direct shear replacing the kinematic restraints inherent in the strain field.

Historic seismic activity includes microseismic events and may include the earthquake of about 1850 reported for the Pyramid Lake area with an estimated Richter magnitude of 7.0. Based on worldwide relations of earthquake magnitude to length of the zone of surface rupture, the Pyramid Lake fault zone is inferred to be capable of generating a 7.0—7.5-magnitude event for a maximum observable length of approximately 6 km and a 6.75—7.25-magnitude event for a half length of approximately 30 km.

572

INTRODUCTION

Purpose

The purposes of this study were (1) to determine the expected location, site, and characteristics of surface faulting within the northern portion of the Walker Lane as exemplified by the Pyramid Lake fault zone, and (2) to evaluate the fault zone in terms of its present stage of evolution and its maximum potential for generating seismic events. Both qualitative and quantitative analyses of available geologic, seismic and historic data were made in an attempt to delineate and evaluate the Pyramid Lake fault zone.

Scope

This investigation of a segment of the northern portion of the Walker Lane is basic to a more extensive regional and intensive site-specific analysis of Late Cenozoic faulting and earthquake hazard assessment in western Nevada. The study was limited to field reconnaissance and the analysis of available literature, satellite imagery, and both conventional (1 : 62,500 scale) and large-scale (1 : 12,000) low-sun angle aerial photography.

Follow-on study will involve field investigation (ground-truth) to verify the results presented in this paper. In addition, data will be obtained on the kinematics of faulting, including (1) data on scarp morphology for assessing the movement history of the zone (Wallace, 1975, 1977) and for predicting future movement, and (2) data on amount of displacement, sense of movement, and trace length for predicting the maximum credible event that could be generated by the zone (Slemmons, 1977a, b; Yeats, 1977).

Investigative techniques

The investigation utilized a multiscale, multiformat approach to interpret and relate regional structural and tectonic elements to site-specific faulting. Regional structural trends were identified and delineated using Landsat imagery. Site-specific faults, local surface manifestations of faulting, and associated tectonic features were identified and delineated, using large-scale (1 : 12,000) low-sun angle aerial photography. Analysis of intermediate scale (1 : 62,500) conventional aerial photography enabled integration of the regional structural and tectonic setting with site-specific features. In addition, the photography served as a base for photogeologic and geomorphic mapping of Late Cenozoic deposits in localized areas.

The results of this systematic analysis of features within the Pyramid Lake fault zone provided the basis for interpreting the local and regional patterns of crustal strain (Tchalenko, 1970) and for inferring the maximum potential capability of the fault zone.

Summary of previous work

The Walker Lane (Fig. 1) was first recognized by Gianella and Callaghan (1934) and named by Locke et al. (1940). This NW-trending zone of right-lateral transcurrent faulting is generally believed to extend for more than 600 km from Las Vegas, Nevada, to the Honey Lake area of California (Gianella and Callaghan, 1934; Locke et al., 1940; Nielson, 1965; Shawe, 1965; Albers, 1967), paralleling the San Andreas fault system to which it may be related (Nielsen, 1965; Shawe, 1965). Previous studies of the Walker Lane have been limited primarily to the southern and central parts of the structure. The southern part of the Walker Lane includes a belt of large-scale sigmoidal bending (Albers, 1967; Stewart, 1967; Stewart et al., 1968) in the vicinity of Tonopah and the Las Vegas shear zone further south (Longwell, 1960; Burchfiel, 1965). The central part of the structure, extending from Tonopah to Walker Lake, is marked by strike-slip faults within the Gillis and Gabbs Valley Ranges and Soda Springs Valley and has been intensively studied (Nielsen, 1964, 1965; Speed, 1975; Hardyman et al., 1975; E.B. Ekren, unpublished data, 1977).

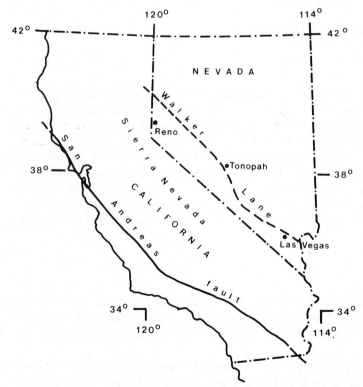

Fig. 1. Index map of the Walker Lane and the San Andreas fault (Nielsen 1965; Albers, 1967).

574

Bonham and Slemmons (1968) suggest that this structural feature, marked by a topographic depression or lane, extends northward from the Cedar Mts., beyond Pyramid Lake to Honey Lake Valley, where the topographic depression is terminated, and where the zone appears to break up into a series of NW-trending faults in Cenozoic volcanic formations. Numerous geomorphic features characteristic of strike-slip faulting (Slemmons, 1977a, b) are exhibited by fault zones within the northern portion of the Walker Lane (Bonham, 1969), including: sag ponds, offset stream channels, recent scarps, spring alignments, and shutter ridges.

The area southwest of Pyramid Lake is a recognized zone of strike-slip movement (Gimlett, 1967; Bonham and Slemmons, 1968; Bonham, 1969; Stewart and Carlson, 1974). Waggoner (1975) reviewed the environmental geologic hazards of the Pyramid Lake basin, which includes the northern portion of the Pyramid Lake fault zone (Nixon 15-min quadrangle). Although Waggoner mapped most of the fault scarps in the area and discussed the potential earthquake faulting hazard, he did not provide a detailed analysis of the fault zone in terms of the local or regional tectonic strain. No systematic or detailed studies exist for the Pyramid Lake fault zone segments of the Walker Lane.

GEOGRAPHIC AND GEOLOGIC SETTING

Geographic setting

The Pyramid Lake fault zone study area (Fig. 2) includes the Sutcliffe, Nixon, Wadsworth, and Two Tips 15-min topographic quadrangles. The area encompasses Warm Springs Valley, the southern portions of Pyramid Lake and Winnemucca Dry Lake, and the lower Truckee River Valley. In addition, it includes portions of the Pah Rah, Truckee, Lake, and Virginia Ranges and the Hot Springs and Dog Skin Mts. Cultural features of the area include the Pyramid Lake Indian Reservation and the towns of Sutcliffe, Nixon, Wadsworth, Fernley, and Hazen. The study area is transected by state highways 34, 40 and 95, interstate highway 80, and the Southern Pacific Railroad.

Geologic setting

Rocks of Tertiary and Quaternary age form approximately 90% of the surface exposures within the study area. Pre-Tertiary rocks, mostly metamorphic and intrusive rocks of Mesozoic age, are exposed in the remainder of the area.

The pre-Tertiary rocks consist of metamorphosed volcanic and sedimentary sequences of Triassic and Jurassic age, and Cretaceous (?) intrusives primarily of granitic composition.

The Tertiary rocks are predominantly volcanic in origin, consisting of

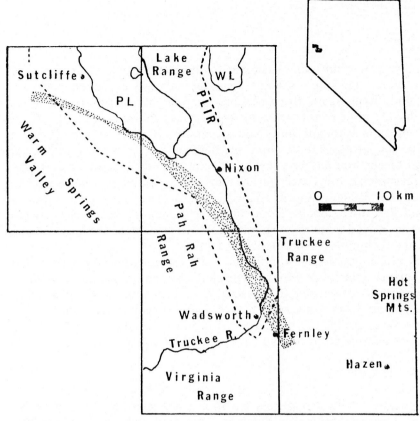

Fig. 2. Index map of the Pyramid Lake fault zone. Approximate extent of fault zone shown by shading. PL = Pyramid Lake; WL = Winnemucca Lake; PLIR = Pyramid Lake Indian Reservation. Area includes the Sutcliffe, Nixon, Wadsworth and Two Tips 15-min quadrangles.

complex, intertonguing piles and sheets of flows and pyroclastics of rhyolitic to basaltic composition. The volcanic sequences contain intercalated lenses of sedimentary rocks. Locally, granitic intrusives are present as stocks, dikes or irregular masses.

The Quaternary deposits consist predominantly of poorly consolidated to unconsolidated alluvial and lacustrine sediments, including the Lake Lahontan sequence (Morrison, 1964; Morrison and Frye, 1965). Locally, eolian deposits are present as a thin veneer overlying both rock and unconsolidated deposits. Volcanic flows and pyroclastic deposits of Pleistocene age are present at several localities within the area.

RESULTS OF THE STUDY

Regional evidence

The Walker Lane fault zone is similar to the San Andreas fault system of California in its right-slip sense of displacement and a general N40°W—N45°W trend. Regionally, the strike-slip components of displacement are indicated by (1) fault offsets from historic surface faulting, (2) fault offsets of Late Cenozoic lithologic units and structures, (3) left-lateral offsets of topography and geologic units along the conjugate NE—trending faults, (4) orientation of chevron and en-echelon structures, and (5) geodetic deformation and seismologic focal mechanisms associated with historic deformation. In general, the Walker Lane in this region is a broad zone of complexly faulted and deformed blocks. Total fault displacements of at least 1.5 km and not more than 16 km are suggested for this portion of the Walker Lane (Bonham and Slemmons, 1968; Bonham, 1969).

Analysis of Landsat imagery indicated a NW-trending lineament (Pyramid Lake lineament) coincident with the Pyramid Lake fault zone and extending approximately 30 km further to the southeast. This suggests the possibility that the Pyramid Lake fault zone may extend further to the south, even though the fault could not be traced with any certainty beyond Fernley where surface expression of the fault is obscured by eolian deposits.

Site-specific evidence

Late Cenozoic strike-slip displacement is evidenced by the observable geomorphic features within the Pyramid Lake fault zone and by the geometry of the shear zone. The geomorphic features characteristic of strike-slip faulting (Slemmons, 1977a) identified during detailed analysis of both conventional (scale 1 : 62,500) and low-sun angle (scale 1 : 12,000) aerial photography of the Pyramid Lake fault zone include sag ponds, en-echelon fault scarps, elongate depressions and troughs, offset stream channels, vegetation alignments, linear gullies, transcurrent buckles, rhombohedral and wedge-shaped enclosed depressions, and recent scarps. These features are conspicuously developed in Late Pleistocene and Holocene deposits and landforms. Examples of these features and their location within the northern portion of the fault zone are shown in Fig. 3.

The Pyramid Lake fault zone has a maximum observable width of 4—5 km and a maximum observable length of approximately 60 km, trending generally N40°W from Fernley. The shear zone is defined principally by Riedel and conjugate Riedel shears (Fig. 4) having maximum observable lengths of 10 and 3 km, respectively (Fig. 5). P-shears have formed symmetrical to the Riedel shears, and the principal displacement shears, or continuous horizontal shears, isolate elongate lenses of essentially passive material (Fig. 5). The principal direction of displacement is inferred to trend approximately

Fig. 3. Geomorphic features characteristic of strike-slip displacement in the northern portion of the Pyramid Lake fault zone.

N35°W—N40°W in the southern portion of the fault zone and approximately N40°W—N45°W in the northern portion of the zone. The inclination angle of the shears to the principal direction of displacement generally ranges from 2 to 8°, with most of the shears inclined at an angle of approximately 4° (Fig. 5). Comparison of the geometry of the shear zone components with the experimental work of Tchalenko (1970; Fig. 6) suggests that the Pyramid Lake fault zone is in an early "Pre-Residual Structure" stage of evo-

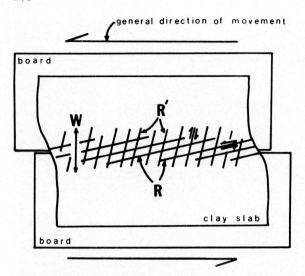

Fig. 4. Diagram of Riedel experiment. R = Riedel shear; R' = conjugate Riedel shear; W = width of shear zone.

lution (Tchalenko, 1970), with the principal deformation mechanism of direct shear replacing the kinematic restraints inherent in the strain field.

Seismic potential

Historic seismic activity in the vicinity of the Pyramid Lake fault zone includes microseismic events (Slemmons et al., 1965a, b; Gimlett, 1967; Ryall, 1977; D. van Wormer, personal communication, 1977), and several events of magnitude 4.0—5.9 (Slemmons et al., 1965a, b; Fig. 7). An earth-

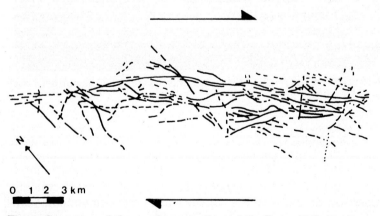

Fig. 5. Geometry of the northern portion of the Pyramid Lake shear zone. Arrows indicate approximate direction and sense of principal displacement.

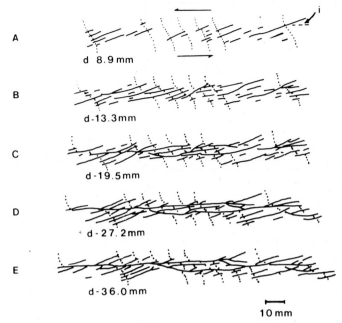

Fig. 6. Sequence of structures in Riedel experiment, d = total board movement. Shears are inclined at angle i at each stage of movement. Stage A appears just before peak shear strength with an average inclination angle of $12°$. Stage B is postpeak shear with some $8°$ shears. Stage C includes some shears at $-10°$ and some Riedel shears connect. Stage D is preresidual structure with the first continuous shears at angles of $0-4°$. Stage E is residual structure with nearly all displacements along a single principal displacement shear (Tchalenko, 1970).

quake in 1852(?) reported near Pyramid Lake, with an estimated Richter magnitude of 7.0 (Slemmons et al., 1965), may be related to the Pyramid Lake fault zone. The location and date of occurrence of this event are based on Paiute Indian accounts of ground cracks, water spouts, and ground shaking observed in the Pyramid Lake and Carson Sink areas. Fig. 7 shows the potential epicentral region of this 1852(?) event based on a possible location error of up to approximately 50 km (D.B. Slemmons, personal communication, 1977); this region includes the Pyramid Lake fault zone. Ryall (1977) reports a date of 1845(?) and places the epicenter in the Stillwater area, based on an 1869 *Gold Hill News* item indicating that the shock knocked people down in the Carson Sink area, that river banks were shaken down in the vicinity of Stillwater, and that the river changed its course.

The seismic potential of the Pyramid Lake fault zone was estimated by comparing the maximum observable length of the zone with world-wide relations of earthquake magnitude to the length of the zone of surface rupture (Slemmons, 1977a; Fig. 8). Based on this analysis and a maximum observable surface rupture zone length of approximately 60 km, the Pyramid Lake

580

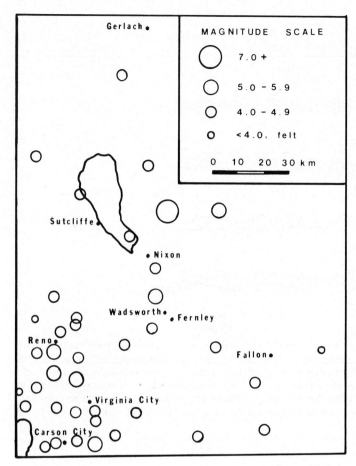

Fig. 7. Historic seismicity in the vicinity of the Pyramid Lake fault zone. The 1852(?) magnitude 7.0 event has a possible location error of 50 km (Slemmons et al., 1965b).

fault zone is inferred to be capable of generating a 7.0—7.5-magnitude event. A fault zone half length of approximately 30 km suggests an event of 6.75—7.25 magnitude. If the Pyramid Lake lineament is representative of the total length of the Pyramid Lake fault zone, then a maximum credible earthquake of 6.9—7.75 magnitude, based on half length (45 km) and total length (90 km), respectively, could be generated by the structure.

CONCLUSIONS RELATING TO THE PYRAMID LAKE FAULT ZONE

(1) The site-specific diagnostic evidence of Late Cenozoic strike-slip faulting includes: sag ponds, en-echelon fault scarps, elongate depressions and troughs, offset stream channels, vegetation alignments, transcurrent buckles, and rhombohedral and wedge-shaped depressions.

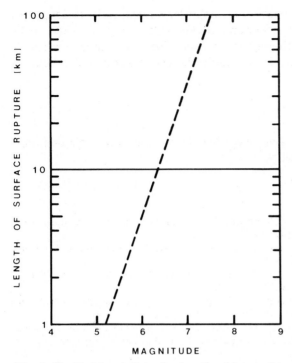

Fig. 8. Worldwide relations for length of zone of surface rupture and earthquake magnitude (Slemmons, 1977a).

(2) The regional and site-specific major fault features disrupt Late Pleistocene—Holocene, and possibly historic deposits and landforms.

(3) The fault zone has a similar orientation to the San Andreas fault system and displays right-lateral, strike-slip displacements.

(4) The shear zone is defined by Riedel shears, conjugate Riedel shears, P-shears, and principal displacement shears. The geometric relationships among these components suggests an early "Pre-Residual Structure" stage of evolution, with the principal deformation mechanism of direct shear replacing the kinematic restraints inherent in the strain field.

(5) The fault zone has a maximum observable width of 4—5 km and a maximum observable length of 60 km. The Pyramid Lake lineament suggests a projection of the fault zone approximately 30 km to the southeast.

(6) The historic seismic activity in the vicinity of the Pyramid Lake fault zone includes microseismic events and several events of magnitude 4.0—5.9. The mid-1800s earthquake reported for the Pyramid Lake—Stillwater area and estimated to have had a Richter magnitude of 7.0 may also be related to this zone. However, reconnaissance examination of fault scarp morphology suggests that displacements along the Pyramid Lake fault zone are neohistoric rather than historic.

(7) Based on worldwide relations of earthquake magnitude to length of the zone of surface rupture, the Pyramid Lake fault zone is inferred to be capable of generating a 7.0—7.5-magnitude event for a maximum observable length of 60 km and a 6.75—7.25-magnitude event for a halflength of approximately 30 km.

ACKNOWLEDGEMENTS

We wish to thank John W. Bell and Chris O. Sanders for their technical and editorial reviews of this article.

REFERENCES

Albers, J.P., 1967. Belt of sigmoidal folding and right-lateral faulting in the western Great Basin. Geol. Soc. Am. Bull., 78: 143—156.

Bonham, H.F., 1969. Geology and mineral resources of Washoe and Storey Counties, Nevada. Nev. Bur. Mines Bull. 70: 140 pp.

Bonham, H.F. and Slemmons, D.B., 1968. Faulting associated with the northern part of the Walker Lane. Geol. Soc. Am. Spec. Pap. 101: 290 pp.

Burchfiel, B.C., 1965. Structural geology of the Specter Range Quadrangle, Nevada, and its regional significance. Geol. Soc. Am. Bull., 76: 175—192.

Gianella, V.P. and Callaghan, E., 1934. The earthquake of December 20, 1932, at Cedar Mountain, Nevada, and its bearing on the genesis of the Basin and Range structure. J. Geol., 42: 1—22.

Gimlett, J.I., 1967. Gravity study of Warm Springs Valley, Washoe County, Nevada, Nev. Bur. Mines Rep., 15: 31 pp.

Hardyman, R.F., Ekren, E.B. and Byers, F.M., 1975. Cenozoic strike-slip, normal and detachment faults in northern part of the Walker Lane, west-central Nevada. Geol. Soc. Am. Abstr. 7 (7): 1100.

Locke, A., Billingsley, P.R. and Mayo, E.B., 1940. Sierra Nevada tectonic pattern. Geol. Soc. Am. Bull., 51: 513—539.

Longwell, C.R., 1960. Possible explanation of diverse structural patterns in southern Nevada. Am. J. Sci., Bradley volume, 258-A: 192-203.

Morrison, R.B., 1964. Lake Lahontan: geology of the southern Carson Desert, Nevada. U.S. Geol. Surv. Prof. Pap., 401: 156 pp.

Morrison, R.B. and Frye, J.C. 1965. Correlation of the middle and late Quaternary successions of the Lake Lahontan, Lake Bonneville, Rocky Mountain (Wasatch Range), southern Great Plains, and eastern Midwest areas. Nevada Bur. Mines Rep., 9: 45 pp.

Nielsen, R.L., 1964. Geology of the Pilot Mountains and Vicinity, Mineral County, Nevada. Ph.D. thesis, Univ. California, Berkeley.

Nielsen, R.L., 1965. Right-lateral strike-slip faulting in the Walker Lane, west-central Nevada. Geol.Soc. Am. Bull., 76: 1301—1308.

Ryall, A. 1977. Earthquake hazard in the Nevada region. Bull. Seismol. Soc. Am., 67 (2) 512—532.

Shawe, D.R., 1965. Strike-slip control of Basin—Range structure indicated by historical faults in western Nevada. Geol. Soc. Am. Bull., 76: 1361—1378.

Slemmons, D.B., 1977a. Determination of design earthquake magnitude from fault length and maximum displacement data: U.S. Army Eng. Waterw. Exp. Stn., Misc. Pap. S-73-1, Rep., 6: 129 pp.

Slemmons, D.B., 1977b. New earthquake magnitude—fault length—maximum displacement relationships. Geol. Soc. Am. Abstr., 9 (4): 501.

Slemmons, D.B., Jones, A.E. and Gimlett, J.I., 1965a. Catalog of Nevada earthquakes 1852—1960. Bull. Seismol. Soc. Am., 55: 539—565.

Slemmons, D.B., Gimlett, J.I., Jones, A.E., Greensfelder, Roger and Koenig, James. 1965b. Earthquake epicenter map of Nevada. Nev. Bur. Mines, Map 29.

Speed, R.C., 1975. Tectonics of the Nevada seismic zone. U.S. Geol. Surv. Grant No. 14-08-0001-G, Tech. Rep., 1: 45 pp.

Stewart, J.H., 1967. Possible large right-lateral displacement in the Death Valley-Las Vegas area, California and Nevada. Geol. Soc. Am. Bull., 78: 131—142.

Stewart, J.H. and Carlson, J.E., 1974. Preliminary geologic map of Nevada. U.S. Geol. Surv. Misc. Field Studies Map MF-609.

Stewart, J.H., Albers, J.P. and Poole, F.G., 1968. Summary of regional evidence for right-lateral displacements in the western Great Basin. Geol. Soc. Am. Bull., 79: 1407—1413.

Tchalenko, J.S., 1970. Similarities between shear zones of different magnitude. Geol. Soc. Am. Bull., 81: 1625—1640.

Waggoner, R.R., 1975. Environmental Geology Problems of Pyramid Lake Basin. M.S. thesis, Univ. Nevada, Reno, 95 pp.

Wallace, R.E., 1975. Geometry and rates of change of fault-generated range fronts, north-central Nevada. Geol. Soc. Am. Abstr., 7 (7): 1310—1311.

Wallace, R.E., 1977. Profiles and ages of young fault scarps, north-central Nevada. Geol. Soc. Am. Bull., 88(9): 1267—1281.

Yeats, R.S., 1977. Evaluation of fault hazard: surface rupture vs. earthquake potential. Geol. Soc. Am. Abstr., 9 (4): 529.

Tectonophysics, 52 (1979) 585—597

RECENT CRUSTAL MOVEMENTS IN THE CENTRAL SIERRA
NEVADA—WALKER LANE REGION OF CALIFORNIA—NEVADA:
PART III, THE OLINGHOUSE FAULT ZONE

CHRIS O. SANDERS and DAVID B. SLEMMONS

*Department of Geological Sciences, Mackay School of Mines, University of Nevada,
Reno, Nev. 89557 (U.S.A.)*

(Revised version accepted June 8, 1978)

ABSTRACT

Sanders, C.O. and Slemmons, D.B., 1979. Recent crustal movements in the Central Sierra
    Nevada—Walker Lane region of California—Nevada: Part III, the Olinghouse fault zone.
    In: C.A. Whitten, R. Green and B.K. Meade (Editors), Recent Crustal Movements,
    1977. Tectonophysics, 52: 585—597.

    The Olinghouse fault zone is one of several NE—ENE-trending fault zones and linea-
ments, including the Midas Trench and the Carson—Carson Sink Lineament, which
exhibit left-lateral transcurrent movement conjugate to the Walker Lane in western
Nevada. The active portion of this fault zone extends for approximately 23 km, from 16
km east of Reno, Nevada, to the southern extent of Pyramid Lake.
    The fault can be traced for most of its length from its geomorphic expression in the
hilly terrain, and it is hidden only where overlain by recent alluvial sediments. Numerous features characteristic of strike-slip faulting can be observed along the fault, including: scarps, vegetation lines, sidehill and shutter ridges, sag ponds, offset stream channels
and stone stripes, enclosed rhombohedral and wedge-shaped depressions, and en-echelon
fractures.
    A shear zone having a maximum observable width of 1.3 km is defined principally by
Riedel shears and their symmetrical P-shears, with secondary definition by deformed conjugate Riedel shears. Several continuous horizontal shears, or principal displacement
shears, occupy the axial portion of the shear zone. The existence of P-shears and principal
displacement shears suggests evolution of movement along the fault zone analogous to the
"Post-Peak" or "Pre-Residual Structure" stage.
    Historic activity (1869) has established the seismic potential of this zone. Maximum
intensities and plots of the isoseismals indicate the 1869 Olinghouse earthquake had a
magnitude of 6.7. Field study indicates the active length of the fault zone is at least 23
km and the maximum 1869 displacement was 3.65 m of left-slip. From maximum fault
length and maximum fault displacement to earthquake magnitude relations, this corresponds to an earthquake of about magnitude 7.

INTRODUCTION

This study was initiated in 1967 when the second author recognized this
young fault scarp during an aerial reconnaissance of the region and learned

from Dr. Vincent Gianella that surface faulting had been observed along the fault zone after the December 28, 1869 earthquake. The primary interest of the first author is in the mechanics and style of surface faulting on this left-oblique-slip fault zone. The study included interpretation of low-sun angle aerial photography over most of the zone, as well as interpretation of conventional photography over the whole zone, in conjunction with detailed field analysis.

METHODS

Small-scale (1 : 62,500) conventional aerial photography and large-scale (1 : 6500) low-sun angle aerial photography were studied to determine the extent and location of the fracturing associated with the fault movement in the Olinghouse zone.

Field studies were carried out concurrent with the photointerpretation to provide quantitative data on the historic fault movement and to check qualitatively the major and minor fractures and features. Much of the ground reconnaissance was carried out in the early morning and late afternoon in order to take advantage of low-sun illumination.

RESULTS

Situated to the southwest of Pyramid Lake, Nevada, the left-oblique-slip Olinghouse fault zone extends approximately 23 km, from 16 km east of Reno to a few kilometers south of Pyramid Lake. The Olinghouse is one of several NE—ENE-trending fault zones and lineaments (Bonham, 1969; Shawe, 1965), including the Midas Trench (Rowan and Wetlaufer, 1974) and the Carson—Carson Sink Lineament (Rogers, 1975), which exhibit left-lateral transcurrent movement conjugate to the right-lateral Walker Lane (Fig. 1). Suspected recent movement along the fault has called attention to the need for a detailed study of the zone and its characteristics. The active length of the fault, the associated fracture pattern, the type of movement and the amount of most recent displacement along the fault all need further definition. This study provides data on these characteristics of the Olinghouse fault zone.

Continuity

The Olinghouse fault zone can be identified along most of its length by its geomorphic expression in the hilly terrain. In some places scarps are conspicuous and are readily visible on small-scale (1 : 62,500) conventional aerial photography. Stream valleys, saddles, and changes in slope all mark the location of the zone which can be traced across the Spanish Springs Valley and Wadsworth 15′ topographic maps (Fig. 2).

Evidence from low-sun angle and conventional aerial photography indicates that the active length of the fault zone may be at least 23 km and pos-

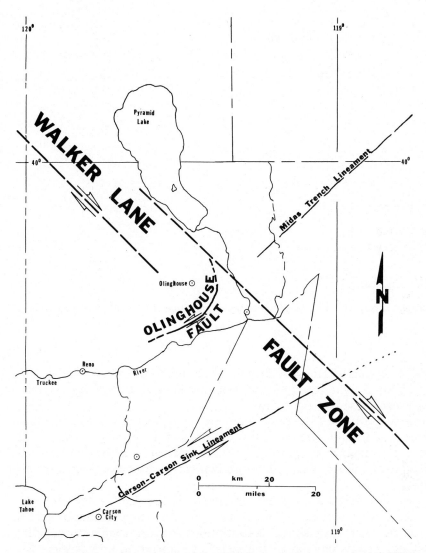

Fig. 1. Map showing the relation of the Olinghouse fault zone and other NE-trending, left-lateral transcurrent structures to the Walker Lane.

sibly greater. At the eastern extent of the zone, recent alluvial deposits obscure most of the evidence of its existence; however, sufficient disturbance can be seen on the aerial photography to be able to ascertain the fault path in this area. This evidence of breakage in the Recent deposits is indicative of fault activity at this eastern extremity. One question arises concerning whether the Olinghouse fault has a connection with the Pyramid Lake zone. Evidence from fracture traces, where the fault cuts the alluvial deposits as it curves toward the north, indicates a trend which follows very close to the

588

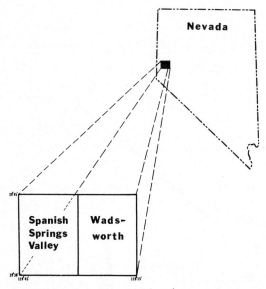

Fig. 2. Location map of the 15' topographic quads in which the Olinghouse fault zone is situated.

hills on the west side of Dodge Flat and which may intersect an active portion of the Pyramid Lake zone. Further evidence is needed to define the relationship between these two fault zones.

In the central portion of the fault zone both conspicuous scarps and abundant fracture traces mark the path of the fault as it extends westward. This portion of the zone exhibits evidence of being the most active, with excellent expression of the shear zone fracture patterns as well as the most prominent scarps.

At the western end the surface expression of the fault zone seems to diminish. The shear fractures become sparse and no long continuous shears or scarps can be seen. Only short, discontinuous fracture segments are indicative of activity in this portion of the zone. In addition, the fault trace can no longer be followed by its expression in the topography, indicating a lack of activity along the Olinghouse zone about 16 km east of Reno.

Fracture patterns

Using large-scale (1 : 6500), low-sun angle aerial photography, a very detailed analysis of the shear zone fracture pattern along the most prominent active portion of the Olinghouse fault was completed. The resulting strip map of the area (Fig. 3) shows fracture patterns diagnostic of strike-slip fault movement. En-echelon Riedel shears, conjugate Riedel shears, P-shears, and continuous horizontal shears or "principal displacement shears" (Tchalenko,

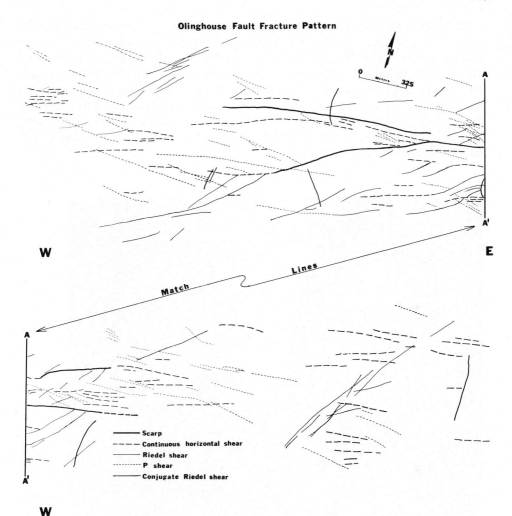

Fig. 3. Strip map of the west-central portion of the Olinghouse zone showing the fracture pattern associated with the left-lateral transcurrent movement in the zone.

1970), define this shear zone. The Riedel and P-shears extend a maximum of 0.65 km from the continuous horizontal shears in the axial portion of the zone, making the maximum width of the shear zone 1.3 km.

The Riedel shears are among the first to appear in this type of shear zone (Tchalenko, 1970), forming a left-lateral en-echelon pattern transverse to the direction of movement at an average angle of 25°. These Riedels are continuous across the width of the zone extending to both the north and south boundaries and exhibiting a maximum, essentially continuous, length of 2.9 km. Some recent offsets are present on the Riedels, and in one location lithologic offset is evident.

Oriented perpendicular to the principal direction of movement, a small number of conjugate Riedel shears represent the deformation caused by continuing movement within the shear zone. Theoretically formed at an angle of 78° to the principal movement vectors (Tchalenko, 1970), the conjugate Riedels quickly lose any movement along them due to their large angle with the direction of movement, and become subject to rotation and deformation. They appear on the strip map (Fig. 3) as generally concave westward, arcuate features nearly orthogonal to the principal displacement shears, and have an average length of 0.4 km.

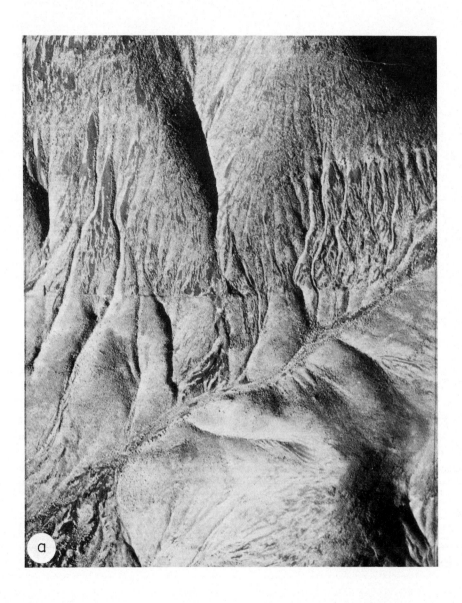

Formation of P-shears has taken place symmetrical to the Riedel shears and ore inclined at an average angle of −13° to the general direction of movement. These shears evolve during the post-peak stage of fault development in response to kinematic restraints which decrease relative displacement on the Riedel and conjugate Riedel shears (Tchalenko, 1970). The Olinghouse P-shears extend the width of the shear zone. The longest P-shear

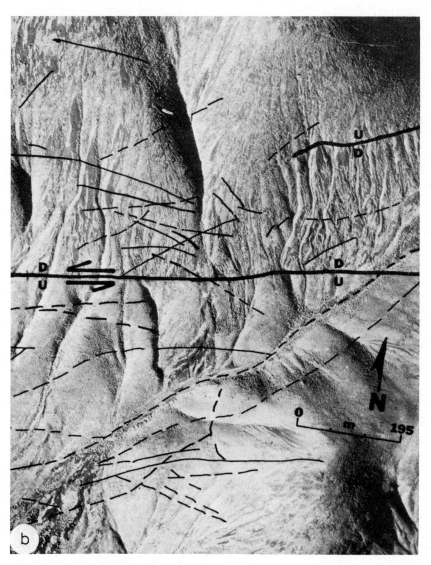

Fig. 4. a. Low-sun angle aerial photograph of the area of most recent faulting. Note offset stone stripes and stream channels. b. Aerial photograph of the area of most recent faulting with fracturing marked.

is 2.25 km long, with the average length being 0.4 km. Like most of the Riedel and conjugate Riedel shears, no discernible recent displacement can be seen along the P-shears.

The most prominent structures are the long principal displacement shears found in the axial portion of the zone. They are not present along the total length of the Olinghouse zone, but are restricted to an area approximately 3.7 km long in the west-central portion of the zone marked by evidence of recent faulting. Similar offsets have occurred along separate segments of these central principal displacement shears. The most prominent scarp exhibits 3 m of dip-slip movement with the upthrown block to the north. Evidence of left-lateral displacement, including offset drainage channels and stone stripes, also exists along the 1.4-km length of this scarp. Similar movement can also be seen along another scarp 0.5 km to the east which has a length of 1.5 km. Between these two horizontal shears lies another less prominent scarp, which exhibits evidence of dip-slip and strike-slip movement. The geomorphic and geologic evidence indicates that movement along this central scarp is the most recent of the three. Furthermore, the southern block is upthrown 0.9 m. The left-lateral movement along this segment may be measured fairly accurately using offset drainage channels and stone stripes, and is a maximum of 3.65 m.

The existence of P-shears, which form during the post-peak stage of movement evolution (Tchalenko, 1970), and of principal displacement shears, which form during the pre-residual stage, suggests a stage in the evolution of the fault zone analogous to either the post-peak or the pre-residual stage or possibly a transitional stage between these two.

Associated movement structures

Along the trace of the more recent scarps geomorphic features have developed which are characteristic of strike-slip fault movement (Slemmons, 1977), including offset stream channels and stone stripes (Figs. 4a and 4b). Stone stripes consisting of cobble- to boulder-size basalt blocks have been displaced left-laterally and provide an excellent means of quantitatively determining offset. In places where the fault cuts large stripes (13—50 m wide) the offset cannot easily be seen. This may be due to the large quantities of material available to fill in and obscure any movement. However, the narrower stripes vary in width from 3 to 6 m and provide invaluable offset markers. In places, streams have been diverted along the sides of stone stripes due to the leveeing action of the stripes. When offset by the fault, these also provide excellent displacement markers.

Much of the length of the historic 1869 scarp (Figs. 4a and 4b) is expressed as a long sidehill ridge, the downhill block having been upthrown relative to the uphill block producing a ridging effect. The strike-slip motion has also produced numerous small sag ponds and shutter ridges along the trace of the scarp. Soil formed in these small (approximately 2-m-wide)

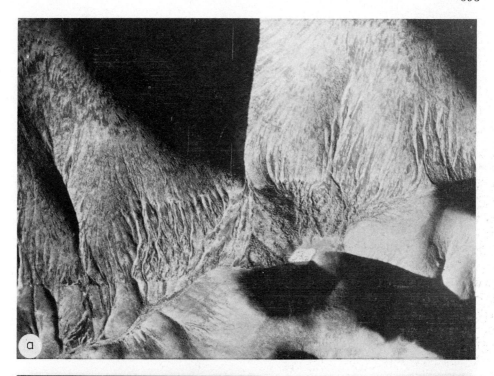

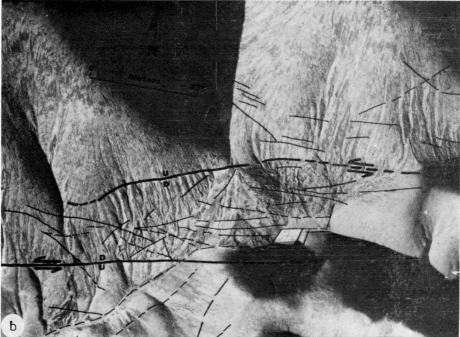

Fig. 5. a. Low-sun angle aerial photograph of rhombohedral depression. b. Aerial photograph of rhombohedral depression with fracturing marked.

594

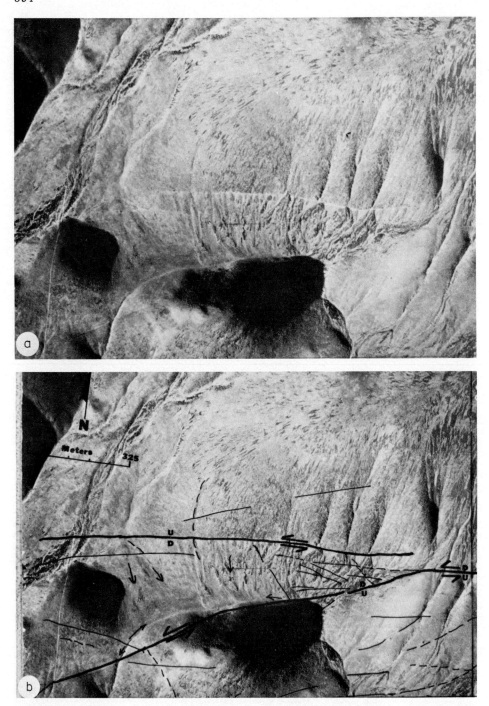

Fig. 6. a. Low-sun angle aerial photograph of wedge-shaped depression. b. Aerial photograph of wedge-shaped depression with fracturing marked.

depressions appears very fresh as it is loosely compacted with no soil horizons discernible, indicating the recency of latest faulting. Vegetational contrasts also occur across the fault.

On a larger scale, two prominent structures have formed in response to the strike-slip movement. Between two of the main en-echelon fault segments a rhombohedral depression has developed (Figs. 5a and 5b). The left-lateral movement along the two fault segments has created a tensional environment in the area between them, resulting in the formation of the rhomb-shaped depression. Due east of this depression lies another similar structure. Here, a wedge-shaped enclosed depression has formed (Figs. 6a and 6b) in response to mechanisms similar to those which formed the rhomb. The left-lateral

Fig. 7. Conventional aerial photograph of central and eastern portion of the Olinghouse fault zone. Note relation to the Walker Lane. Insets show locations of Figs. 3—6.

movement on the north- and south-bounding faults has again created a tensional environment mainly responsible for producing the depression. One other mechanism is also involved. At the eastern edge of the wedge the south-bounding fault curves north and then east to become aligned with the north-bounding fault. This curve in the strike of the south-bounding fault, in conjunction with the left-lateral transcurrent motion, appears to have produced additional tension resulting in the drop of the block in this area. Dip-slip offset along this curved portion of the fault varies from 0 to 4.3 m. At one point along this curve a former SE-running drainage has been offset so that it now slopes northwest.

Recent movement

Historic activity has established the seismic potential of this zone. Maximum intensities and plots of the isoseismals indicate that the Olinghouse earthquake of December 27, 1869, had a magnitude of 6.7 (Slemmons, 1977). Reports from a prospector in the area tell of faulting extending from the west-central portion of the fault zone to the road leading to Olinghouse.

Field mapping of the main west-central portion of the zone reveals a segment of recent ground breakage 2.3 km long. The distance from this area of breakage to the area of breakage reported by the 1869 witness is approximately 14 km. Aerial photographic studies indicate that additional fractures occur further to the east, where the fault curves northward, and to the west, extending the active fault length 9 km, or a total length of 23 km.

By relating maximum fault length and maximum displacement to earthquake magnitude (Slemmons, 1977), a design earthquake value may be calculated. Worldwide and North American relations suggest this structure capable of about a magnitude 7 event.

CONCLUSIONS

(1) The Olinghouse fault is an active left-oblique-slip fault zone conjugate to the Walker Lane with a length of at least 23 km.

(2) The fault is characterized by scarps, offset stream channels, sidehill and shutter ridges, sag ponds, rhombohedral and wedge-shaped depressions, and en-echelon fractures.

(3) The zone has similar development of Riedel and P-shears and principal displacement shears to that of Tchalenko (1970) for "Post-Peak" and "Pre-Residual Structure" stages of development, indicating a prolonged period of prehistoric offsets.

(4) The historic offsets from the December 27, 1869 earthquake included a rupture length of 23 km and a maximum left-slip offset of 3.65 m. The earthquake is inferred to have a magnitude of about 6.7.

ACKNOWLEDGEMENTS

We are grateful for assistance provided by David Findley, who assisted in the field work and critically reviewed this paper, and for the many helpful suggestions from Elaine Bell, who also participated in many informative discussions. We also wish to express our gratitude to Dean Arthur Baker, Mackay School of Mines, who provided financial assistance for our work.

REFERENCES

Bonham, H.F., 1969. Geology and mineral deposits of Washoe and Storey Counties, Nevada. Nev. Bur. Mines Bull., 70.

Rogers, D.K., 1975. The Carson Lineament: its influence on Recent left-lateral faulting near Carson City, Nevada. (Abstr.) Geol. Soc. Am. Annu. Meet., S.L.C.

Rowan, L.C. and Wetlaufer, P.H., 1974. Structural Geologic Analysis of Nevada Using ERTS-1 Images: A Preliminary Report. NASA SP-327, pp. 413—423.

Shawe, D.R., 1965. Strike-slip control of Basin-Range structure indicated by historical faults in western Nevada. Geol. Soc. Am. Bull., 76: 1361—1378.

Slemmons, D.B., 1977. State-of-the-art for assessing earthquake hazards in the United States, Report 6, Faults and Earthquake Magnitude, Office, Chief of Engineers, U.S. Army.

Tchalenko, J.S., 1970. Similarities between shear zones of different magnitudes. Geol. Soc. Am. Bull., 81: 1625—1640.

Tectonophysics, 52 (1979)599

599

© Elsevier Scientific Publishing Company, Amsterdam — Printed in The Netherlands

STRAIN PATTERN REPRESENTED BY SCARPS FORMED DURING THE EARTHQUAKES OF OCTOBER 2, 1915, PLEASANT VALLEY, NEVADA

R.E. WALLACE

U.S. Geological Survey, Menlo Park, Calif. 94025 (U.S.A.)

(Accepted for publication, June 8, 1978)

ABSTRACT

The pattern of scarps developed during the earthquakes of October 2, 1915, in Pleasant Valley, Nevada, may have formed as a result of a modern stress system acting on a set of fractures produced by an earlier stress system which was oriented differently. Four major scarps developed in a right-stepping, en-echelon pattern suggestive of left-lateral slip across the zone and an extension axis oriented approximately S85°W. The trend of the zone is N25°E. However, the orientation of simple dip-slip on most segments trending approximately N20—40°E and a right-lateral component of displacement on several N- and NW-trending segments of the scarps indicate that the axis of regional extension was oriented between N50° and 70° W, normal to the zone.

The cumulative length of the scarps is 60 km, average vertical displacement 2 m, and the maximum vertical displacement near the Pearce School site 5.8 m. Almost everywhere the 1915 scarps formed along an older scarp line, and in some places older scarps represent multiple previous events. The most recent displacement event prior to 1915 is interpreted to have occurred more than 6600 years ago, but possibly less than 20,000 years ago. Some faults expressed by older scarps that trend northwest were not reactivated in 1915, possibly because they are oriented at a low angle with respect to the axis of modern regional extension.

The 1915 event occurred in an area of overlap of three regional fault trends oriented northwest, north, and northeast and referred to, respectively, as the Oregon—Nevada, Northwest Nevada, and Midas—Battle Moutain trends. Each of these trends may have developed at a different time; the Oregon—Nevada trend was possibly the earliest and developed in Late Miocene time (Stewart et al. 1975). Segments of the 1915 scarps are parallel to each of these trends, suggesting influence by older sets of fractures.

REFERENCE

Stewart, J.H., Walker, G.W. and Kleinhampl, F.J., 1975. Oregon—Nevada lineament. Geology, 5 (2): 265—268.

Tectonophysics, 52 (1979) 600

FAULT-CROSSING P DELAYS, EPICENTRAL BIASING, AND FAULT BEHAVIOR IN CENTRAL CALIFORNIA

S.M. MARKS and C.G. BUFE

U.S. Geological Survey, Menlo Park, Calif. 94025 (U.S.A.)

(Accepted for publication June 8, 1978)

ABSTRACT

The P delays across the San Andreas fault zone in central California have been determined from travel-time differences at station pairs spanning the fault, using off-fault local earthquake or quarry blast sources. Systematic delays as large as 0.4 sec have been observed for paths crossing the fault at depths of 5—10 km. These delays can account for the apparent deviation of epicenters from the mapped fault trace. The largest delays occur along the San Andreas fault between San Juan Bautista and Bear Valley and Between Bitterwater Valley and Parkfield. Spatial variations in fault behavior correlate with the magnitude of the fault-crossing P delay. The delay decreases to the northwest of San Juan Bautista across the "locked" section of the San Andreas fault and also decreases to the southeast approaching Parkfield. Where the delay is large, seismicity is relatively high and the fault is creeping.

Tectonophysics, 52 (1979) 601—602

A CHANGE IN FAULT-PLANE ORIENTATION BETWEEN FORESHOCKS AND AFTERSHOCKS OF THE GALWAY LAKE EARTHQUAKE, M_L = 5.2, 1975, MOJAVE DESERT, CALIFORNIA

G.S. FUIS and A.G. LINDH

U.S. Geological Survey, at the Seismological Laboratory, Caltech, Pasadena, Calif. 91125 (U.S.A.)
U.S. Geological Survey, Menlo Park, Calif. 94025 (U.S.A.)

(Accepted for publication June 8, 1978)

ABSTRACT

A marked change is observed in P/SV amplitude ratios, measured at station TPC, from foreshocks to aftershocks of the Galway Lake earthquake. This change is interpreted to be the result of a change in fault-plane orientation occurring between foreshocks and aftershocks.

The Galway Lake earthquake, M_L = 5.2, occurred on June 1, 1975. The first-motion fault-plane solutions for the main shock and most foreshocks and aftershocks indicate chiefly right-lateral strike-slip on NNW-striking planes that dip steeply, 70—90°, to the WSW. The main event was preceded by nine located foreshocks, ranging in magnitude from 1.9 to 3.4, over a period of 12 weeks, starting on March 9, 1975. All of the foreshocks form a tight cluster approximately 1 km in diameter. This cluster includes the main shock. Aftershocks are distributed over a 6-km-long fault zone, but only those that occurred inside the foreshock cluster are used in this study.

Seismograms recorded at TPC (Δ = 61 km), PEC (Δ = 93 km), and CSP (Δ = 83 km) are the data used here. The seismograms recorded at TPC show very consistent P/SV amplitude ratios for foreshocks. For aftershocks the P/SV ratios are scattered, but generally quite different from foreshock ratios. Most of the scatter for the aftershocks is confined to the two days following the main shock. Thereafter, however, the P/SV ratios are consistently half as large as for foreshocks. More subtle (and questionable) changes in the P/SV ratios are observed at PEC and CSP.

Using theoretical P/SV amplitude ratios, one can reproduce the observations at TPC, PEC and CSP by invoking a 5—12° counterclockwise change in fault strike between foreshocks and aftershocks. This interpretation is not unique, but it fits the data better than invoking, for example, changes in dip or slip angle. First-motion data cannot resolve this small change, but they permit it. Attenuation changes would appear to be ruled out by the fact that changes in the amplitude ratios, P_{TPC}/P_{PEC} and P_{TPC}/P_{CSP}, are observed, and these changes accompany the changes in P/SV.

Observations for the Galway Lake earthquake are similar to observations for the Oroville, California, earthquake ($M_L = 5.7$) of August 1, 1975, and the Brianes Hills, California, earthquake ($M_L = 4.3$) of January 8, 1977 (Lindh et al., Science Vol. 201, pp. 56—59).

A change in fault-plane orientation between foreshocks and aftershocks may be understandable in terms of early en-echelon cracking (foreshocks) giving way to shear on the main fault plane (main shock plus aftershocks). Recent laboratory data (Byerlee et al., Tectonophysics, Vol. 44, pp. 161—171) tend to support this view.

Tectonophysics, 52 (1979) 603

© Elsevier Scientific Publishing Company, Amsterdam — Printed in The Netherlands

EARTHQUAKE RECURRENCE ON THE CALAVERAS FAULT EAST OF SAN JOSE, CALIFORNIA

CHARLES G. BUFE, PHILIP W. HARSH and ROBERT O. BURFORD

U.S. Geological Survey, Menlo Park, Calif. 94025 (U.S.A.)

(Accepted for publication June 8, 1978)

ABSTRACT

Occurrence of small ($3 \leqslant M_L < 4$) earthquakes on two 10-km segments of the Calaveras fault between Calaveras and Anderson reservoirs follows a simple linear pattern of elastic strain accumulation and release. The centers of these independent patches of earthquake activity are 20 km apart. Each region is characterized by a constant rate of seismic slip as computed from earthquake magnitudes, and is assumed to be an isolated locked patch on a creeping fault surface. By calculating seismic slip rates and the amount of seismic slip since the time of the last significant ($M \geqslant 3$) earthquake, it is possible to estimate the most likely date of the next ($M \geqslant 3$) event on each patch. The larger the last significant event, the longer the time until the next one. The recurrence time also appears to be increased according to the moment of smaller ($2 < M_L < 3$) events in the interim. The anticipated times of future larger events on each patch, on the basis of preliminary location data through May 1977 and estimates of interim activity, are tabulated below with standard errors. The occurrence time for the southern zone is based on eight recurrent events since 1969, the northern zone on only three. The 95% confidence limits can be estimated as twice the standard error of the projected least-squares line. Events of $M \geqslant 3$ should not occur in the specified zones at times outside these limits. The central region between the two zones was the locus of two events ($M = 3.6, 3.3$) on July 3, 1977. These events occurred prior to a window based on the three point, post-1969 slip-time line for the central region.

Latitude	Longitude	Depth	Mag.	Target date	Standard error (days)
$37°17' \pm 2'$N	$121°39' \pm 2'$W	5.0 ± 2 km	3.0—4.0	7—22-77	22.3
$37°26' \pm 2'$N	$121°47' \pm 2'$W	6.0 ± 2 km	3.0—4.0	9—02-77	8.0 *

* The remarkable degree of linearity in this zone may be fortuitous as it is based on only three occurrences. The standard error associated with the southern zone (eight occurrences) may be more representative of uncertainty in occurrence dates for both zones.

Tectonophysics, 52 (1979) 604

EARTHQUAKES AND FAULT CREEP ON THE NORTHERN SAN ANDREAS FAULT

R. NASON

U.S. Geological Survey, Menlo Park, Calif. 94025 (U.S.A.)

(Accepted for publication June 8, 1978)

ABSTRACT

At present there is an absence of both fault creep and small earthquakes on the northern San Andreas fault, which had a magnitude 8 earthquake with 5 m of slip in 1906. The fault has apparently been dormant after the 1906 earthquake. One possibility is that the fault is 'locked' in some way and only produces great earthquakes. An alternative possibility, presented here, is that the lack of current activity on the northern San Andreas fault is because of a lack of sufficient elastic strain after the 1906 earthquake. This is indicated by geodetic measurements at Fort Ross in 1874, 1906 (post-earthquake), and 1969, which show that the strain accumulation in 1969 ($69 \cdot 10^{-6}$ engineering strain) was only about one-third of the strain release (rebound) in the 1906 earthquake ($200 \cdot 10^{-6}$ engineering strain).

The large difference in seismicity before and after 1906, with many strong local earthquakes from 1836 to 1906, but only a few strong earthquakes from 1906 to 1976, also indicates a difference of elastic strain.

The geologic characteristics (serpentine, fault straightness) of most of the northern San Andreas fault are very similar to the characteristics of the fault south of Hollister, where fault creep is occurring. Thus, the current absence of fault creep on the northern fault segment is probably due to a lack of sufficient elastic strain at the present time.

Tectonophysics, 52 (1979) 605—611

Experimental and Theoretical Models

INTERFEROMETRIC METHODS FOR DISTANCE MEASURING IN THE STUDY OF RECENT CRUSTAL MOVEMENTS

M.T. PRILEPIN, A.N. GOLUBEV and A.S. MEDOVIKOV

Institute of Geodesy, Aerial Surveying and Cartography, Moscow (U.S.S.R.)

(Accepted for publication June 8, 1978)

ABSTRACT

Prilepin, M.T., Golubev, A.N. and Medovikov, A.S., 1979. Interferometric methods for distance measureing in the study of recent crustal movements. In: C.A. Whitten, R. Green and B.K. Meade (Editors), Recent Crustal Movements, 1977. Tectonophysics, 52: 605—611.

In solving a series of problems in the study of recent crustal movements, interferential methods may prove to be more effective than electronic distance measurements or other methods. The interferential methods now in existence have substantial shortcomings impeding their wide used in solving the above problems. The possibilities of the interferometric method based on the employment of the light source with variable coherence time are discussed.

INTRODUCTION

There is a series of geodynamic problems whose solution requires the high-precision measurement of short distances (of the order of several hundred metres). By way of example, we shall mention only one problem relating to the study of the regime of horizontal movements directly in zones of faults. Nowadays electronic rangefinders and interferometers can be used for high-precision distance measurement.

METHOD OF RANGEFINDER

None of the approximately 40 different types of phase rangefinders now at the disposal of geodesists can be used for this purpose because the actual root-mean-square error for distances of the order of few hundred metres is at best 1 mm. Such precision is obviously insufficient for the solution of the problems in question, for here the movements to be determined are smaller by one order at least (Boulanger, 1971). Since higher precision for the existing rangefinders probably depends on future development, it is advisable to turn to the interferometric method.

606

INTERFEROMETRIC METHOD

Today, Michelson's interferometer (or some modification) with a continuous He—Ne laser is used for high-precision absolute measurements, the distance being measured directly in laser radiation half-waves by continuously moving along the line being measured. However, this method is not feasible for the solution of the problems under discussion because of the difficulties involved in constructing long guidelines in the field. Differential interferometric measurements of long bases in vacuum tubes (Vali et al., 1968) have been used for geodynamic ranging; but these differential methods are not discussed in this paper.

The relative method or the method of optical multiplication has found

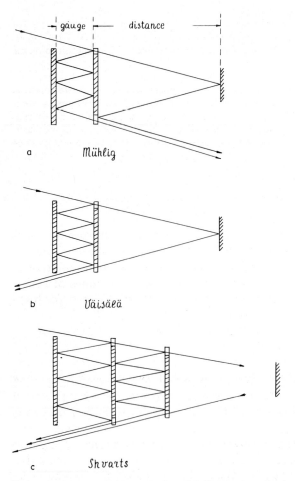

Fig. 1. Diagrams of the optical multiplication method.

much wider application in linear measurements for purely geodetic purposes (Honkasalo, 1950). Some diagrams of relative interferometers are shown in Fig. 1. Fundamentally, all relative measurements can only be made by using the light source with a very small coherence length. Although the He—Ne laser is ideal in almost every respect, it cannot be used here precisely because of its too great coherence length.

The multiplication method has two substantial drawbacks. The first consists in that lines several hundred metres long must be measured in several consecutive stages. Moreover, rather bulky intermediate supports are required for the mirrors, which in some cases may prove not to be feasible because of the land profile along the lines which are of interest from the geodynamic viewpoint. The second drawback is connected with the need to employ a thermal source with a small coherence length. This greatly complicates the measuring process, because we have to preliminarily equate the multiplied length of the gauge and the line being measured to an accuracy of within a few micrometres. These drawbacks are responsible for the fact that interferometric measurements of geodetic bases are made only in exceptionally rare cases.

METHOD OF INTERFEROMETRIC RANGE-FINDING

The interferometric method of linear measurement needs substantial improvement to become suitable for geodynamic measurements. Here we have to endeavour to construct the interferometric scheme of measurements according to the rangefinder principle, which requires no intermediate points. The block diagram of such an interferometer, which is hereafter called an interferometric rangefinder, is given in Fig. 2. A light source with a variable coherence length (block A) is the main element of such a rangefinder. The purpose of all the other blocks can be seen from the figure. The reference arm is made in the form of a multipath delay line. Calculations show that the reference arm up to 1 km long can be achieved when its constructional length (ℓ) is about 1.5 m, the diameter of the spherical mirrors 10 cm, and the loss factor not more than 95%.

The rangefinder under consideration operates on the following principle. Suppose, according to Fig. 3, we have a light source of constant intensity within some spectral interval, or two identical spectral components separated by the value $\Delta\Omega$. In both cases the contrast of the interference pattern will have specific points easy to register. For these points, the source coherence length L ($L_{1,2}$) and the optical path difference δ, which is to measure, are equal. The relationship between the errors in measuring the optical path difference and contrast is expressed by the following equations:

$$d\delta = L \cdot dk \tag{1}$$

$$d\delta = \exp\left(\lambda \frac{\delta}{L} \cdot \frac{L_{1.2}}{\lambda}\right) dk \tag{2}$$

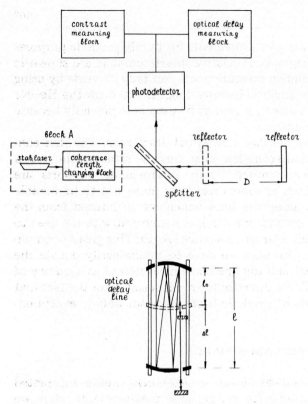

Fig. 2. Block diagram of an interferometric rangefinder.

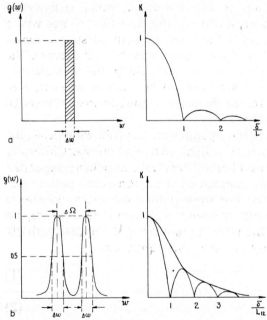

Fig. 3. Relation between the interferometric pattern contrast and the coherence length of the source.

where dk is the error of contrast measurement. Thus, if we have a source with a variable coherence length, then the optical path difference can be consecutively specified by reducing the coherence length. At present the photoelectrical method of contrast measurement allows a precision in the order of 0.01—0.001 (Kolomiytsov, 1976). Taking the error of the optical path difference measurement as 0.01 mm, it is advisable to choose a value of the order of 10 cm for the upper limit of the coherence length, and the coherence length of the thermal source for the lower limit. The problem of creating a light source with a variable coherence length can be solved in two ways: on the one hand, we can try to increase the coherence length for white light, which is about 2—3 μm, and on the other we can reduce the laser coherence, which is several hundred metres long. The simplest way of increasing the coherence length is by radiation monochromatisation, but there is only a limited possibility of using this method here owing to reduced radiation energy.

In employing lasers, we may point out the following possibilities: (1) alteration of the radiation frequency by means of intraresonator modulation; (2) tuning of the radiation frequency by using the Zeeman effect; (3) use of the laser pulse regime; and (4) change in the supply of a laser emitting several lines.

Operating range of an interferometric rangefinder

Insofar as in the rangefinder under discussion the object of measurement is the contrast of the interference pattern, the operating range of the instrument will depend on optical path fluctuations. For the atmospheric conditions in which the Kholmogorov—Obukhov "low $\frac{2}{3}$" (Tatarsky, 1967) holds we have obtained the following equation for the operating range:

$$D_{max} \geqslant 0.04 \frac{b^2}{C_n^2(UT)^{5/3}} \tag{3}$$

$b = \lambda/4$ for the laser

$b = 2\mu$ for the white light

where C_n is the structure parameter, U the wind speed, and T the observation time (averaging interval).

On the basis of (3), we have plotted a graph for the white light which shows that, in the case of weak turbulence ($C_n = 2 \cdot 10^{-16}$ cm$^{-2/3}$), the operating range can reach 1 km with $UT = 1$. In this calculation it is assumed that only one contrast measurement is made in the interval T (see Fig. 4). The range of 860 m has already been achieved in experiment (Honkasalo, 1950).

The dispersion method for determining the mean integral index of refractivity in interferometric measurements

By calculation it can be shown that, in the case of phase rangefinders, the dispersion method cannot give the required accuracy ($1 \cdot 10^{-7}$) for distances

610

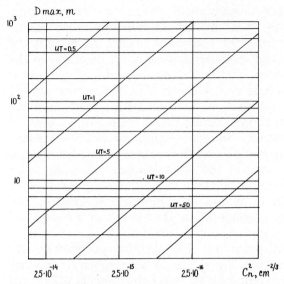

Fig. 4. The operating range of the interferometric rangefiner as a function of the structural parameter C_n.

of several hundred metres because the optical path difference should be measured with an error of only a few micrometres (Prilepin, 1957; Thompson, 1968).

By using the dispersion method in an interferometric rangefinder, the phase mean integral index of refractivity can be determined with an accuracy of 10^{-7}, for here we achieve the required accuracy of measuring the dispersion difference (for wavelengths of 0.44 and 0.63 μm the error should not exceed 3 μm for a distance of 500 m).

A number of problems currently facing geodesists concern the elaboration of the dispersion method for measuring short and, incidentally, long distances connected with the study of the difference between the actual composition of the atmosphere at the moment of measurement and its standard composition, for which the coefficients of dispersion are determined in laboratories.

One of the variants of a laser refractometer intended to determine the absolute value of the refractivity index in the field and to study air dispersion is described below. The block diagram of such a refractometer is given in Fig. 5. This refractometer differs from the usual types in that the required accuracy $(1 \cdot 10^{-7})$ can be achieved with the aid of a vacuum chamber about 100 mm long, using the integration effect. If the reflectors of the vacuum and the working channel are synchronously displaced to and fro several times by the length S, then the index of air refractivity in the working

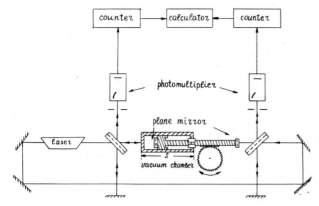

Fig. 5. Block diagram of a laser refractometer.

channel can be found from the following equations:

$$\frac{2S}{\lambda_0} = N_{\mathrm{v}}, \qquad \frac{2S \cdot n}{\lambda_0} = N_{\mathrm{w}}, \qquad n = \frac{N_{\mathrm{w}}}{N_{\mathrm{v}}} \tag{4}$$

where N_v is the number of interference fringes counted in the vacuum channel, and N_w is the number of fringes counted in the working channel. As can be seen from eq. 4, a frequency-nonstabilized laser can be used in measuring the index of air refractivity; λ_0 is the laser wavelength in vacuum.

REFERENCES

Boulanger Yu.D. (Editor), 1971. Proceedings of the 4th International Symposium on Recent Crustal Movements, Valgus, Tallin 1975, 249 pp.

Honkasalo, T., 1950. Measuring of the 864-m-long base line. Veröff. Finn. Geod. Inst., No. 37, Helsinki.

Kolomiytsov Yu.V., 1976. The Interferometers. Mashinostroyeniye, Leningrad (in Russian).

Prilepin, M.T., 1957. Proceedings of the Central Research Institute of Geodesy, Aerial Surveying and Cartography, Vol. 114. Moscow (in Russian).

Tatarsky, V.J., 1967. Wave Propagation in Turbulent Medium. Nauka, Moscow (in Russian).

Thompson Jr., M.C., 1968. Space averages of air and water vapor densities by dispersion for refractive correction of electro-magnetic measurements. J. Geophys. Res., 73 (10).

Vali, V. and Bostrom, R., 1968. One thousand-meter laser interferometer, Rev. Sci. Instr., 39 (9).

Tectonophysics, 52 (1979) 613—626

© Elsevier Scientific Publishing Company, Amsterdam — Printed in The Netherlands

AGING AND STRAIN SOFTENING MODEL FOR EPISODIC FAULTING

WILLIAM D. STUART

U.S. Geological Survey, Menlo Park, Calif. 94025 (U.S.A.)

(Submitted December 10, 1977; Revised manuscript accepted for publication July 7, 1978)

ABSTRACT

Stuart, W.D., 1979. Aging and strain softening model for episodic faulting. In: C.A. Whitten, R. Green and B.K. Meade (Editors), Recent Crustal Movements, 1977. Tectonophysics, 52: 613—626.

Episodic slip on shallow crustal faults can be qualitatively explained by postulating a fault constitutive law that is the superposition of two limiting material responses: (1) strain softening after peak stress during large strain rates, and (2) strength (peak stress) recovery during aging at small strain rates. A single law permits a variety of seismic and aseismic phenomena to occur over a range of space and time scales. Specific cases are determined by the spatial variation of material constants, recent deformation history, crustal rigidity, and remote forcing.

INTRODUCTION

During time intervals of decades and less, ground surface deformation associated with shallow crustal faulting can range from episodic to nearly steady. For longer time intervals, the same is probably true, but observations are scarce and only suggestive. Rates of episodic deformation caused by faulting may be high, as during an earthquake, or low, as with aseismic creep. Duration and recurrence times of episodes and clusters of episodes also vary widely, as do the space scales. Some fairly well documented examples of episodic ground deformation and faulting are the once-repeated Palmdale uplift and collapse in southern California (Castle et al., 1976, 1977), periodic moderate to great earthquakes (McEvilly et al., 1967; Fedotov, 1968), pre-earthquake seismic gaps (Kelleher and Savino, 1975), repetitive pulses of enhanced regional seismicity (see Fig. 7), earthquake migration in time (Kagan and Knopoff, 1976), and recurring creep events on the San Andreas fault (Tocher, 1960). In fact, the two most striking features of crustal deformation are its variability in time and its tendency to localize into thin bands called faults.

Suppose that a number of thin $(0(m))$ potential fault zones are embedded

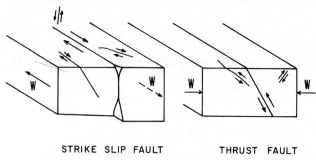

STRIKE SLIP FAULT   THRUST FAULT

Fig. 1. Model geometry.

in a linearly elastic crust subject to a growing regional displacement W as shown in Fig. 1. In principle, the stress and faulting history are then completely determined by fault constitutive properties, possibly space and time dependent, and W. During a definite short time interval, only a few fault zones are likely to have strength and orientation favorable to failure. If the fault zones are thin enough, a friction law on a fault plane is an acceptable mathematical substitute for a constitutive law. Hereafter the term "fault plane" refers to these individual thin tabular zones and the term "fault zone" is reserved for a wider, dense collection of fault planes, e.g., the San Andreas fault zone. This paper attempts to explain certain episodic phenomena with a qualitative model containing single or interacting fault planes which obey a nonlinear friction law.

CONSTITUTIVE PROPERTIES OF FAULTS

Numerous laboratory experiments, field observations, and theoretical considerations leave little doubt that seismic faulting in the upper crust is due to release of elastic potential energy during progressive weakening of rock. The weakening is assumed here to come from strain softening, but other mechanisms such as strain rate softening may exist. As will be seen, accelerated fault slip rates, though subsonic, can have a similar physical origin; geometric and constitutive conditions determine whether the slip rate will be inertia-limited. But strain softening alone can account for no more than a single earthquake or aseismic slip event. Recurrent transient deformation seems to require healing or aging of fault planes such that they regain their strength. An alternative to aging, not considered here, may be a slight geometric rearrangement of fault structure such that formerly freely sliding fragments become interlocked. Physical mechanisms for strain softening are probably grain fracture and intergrain relative displacement and rotation. Aging may be a time dependent interpenetration of grains, cementation, or recrystallization.

We shall postulate that fault materials are initially strain hardening and then become strain softening after a peak stress is reached. Furthermore, we

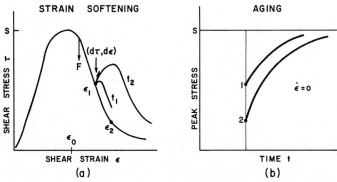

Fig. 2. Proposed constitutive law for faults. a. Strain hardening and softening at large strain rate, b. Recovery of peak stress during aging at small strain rate.

postulate that the material can recover its strength by aging at low strain rates such that the hardening/softening cycle can repeat itself. Although the available laboratory data tend to support the above postulates, we do not wish the model to be too severely constrained by such data because of scaling uncertainties. Fig. 2a shows a typical stress—strain curve for rock deforming brittlely at relatively high strain rate. A positive slope obtains during strain hardening, a peak stress S is reached, and then strain softening occurs during the approach to a residual stress at large strain. In the laboratory, sudden failure marked F commonly occurs; however, it is a fundamental property not of the rock, but of the combined sample plus testing machine system and can be prevented by increasing the machine stiffness. If the sample were under seismically appropriate pressure and temperature conditions, a test terminated and maintained for a time at post-peak strains, say ϵ_1 or ϵ_2 ($\dot{\epsilon} = d\epsilon/dt = 0$ at ϵ_1 or ϵ_2 in this example), might allow the sample to recover partially its peak stress (Fig. 2b) so as to follow subdued versions of its original curve upon restraining. Curves marked t_1 and t_2 in Fig. 2a are paths showing such continued deformation where curve t_2 represents greater aging than curve t_1. Otherwise, an immediately restarted test would proceed along the original curve. Little is known about the recovery rate of S. It might, for example, be proportional to the logarithm of time, as Dieterich (1972) observed in friction experiments on saw-cut rock surfaces at room temperature, and depend on the loss of peak stress $(S - \tau)$.

Equation 1 is a hypothetical differential form for the next deformation increment $(d\tau, d\epsilon)$ from (τ, ϵ), a general post-peak state such as ϵ_1 in Fig. 2, and is obtained by summing hardening/softening, $d\tau^S$, and aging, $d\tau^A$, terms:

$$d\tau = d\tau^S + d\tau^A = \frac{\partial \tau^S}{\partial \epsilon} d\epsilon + \frac{\partial \tau^A}{\partial \epsilon} d\epsilon$$

$$\tau^S = S \exp\left[-\left(\frac{\epsilon - \epsilon_0}{\alpha}\right)^2\right]$$

$$\tau^A = [\tau + \eta(S - \tau)] \exp\left[-\left\{\frac{\eta(\epsilon_0 - \epsilon)}{\alpha}\right\}^2\right]$$

$$\eta = 1 - \exp(-\beta t) \tag{1}$$

Aging during time t is assumed to transpire only after strain softening, viz. $\epsilon > \epsilon_0$. The first bracket of the τ^A definition specifies the recovery of peak stress, and the η in the exponential argument moves the tail of vector $(d\tau, d\epsilon)$ to the hardening side of the aged curve. Material constants for a pristine or thoroughly aged medium are the strain at peak stress ϵ_0, the curve "width" α, and peak stress S. Later S will be presumed to be a function of position on the fault plane. With no aging, ϵ_0, α, and S describe a Gaussian stress–strain curve. Aging factor η ranges from $\eta = 0$ for no aging to $\eta = 1$ for complete aging; β is the aging time constant.

Two limiting cases are immediately evident. For short times compared to β (large $\dot{\epsilon}$, small η), little aging occurs. Consequently, the original strain softening curve is closely followed. For long times compared to β (small $\dot{\epsilon}$, large η), the material "forgets" its strain softening character according to $(1 - \eta)$ while recovering its former peak stress; the response is strain hardening. Intermediate η at constant $\dot{\epsilon}$ yields a peak stress that varies inversely with $\dot{\epsilon}$. In contrast to Newtonian viscosity and certain creep laws, such material appears stronger the slower it deforms. In related work, Dieterich (1977) proposes a velocity-weakening law to explain frictional sliding between rock surfaces. His law accommodates observed time dependent friction (Dieterich, 1972) and reconciles the inequality of static and kinetic friction coefficients.

Equation 1 will surely need revision when more data are available. If aging proves to be important, a multiple integral expansion within the axiomatic theory of constitutive laws may be worth developing (e.g. Malvern, 1969; Strauss, 1974).

Constitutive law (1) is assumed to apply within a thin fault layer undergoing finite shear strain. Since neither the width nor the strain is known, the friction law version of eq. 1 is used in model calculations. The friction law is obtained by integrating strain across the layer width, in which case the form of eq. 1 is preserved with fault slip $2w$ replacing strain. Use of a friction law, however, does not necessarily imply an exact correspondence with phenomena observed in laboratory friction experiments.

STRAIN SOFTENING ON A LONG STRIKE–SLIP FAULT

Computed results for a single earthquake cycle with a slip hardening/softening law, but no aging, are described in this section. That is, deformation is assumed to be fast enough to preclude significant aging. Details and

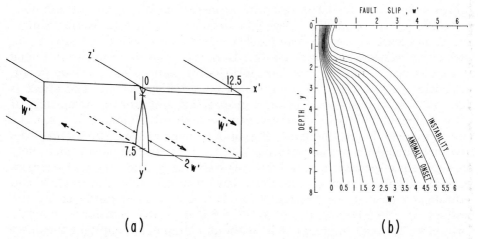

(a) (b)

Fig. 3. a. Geometry of strain softening model of a long, vertical, strike-slip fault. b. Solutions for fault slip w' vs. depth y' at successive boundary displacement W'.

additional results are given in Stuart (1978). In the following sections, the model results are qualitatively extended to infer the consequences of aging and peak stress variation on a fault plane.

Approximate solutions for quasi-static deformation near a long, vertical, strike-slip fault (Fig. 3a) have been obtained using the finite element method. Material surrounding the fault is assumed to be linearly elastic. The assumed fault friction law (eq. 2) is Gaussian in depth y and total fault slip $2w$. Thus the peak stress first increases, then decreases with depth.

$$\tau = S \exp\left[-\left(\frac{2w - 2w_0}{a}\right)^2\right] \exp\left[-\left(\frac{y - y_0}{b}\right)^2\right] \tag{2}$$

At each depth the fault is initially slip hardening before becoming slip softening as the peak stress is surpassed. In the earth, these frictional stresses are superimposed on an unknown stationary background stress and are presumably restored after the earthquake by aging. Increasing displacement W at $x = \pm 12.5\, y_0$ and $y = 7.5\, y_0$ simulates growing regional displacement which is balanced by frictional resistance mainly near y_0 ($y' = 1$).

Scaling Hooke's law, the friction law, equilibrium equations, strain definition and boundary conditions with a reference distance y_0, displacement a, and stress S reveals that two primed dimensionless parameters $\mu' = \mu a/Sy_0$, $b' = b/y_0$ sufficiently specify mechanically similar boundary value problems. μ is crustal rigidity, and S, a, y_0 and b are friction law coefficients ($w_0 = 0$ for convenience). μ' and b' thus determine if an inertia-limited instability is possible (an earthquake analog) or only temporarily rapid quasi-static slippage. Conversion to dimensional variables (unprimed) follows from $(x, y) = (x', y')y_0$, $(w, W) = (w', W')a$, and $\tau = \tau'S$.

618

Figure 3b shows fault slip response to increasing W' for an unstable case $\mu' = 0.25$, $b' = 1$. Fault slip w' lags W' most near $y' = 1$ where friction is maximum, and least at $y' = 7.5$. The fault becomes slip softening first at $y' = 7.5$, later at successively shallower depths. During the transition to slip softening at a given depth, the slip rate increases sharply as the nearby elastic potential energy stored during slip hardening is partially released. The effect is that fault slip appears as an upward propagating and growing creep wave of screw dislocations which concentrates stress upward. At about $W' = 4.5$, the rate of increase is fast enough that a rapid increase in shear strain rate at the free surface appears as shown in Fig. 4. This slope change, most evident near the fault trace (small x'), may be recognizable as an anomaly precursory to an earthquake and is labeled as such. Saw-tooth oscillations at small W' and cusp-shaped oscillations at large W' in Figs. 4 and 5 are produced by the method of numerical solution. At $W' = 6.0$ no stable solution is possible for arbitrarily small increases of W' without a jump in w', implying the onset of an inertial instability. Immediately before the instability, the greatest fault slip rate and acceleration are in the depth interval $1 < y' < 2$. This position is the probable counterpart of an earthquake focus, though further analysis is needed. No geodetic data are available to test the model, but ground breakage two weeks prior to the $M_L = 5.5$, 1966 Parkfield, California (Allen and Smith, 1966) and rapid epicentral uplift during the two years

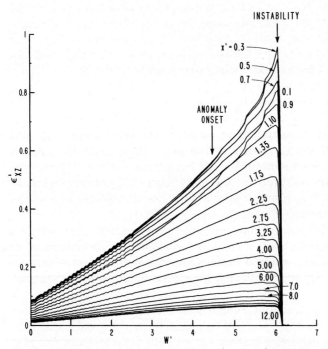

Fig. 4. Solutions for surface shear strain ϵ'_{xz} vs. W' at various distances x' from fault trace.

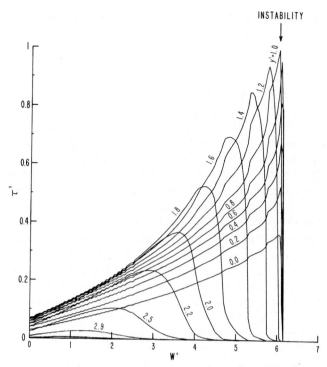

Fig. 5. Solutions for fault stress τ' vs. W' at various depths y'.

preceding the M_L = 6.4, 1971 San Fernando earthquake (Castle et al., 1974; Thatcher, 1976) are compatible with accelerating precursory fault slip.

Fault stress τ' vs. W' is shown in Fig. 5. The peak stress is reached progressively later at shallower depth for $y' > 1$. at the instability, it is clear that much of the fault resides at stress levels much lower than at earlier times at the same position. Only the fault close to $y' = 1$ is near peak stress at instability onset. Prior to instability, the average fault stress has reached a maximum and is declining. Instability occurs, of course, when the average fault stress decreases with fault slip faster than the nearby elastic stress, thus preventing static equilibrium.

The space (μ', b') has been partially explored to determine the boundary between inertially stable and unstable modes by Stuart (1978). Stuart and Mavko (1979) show that with both a stress-free and displacement boundary condition at the plate bottom, lowering b' or raising μ' tends to replace unstable with stable deformation. That inertially stable modes must exist is easily seen by considering the limit of a very rigid crust or slowly softening fault corresponding to large μ'. Inertially stable modes are taken to be analogs of aseismic slip episodes because rapid, but quasi-static, fault slip occurs during strain softening. Shear strain and fault stress curves, analogous to Figs. 4 and 5, for stable cases resemble Gaussian functions and show no slope discontinuities.

HETEROGENEOUS FAULT

Similar calculations for arbitrary fault dip, slip, and peak stress variation are not so easy, but qualitatively it is reasonable to expect fault slip rate to increase during strain softening near the greatest peak stress positions. As above, the form of the friction law and crustal rigidity will decide if a slip episode will culminate in an inertial instability. With the central San Andreas fault in mind, suppose that contours of peak stress on the side view of a strike-slip fault plane (Fig. 6) have been somehow established. Suppose, furthermore, that any earthquake slip covers the shaded area roughly, but not identically, spanning the largest peak stress contours. While the regional displacement gradually increases, perhaps by tectonic plate motions, accelerated precursory fault slip will occur mainly below the earthquake in patch A, to the right in B, and above in C, if the patches respond independently. The reason is that the fault near the greatest peak stress position softens last. Smaller closed contours within the patches might fail as smaller earthquakes, possibly foreshocks. Again, in all cases seismic failure may be replaced by a rapid aseismic slip episode according to the constitutive law, etc. However, sufficiently close patches will actually interact because failure of one will cause increased loading of its neighbor, adding to the complexity of the deformation. Wesson et al. (1973) invoke a similar patch concept, but without reference to a specific fault constitutive law.

A particular patch interaction that is basic to later arguments is now considered. For convenience, divide the fault into a shallow seismically active

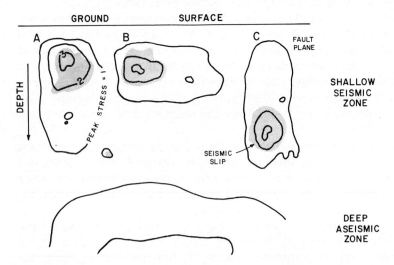

Fig. 6. Side view of a vertical fault plane with hypothetical peak stress contours.

zone containing patches such as A, B, and C in Fig. 6 (ca. 0—15 km for the San Andreas fault) and an underlying aseismic zone, called the deep zone. The deep zone, extending perhaps another 15 km, is presumed also to undergo episodic deformation by a strain softening mechanism, but only quasi-statically. Instability would be suppressed in the deep zone because the brittle component of deformation is dominated more by the ductile component than at shallower depths. The shallow zone may at one time or another have a range of earthquake magnitudes, all smaller than the occasional largest earthquakes which coincide with patches such as A. We wish to infer the interaction of shallow patches with the underlying deep zone during a cycle defined by the repeat time of the largest earthquakes occurring at A, B, or C.

Seismic and aseismic failure in the shallow zone will be encouraged by strain softening in the deep zone. This is so because more of the regional load must then be taken up by the shallow zone. Unless the shallow zone also fails completely, deep zone failure will be accompanied by pulse-like transfer of elastic potential energy upward. A comparable interaction may occur between en echelon and laterally adjacent fault segments if, as is usual, no continuous single fault plane occupies the fault zone.

How might aging modify the failure interactions? At a high shear strain rate it was asserted that the constitutive law was either strain hardening or softening, and at a small strain rate the fault material tended to recapture its peak stress. At intermediate strain rates, peak stress depended on strain rate. Episodic deformation arises naturally with this constitutive law when the coefficients are space dependent, and is in fact difficult to prevent. On a freshly faulted surface, for example, a small area of relatively high strength will tend to retard the local strain rate, which in turn will promote aging and further strain-rate retardation. Near such an isolated high strength position the area of fault-slip retardation might enlarge laterally with time as aging proceeds. During failure, the process would reverse as a ring of high slip rate collapsed on the high strength core. These events also imply that failure modes and rates at the same location may be different at different times according to the deformation history.

Referring to Fig. 6, if patch A fails long before B, A may have recovered enough of its strength during small $\dot{\epsilon}$ to delay failure of B by sharing the tectonic load. Similarly, aging will proceed after slip associated with shallow seismic and aseismic failure which was stimulated by a more extensive, deep failure. The deep failure site itself will also age after slippage. Renewed aging in the deep zone permits a new episode of potential energy accumulation in the adjacent elastic rocks followed by another upward transfer of potential energy during a future deep slip episode. Several repetitions ultimately will have transferred enough potential energy to the seismic zone for the largest earthquake to happen. The high slip rates accompanying deep failure during the final episode merge into high precursory slip rates (as suggested by Fig. 3b) of the largest earthquake.

From static dislocation theory one can argue that each of the deep slip episodes serves to smooth (the "priming" of Weertman, 1964) the spatial irregularities of stress in the shallow zone by means of earthquakes and aseismic slip. Smoothness implies a nearly uniform $\dot{\epsilon}$ field and therefore favors more uniform and widespread aging if $\dot{\epsilon}$ is also small. Consequently, the first few stress pulses from the deep zone after the largest earthquake will be weak because most of the shallow zone will have had insufficient time to age, thus preventing much aging of the deep zone. One might expect relatively small magnitude seismicity and a large fault-slip rate for a time after the largest events. Later pulses will be more intense and seismic because of longer net aging time, both shallow and deep, and greater fault area subject to aging. Average fault-slip rate in the shallow zone should be retarded most just prior to the high slip rates during the final deep failure. For certain great earthquakes like the 1906 San Francisco event (M = 8.3) whose seismic zone appears to have been locked since 1906, episodic priming and aging may be most important at the locked zone periphery.

Thus, a picture emerges of repeated stress pulses applied to the seismic zone in the time interval between the largest earthquakes. During each pulse, seismicity may increase, but then it will diminish as aging again dominates on newly slipped faults. Regions of such reduced seismicity may appear as seismic gaps. The last pulse leads into the largest earthquake. Small fault patches may have slipped episodically and rehealed a number of times, whereas the largest patches only once or twice.

SEISMICITY PATTERNS AND FAULT CREEP

Consider now the central San Andreas fault where both continuous small to moderate seismicity and slow surface faulting (creep) occur. The model implies that episodic seismicity will be caused by rapid loading associated with nearby faulting. In other words, patches of recurrent high slip rate are likely to have seismically active perimeters. Prior to rapid patch failure, slip rates within the patch will be unusually low, and during and after, unusually high. These assertions permit us to use observed seismicity to infer crudely some of the fault slip history.

Figure 7 displays all $M_L > 2.5$ seismicity within 6 km of the central San Andreas fault from 1969 through 1977. The ordinate is the position of the orthogonally projected epicenter onto the fault trace, and the abscissa is time. Symbol centers and epicenters coincide, and symbol height approximately equals rupture length (km) via $L = 10 \cdot 5^{M_L - 5}$ (approximate fit to source parameters cited by Dieterich, 1974, p. 296). Thus a magnitude 5 earthquake has a rupture length of 10 km, a magnitude 3 has a length of 400 m. For legibility, symbols of height appropriate for magnitude 3 are plotted for all events of magnitude less than 3. Symbol value provides hypocenter depth d interval, e.g., 2: $2 \leqslant d < 3$ km, ...; A: $10 \leqslant d < 11$, ...

Several features, some recurrent, can be extracted from the seemingly ran-

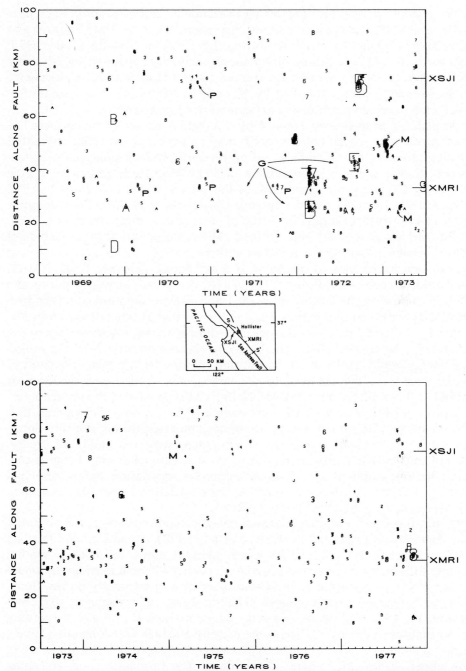

Fig. 7. Space-time seismicity within 6 km of the San Andreas fault, M > 2.5 1969—1977. Earthquake data from the U.S. Geological Survey central California seismometer net. Symbol value is truncated hypocenter depth, height is approximate rupture length. *P* = pulses of enhanced seismicity, *M* = seismicity migrations, *G* = seismic gaps. XSJ1 and XMR1 mark creepmeter locations. *S—S'* on index map marks fault length of seismicity plot.

dom pattern. They are: (1) average seismicity is persistently high near XSJ1 and XMR1 creepmeters where, (2) several M_L = 4—5 earthquakes and associated aftershocks occurred during 1971—1973 (Wesson et al., 1973; Ellsworth, 1975), (3) pulses of enhanced seismicity separated by relatively quiet intervals occur — some are marked by P, (4) migrations of seismicity in space-time appear — marked by M, and (5) seismic gaps, sometimes fulfilled by a dominant earthquake are common — marked by G.

Explaining observations *1* and *4* by a judicious choice of constitutive law coefficients and spacial heterogeneity may be possible, but is not warranted at present owing to sparse data and no relevant model simulations. As mentioned above, however, the increase of small magnitude seismicity during the years following the large earthquakes (symbols *7* and *B*) in early 1972 near XMR1 is implied by the model. Observations *2*, *3*, and *5* harmonize more clearly with the model by virtue of independent creepmeter data.

Burford et al. (1973) and Burford (1976) point out that surface fault creep became progressively retarded with respect to the secular rate during one to two years prior to the M_L = 4—5 earthquakes close to XSJ1, XMR1, and other creepmeters during 1971—1973. Creep rate increased during the months following the earthquakes. Burford (private communication) conjectures that the creep retardation was due to a relative locking of the coseismic sections of the fault. Since seismic slip during an M_L = 4—5 earthquake is roughly 1—10 cm (cf. Dieterich, 1974), it appears that much of the potential energy could have been stored during the one to two years of observed 5—10 mm/yr relative lag prior to the earthquakes. Earlier earthquakes near XMR1, M_L = 4.9, 5.0 in 1951 (McNally, 1976), suggest a recurrence interval of about 20 years which is long compared to 2 years. This implies that the traditional notion of steady strain accumulation during the time interval between co-located recurrent earthquakes may not hold universally. Completely locked faults failing during great earthquakes may be an exception. In the proposed model, both increasing retardation before principal earthquakes and the dominance of the last retardation interval are explained by extensive aging of a well-primed patch.

Rapid but brief creep rate increases called creep events are common in the creepmeter data presented in Burford et al. (1973), Yamashita and Burford (1973), and Schulz et al. (1976). These steplike jumps, recurring at weekly to monthly intervals, appear as partial recoveries of creep retardation. Many of the creep events occur a few weeks after nearby seismicity pulses. Some good examples are creep events at XMR1 in April, 1970; January, 1971; and November, 1971 following seismicity pulses marked P in Fig. 7. According to an analysis of creep, tilt, and strain data by McHugh and Johnston (1976), some of the creep events can be explained by propagating dislocations along the upper several kilometers of the fault. The proposed aging—softening model implies that the underlying seismicity pulses are associated with nearby high fault slip rate during strain softening. Thus if creep events are due to shallow dislocations, the model indicates that they are forced into motion by deeper slip. In general, the model associates aging of the fault with creep retardation and diminished seismicity and associates strain softening with creep rate recovery and enhanced seismicity.

IMPLICATIONS

Although certain features of space—time seismicity and fault creep can be reconciled with the theoretical model invoking a strain softening and aging fault constitutive law, overall understanding is far from satisfactory. This may be due to an intricate spatial variation of constitutive properties instead of inadequate mechanical postulates. Even if the correct and complete form of the constitutive law for a fault were known, the distribution of material constants, especially the shorter wavelengths, may be practically unknowable because of inaccessibility and sparse and imprecise crustal deformation data. A related complication may be an extreme sensitivity of fault motions to small stress or slip perturbations such that two initially close states can substantially diverge with time.

When the form of the fault constitutive law is known, understanding episodic faulting means discerning the distribution of material constants on the fault and in the surrounding media. This task, at least formally, can be posed as an inversion of the data history. If successful, it may be possible to extrapolate a short time forward to estimate future deformation.

ACKNOWLEDGEMENTS

This paper arose from discussions with A.G. Lindh. I thank R.O. Burford for comments and A. Steppe for earthquake data.

REFERENCES

Allen, C.R. and Smith, S.W., 1966. Pre-earthquake and post-earthquake surficial displacement. In: Parkfield earthquakes of June 27—29, 1966, Monterey and San Luis Obispo Counties, California, Preliminary Report, Bull. Seismol. Soc. Am., 56: 966—967.

Burford, R.O., 1976. Fluctuations in rates of fault creep associated with moderate earthquakes along the central San Andreas fault. EOS, Trans. Am. Geophys. Union, 57: 1012.

Burford, R.O., Allen, S.S., Lamson, R.J. and Goodreau, D.D., 1973. Accelerated fault creep along the central San Andreas fault after moderate earthquake during 1971—1973. In: R.L. Kovach and A. Nur (Editors), Proceedings of the Conference on Tectonic Problems of the San Andreas Fault System. Stanford University, Stanford, Calif., pp. 268—274.

Castle, R.O., Alt, J.N., Savage, J.C. and Balazs, E.I., 1974. Elevation changes preceeding the San Fernando earthquake of February 9, 1971. Geology, 2: 61—66.

Castle, R.O., Church, J.P. and Elliott, M.R., 1976. Aseismic uplift in southern California. Science, 192: 251—253.

Castle, R.O., Elliott, M.R. and Wood, S.H., 1977. The southern California uplift. EOS, Trans. Am. Geophys. Union, 58: 495.

Dieterich, J.H., 1972. Time-dependent friction in rocks. J. Geophys. Res. 77: 3690—3697.

Dieterich, J.H., 1974. Earthquake mechanism and modeling. Annu. Rev. Earth Planet. Sci., 2: 275—301.

Dieterich, J.H., 1978. Time-dependent friction and the mechanics of stick-slip. Pure Appl. Geophys., 11: 790—806.

Ellsworth, W.L., 1975. Bear Valley, California, earthquake sequence of February—March 1972. Bull. Seismol. Soc. Am., 65: 483—506.

Fedotov, S.A., 1968. The seismic cycle, quantitative seismic zoning, and long term seismic forecasting. In: S.V. Medvedev (Editor), Seismic Zoning of the USSR. Nauka, Moscow (in Russian). English Transl. by Israel Progr. for Sci. Transl., 1976 — U.S. Dept. of Commerce, Natl. Tech. Inf. Serv., Springfield, Va., pp. 133—166.

Kagan, Y. and Knopoff, L., 1976. Statistical search for non-random features of the seismicity of strong earthquakes. Phys. Earth Planet. Inter., 12: 291—318.

Kelleher, J. and Savino, J., 1975. Distribution of seismicity before large strike slip and thrust-type earthquakes. J. Geophys. Res., 80: 260—271.

Malvern, L.E., 1969. Introduction to the Mechanics of a Continuous Medium. Prentice-Hall, Englewood Cliffs, N.J., 713 pp.

McEvilly, T.V., Bakun, W.H. and Casaday, K.B., 1967. The Parkfield, California earthquakes of 1966. Bull. Seismol. Soc. Am., 57: 1221—1224.

McHugh, S. and Johnston, M.J.S., 1976. Short-period nonseismic tilt perturbations and their relation to episodic slip on the San Andreas Fault in central California. J. Geophys. Res., 81: 6341—6346.

McNally, K., 1976. Spatial, Temporal, and Mechanistic Character in Earthquake Occurrence: a Segment of the San Andreas Fault in Central California. Ph.D. Thesis, University of California, Berkeley, Calif., 140 pp.

Schulz, S.S., Burford, R.O. and Nason, R.D., 1976. Catalog of creepmeter measurements in central California from 1973 through 1975. U.S. Geol. Surv., Open-file Rep. 77—31, 480 pp.

Strauss, A.M., 1974. Continuum theory of time-dependent hysteretic earth materials. J. Geophys. Res., 79: 351—356.

Stuart, W.D., 1978. Strain softening prior to two-dimensional, strike-slip earthquakes. J. Geophys. Res., 83: (in press).

Stuart, W.D. and Mavko, G.M., 1979. Earthquake instability on a strike-slip fault. J. Geophys. Res., 84: (in press).

Thatcher, W., 1976. Episodic strain accumulation in southern California. Science, 194: 691—695.

Tocher, D., 1960. Creep on the San Andreas fault: creep rate and related measurements at Vineyard, California. Bull. Seismol. Soc. Am., 50: 396—404.

Weertman, J., 1974. Continuum distribution of dislocations on faults with finite friction. Bull. Seismol. Soc. Am., 54: 1035—1038.

Wesson, R.L., Burford, R.O. and Ellsworth, W.L., 1973. Relationship between seismicity, fault creep, and crustal loading along the central San Andreas fault. In: R.L. Kovach and A. Nur (Editors), Proceedings of the Conference on Tectonic Problems of the San Andreas Fault System, Stanford University, Stanford, Calif., pp. 303—321.

Yamashita, P.A. and Burford, R.O., 1973. Catalog of preliminary results from an 18-station creepmeter network along the San Andreas fault system in central California for the time interval June 1969 to June 1973. U.S. Geol. Surv., Open-file Rep., 215 pp.

Tectonophysics, 52 (1979) 627   627

THEORETICAL MODELS OF ANELASTIC CRUSTAL DEFORMATION

J.B. RUNDLE

Department of Earth and Space Sciences, U.C.L.A., Los Angeles, Calif. 90024 (U.S.A.)
(Accepted for publication June 8, 1978)

ABSTRACT

Recent geodetic data indicate that the earth exhibits systematic long-term nonseismic deformation. Numerous examples of this behavior can be found in both local precursory and postseismic displacements, and in glacial rebound. In many cases, the best way to explain the observed motion is with the use of earth models consisting of a brittle region overlying a more ductile, easily deformable zone. A widely used method for construction of theoretical models to explain the observed deformation is the Green's function technique. With this method, either the displacements or stresses over surfaces of displacement can be readily specified. Modern data-inversion theory can be employed and the values of physically meaningful parameters in the earth model can be determined.

An example of such an anelastic earth model is presented. The model consists of an elastic layer over a viscoelastic half-space. It is shown how the quasistatic viscoelastic response to a fracture occurring in the elastic layer can be computed by two mutually comparible techniques, both of which are accurate and convenient to use. As an example, a model of the 1906 San Francisco earthquake is presented and it is shown that the model fits the data well.

Tectonophysics, 52 (1979) 629—641 629

RHEOLOGY OF THE CRUST BASED ON LONG-TERM CREEP TESTS OF ROCKS

HIDEBUMI ITÔ

University of Osaka Prefecture, Sakai-City, Osaka (Japan)

(Accepted for publication June 8, 1978)

ABSTRACT

Itô, H., 1979. Rheology of the crust based on long-term creep tests of rocks. In: C.A. Whitten, R. Green and B.K. Meade (Editors), Recent Crustal Movements, 1977. Tectonophysics, 52: 629—641.

For 20 years Kumagai and Itô have carried out creep tests on granite beams each of $215 \times 12.3 \times 6.8$ cm. They have concluded that granite flows plastically with a vanishingly small yield stress, that is, viscous flow with a viscosity of $3-6 \cdot 10^{20}$ P. Itô and Sasajima have carried out more accurate creep tests by bending three granite test-pieces (each $21 \times 2.5 \times 2.0$ cm) and three gabbro test-pieces ($16 \times 2.0 \times 1.5$ cm). Deformations were measured using a light-interference technique. The results obtained over a three-year period show that the creep in granite is comparable with that found formerly, and that gabbro has a slower rate of creep than granite.

Consequently, over a long period of time, rock may be represented rheologically as a Maxwell liquid. We may thus imagine the crust to be modelled by a Maxwell liquid, which floats on a more fluidal underlying layer supporting the crust isostatically. Based on this conception, subsidence of guyots and atolls has been investigated and the viscosity of the ocean crust has been found to be 10^{25-26} P. Analysing the Quaternary crustal movements in the Himalayas and in southwest Japan, the viscosity of the orogenic crust has been estimated to be 10^{22} P. The viscosity of the cratonic crust seems to be greater than that of the orogenic crust.

From the viscosities estimated above and the generally accepted value of rigidity of the crust, we can draw the stress—strain diagrams of the crust for different strain rates. For example, the diagram obtained for the orogenic crust suggests that the crust would behave elastically for times shorter than the relaxation time of 3000 years and would flow as Newtonian liquid for longer times, and if a shearing strength of the crust were greater than 10 bar and the shearing strain rate were smaller than 10^{-15}/sec, the crust would flow without fracture.

INTRODUCTION

The author once thought that geological phenomena occurring over a very long period of time might be not explainable by the usual laws as discovered from laboratory experiments. For example, he doubted if an iron bridge would constantly keep its elastic deflection over geological time. Since then he intended to devote his work to the bending of rock beams from a geological viewpoint, and started an experiment of bending large granite beams on August 7, 1957 with N. Kumagai. The experiment lasted for about 10 years and showed that granite beams would flow as viscous liquid.

Conceptionally, we classify flow as two kinds: a viscous flow and a plastic flow. However small the stress may be, a viscous material flows as does a liquid. But a plastic material does not flow under stresses smaller than the yield stress. Griggs (1940) carried out creep tests for 500 days on seven alabaster specimens immersed in water and loaded under different compressive stresses, the lowest stress being 103 kg/cm². He then found from the relation between stress and strain rate of steady (secondary) creep that the strain rate would become zero at a stress of 92 kg/cm². This is only the yield stress. However, this was not confirmed by creep tests of specimens loaded less than 92 kg/cm². Determination of the value of the yield stress of rock should be of considerable importance to geology and geophysics.

In the first part of this paper two experiments are presented on long-term creep of rocks: the experiment by Kumagai and Itô and that by Itô and Sasajima. From the experiments, it is concluded that hard rocks may have zero yield stress. Based on this conclusion, the author then discusses the rheology of the crust.

EXPERIMENT BY KUMAGAI AND ITÔ

In Fig. 1 the test-pieces used in the experiment by Kumagai and Itô (1965, 1968, 1970, 1971) are shown. The test-pieces were cut from a granite block from Akasaka, Hiroshima Prefecture, Japan and shaped into a straight beam (215 × 12.3 × 6.8 cm). One of these pieces, bending under its own weight, is called the unloaded beam, and the other, bending under its own weight plus a center-load is called the center-loaded beam. The maximum bending stress in the unloaded beam is 12.8 kg/cm² and that in the center-loaded beam 24.8 kg/cm². The experiment started in the basement laboratory of the Geological and Mineralogical Institute of Kyoto University. However, because of the reconstrucion of the institute buildings, the test-pieces were moved very carefully on October 14, 1967 to a distant laboratory of the First Gravity Station of Kyoto University.

The chief instrument used to measure the deformation of the test-piece is a dial gauge (0.01 mm/div.) which slides along a straight steel beam 103 cm long (Fig. 1). When measuring, the straight steel beam is placed in the middle portion of the upper surface of the test-piece. A precision level placed at one end of the test-piece is used to measure the additional small elastic deformation of the test-piece caused by mounting the straight steel beam. Using these instruments we can find the vertical displacement of measuring points marked at the upper surface and thereby construct the most probable deflection curve of the beam.

Fig. 1. Test-pieces under the experiment by Kumagai and Itô. The left-hand test-piece is the unloaded beam and the right-hand one the center-loaded beam.

The deflection curve of the beam at the beginning is given from the theory of elasticity by $y = 1/E \cdot X(x)$, where E (dyn/cm^2) is Young's modulus. Since the deflection curve changes with time, let us assume:

$$y = T(t) X(x) \tag{1}$$

where $T(t)$ is a function of time t, having the dimensions of the reciprocal Young's modulus, with $T(0) = 1/E$. A numerical value of $T(t)$ is determined from the most probable deflection curve. Let $S(t)$ denote the sag of the middle point, where $x = l/2$. Moreover, let us denote by ϵ and ϵ_T the mean-square errors respectively of a single measurement of the vertical displacement and the adjusted value of $T(t)$.

In Fig. 2 the time-dependent change of $T(t)$, $S(t)$, ϵ_T and ϵ obtained over about 20 years are shown, the changes of room temperature and humidity being given at the bottom of the figure. Since the second year, the temperature was kept constant (23–25°C), but the humidity was left uncontrolled.

Here we have to explain that ϵ or ϵ_T includes a systematic error. The measured value of the vertical displacement does not deviate at random, but rather systematically from the most probable deflection curve for all measurements exept those for the first 50 days. Kumagai and Itô have noted that after about 50 days minute folds develop in the upper surface of beam and

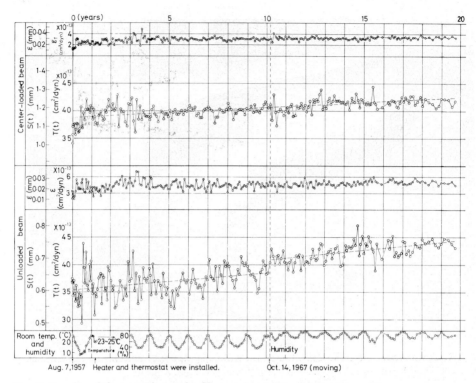

Fig. 2. Results of the experiment by Kumagai and Itô, obtained over 20 years.

gradually grow with the passage of time. Fig. 2 shows ϵ slightly increasing. Values of ϵ for the first 50 days are nearly a constant, equal to about 0.014 mm for the both beams (unloaded and center-loaded). This value of 0.014 mm is regarded merely as an accidental error of measurement. Therefore the accidental error of $T(t)$ derived from ϵ may be $1.5 \cdot 10^{-13}$ cm²/dyn for the center-loaded beam and $2.7 \cdot 10^{-13}$ cm²/dyn for the unloaded one.

Looking at the change of $T(t)$ as a whole, it tends to increase for both the beams. If we assume that $T(t)$ increases linearly, the approximate straight line becomes the broken one, as shown in Fig. 2. However, the graph has been divided in two parts because of the moving referred to above for the following two reasons: the humidities in the old and new laboratories are different and the distance (210 cm) between knife-edges before and after moving may not be equal. The approximate straight lines divided in the two parts show nearly the same inclination for each of the beams. Therefore, we may conclude that the creep rate is constant for 20 years. The maximum bending stress is 12.8 kg/cm² for the unloaded beam, which is small compared with a tensile strength of granite which is usually about 100 kg/cm². The stresses distributed in the beam are considerably smaller. If granite had some finite yield stress we could not explain the experimental results.

Thus, Kumagai and Itô have concluded that granite does flow plastically with a vanishingly small yield stress, i.e., the flow is viscous.

Accordingly, let us assume granite to be a Maxwell liquid, and $T(t)$ (eq. 1), is given by:

$$T(t) = 1/E + t/3\eta \tag{2}$$

where η is the viscosity. The value of the viscosity as obtained from the inclination of the approximate straight line in Fig. 2 is $5.7 \cdot 10^{20}$ P for the center-loaded beam and $3.2 \cdot 10^{20}$ P for the unloaded beam. Therefore, it may be concluded that the viscosity of granite is $3{-}6 \cdot 10^{20}$ P.

Looking at $T(t)$ and considering the experimental error mentioned above, the mean curve of the change of $T(t)$ should oscillate. The oscillation correlates with humidity changes in the old laboratory, but it is independent of humidity in the new laboratory where the annual change of humidity is small. Even if $T(t)$ was corrected for humidity, the mean curve of $T(t)$ would show some oscillations with relatively long periods. This phenomenon will be mentioned again later.

EXPERIMENT BY ITÔ AND SASAJIMA

Since August 1974 Itô and Sasajima (1977) have carried out more accurate creep tests by bending three granite test-pieces (each 21 × 2.5 × 2.0 cm) and three gabbro test-pieces (16. × 2.0 × 1.5 cm). The laboratory occupies a vacant adit of Kisenyama Underground Power Station of Kansai Electric Power Inc. in Uji City, Kyoto Prefecture, where the constant room temperature (18°C) and humidity (nearly 100%) is maintained naturally.

The fine-grained granite is from Aji, Kagawa Prefecture, Japan, and the fine-grained gabbro is from Sweden. The test-piece was shaped into a beam with its upper surface highly polished. The test-piece has been bent convex upwards as shown in Fig. 3; each granite test-piece has undergone the maximum bending stress, 19.5 kg/cm² (l = 20.0 cm, l_2 = 8.0 cm, W = 5.25 kg), and for each gabbro test-piece, 20.4 kg/cm² (l = 14.9 cm, l_2 = 6.0 cm, W = 3.30 kg). While measuring, an optical flat is set above the polished surface to produce interference fringes with monochromatic light (Na–D ray of wavelength 0.5893 μm), its principle being based on Newton's rings. From an analysis of the interference fringes, we can determine the profile of the upper surface with an accuracy of better than one-tenth of a wavelength.

A bending of the test-piece should be reduced to a difference between the profile at the measurement and the initial profile before loading. At present, let us represent the bending by a displacement of a middle point of the test-piece. However, in future the value of $T(t)$ for the deflection curve assumed by eq. 1 will be calculated.

In Fig. 4 the experimental results obtained over three years are shown. For the results from the granite we can draw a common mean curve, as shown by a broken line. It seems to show both a primary and a secondary

634

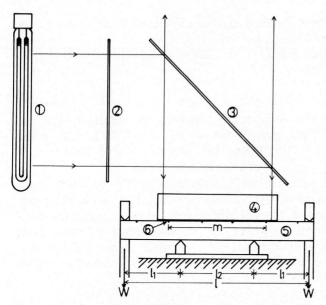

Fig. 3. Figure showing the method of the experiment by Itô and Sasajima. *1* = Sodium lamp, *2* = frosted glass, *3* = plane parallel plate of glass, *4* = optical flat with 13-cm diameter, *5* = test-piece, *6* = marks carved on a polished surface of the test-piece.

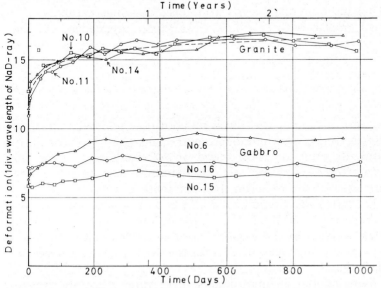

Fig. 4. Results of the experiment by Itô and Sasajima, obtained for about three years. *10*, *11* and *14* = granite test-pieces, *6*, *15* and *16* = gabbro test-pieces.

creep. The viscosity of the secondary creep is calculated to be $1 \cdot 10^{20}$ P. This value may be subject to some change, but is comparable with that found by the experiment of Kumagai and Itô. For gabbro results we cannot draw a common mean curve, but after 180 days there is some parallelism in the bending rates of the three pieces. We cannot yet determine the secondary creep. However, it can be said that gabbro does creep more slowly than granite.

In this experiment we also see, as did Kumagai and Itô, that rock creep does not steadily progress but sometimes regresses for a relatively long period. This experiment has been done in constant humidity and with the high accuracy mentioned above. Therefore, we must accept the phenomenon of regression of creep as a fact. This presents a new problem in creep studies.

VISCOSITY OF THE OCEANIC CRUST ESTIMATED FROM SUBSIDENCE OF GUYOTS AND ATOLLS

If we consider the creep of rocks over geological time, we may ignore both the regression of creep and the primary creep. Based on the suggestion by Kumagai and Itô that rock has a vanishingly small yield stress, we may employ the Maxwell liquid as a rheological model of rock. We may imagine that a viscous crust, modeled by a Maxwell liquid, floats on an underlying, more fluidal, layer in the state of isostasy. If there were no orogenic forces, mountains would flow down and trenches would rise up by intercrustal viscous flow. How would the crust flow to flatten the topography? This problem was solved (Itô, 1972a) to explain the subsidence of guyots and atolls (Itô, 1972b).

Let us consider the subsidence of a large idealized seamount whose topographical elevation $\zeta(r, t)$ has radial symmetry as shown in Fig. 5, the initial

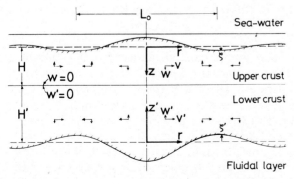

Fig. 5. Diagram to explain the decay of an elevation of topography of the viscous crust under the condition of isostasy.

elevation at $t = 0$ being given by the series:

$$\zeta(r, 0) = aJ_0(\kappa r) \tag{3}$$

where J_0 is the Bessel function. If we introduce L_0 as shown in Fig. 5 we obtain:

$$\kappa = 2 \times 3.83/L_0 \tag{4}$$

because $J_0(\kappa r)$ has its first minimum at $\kappa r = 3.83$. Here, L_0 may be approximately regarded as a diameter of the seamount bottom. From the assumption that the crust always flows in the state of isostasy, the vertical component of flow must change its direction at a certain level $z = H$. Then the crust at this level is divided into the upper crust and the lower crust. Analyzing the flow in the upper crust, we obtain the horizontal velocity $v(r, z, t)$ and the elevation $\zeta(r, t)$, both of which include unknown constants. The vertical velocity w is of less importance compared with the horizontal one, since $\zeta \ll L_0$ and $H < L_0$. Also under the assumption that the viscosity of the fluidal layer is considerably smaller than that of the crust, the analysis of the flow in the lower crust is carried out to obtain the horizontal velocity v' and the elevation of the crustal bottom ζ', both of which include other unknown constants. The above-mentioned unknown constants are determined by the condition of isostasy between ζ and ζ' and the condition that v and v' are continuous at $z = H$. Thus we obtain:

$$\zeta(r, t) = aJ_0(\kappa r) \exp\left[-\frac{(\rho - 1)gH^3\kappa^2}{3\eta}\right]t \tag{5}$$

$$H = \frac{\text{average thickness of the crust}}{1 + \sqrt[3]{\rho/(\rho' - \rho)}} \tag{6}$$

when η is viscosity of the crust and ρ and ρ' are densities of the crust and the fluidal layer (sea-water density has been taken to be 1 g/cm^3). The author (Itô 1972a) proposed a new technical term "average life of topography", which is equal to the time in which the elevation of the topography decreases to $1/e$ of the initial one. In this case the average life τ is given by:

$$\tau = 3\eta L_0^2/4 \times 3.83^2(\rho - 1)gH^3 \tag{7}$$

The time when the flat top of a guyot existed at sea level is estimated from fossils of reef fauna or age determination of basalts dredged on the top. A chronological process of subsidence of an atoll is known from deep drilling of the atoll. If the weight of sediments of reef corals is disregarded, the subsidence of an atoll may be treated in the same manner as the subsidence of a guyot. Studying three guyots (Suiko seamount, Hess guyot and Cape Johnson guyot) and four atolls (Funafuti atoll, Eniwetok atoll, Bikini atoll and Midway atoll), the results in Fig. 6 are obtained. Here $\zeta(r, t)$ is adopted as the height to a seamount relative to an ocean basin around it and $\zeta(r, 0)$ a depth of the ocean basin which may be regarded as the height of the sea-

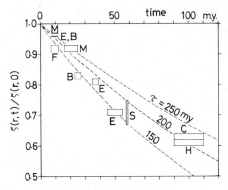

Fig. 6. Time-dependent subsidence of guyots and atolls after Itô (1972). M = Midway atoll, E = Eniwetok atoll, B = Bikini atoll, F = Funafuti atoll, S = Suiko seamount, C = Cape Johnson guyot, H = Hess guyot. In this figure the Suiko seamount has been altered by the new data of age determination by Saito and Ozima (1977).

mount at $t = 0$. At the same time, an average diameter of the seamount foot is examined to give L_0. Broken lines are theoretical curves for different τ. The atolls show a tendency of faster sinking than in the theoretical case. This is reasonable, since it is caused by the weight of sediments of reef corals. Thus the average life of a seamount with $L_0 = 100$ km would be 200—250 m.y.

In this case the crust is one floating on the fluidal layer and may be equated with the lithosphere. Assuming a thickness of the oceanic crust to be 50—75 km and also $\rho = 3.1$ and $\rho' = 3.5$ g/cm^3, its viscosity is obtained from eqs. 6 and 7 to be 2—$7 \cdot 10^{25}$ P. Let us adopt 10^{25-26} P for it.

VISCOSITY OF THE OROGENIC CRUST ESTIMATED FROM QUATERNARY CRUSTAL MOVEMENT IN HIMALAYA AND IN SOUTHWEST JAPAN

Yamashita (1971) discussed the geological process of the uplift of the Himalayas. The author (Itô 1971) considered that the crust of the Himalayas has been compressed laterally to form the mountain range and in doing so the Himalayas have been uplifted. In Fig. 7 a sectional profile is shown along the transverse valley of the Kali Gandaki River which cuts the Himalaya Mts. between Mt. Daulagiri and Mt. Annapurna (after Yamashita, 1971, Fig. 4). Here the upper and lower lines represent the profile of mountains and hills on both sides of the transverse valley and that of the river bed. Let us take the mean height of the Great Himalaya and Tibet zones to be 5000 m. These zones are estimated geologically to have occupied the mean height 1500 m at the beginning of the Quaternary. In Fig. 7 there is shown schematically the crust of these zones at the present ($ABCD$) and that at the beginning of the Quaternary ($A'B'C'D'$). The Himalayas are a birthplace of the theory of isostasy. If the crust has been deformed isostatically, the

638

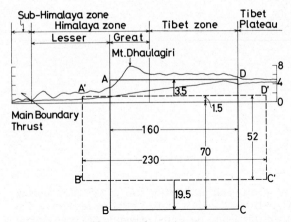

Fig. 7. Schematical diagram to explain the uplift of the Himalayas caused by lateral compression of the crust.

thickness of the crust has increased 3.5 + 19.5 km during the Quaternary, assuming that the mean density of the crust is 2.8 g/cm^3 and that of the upper mantle is 3.3 g/cm^3. Assuming the depth of Moho-discontinuity to be 70 km, the thickness of $A'B'$ would have been 52 km. Let us adopt a mean width of 160 km for the Great Himalaya and Tibet zones. Since there is no Quaternary volcanic activity in the Himalayas, the addition of materials from the mantle to the crust has not taken place (i.e., the volume of the crust would have been kept nearly constant during the Quaternary). Hence the width $A'D'$ would have been initially 230 km. Thus the Great Himalaya and Tibet zones would have contracted laterally from 230 km to 160 km during 2 m.y. Therefore, we obtain a mean horizontal strain rate of:

$$\dot{e}_n = 5 \cdot 10^{-15}/\text{sec} \tag{8}$$

This value equals an ordinary geodetic strain rate.

Between the strain rate $\dot{e}_n$, the tectonic stress p_n and the viscosity of the crust η, there is a simple relation:

$$p_n = 3\eta\dot{e}_n \tag{9}$$

If the tectonic stress were about 100 bar, the viscosity should have been about 10^{22} P.

In Kinki and Chubu (southwest Japan) are distributed abundant active faults cutting the pre-Neogene basement rocks. The strike-slip faults consist of conjugate sets of NW—SE (left-lateral) and NE—SW (right-lateral) trends, and the faults of N—S trend are thrust type. Such a fault system indicates that the areas have been under a tectonic stress state of horizontal compression in the E—W direction. This indication is also supported by seismological and geodetic data. Recently, in-situ rock stress measurements have revealed that rock stresses are closely connected with tectonic stress (Hiramatsu et al., 1973; Itô et al., 1976; Oka et al., 1977), it is the tectonic stress only that has caused the Quaternary crustal movements of these areas. Based on neotectonical studies of southwest Japan (Huzita 1969), the author and Huzita (Itô and Huzita, 1974) carried out the same analysis for Himalaya and obtained a strain rate of $\dot{e}_n = 5 \cdot 10^{-15}$/sec. The recent tectonic stress in these areas is estimated from the in-situ rock stress measurements and stress drops (releases) of earthquake faults to be about 100 bar. Thus, we again obtain a viscosity of about 10^{22} P.

This viscosity (10^{22} P) obtained for the orogenic crust is significantly smaller than the 10^{25-26} P viscosity for the oceanic crust obtained in the preceding section. The author cannot as yet estimate the viscosity of the cratonic crust as he has not the appropriate geological data necessary. However, it is thought that it may be greater than the viscosity of the orogenic crust, because, for example, the Indian craton being adjacent to the Himalayas has not been deformed, although it would also have been compressed laterally as were the Himalayas. Therefore the orogenic belts would be mobile when compared with cratons and ocean floors.

The viscosity of granite, $3-6 \cdot 10^{20}$ P, obtained by the experiment of Kumagai and Itô may be somewhat different from that under a natural condition, but it is comparable with the 10^{22} P viscosity estimated for the orogenic belts which mainly consist of granitic rocks. The experimental confirmation by Itô and Sasajima that gabbro creeps more slowly than does granite, is consistent with the above estimation that the oceanic crust consisting of basic rocks has a viscosity larger than that of the orogenic crust.

STRESS–STRAIN DIAGRAM OF THE CRUST – CONCLUDING DISCUSSION

As already mentioned, the author has considered the viscous crust modeled by a Maxwell liquid. The rheological equation of a Maxwell liquid is given by:

$$\dot{e}_t = \dot{p}_t/\mu + p_t/\eta \tag{10}$$

where e_t is the shearing strain, p_t the shearing stress and μ the rigidity. Let us make the assumption of a constant strain rate:

$$\dot{e}_t = c \text{ (const.)} \quad \text{or} \quad e_t = ct \tag{11}, (11')$$

and an initial condition:

$$(p_t)_{t=0} = 0 \tag{12}$$

Solving eq. 10 subject to eqs. 11 and 12, we obtain:

$$p_t = \eta c \{1 - \exp(-\mu/\eta \cdot t)\} \tag{13}$$

The pair of eqs. 13 and 11' represents a relation between stress p_t and strain e_t with a parameter of time, t (Itô, 1970). Therefore a stress–strain diagram can be drawn, if the viscosity, the rigidity and the strain rate c are known.

640

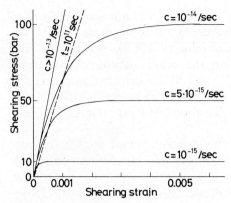

Fig. 8. Stress—strain diagrams of the orogenic crust for different strain rates.

For the orogenic crust we employ:

$$\eta = 10^{22} \text{ P}, \quad \mu = 10^{11} \text{ dyn/cm}^2 \quad \text{and} \quad \eta/\mu = 10^{11} \text{ sec} \tag{14}$$

where η/μ is a relaxation time of the Maxwell liquid. The static rigidity μ has been estimated as a little smaller than that seismologically accepted. Thus, in Fig. 8 the stress—strain diagrams of the orogenic crust are drawn for different strain rates. This figure suggests that if the shearing strength of the crust were greater than 10 bar and the strain rate were smaller than 10^{-15}/sec, the crust would flow without fracture.

If t is much smaller than the relaxation time, the pair of equations (13) and (11′) becomes:

$$p_t = \mu c t = \mu e_t \quad (t \ll \eta/\mu). \tag{15}$$

This is the equation of Hooke on solids. If t is much larger than the relaxation time, then:

$$p_t = \eta c = \eta \dot{e}_t \quad (t \gg \eta/\mu) \tag{16}$$

This is the equation of Newtonian liquid, which corresponds to the part shown by a straight line parallel to the abscissa for each curve in Fig. 8.

When t is equal to the relaxation time, we have the relation:

$$p_t = (1 - e^{-1}) \mu e_t \quad (t = \eta/\mu) \tag{17}$$

which is shown by the broken straight line in Fig. 8. Therefore, Fig. 8 also suggests that the crust would behave elastically for times shorter than the relaxation time (approx. 3000 years) and would flow as the Newtonian liquid for longer times. And although the strength of the crust is different from place to place, if the crust were fractured, the fracture would have occurred within a period comparable to 3000 years.

REFERENCES

Griggs, D., 1940. Experimental flow of rocks under conditions favoring recrystallization. Bull. Geol. Soc. Am., 51: 1001—1022.

Hiramatsu, Y., Oka, Y., Itô, H. and Tanaka, Y., 1974. The correlation of the rock stress measured in situ and the tectonic stress inferred from geological and geophysical studies. J. Soc. Mater. Sci. Jap., 23: 380—386 (in Japanese, with English abstract).

Huzita, K., 1969. Tectonic development of southwest Japan in the Quaternary period. J. Geosci. Osaka City Univ., 12: 53—70.

Itô, H., 1970. Imaginary experiments by a testing machine with geologic strain rate. The Shizen Kagaku Ronso (J. Kyoto Women's Univ.), 2: 15—22 (in Japanese).

Itô, H., 1971. Viscosity of the earth's crust in an orogenic belt estimated from the uplift of Himalayas. Autumn Meet. Seismol. Soc. Japan, Prepr., p. 53. (in Japanese).

Itô, H., 1972a. Change of topography due to flow of the crust under the condition of isostasy. J. Geol. Soc. Jap., 78: 29—38 (in Japanese, with English abstract).

Itô, H., 1972b. On the subsidence of guyots and atolls. Tsukumo Earth Sci., 7: 37—54 (in Japanese).

Itô, H. and Huzita, K., 1974. The flow of the earth's crust considered from the Quaternary crustal movements in Southwest Japan. Rock Mech. Jap., 2: 181—183.

Itô, H. and Sasajima, S., 1977. Creep tests of rocks measured by making use of interference fringes of light — experimental results obtained for the first 2 years. Proc. 5th Natl. Symp. Rock Mech., Jap., pp. 49—54 (in Japanese, with English abstract).

Itô, H., Oka, Y. and Huzita, K., 1976. Shrinking islands of Japan — rheology of the earth's crust from the viewpoints of laboratory experiments, in-situ measurements and field geology. Kagaku, 46: 745—754 (in Japanese).

Kumagai, N. and Itô, H., 1965. Method to find secular bending of big granite beams and results obtained for the first seven years. J. Soc. Mater. Sci. Jap., 14: 507—519 (in Japanese, with English abstract).

Kumagai, N. and Itô, H., 1968. Results of experiments of secular bending of big granite beams extending for 10 years and their analyses. J. Soc. Mater. Sci. Jap., 17: 925—932 (in Japanese, with English abstract).

Kumagai, N. and Itô, H., 1970. Creep of granite observed in a laboratory for 10 years. In: S. Onogi (Editor), Proc. 5th Int. Congr. Rheol., 2: 579—590.

Kumagai, N. and Itô, H., 1971. The experimental study of secular bending of big granite beams for a period of 13 years with correction for change in humidity. J. Soc. Mater. Sci. Jap., 20: 185—189 (in Japanese with English abstract).

Oka, Y., Hiramatsu, Y. and Kameoka, Y., 1977. Investigations into the method of determining rock stress by the stress relief technique accompanied by applications of the method obtained. Proc. 5th Natl. Symp. on Rock Mech., Jap., pp. 187—192 (in Japanese, with English abstract).

Saito, K. and Ozima, M., 1977. ^{40}Ar—^{39}Ar geochronological studies on submarine rocks from the western Pacific area. Earth Planet. Sci. Lett., 33: 353—369.

Yamashita, N., 1971. Development of the Himalayas, Kagaku, 41: 221—230 (in Japanese).

Tectonophysica, 52 (1979) 643
© Elsevier Scientific Publishing Company, Amsterdam — Printed in The Netherlands

MONITORING MASSIVE FRACTURE GROWTH AT 2-KM DEPTHS USING SURFACE TILTMETER ARRAYS

M.D. WOOD

U.S. Geological Survey, 345 Middlefield Road, Menlo Park, Calif. 94025 (U.S.A.)

(Accepted for publication June 8, 1978)

ABSTRACT

Tilt due to massive hydraulic fractures induced in sedimentary rocks at depths of up to 2.2 km have been recorded by surface tiltmeters. Injection of fluid volumes up to $4 \cdot 10^5$ liters and masses of propping agent up to $5 \cdot 10^5$ kg is designed to produce fractures approximately 1 km long, 50—100 m high and about 1 cm wide. The surface tilt data adequately fit a dislocation model of a tensional fault in a half-space. Theoretical and observational results indicate that maximum tilt occurs at a distance off the strike of the fracture equivalent to 0.4 of the depth to the fracture. Azimuth and extent of the fracture deduced from the geometry of the tilt field agree with other kinds of geophysical measurements. Detailed correlation of the tilt signatures with pumping parameters (pressure, rate, volume, mass) have provided details on asymmetry in geometry and growth rate. Whereas amplitude variations in tilt vary inversely with the square of the depth, changes in flow rate or pressure gradient can produce a cubic change in width. These studies offer a large-scale experimental approach to the study of problems involving fracturing, mass transport, and dilatancy processes.

Tectonophysics, 52 (1979) 644—645

CONTRIBUTION TO EARTHQUAKE PREDICTION BY THE DATA OF RECENT CRUSTAL MOVEMENT ANOMALIES

A.A. NIKONOV

Institute of Physics of the Earth, USSR Academy of Sciences, Moscow 123242 (U.S.S.R.)

(Accepted for publication June 8, 1978)

ABSTRACT

The changes in velocity and sign of Recent crustal movements in the epicentral zones of earthquakes have been revealed in different seismically active areas (Japan, California, and Middle Asia) from sea-level observations and releveling data (Mescherikov, 1973; Rikitake, 1969; Hofmann, 1970; Lensen, 1971; Nikonov, 1971; Fujii, 1974; Wyss, 1976).

Earthquakes are often preceded by movement acceleration in the epicentral zones; the greater the earthquake and the lesser the epicentral distance, the larger is the amplitude of altitude changes (from several centimeters to a few meters); the period of the movement anomaly before the earthquake ranges from some hours to tens of years, in proportion to the earthquake intensity.

Four main phases of the movements can be found: secular or background movements, the movements preceding the earthquake, and movements accompanying and following it. The phase of the preceding movements is the most interesting from the point of view of earthquake prediction.

After a retrospective analysis of the available data and in elaboration of the scientific works begun in Japan, the author improved some relations between the anomalous movements duration in years (t) and the corresponding earthquake magnitude (M).

The relation for the Pacific seismic belt (P) from 18 events is $\log t_{\mathrm{P}} = 0.57M - 3.16$. The relation for the Mediterranean—Central Asiatic seismic belt (M—CA) is $\log t_{\mathrm{M-CA}} = 0.56M - 2.88$. They are seen to be similar.

The author has attempted eartquake prediction from some geodetic data (Castle et al., 1976) in the territory of Central California. If the surface bulge which has been in progress since 1959 precedes an earthquake, the earthquake magnitude is to be evaluated at 7.7 for 1977. An earthquake with $M \simeq 8$ (such as the great earthquake of 1857) can be expected by 1985. Therefore, the magnitude of this forthcoming earthquake in the territory of Central California can be expected to range within 7.5—8, i.e., it will be very strong.

REFERENCES

Castle, R.O., Church, J.P. and Elliott, M.R., 1976. Aseismic elevation in South California. Science, 192 (4236): 127—131.

Fujii, Y., 1974. Relation between duration period of the precursory crustal movement and magnitude of the earthquake. J. Seismol. Soc. Jpn., 27(3): 197—214.

Hofmann, R.B., 1970. Earthquake prediction from fault movement and strain precursors in California. In: Earthquake Displacement Fields and Rotation of the Earth. Reidel, Dordrecht, pp. 234—245.

Lensen, G.J., 1971. Phases nature and rates of Earth deformation. In: Recent Crustal Movements. Bull. R. Soc. N. Z., 9: 97—105.

Mescherikov, Yu., A., 1973. On the crustal movements as forerunners of earthquakes. In: The Earth's Crust of Seismically Active Zones. Upper Mantle, 11. Nauka, Moscow (in Russian).

Nikonov, A.A., 1971. On recent vertical movements of the Earth's crust in seismically active areas of Middle Asia. Tectonophysics, 12(2): 119—127.

Rikitake, T., 1969. An approach to prediction of magnitude and occurrence time of earthquakes. Tectonophysics, 8(1): 81—95.

Wyss, M., 1976. Local changes of sea level before large earthquakes in South America. Bull. Seismol. Soc. Am., 66(3): 903—914.

Tectonophysics, 52 (1979) 647—663
© Elsevier Scientific Publishing Company, Amsterdam — Printed in The Netherlands

MODELING THE LOCAL STRESS FIELD AND KINEMATICS OF THE SAN ANDREAS FAULT SYSTEM

D.N. OSOKINA, A.A. NIKONOV and N.YU. TSVETKOVA

Institute of Physics of the Earth, Academy of Sciences, Moscow 123242 (U.S.S.R.)

(Accepted for publication June 8, 1978)

ABSTRACT

Osokina, D.N., Nikonov, A.A. and Tsvetkova, N.Yu., 1979. Modeling the local stress field and kinematics of the San Andreas fault system. In: C.A. Whitten, R. Green and B.K. Meade (Editors), Recent Crustal Movements. 1977. Tectonophysics, 52: 647—663.

The present paper is a continuation of the previous work on modeling the local stress field induced by the San Andreas fault system (Nikonov et al., 1975). This system has been simulated on plane elastic models made of optically sensitive material, the models being under homogeneous uniaxial compression. The photoelastic method has been used to study the redistribution of τ_{max} around the fault system with sides closed under compression.

Three main features emerge in the kinematics of fault-system modeling. The first is a peculiar distortion of an originally rectangular grid, reflecting right-lateral movements on the San Andreas fault. This is especially noticeable in its central part. The second is the appearance and spreading of tear breaks near the ends of the zone nearly normal to the strike of the ends of the master fault. The third feature is separation of fault wings in certain sections of the San Andreas fault in the model. All these features are in general correspondence with the phenomena actually observed in the San Andreas fault system.

INTRODUCTION

The San Andreas fault system in California is a phenomenon that excites the continued interest of geologists and geophysicists worldwide, in view of its great extension, peculiarity of structure, and high tectonic and seismic activity. There are several features of this zone that combine to make it a particularly suitable object on which to model local stress fields by the method developed in the Tectonophysical Laboratory, Institute of Physics of the Earth, Moscow (Gzovsky, 1975; Osokina et al., 1976a, b). Those features are as follows:

(1) The fault system and the character of fault movements there are very well known, as also is the crustal structure in California.

(2) The fault plane in the main faults is vertical.

(3) Displacement vectors are horizontal.

(4) The character of the regional stress field is favorable for model studies.

Some preliminary results in model studies on the local τ_{max} field of the San Andreas fault system have been published (Nikonov et al., 1975).

TECTONIC FEATURES, RECENT MOVEMENTS AND REGIONAL STRESS FIELD OF THE SAN ANDREAS FAULT SYSTEM

The San Andreas fault system extends in a NW—SE direction for over 900 km. In the northern part it is represented by a single major fault with a few branchings and some small faults accompanying it, whereas the southern part consists of a set of parallel faults and faults inclined to the general strike of the zone, all about the same size, lying in a strip 100 km or more in width.

The San Andreas fault proper is a zone ranging between a few meters and several kilometers in width, and with a dip of $90°$. Estimates of depths to which the fault penetrates the crust, based on earthquake hypocenter depths, are 10—15 km, reaching 25 km for the end portions of the fault. All of the transverse and some of the longitudinal side faults show an inclination of the fault plane of 30—$45°$, rarely less, the angle of inclination increasing as we go away from the Transverse Ranges and down the fault plane.

American scientists consider the San Andreas fault system as a present conservative boundary of lithospheric plates. According to these concepts (Dickinson and Grantz, 1968; Silver, 1971; Elders et al., 1972; Kovach and Nur, 1973), the San Andreas is a giant transform fault along which during the last several million years the Pacific Plate is being displaced with respect to the North American Plate towards the northwest (Fig. 1) Evidence for this has recently been summarized in Nikonov (1975).

The amounts and rates of vertical movements on the fault are several orders of magnitude less than the corresponding values for horizontal, strike-slip displacements. The same is true with regard to recent movements.

Recent horizontal movements in the San Andreas fault system have been studied by different methods over a great many years (Burford, 1966; Pope et al., 1966; Howard, 1968; Hofmann, 1970; Whitten, 1970; Meade, 1971; Savage and Burford, 1973; Nikonova and Nikonov, 1973; Nikonov, 1977). The different instrumental observations as well as geological data, all point to a continuing right-lateral movement, both along the main San Andreas fault and the associated longitudinal feather faults. The rate of present-day displacement at the walls measured on separate faults is 1—3 cm yr^{-1}. When estimated across the whole fault zone 100—200 km wide (in the southern part), it may reach 5—8 cm yr^{-1}.

An important feature of recent strike-slip movements is that they are not uniform in time and along the fault length (Fig. 2), and also their peculiar connection with seismic activity. The movement is continuous or is punctuated by short stops over areas of continuous seismicity, while areas of sporadic seismicity may show no movement over many decades (Allen, 1968;

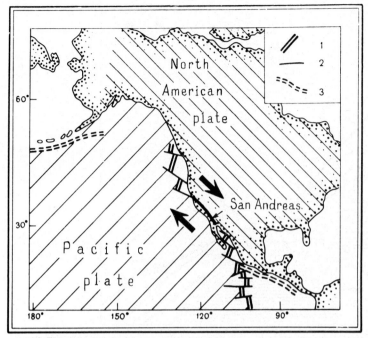

Fig. 1. Position of the San Andreas fault system at the junction of the Pacific and North American plates. *1* = mid-ocean ridge, *2* = transform faults, *3* = deep-ocean trenches.

Richter, 1971). Fig. 3 shows a classification of the San Andreas fault system into areas of continuous seismicity (I, III, V) and those of sporadic seismicity (II, IV) after Allen (1968).

The directions of axes of the present regional stress field in California can be determined using seismological, geodetic, and strainmeter data.

Focal mechanism studies of large Californian earthquakes have shown that the regional compressional axes are meridional (Ritsema, 1961; Hodson, 1962; Scheidegger, 1965). Seismological data over parts of the Transverse Ranges have established submeridional compression there too (Richter et al., 1958, Elsworth et al., 1973, Whitcomb et al., 1973), as also for the Gorda Escarpment (Seeber et al., 1970, Silver, 1971). The same method has been used to find a northwestern (290—305°) regional horizontal crustal extension in parts of Nevada adjoining California (Ryall and Malone, 1971). A NW—SE extension has also been found from seismological data for the rift parts of submarine elevations in the Pacific north and south of the San Andreas fault (Bolt et al., 1968; Elders et al., 1972).

Secular strain observations made with three-component horizontal strainmeters at the Stone Canyon Observatory (30 km south of Hollister) and at several other points, southern Nevada included (Smith and Kind, 1972; Bufe et al., 1973), show that the major axes of the strain ellipses have a NW—SE

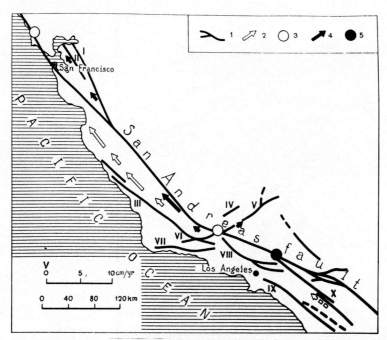

Fig. 2. Recent horizontal movements on the San Andreas fault system (after Hofmann, 1970). *1* = active faults (denoted by Roman numerals): *I* — Calaveras, *II* — Hayward, *III* — Nacimento, *IV* — White Wolf, *V* — Garlock, *VI* — Big Pine, *VII* — Santa Ines, *VIII* — San Gabriel, *IX* — Englewood, *X* — San Jacinto; *2* = mean displacements on the San Andreas fault for the years 1959—1968, *3* = portions of the fault with no displacements during this period; *4* = mean displacements on the San Andreas fault for the period 1969—1971; *5* = portions of the fault with no displacements for this period.

direction in nearly all cases. The WNW direction of the major axis of the strain ellipse found at the Stone Canyon Observatory supports a submeridional compression and a sublatitudinal tension (Bufe et al., 1973).

Geodetic data (Whitten, 1970; Meade, 1971; Savage and Burford, 1973) concerning right-lateral strike-slip movement on NW—SE faults and a left-lateral one on NE—SW faults indicate submeridional compression or sublatitudinal extension. Submeridional compression in the Transverse Ranges area is indicated by geodetic observations of thrust movements on sublatitudinal reverse faults (Wilt, 1958; Jungels and Frazier, 1973). Precise triangulation work in the San Andreas fault zone north of the Transverse Ranges has determined the direction of greatest compression for the years 1932—59 as N—S with a NNW deviation (Howard, 1968), indicating meridional compression and latitudinal extention at the boundary between the San Andreas and Garlock faults (Meade, 1971). An analysis of deformations occurring during 1930/32—1951 in the triangulation networks that cut across the fault zone as strips 70 km long near Hollister and 110 km near Cholame, has revealed (Burford, 1966) a regional compression

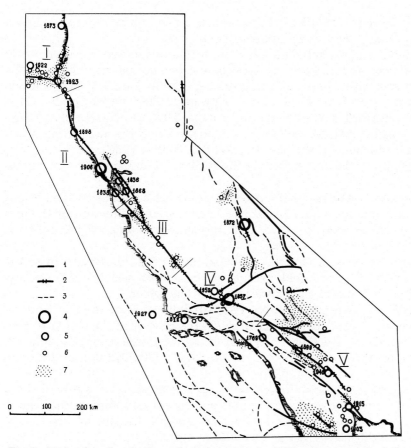

Fig. 3. Map of major faults and earthquakes in California. *1—3* = faults: *1* — faults of Nevada and the San Andreas system that were active in late Cenozoic era; *2* — the same, active at the time of known earthquakes; *3* — other faults. *4—7* = earthquake epicenters (the numerals denote the year of occurrence): $4 - M \geqslant 8$; $5 - 7 < M < 8$; $6 - 6 < M < 7$; *7* — areas with numerous small earthquakes. *I—V* = portions of the San Andreas fault system which have different levels of seismicity.

of the order of 10^{-4} with azimuth NE35° that becomes meridional in a strip 15—20 km from the fault.

Geodimeter observations in 1959—1969 also demonstrated meridional compression and latitudinal extension north of the Garlock fault, and a similar compression without the corresponding extension further southeast, between San Bernardino and Palm Springs (Hofmann, 1970). The axis of greatest compression is NE13° over the northern of the two areas mentioned and NW7° over the southeastern (Savage et al., 1973). Changes in the lengths of a number of differently directed lines observed in northern Central California during 1959—1970 show that the σ_3-axis has a NE11°—18° direction (Scholz and Fitch, 1970; Savage and Burford, 1973). For southern California

Savage and Prescott (1976) establish the existence of a meridional compression and possibly a very weak latitudinal extension.

The general regional submeridional compression is relived in places — near faults in a strip of 1—2 km — by local extensions normal or oblique to the fault line; such are observed, for instance, around Parkfield and the San Andreas Lake (see Pope et al., 1966; Howard, 1968; Burford et al., 1969; Raleigh and Burford, 1969; Cherry and Savage, 1972). Such areas may correspond to longitudinal depressions and troughs along the fault zone which have given it the name of the San Andreas Rift (Willis, 1938).

Our general conclusions regarding the stress field may now be summed up as follows:

(1) The present-day regional stress field is characterized by the conditions of active compression and a horizontal submeridional direction of the axis of greatest compression, σ_3. The most probable direction of the σ_3-axis is NNE 0—15°.

(2) There are areas where the principal axes of stress deviate from the direction of the regional field axes; such areas exist near the San Andreas fault itself as well as near some other faults in the system.

(3) A NW—SE extension is observed near both ends of the San Andreas fault system in the rifts of mid-ocean ridges (northwest of the Gorda Escarpment and south of the Baja California extremity), in a transition zone (in the trough of the Salton Sea and in the Gulf of California), and also in Nevada.

These conclusions about the character of the present-day stress field are in agreement with the geologic features and recent structure of the region, thus indicating continuity in its development and a constancy of direction of tectonic forces during the last several million years.

We can regard changes in the orientation of stress axes near the fault itself, as well as the presence of areas of tensile stresses directed along the fault system (around its ends), as a local phenomenon, being due to the regional field being disturbed by a major strike-slip fault, as described by Osokina et al. (1976a, b).

Unfortunately the results of nearly all papers mentioned permit no unique conclusions to be made concerning the sign of the second principal stress lying in the horizontal plane. A latitudinal or a NWW extension may be found both for uniaxial and biaxial compression, provided in latter case that the sublatitudinal compression (σ_1) is smaller than the submeridional compression (σ_3) by a factor of 4.

Measurements in mines and tectonophysical investigations (Hast, 1969; Kropotkin et al., 1973; Gzovsky, 1975; Gushchenko et al., 1977) relating mostly to Eurasia, lead to the conclusion that, except in the case of rift zones, all three principal stresses are as rule compressive. This permit us to suppose that the second principal horizontal stress of the regional field, external with regard to the San Andreas fault system, must be either compressive or nearly zero. It should be noted that Willis (1938) and Moody and

Hill (1960) consider this stress to be the algebraically greatest principal stress, i.e., σ_1.

FAULT SCHEME AND THE EXPERIMENTATION CONDITIONS

The San Andreas fault system that has been simulated in our experiments is that shown in Fig. 3. The fault map of California given in Dickinson and Grantz (1968) has been taken as a basis. For the purposes of model imitation some simplifications were necessary — some faults have been united, a number of short ones discarded, and all faults have been simulated by vertical cuts.

In accordance with the preceding discussion the regional stress field has been assumed to correspond to the conditions of active horizontal compression, the axis of greatest compression, σ_3, having a meridional or submeridional orientation. The second principal stress in the horizontal plane has been taken to be zero. From the experiments it appears that adding a second principal compressive stress σ_1 two to three times less than σ_3 would not produce any significant changes in the local τ_{max} field over the fault zone.

The original stress field in an undisturbed model corresponds to the action on it of forces that are constant in time, so that a redistribution of that field also gives a static picture of the local field. However, the crustal area under investigation is a boundary between two continuously moving blocks or lithospheric plates, which apparently corresponds to a constant (on the average) velocity of movement. This must lead to accumulation of strain in time accompanied by stress changes. From time to time local disturbances of the process occur, observed as slips along already existing faults, or formation of new ones. The disturbances are accompanied by considerable displacements and seismic vibrations following the release of energy. Thus, the process of crustal deformation can roughly be represented as a stable creep process that is replaced, locally in time and space, by dynamic phenomena, and then again for a span of time goes on statically. As we are not able to construct an adequate model of the natural process, we must limit ourselves to static model investigations only. The static picture obtained on the model may be regarded as a stage in or a single still taken of the dynamical process for a system of faults close to the natural one.

Models have been studied for two orientations of the σ_3-axis, a meridional ($\varphi = 0$) and a NNE one (azimuth $11°$, $\varphi = 11°$). In addition to this, to study the effect of the orientation of the σ_3-axis on stress distribution we took the extreme possible position of the σ_3-axis, that with the N35°E azimuth ($\varphi = 35°$).

A homogeneous uniaxial compression was created in the middle portions of undisturbed models. The stresses were increased until a sensitive colour appeared in the model. Call this value $\tau_{max}^°$. After this, vertical cuts were made one by one in the model. The cut simulating the San Andreas fault proper was made at the first stage, the Garlock with its western continuation at the second stage; the stages following simulated successively the remaining faults. At each stage we recorded the isochrome pattern arising in the model and the displacements of fault sides, and in some cases the distortions produced in a grid which had been made upon the surface of the model prior to the experiment, and the growth of tear faults appearing around the ends of the fault system.

DISPLACEMENTS IN THE FAULT-ZONE MODEL

Models have been studied in two ways. The first was to reproduce the fault zone to a scale of $1 : 6.0 \cdot 10^6$. A rectangular grid was marked upon the surface of an unstrained model. Taking a photo of the loaded model, we could observe the distortion of the grid and the displacements of its lines along the fault zone. Under small stresses ($\tau_{max} < 0.5\,\tau^\circ_{max}$), displacements on fractures imitating the faults are not observed visually, while at the normal stress level ($\tau_{max} = \tau^\circ_{max}$), the grid is distorted over the whole of the fault zone (Fig. 4). "Latitude" lines are curved even north of the "Pt. Mendosino". * The greatest displacements have been observed on the fault along the line "Cholame—San Francisco", their magnitude reaching the maximum around "Parkfield" (5—6 mm). South of the "Garlock" there is curvature to be observed in the "latitude" lines, but displacements along separate faults are not noticeable visually. The curvature is distributed over a wide band (about 100 km wide in nature) west of the "San Andreas fault". The greatest curvature is observed around a line passing south of the "Garlock fault" (Fig. 4).

We see that the same segment of the San Andreas fault where significant displacements are observed in the model, a single nearly straight line extending from the Garlock fault up to San Francisco Bay, is also one where active recent movements are observed. South of the "Garlock fault", in the model as well as in nature, deformations occur over a wide zone, and displacements along separate faults are not observed. Thus, areas having the greatest displacements, on the fracture in the model and on the fault in nature, have approximately similar positions. It should be noted that fault displacements in nature are determined by elastic, plastic, viscous, and quasi-viscous (displacements along small fractures) deformation of blocks, while in the model, by elastic deformation alone. Therefore, comparison of natural and modeled displacements can be qualitative only.

In the second type of investigation mentioned above, the San Andreas fault system is modeled to a large scale ($1 : 4.9 \cdot 10^6$). In that case the strike-directed tensile local stress σ_1 over tension sectors around the ends of the zone exceeded the strength of the material, leading to tear breaks developing

---

* To avoid repetition, inverted commas are used when referring to parts in the model corresponding to areas, faults, and directions in nature.

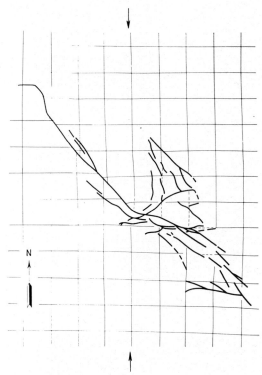

Fig. 4. Distortions of a rectangular grid upon the surface of the model after applying uni-axial horizontal "meridional" compression, and displacements along fractures simulating the San Andreas fault system.

around the ends of the zone. Also, in some portions of the fracture, there appeared a separation of the fault sides, referred to below as fracture opening. Fig. 5 shows the separate stages of the model experiment mentioned above and the tear breaks and fracture-opening portions.

In stage I a cut simulating the part of the San Andreas fault between Pt. Gorda and Garlock was made. There appeared an area of strong concentration of τ_{max} at both ends of the fault, and one of strong decrease along the fault. In stage II the "San Andreas" fault was cut towards its "southern" end. The maximum stress moved to this end, and virtually at this moment a tear break began to develop about this end of the fault and perpendicularly to it. Two fracture-opening portions on the "San Andreas" appeared — one along a "submeridional" part near the "northern" end, between "Pt. Mendocino" and Pt. Arena ("A"), and the other along a portion where the strike is changed in the "southern" part of the "San Andreas" ("D") (Fig. 5a).

In stage III cuts were made simulating the Garlock fault and its western continuation, Big Pine. The maxima of τ_{max} around the ends of the "San Andreas" fault now increased considerably, and the tear break at the "south-

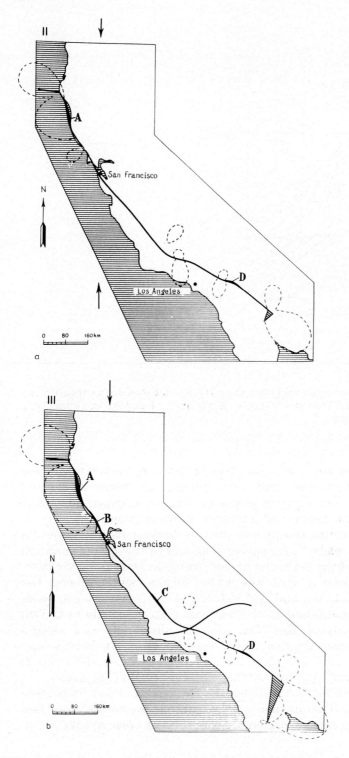

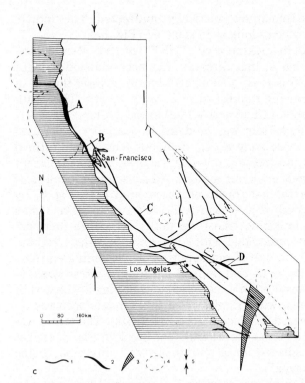

Fig. 5. Portions with fracture opening and the formation of tear breaks in a model simulating the San Andreas fault system. a—c. Different stages (II, III, V) in modelling the faults of the San Andreas system. Shown here are the appearance of fracture-opening segments and development of tear breaks around the ends of the fault zone; 1 = faults; 2 = fracture opening in portions of the master fault; 3 = tear breaks which begin to grow at the ends of the fault zone; 4 = lines marking areas of higher τ_{max}; 5 = σ_3-axis of the regional stress field.

eastern" end of the "San Andreas" doubled in length. Simultaneously two more fracture-opening zones appeared, near "San Francisco Bay" (B) and "north" of "Cholame" (C) (Fig. 5b).

In stage IV we reproduced all longitudinal faults of the San Andreas fault system and the adjoining parts of Nevada that lie north of the Garlock. This has led to a further increase of the end maxima; the tear break at the "south-eastern" end nearly doubled in length compared with stage III, and it became apparent that it bends downward ("southward"). In addition, a new tear break appeared at the "northwestern" end of the "San Andreas". Since this portion of the fault is "latitudinal", the tear break near its end which developed normal to it, turned out to be "meridional". The "A" zone of fracture opening gaped wider, the rest remaining the same.

In stage V all the remaining faults south of the "Garlock" were added. This resulted in all the fracture-opening zones on the "San Andreas" and its "southeastern" tear break remaining practically unaffected. The "northwestern" tear break became twice as long as in stage IV (Fig. 5c).

Finally, stage VI consisted in making a cut "NNE" of "Pt. Mendocino" along the coast. A fault striking in that direction has been claimed to exist there by some geologists. As this is an area of high tensile stresses the cut at once became a tear break, growing rapidly to 5—7 cm in length. This result shows that there is little likelihood of an active fault existing there.

When the San Andreas fault system was modeled to a smaller scale, used in most of our experiments, there were no spontaneous tear breaks formed at the ends of the zone. Also, no fracture opening has been observed, except in zone A, which could be seen in all our models. All subsequent small-scale models had the fault zone extended southward compared with Fig. 5, by attaching to the "southern" end of the "San Andreas" a set of three small fractures parallel to it, thus imitating transform faults. In nature there are displaced portions of the midocean ridge with a rift zone. In order to stimulate the growth of a tear break, in one of our experiments we simulated these transverse parts of the rifts by cuts, and applied an original stress twice as great as the normal one. This resulted in a tear break developing at the end of the last of these three fractures, while significant displacements occurred on all, and gaping openings appeared in place of the cuts (rifts). The tear break had the same strike and shape as the "southern" tear break in Fig. 5, except that it was considerably displaced in a "southeastern" direction, being situated in the "Gulf of California".

We thus see that the position of a tear break is determined by that of an end of the active portion of a fault system. Development of a tear break near the southern end of the fault system in the model corresponds in position and shape to the large natural tension structure of the Gulf of California (Kovach and Nur, 1973; Elders et al., 1975). The "meridional" tear break near the "northwestern" end of the fault zone in the model may correspond to the NNE middle rift within the Gorda Escarpment in nature. Thus, both of the large tear breaks near the ends of the fault zone in the model can be compared with tension (and sinking) structures in nature.

To make a similar comparison possible for intermediate fracture-opening portions of the fault, one must first establish the existence and positions of young longitudinal grabens near the fault. The combined geomorphic, geological, and geophysical data make it clear that narrow longitudinal grabens along the San Andreas fault zone do in fact exist, although in each case they do not have a morphological expression, and their existence is proved with varying degrees of confidence.

Geomorphic and geological data indicate with fair certainty the existence of narrow depressions along the San Andreas fault line south and north of San Francisco Bay (with San Andreas Lake and Tomales Bay) (Fig. 6). Although it is not clear whether these depressions are interconnected, their bilateral ends are close to the ends of the fracture opening on the "B" line in the model (Fig. 5). There are indications from gravity anomaly data that

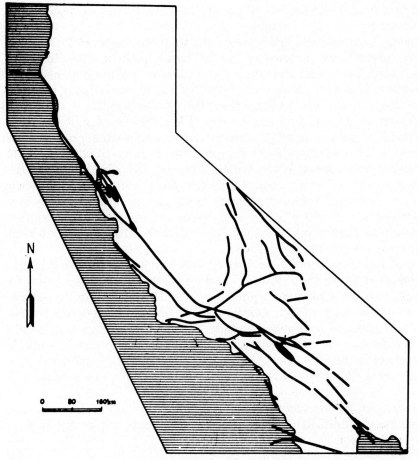

Fig. 6. Map of known tension structures in the San Andreas fault system.

a narrow longitudinal graben exists on the same portion of the San Andreas fault system, along the Calaveras fault which is parallel to the San Andreas (Robbins, 1971). It is filled with sediments 4 km thick, but has an almost indiscernible surface expression.

Morphological and geological data indicate with reasonable certainty the existence of a narrow longitudinal graben in the southern part of the fault system along the middle portion of the active San Jacinto fault (Cheatum and Combs, 1973). The width of the graben is 3—4 km and the basement is at least 2 km below the surface. There are proofs that the graben was formed contemporaneously with the strike-slip movement on the fault. The graben lies on the exact part of the fault system as the "D" segment in the model, although along a parallel fault.

As to the existence of a graben north of the Transverse Ranges on the Carrozo Plain that would correspond to part "C" in the model, no indications of this could be found in the literature.

We found no longitudinal depression on the shelf area between Pt. Mendocino and Pt. Arena corresponding to the "A" zone in the model. However, it may have been filled with river deposits. A narrow, longitudinal, negative-gravity anomaly west of the San Andreas fault (Griscom, 1973) may reflect a graben-like structure like that existing along the Calaveras fault.

There is neither geomorphic nor geological evidence to suggest the presence of tension structures in the Transverse Ranges area where, on the contrary, evidence for a submeridional compression is quite convincing (Wilt, 1958; Dickinson and Grantz, 1968; Kovach and Nur, 1973; Nikonov, 1975).

Therefore, the above comparison, in the first place, confirms the presence of narrow near-to-fault tension structures along the San Andreas fault system, similar to those found in the model. Secondly, the positions of known grabens along the fault zone does not contradict the distribution of local fracture-opening portions in the model.

In most of our experiments we noticed that the "southern" wall of the fault was elevated above the "northern" one over the area corresponding to the section along the Gorda Escarpment to San Francisco Bay. In nature, too, the southern wall of the Gorda Escarpment is elevated above the northern. This effect in the model is easily explained by a thickening of the model at the fault ends where the material is under compression, while it becomes thinner in tension areas. The possibility cannot be excluded that there is the same effect operating at the ends of a large natural strike-slip fault.

Model results concerning local stress fields and their connection with seismicity have been published (Nikonov et al., 1975) and will be reported in more detail elsewhere.

CONCLUSIONS

Analysis of a large number of seismological, geodetic, and strainmeter observations enables us to infer a horizontal submeridional axis of greatest compression (σ_3) for the present-day regional stress field in California and adjacent areas, and to suggest a horizontal sublatitudinal direction of the σ_1-axis.

Investigation of elastic models with fractures made in simulation of the San Andreas fault system, has shown general correspondence in the character of displacements in nature and in the model. In particular, displacements on model fractures are more pronounced in the same part of the fracture zone where the most intensive recent movements are observed along the San Andreas fault.

The main features in the morphology of the model fracture zone are as follows: formation of tear breaks around both ends of the zone and their peculiar character; formation of (four) fracture-opening segments situated along the master fault. These features correspond to a number of morphological features at the surface of the crust within the San Andreas fault system, including the presence of major tension structures about both ends of the fault zone.

REFERENCES

Allen, C.R., 1968. The tectonic environments of seismically active and inactive areas along the San Andreas fault system. In: W.R. Dickinson and A. Grantz (Editors), Proceedings of the Conference on Geologic Problems of the San Andreas Fault System. Stanford Univ. Publ. Geol. Sci., 11: 70—82.

Bolt, B.A., Lomnitz, C. and McEvilly, T.V., 1968. Seismological evidence of the tectonics of central and northern California and the Mendocino Escarpment. Seismol. Soc. Am. Bull., 58: 1725—1767.

Bufe, Ch.G., Bakun, W.H. and D. Tocher, 1973. Geophysical studies in the San Andreas fault zone at the Stone Observatory, California. Proceedings of the Conference on Tectonic Problems of San Andreas Fault System. Stanford Univ. Publ. Geol. Sci., 13: 86—93.

Burford, R.O., 1966. Strain analysis across the San Andreas fault and Coast Ranges of California. Ann. Acad. Sci. Fenn., A III, 90: 99—110.

Burford, R.O., Eaton, J.P. and Pakiser, L.C., 1969. Crustal strain and microseismicity investigations at the National Center for earthquake research of the United States Geological Survey. In: Yu. Boulanger and Yu. Mescheryakov (Editors), Problems of Recent Crustal Movements, Moscow, pp. 370—377.

Cheatum, C. and Combs, J., 1973. Microearthquake study of the San Jacinto Valley, Riverside County, California. Proceedings of the Conference on Tectonic Problems of the San Andreas Fault System. Stanford Univ. Publ. Geol. Sci., 13: 1—10.

Cherry, J.T. and Savage, J.C., 1972. Rock dilatancy and strain accumulation near Parkfield, California. Bull. Seismol. Soc. Am., 62(5).

Dickinson, N.R. and Grantz, A. (Editors), 1968. Proceedings of the Conference on Geological Problems of the San Andreas Fault System. Stanford Univ. Publ. Geol. Sci., 11: 374 pp.

Elders, W.A., Rex, R.W., Meidav, T., Robinson, P.T. and Bichler, Sh., 1972. Crustal spreading in southern California. Science, 178 (4056): 15—24.

Elsworth, W.L., Campbell, R.H., Hill, D.P. et al., 1973. Point Mugu, California, earthquake of 21 February 1973 and its aftershocks. Science, 182 (4117): 1127—1129.

Griscom, A., 1973. Tectonics of the junction of the San Andreas fault and Mendocino fracture zone from gravity and magnetic data. Proceedings of the Conference on Tectonic Problems of the San Andreas Fault System. Stanford Univ. Publ. Geol. Sci., 13: 383—390.

Gushchenko, O.I., Stepanov, V.V. and Sim, L.A., 1977. Directions of megaregional tectonic stresses in seismically active areas of the southern Eurasia. Dokl. Akad. Nauk S.S.S.R., 234 (3): 556—559 (in Russian).

Gzovsky, M.V., 1975. Principles of Tectonophysics. Nauka, Moscow, 536 pp. (in Russian).

Hast, N., 1969. The state of stress in the upper part of the Earth's crust. Tectonophysics, 8 (2): 169—211.

Hodgson, J.H., 1962. Movements in the earth's crust as indicated by earthquakes. In: S.K. Runcorn (Editor), Continental Drift. Academic Press, London.

Hofmann, R.B., 1970. Earthquake prediction from fault movement and precursors in California. In: L. Masina (Editor), Earthquake Displacement Fields and Rotation of the Earth. Reidel, Dordrecht, pp. 234—245.

Howard, I.N., 1968. Recent deformation of the Cholame and Taft—Maricopa areas, California. Proceedings of the Conference on Geological Problems of the San Andreas Fault System. Stanford Univ. Publ. Geol. Sci., 11.

Jungels, P.H. and Frazier, G.A., 1973. Finite element analysis of the residual displacements for an earthquake rupture: source parameters for the San Fernando earthquake. J. Geophys. Res., 78 (23).

Kovach, R.L. and Nur, A. (Editors), 1973. Proceedings of the Conference on Tectonic Problems of the San Andreas Fault System. Stanford Univ. Publ. Geol. Sci. 13: 494 pp.

Kropotkin, P.N., Nesterenko, G.T. and Bulin, N.K. (Editors), 1973. Stresses in the Earth's Crust. Nauka, Moscow, 186 pp. (in Russian).

Meade, B.K., 1971. Horizontal movement along the San Andreas fault system. In: B.W. Collins and R. Fraser (Editors), Recent Crustal Movements. Bull. R. Soc. N.Z., 9: 175—179.

Moody, J.D. and Hill, M.J., 1956. Wrench-fault tectonics. Bull. Geol. Soc. Am., 67 (9): 1207—1246.

Nikonov, A.A., 1975. Cenozoic tectonic movements along the San Andreas fault system in California. Geotektonika, (2): 98—113 (in Russian).

Nikonov, A.A., 1977. Recent Crustal Movements of the Crust. Nauka, Moscow, 240 pp. (in Russian).

Nikonova, K.I. and Nikonov, A.A., 1973. Study of recent crustal movements on the San Andreas fault system (a review of American work). In: Yu. Boulanger (Editor), Recent Crustal Movements, No. 5, Tartu, pp. 643—650 (in Russian, English summary).

Nikonov, A.A., Osokina, D.N., and Tsvetkova, N.Yu., 1975. Recent movements and stress field in the San Andreas fault system by the results of modelling. In: N. Pavoni and R. Green (Editors), Recent Crustal Movements. Tectonophysics, 29 (1—2): 153—160.

Osokina, D.N., Grigoryev, A.S. and Tsvetkova, N.Yu, 1974. The method and results of modelling redistribution of regional field and formation of local tectonic stress fields in the vicinity of tectonic faults. In: A.N. Kozakov (Editor), Mechanics of the Lithosphere, pp. 16—18 (in Russian).

Osokina, D.N., Grigoryev, A.S., Gushchenko, O.I. and Tsvetkova, N.Yu., 1976a. Potentialities of tectonophysical method in the study of present-day stress field in connection with the problem seismic hazard prediction. In: E. Savarensky (Editor), Searching Earthquake Forerunners. F.A.N., Tashkent, pp. 193—199 (in Russian, English summary).

Osokina, D.N., Tsvetkova, N.Yu. and Smirnov, L.A. (1976b). Tectonophysical investigation of regularities in the formation of local stress field anomalies and secondary faults in the vicinity of faults. In: A. Kovalev (Editor), Geodynamics and Mineral Deposits. Moscow, pp. 43—45 (in Russian).

Pope, A.J., Stearn, J.L. and Whitten, C.A., 1966. Survey for crustal movement along the Hayward fault. Bull. Seismol. Soc. Am., 56 (2): 311—323.

Raleigh, C.B. and Burford, R.O., 1969. Tectonics of the San Andreas fault system, Strain studies. EOS, Trans. Am. Geophys. Union, 50 (5): 380—381.

Richter, C.F., 1971. Sporadic and continuous seismicity of faults and regions. In: B.W. Collins and R. Fraser (Editors), Recent Crustal Movements. R. Soc. N.Z. Bull., 9: 171—173.

Richter, C.F., Allen, C.R. and Nordquist, J.M., 1958. The Desert Hot Springs earthquakes and their tectonic environment. Bull. Seismol. Soc. Am., 48 (4): 315—337.

Ritsema, A.R., 1961. Further focal mechanism studies. Publ. Dom. Obs., 24 (10): 355—358.

Robbins, St.L., 1971. Gravity and magnetic data in the vicinity of the Calaveras, Hayward, and Silver Creek faults near San Jose, California. Geol. Surv. Res., 750-B, Washington.

Ryall, A. and Malone, S.D., 1971. Earthquake distribution and mechanism of faulting in the Rainbow Mountain—Dixie Valley—Fairview Peak area, Central Nevada. J. Geophys. Res., 76 (29): 241—248.

Savage, J.C. and Burford, R.C., 1973. Geodetic determination of relative plate motion in Central California. J. Geophys. Res., 78(5): 832—845.

Savage, J.C. and Prescott, W.H., 1976. Geodolite measurement of deformation across the Salton Trough. Earthquakes Notes. East. Sect. Seismol. Soc. Am., 47 (2) 46—47.

Savage, J.C., Prescott, W.C. and Kinoshita, W.T., 1973. Geodimeter measurements along the San Andreas fault. Proceedings of the Conference on Tectonic Problems of the San Andreas Fault System. Stanford Univ. Publ. Geol. Sci., 13: 44—53.

Scheidegger, A.E., 1965. The tectonic stress and tectonic motion direction in the Pacific and adjacent areas as calculated from earthquake fault plane solutions. Bull. Seismol. Soc. Am., 55 (1): 147—152.

Seeber, L., Barasangi, M. and Nowroozi, A., 1970. Microearthquake seismicity and tectonics of coastal northern California. Bull. Seismol. Soc. Am., 60 (5): 1669—1700.

Scholz, C.H. and Fitch, T.I., 1970. Strain and creep in Central California. J. Geophys. Res., 75 (23).

Silver, E.A., 1971. Tectonics of the Mendocino Triple Junction. Geol. Soc. Am. Bull., 82 (11): 2965—2978.

Smith, S.W. and Kind, R., 1972. Regional secular strain fields in southern Nevada. Tectonophysics, 14 (1): 57—70.

Whitcomb, J.H., Allen, C.R. and Garmany, J.D., 1973. San Fernando earthquake series, 1971. Focal mechanisms and tectonics. Rev. Geophys. Space Phys., 11 (3): 693—730.

Whitten, C.A., 1970. Crustal movement from geodetic measurements. In: L. Mansina (Editor), Earthquake Displacement Field and the Rotation of the Earth. Reidel, Dordrecth, pp. 255—267.

Willis, B., 1938. San Andreas Rift, California. J. Geol., 46 (6): 793—827; (8): 1017—1057.

Wilt, J.W., 1958. Measured movement along the surface trace of an active thrust fault in the Buena Vista Hills, Kern County, California. Bull. Seismol. Soc. Am., 48 (9): 169—176.